AN INTRODUCTION TO

APPLIED MATHEMATICS

AN INTRODUCTION TO APPLIED MATHEMATICS

BY

J. C. JAEGER

AND

A. M. STARFIELD

SECOND EDITION

OXFORD
AT THE CLARENDON PRESS
1974

Oxford University Press, Ely House, London W. 1

GLASGOW NEW YORK TORONTO MELBOURNE WELLINGTON
CAPE TOWN IBADAN NAIROBI DAR ES SALAAM LUSAKA ADDIS ABABA
DELHI BOMBAY CALCUTTA MADRAS KARACHI LAHORE DACCA
KUALA LUMPUR SINGAPORE HONG KONG TOKYO

Casebound ISBN 0 19 8531516
Paperback ISBN 0 19 8531540

First Edition 1951
Second Edition 1974

Printed in Great Britain
at the University Press, Oxford
by Vivian Ridler
Printer to the University

PREFACE TO THE SECOND EDITION

THE first edition of this book was an attempt to give an introduction to the basic mathematics needed in physics and engineering. Only a knowledge of the principles of statics and dynamics and of the calculus was assumed; other techniques were developed *ab initio*, but as rigorously as possible, and an attempt was made to develop mathematical skill through the solution of a large number of problems of technical significance. To this end, in addition to conventional statics and dynamics, the book covered much of what is now described as 'mathematical methods', including vector analysis, numerical analysis, ordinary and partial differential equations, special functions, Fourier series, and Fourier and Laplace Transforms. In dynamics it included a long chapter on mechanical vibrations and another on electric circuit theory. Boundary value problems were introduced through the theory of bending of beams.

This new edition has been prepared largely by Professor Starfield. It has been found that most of the old material is still needed. In some places the examples have been modernized, for example solid-state devices replace vacuum tubes. Other areas have been extended; for instance servomechanisms are discussed in terms of the transfer function, and the applications of Fourier transforms have been widened to include ideas in communication theory. Also, over the last few years, the application of mathematics to biological and economic problems has greatly increased, and discussion and examples relating to both of these topics have been added.

The great change over the past two decades has been the development of computer techniques which have not only enormously widened the range of problems which can be solved, but have changed habits of thought in the formulation of problems and demanded new approaches in their solution. The emphasis has to some extent changed from problems leading to differential equations to those involving difference equations or an algorithmic approach. Moreover, the numerical solution of differential equations involves their expression in terms of difference equations and creates a new concern with questions of accuracy and the stability of solution. In more complicated cases, such as systems of equations and the numerical solution of partial differential equations, matrices have become an essential computing tool.

To cope with these changes two new chapters have been added; one on difference equations and the numerical solution of differential equations, and the other on matrices. The last two chapters on partial differential equations have also been rewritten and extended and a number of examples have been added for numerical solution on a computer.

J. C. J.
A. M. S.

Tasmania
University of the Witwatersrand, South Africa
May 1973

CONTENTS

X. BOUNDARY VALUE PROBLEMS

XI. FOURIER SERIES AND INTEGRALS

XII. ORDINARY LINEAR DIFFERENTIAL EQUATIONS WITH VARIABLE COEFFICIENTS

XVI. NUMERICAL METHODS FOR PARTIAL DIFFERENTIAL EQUATIONS

I

MATHEMATICAL MODELS AND DIFFERENTIAL AND DIFFERENCE EQUATIONS

1. Introductory

THERE are three essential steps in the solution of a problem in applied mathematics. In the first step, the problem is stated in mathematical terms. This means that the relevant variables are identified and that mathematical relationships are established between them, either by using physical laws or empirical evidence, or by hypothesis. The second step consists of the solution of the mathematical relationships, either by standard mathematical techniques, or, if these prove intractable, by numerical methods with the aid of a computer. Finally, in the third step, the solution is expressed in a form which enables one to interpret it and draw physical conclusions from it.

Most of the problems with which we shall be concerned will lead to mathematical relationships involving either differential, or, occasionally, difference equations. The subsequent chapters of this book are therefore largely concerned with the solution and applications of these equations. In this chapter we shall use simple examples to show how differential and difference equations arise, and will lay the groundwork for both the mathematical and numerical methods of solving them.

2. Mathematical models

To illustrate the first step in the solution procedure, that is, the development of a mathematical model, we discuss the problem of predicting the growth of a population.

In the simplest case, this problem involves only two variables; the independent variable is the time t, and the dependent variable is the size of the population x. The mathematical relationship between x and t is then determined by the specific conditions relating to the population we are studying. For a large but uncrowded human population, with an unlimited food supply, we could argue that both the birth-rate and mortality rate at any time are proportional to the size of the population at that time,

$$\text{birth-rate} = bx$$

and

$$\text{mortality rate} = cx,$$

where b and c are constants. The rate of increase of the population is then

$$\frac{dx}{dt} = bx - cx, \tag{1}$$

which is an *ordinary differential equation.*

Alternatively, if we consider an insect population where one generation dies out before the next generation hatches, it is unsatisfactory to think of the population as a continuous function of time. We therefore introduce the notation $x_1, x_2, x_3,...,$ etc., for the size of generation $1, 2, 3,...,$ etc., and postulate that the increase in population from generation r to generation $r+1$ should be directly proportional to the size of the rth generation,

$$x_{r+1} - x_r = kx_r, \tag{2}$$

where k is a constant. This is an example of a *difference equation.*†

Both the previous models are deterministic in the sense that statistical fluctuations are ignored. If the population under study is sufficiently large, it is reasonable to assume that chance effects can be neglected. This is not true of small populations, where one must develop a probabilistic or stochastic model which includes an element of chance.

To take an extreme example, suppose that we start with a single cell at time $t = 0$, and that we wish to predict its subsequent division and subdivision. Suppose further that empirical evidence indicates that there is a probability q that the cell will divide in time T. We then cannot say how many cells we will have at time T; only that there is a probability q that we will

† Ex. 4 of § 112 illustrates some of the differences between a continuous (differential equation) and discrete (difference equation) formulation of the same problem.

have two cells and a probability $1-q$ that we will still have one cell. If we introduce the notation $p_{m,n}$ to denote the probability that we will have precisely m cells at time nT, then

$$\left.\begin{aligned} p_{1,0} = 1, \quad p_{r,0} = 0 \quad \text{for } r \geqslant 2 \\ p_{1,1} = 1-q, \quad p_{2,1} = q, \quad p_{r,1} = 0 \quad \text{for } r \geqslant 3 \end{aligned}\right\}. \tag{3}$$

Using the theory of probability, we can then argue that at time $2T$

$$p_{1,2} = (1-q)^2, \quad p_{2,2} = q(1-q)(2-q), \quad p_{3,2} = 2q^2(1-q),$$
$$p_{4,2} = q^3, \quad p_{r,2} = 0 \quad \text{for } r \geqslant 5,$$

and so on. For example the argument leading to $p_{4,2}$ is that there can only be four cells at time $2T$ if the first cell divides at time T and both halves subsequently divide again. The probability of the former is q, of the latter q^2, and so the probability of both events occurring is q^3.

The probability distribution $p_{m,n}$ at each time nT gives a full statistical picture of the process of cell division. To compare this with a deterministic model, we calculate the average size or *expected value* of the population at nT. This is defined as

$$\bar{p}_n = \sum_{m=1}^{\infty} m p_{m,n}, \tag{4}$$

and it can be shown that, in this example,

$$\bar{p}_{n+1} = (1+q)\bar{p}_n. \tag{5}$$

Comparing (2) and (5) we see that, for this particular example, the deterministic model describes the average behaviour of the stochastic model.

Deterministic models will be perfectly adequate for most of the problems that we shall study. It is, however, important to bear in mind that some problems (see, for example, § 113 Ex. 1, and Ex. 6 at the end of Chapter XIV) can properly be described only by a stochastic model.

3. Solution and interpretation of results

The second step in the solution of a problem involves the actual solution of the mathematical model. In the example of the previous section, both the differential equation § 2 (1) and the difference equation § 2 (2) can be solved by the methods of Chapters III and XIV. The solution to the differential equation is

$$x(t) = Ae^{(b-c)t}, \tag{1}$$

where A is the size of the population at $t = 0$,while the solution to the difference equation is

$$x_r = x_1(1+k)^{r-1}, \tag{2}$$

which expresses subsequent generations in terms of the size of the first generation x_1.

We will not attempt to find the general form of the probabilities $p_{m,n}$ of § 2 (3). However, we notice that there is a definite argument which enables one to proceed from the probabilities at time nT to those at time $(n+1)T$, and this argument could be developed into a computer program which would calculate and print out tables of $p_{m,n}$ for different values of q.

Similarly, if we did not know how to solve the difference equation § 2 (2), we notice that writing it in the form

$$x_{r+1} = (1+k)x_r \tag{3}$$

suggests a direct method of calculating x_2 from x_1, x_3 from x_2, and so on, for a given value of k. This simple correspondence between difference equations and computer routines is one of the reasons why difference equations are important. In fact, in a more realistic model of population growth, k in (3) might well not be a constant but rather a complicated function of r and x_r, depending on environmental and other effects. It is then unlikely that a mathematical solution of (3) could be found, and the only way to solve the problem is to compute x_2, x_3,... etc., step by step.

The final step in the solution of a problem is the interpretation of the results. Apart from presenting the results as graphs or tables for various values of the constants, it is possible to draw some definite conclusions from solutions such as (1) and (2). For example, from (1) we see that the population will increase with time if $b > c$ (which we could have concluded from the differential equation § 2 (1) without even solving it). A less trivial result is that, if $b > c$, the population will double in a time period $(\ln 2)/(b-c)$† and will double again in a similar time period. In the case of the discrete model (2), the population doubles every $1+\{\ln 2/\ln(1+k)\}$ generations.

The preceding discussion indicates the importance of differential equations, difference equations, and computer methods in the solution of problems. In the next two sections we shall introduce the terminology of differential and difference equations, and discuss computer methods in greater detail.

4. Differential equations and difference equations. Definitions

Any relation between the independent‡ variable x, the dependent variable y, and its successive derivatives $dy/dx, d^2y/dx^2,\ldots$, is called an *ordinary*§ *differential equation*. The *order* of a differential equation is the order of the highest differential coefficient occurring in it. Thus, for example,

$$\frac{d^2y}{dx^2}+xy = 1, \tag{1}$$

$$\frac{d^2y}{dx^2}+x\left(\frac{dy}{dx}\right)^3+y = 0, \tag{2}$$

and

$$\left(\frac{d^2y}{dx^2}\right)^2+y\left(\frac{dy}{dx}\right)^3+y = 0, \tag{3}$$

are all second-order differential equations.

All the differential equations we shall need will contain only rational integral algebraic functions of the differential coefficients (fractional powers of x and y may sometimes occur), and in such cases the degree of the highest differential coefficient is called the *degree* of the equation. Thus (1) and (2) are both of the second order and the first degree, while (3) is of the second order and second degree.

† The notation $\ln x$ for $\log_e x$ will always be used.

‡ When discussing the theory of differential equations we shall take the independent variable to be x and the dependent variable y. In applications the symbols are determined by the problems. It is assumed throughout that y has derivatives of all the orders involved for all values of x.

§ If there are two or more independent variables, the equation is a partial differential equation; these will be discussed in Chapter XV. Until then the word 'ordinary' will usually be omitted.

By far the most important type of differential equation with which we shall be concerned is that in which all terms are of at most the first degree in y and its derivatives. This is called an ordinary *linear* differential equation, and its general form for the nth order is

$$a_0(x)\frac{d^ny}{dx^n}+a_1(x)\frac{d^{n-1}y}{dx^{n-1}}+\ldots+a_{n-1}(x)\frac{dy}{dx}+a_n(x)y = \phi(x). \quad (4)$$

The quantities $a_0(x), a_1(x),\ldots, a_n(x)$ are called the coefficients; if these are all constants, the equation is referred to as an ordinary linear differential equation *with constant coefficients*, otherwise it is a differential equation *with variable coefficients.*

Equation (4) is called an *inhomogeneous* equation to distinguish it from the corresponding equation with $\phi(x) = 0$,

$$a_0(x)\frac{d^ny}{dx^n}+a_1(x)\frac{d^{n-1}y}{dx^{n-1}}+\ldots+a_{n-1}(x)\frac{dy}{dx}+a_n(x)y = 0, \quad (5)$$

which is called a *homogeneous equation.*†

Equations (4) and (5) have fundamental properties which distinguish them from all other types of differential equations. Considering the homogeneous equation (5) first, suppose that y_1 and y_2 are two different solutions of it, so that

$$a_0(x)\frac{d^ny_1}{dx^n}+a_1(x)\frac{d^{n-1}y_1}{dx^{n-1}}+\ldots+a_n(x)y_1 = 0 \quad (6)$$

and $$a_0(x)\frac{d^ny_2}{dx^n}+a_1(x)\frac{d^{n-1}y_2}{dx^{n-1}}+\ldots+a_n(x)y_2 = 0. \quad (7)$$

If c_1 and c_2 are constants, it follows, by adding c_1 times (6) to c_2 times (7), that $c_1 y_1+c_2 y_2$ also satisfies (5). That is, if we know two solutions of (5), any linear combination of these is also a solution. Similarly, if $y_1, y_2,\ldots, y_n$ are n different solutions of (5), the general linear combination

$$c_1 y_1+c_2 y_2+\ldots+c_n y_n,$$

where $c_1, c_2,\ldots, c_n$ are any constants, is also a solution. This result does not hold for the inhomogeneous equation (4).

† The term 'homogeneous' is also used in a different context for certain special types of differential equations; cf. § 25. These are not of much importance in applied mathematics and no confusion is likely to arise.

The important result for the inhomogeneous linear equation (4) is that if y_1 is a solution of it with a function $\phi_1(x)$ on the right-hand side, and y_2 a solution with $\phi_2(x)$ on the right-hand side, and so on, then $y = y_1+y_2+...+y_n$ satisfies (4) with $\phi(x) = \phi_1(x)+\phi_2(x)+...+\phi_n(x)$ on the right-hand side. This can be confirmed by adding the equations of type (4) for y_1 to y_n. In many applications we will find that $\phi(x)$ refers to the cause and y describes the effect. The above result thus implies that, *if the equations governing the problem are linear*, the effects of a number of superposed causes can be added. This is known as the Principle of Superposition.

Exactly the same terminology and results apply to difference equations. Any relation between the terms $y_r, y_{r+1},..., y_{r+n}$ of a sequence is called a *difference equation* of order n. The *linear* difference equation of order n is

$$a_0 y_{r+n}+a_1 y_{r+n-1}+...+a_{n-1} y_{r+1}+a_n y_r = \phi(r), \tag{8}$$

and if the coefficients $a_0, a_1,..., a_n$ are independent of r we call (8) a linear difference equation with constant coefficients. If $\phi(r) = 0$, (8) is homogeneous, otherwise it is inhomogeneous. It can easily be shown that the general linear combination of n different solutions of the linear homogeneous equation is also a solution of the homogeneous equation, and that the Principle of Superposition holds for linear inhomogeneous difference equations.

However, neither of these results hold for non-linear differential or difference equations. For example, if y_1 and y_2 both satisfy

$$\frac{d^2y}{dx^2}+y\frac{dy}{dx}+y = 0, \tag{9}$$

neither $c_1 y_1+c_2 y_2$ nor even $c_1 y_1$ satisfies (9) because of the non-linear term $y\,dy/dx$.

The distinction between linear and non-linear differential or difference equations is of fundamental importance. Broadly speaking, it will become apparent that, for both differential and difference equations, the solution of linear equations with constant coefficients is relatively straightforward; linear equations with variable coefficients are more difficult to solve, but special equations of importance have been extensively studied. However, much less is known about non-linear equations, and apart from a number of special cases, these must be attacked either by a method of successive approximation, or by a more direct numerical method on a computer.

5. Computers and algorithms

The mathematical solution of a problem is accomplished by a sequence of mathematical operations; a problem is solved on a computer by a sequence of numerical calculations. An explicit, unambiguous statement of how to proceed in order to solve a particular class of problems is called an *algorithm*. For example, an algorithm for the numerical solution of $x_{r+1} = (1+k)x_r$, § 3 (3), could be the following:

(i) start with numerical values of x_1 and k;
(ii) put $r = 1$;
(iii) calculate $x_{r+1} = (1+k)x_r$;
(iv) if r is large enough, stop:
(v) otherwise replace r by $r+1$ and go back to (iii).

It is a small step from this to a formal computer program.

Numerical methods have been developed for finding approximate solutions to most types of mathematical problems. We shall see in Chapter XIV that differential equations can be approximated by difference equations which can then be solved on the computer using algorithms similar to the one above. In this section we illustrate the numerical approach by considering the somewhat simpler problem of finding a root of an ordinary equation.

Ex. *Find the root of* $$\cos x = \sin 2x \qquad (1)$$ *that lies between* 0 *and* $\frac{1}{2}\pi$.

The mathematical solution uses the result $\sin 2x = 2\sin x\cos x$ to reduce (1) to the equation $\cos x(1-2\sin x) = 0$, which leads directly to the result $\sin x = \frac{1}{2}$ or $x = \frac{1}{6}\pi$.

A numerical method for solving (1) might proceed as follows. If $f(x) = \cos x - \sin 2x$, we notice that $f(x_1) > 0$ if $x_1 = 0$, and $f(x_2) < 0$ if x_2 is just less than $\frac{1}{2}\pi$. We therefore know that a root lies between x_1

and x_2. With these values as a starting-point, we then apply the following algorithm:

(i) let $x_3 = (x_1+x_2)/2$ and calculate $f(x_3)$;
(ii) if $f(x_3) > 0$ replace x_1 by x_3;
(iii) if $f(x_3) < 0$ replace x_2 by x_3;
(iv) if $|x_1-x_2|$ is sufficiently small, stop;
(v) otherwise go back to (i)

This method is known as 'halving the interval'; each run through the algorithm halves the distance between x_1 and x_2 and ensures that the root we are looking for lies between x_1 and x_2. When the distance between x_1 and x_2 is sufficiently small, either x_1 or x_2 is acceptable as an answer.

Obviously, in the above example, the mathematical method is far quicker and easier to use than the numerical method. However, the mathematical solution depends on the relation $\sin 2x = 2\sin x\cos x$, and if a similar 'trick' cannot be found in a different example, the mathematics breaks down. For example, $\tan x = 2/x$ is an equation that cannot be solved mathematically, but is easily solved by halving the interval.

An alternative method for finding a root of the equation $f(x) = 0$ is an algorithm due to Newton. Starting with an initial guess x_0, the algorithm defines successive approximations x_1, x_2,..., etc., by the formula

$$x_{r+1} = x_r - f(x_r)/f'(x_r), \tag{2}$$

where we have written $f'(x_r)$ for df/dx evaluated at $x = x_r$.

Newton's method applied to (1) leads to the formula

$$x_{r+1} = x_r + (\cos x_r - \sin 2x_r)/(\sin x_r + 2\cos 2x_r). \tag{3}$$

Starting with the guess $x_0 = 0$, and working to four decimal places, (3) gives

$$x_1 = 0{\cdot}5000, \quad x_2 = 0{\cdot}5231, \quad x_3 = 0{\cdot}5236 = x_4 = x_5 \ldots$$

The required root is therefore $x = 0{\cdot}5236$.

A method which involves the repeated application of a formula such as (2), or the repetition of a set of operations, such as the algorithm for halving the interval, is called an *iterative method*. If an iterative method leads to a definite result, we say that it *converges*; otherwise it *diverges*. The method of halving the interval always converges provided $f(x_1)$ and $f(x_2)$ have opposite signs to begin with. Whether or not Newton's method converges depends on the function $f(x)$ and the initial guess x_0.

Whenever one solves a problem numerically on a computer, it is of the utmost importance to estimate the accuracy of the results. The major source of error on the computer is the fact that all numbers are stored as a fixed number of digits, called the 'wordlength' plus an exponent. To illustrate what we mean by this, let us consider a computer with a wordlength of only 4 decimal digits. This computer would store the number $a = 23{\cdot}1$ as $\cdot 2310 \times 10^2$, and $b = 0{\cdot}054$ as $\cdot 5400 \times 10^{-1}$. Now $a+b = 23{\cdot}154$, but since the computer can only store 4 digits, the sum $a+b$ is recorded as $\cdot 2315 \times 10^2$. The loss of the last digit is called a *rounding error*, and in this case is about 0·02 per cent. However, if at some later stage in the calculation we were to subtract $c = 23{\cdot}14$ from $a+b$, our answer would be 0·010 instead of 0·014, an error of 29 per cent. Computers typically have a wordlength of about 8 decimal digits.

Because of the large number of calculations that are usually performed during the numerical solution of a problem, the accumulation of rounding errors can become a serious matter. For example, if $J_n(x)$ is Bessel's function of order n, it is shown in § 92 (20) that

$$J_{n+1}(x) = \frac{2n}{x} J_n(x) - J_{n-1}(x). \tag{4}$$

Starting with values of $J_0(x)$ and $J_1(x)$, it should be possible to use (4) to generate $J_2(x)$, $J_3(x)$,..., etc. This implies that provided one has accurate tables for J_0 and J_1, it is not necessary to tabulate the higher order Bessel functions. What happens in practice for

TABLE I

n	*Correct* $J_n(x)$	*Computed* $J_n(x)$
0	0·96039823	
1	0·19602658	
2	0·01973470	0·01973462
3	0·00013201	0·00013196
4	0·00000661	0·00006008
5	0·00000026	−0·00011802

the case $x = 0{\cdot}4$ is shown in Table I. The rounding errors introduced at one step are magnified in the succeeding steps of the calculation until the accumulated error is far larger than the numbers we are trying to calculate. We shall see in Chapter XIV that this sort of accumulation of errors has to be guarded against in the numerical solution of differential equations.

6. The differential equations of particle dynamics and electric circuit theory

In this and the following two sections we shall look at some of the differential and difference equations that arise in various branches of applied mathematics.

The simplest equations of particle dynamics arise in problems on the motion of a particle of mass m in a straight line. If y is the position of the particle at time t, Newton's second law states that

$$m\frac{d^2y}{dt^2} = F, \tag{1}$$

where the force F is, in general, a complicated function of y, t, and the velocity dy/dt. In many cases of practical importance F simplifies into an expression of the form

$$F = f(y)+g\left(\frac{dy}{dt}\right)+h(t), \tag{2}$$

so that (1) becomes

$$m\frac{d^2y}{dt^2} = f(y)+g\left(\frac{dy}{dt}\right)+h(t). \tag{3}$$

(3) will be a linear inhomogeneous second-order equation if f and g are linear functions of y and dy/dt respectively. The case $f = ky$, where k is a constant, occurs in the theory of small oscillations, and one often makes the approximation $g = k'dy/dt$, where k' is also a constant, although this is not always very realistic. In general, therefore, the equations of particle dynamics are non-linear, but linear equations are encountered in approximations to important problems.

On the other hand, the differential equations associated with simple electric circuits are linear to a high degree of accuracy. The standard equation for a closed circuit containing a charge Q

on a condenser of capacitance C, an inductance L, a resistance R, and a voltage source V is

$$L\frac{d^2Q}{dt^2}+R\frac{dQ}{dt}+\frac{1}{C}Q = V, \tag{4}$$

where L, R, and C are constants. (4) is a linear second-order equation with constant coefficients.

Non-linear equations do arise in electric circuit theory, for instance when the circuit contains transistors, or if the inductance has an iron core. In the latter case L is a complicated function of the current dQ/dt. These and other problems in particle dynamics and electric circuit theory are discussed in Chapters IV, V, and VII.

7. Biological examples

The laws of particle dynamics and electric circuit theory are well established. The mathematical models of biological processes tend, except in certain fields such as genetics, to be more speculative. We have seen examples of this kind in the population growth models of § 2. Some more complex examples are the following.

Ex. 1. *Bacterial growth.*

Let x be the concentration of cells in a bacterial culture at time t, and let s be the concentration of the nutrient on which the cells feed. The linear inhomogeneous equation

$$\frac{dx}{dt}+ax = b\frac{ds}{dt}, \tag{1}$$

where a and b are constants, accords well with experimental evidence. In any particular example, the feed rate ds/dt is a known function of time.

Ex. 2. *A predator–prey model.*

Let x and y be the respective populations of a certain species and the predator that preys on it. The Lotka–Volterra model† for the dynamics of the interaction between predator and prey is the pair of differential equations

$$\left.\begin{aligned}\frac{dx}{dt} &= px-qxy\\ \text{and}\qquad \frac{dy}{dt} &= rxy-sy\end{aligned}\right\} \tag{2}$$

† See Lotka, *Elements of Mathematical Biology* (Dover).

where p, q, r, and s are positive constants. This model leads to cyclic changes in the two populations.

Ex. 3. *Competition between two species.*

A similar pair of differential equations arises in the study of two species, x and y, competing for the same food resource,

$$\left.\begin{aligned}\frac{dx}{dt} &= (a-bx-cy)x\\ \frac{dy}{dt} &= (e-fx-gy)y\end{aligned}\right\} \tag{3}$$

where a, b, c, e, f, and g are all positive constants.

The equations in the last two examples are all non-linear, as are most of the more realistic models of biological systems. It is sometimes possible to work with approximate equations that are linear in the neighbourhood of an equilibrium state, as in the following example.

Ex. 4. *Glucose and insulin levels in the bloodstream.*

Let u and v denote deviations from the normal levels of glucose and insulin in the bloodstream. If these deviations are small, it can be argued that

$$\left.\begin{aligned}\frac{du}{dt} &= I-au-bv\\ \text{and}\qquad \frac{dv}{dt} &= ru-sv\end{aligned}\right\} \tag{4}$$

where I is the rate at which glucose is injected into the bloodstream, and a, b, r, and s are constants.

Both difference and differential equations occur in the study of selection processes and mutation rates in genetics.

Ex. 5. *Selection against a recessive gene.*

Consider the three gene types AA, Aa, and aa. If the first two have the same characteristics, then the third gene aa is called recessive. Suppose that the recessive gene is fatal to its bearer before maturity. Successive generations will then only be able to inherit the aa gene from two Aa parents. It can be shown that if p_n is the proportion of individuals born with the recessive gene in generation n, the proportion in generation $n+1$ is

$$p_{n+1} = \frac{p_n}{1+p_n}, \tag{5}$$

a non-linear difference equation.

8. Other examples of differential and difference equations

In this section we consider a number of examples which illustrate how differential or difference equations can be written down from the statement of a problem.

Ex. 1. *Radioactive decay.*

The atoms of radioactive substance A change into atoms of a substance B at a rate k times the number of atoms of A present. If n is the number of atoms of A present at time t, this statement is equivalent to the equation

$$\frac{dn}{dt} = -kn. \tag{1}$$

If substance B is also radioactive and changes into substance C at a rate k_1 times the number of atoms n_1 of B present, then

$$\frac{dn_1}{dt} = kn-k_1 n_1. \tag{2}$$

Ex. 2.

A mass M of water (unit specific heat) is so well stirred that its temperature T is constant throughout the mass. Heat is lost at a rate H times the difference between the water temperature and the temperature T_0 of its surroundings. The differential equation is

$$M\frac{dT}{dt} = -H(T-T_0). \tag{3}$$

If the heat loss is primarily by radiation, heat is lost at a rate H' times the difference $T^4-T_0^4$. Then

$$M\frac{dT}{dt} = -H'(T^4-T_0^4), \tag{4}$$

which is non-linear.

Ex. 3.

Water at temperature T_1 flows into a bath at the rate of m cm^3 s^{-1}. If T is the temperature of the water in the bath and T_2 that of its surroundings, heat is lost from the bath at the rate $H(T-T_2)$.

Suppose M is the mass of water in the bath at time t. Then

$$\frac{dM}{dt} = m. \tag{5}$$

The rate of change of the quantity of heat MT is

$$\frac{d}{dt}(MT) = mT_1-H(T-T_2) \tag{6}$$

or, by (5)

$$M\frac{dT}{dt}+mT = mT_1-H(T-T_2). \tag{7}$$

The pair of equations (5) and (7) are the differential equations of the problem.

Ex. 4. *A 'car-following' model.*

The position of car A on a road is a known function y of the time t. Let $x(t)$ be the position of car B which is following car A. Suppose that the driver of car B tries to keep a fixed distance c behind A, that is, he accelerates when $y-x > c$ and decelerates when $y-x < c$. Then

$$\frac{d^2x}{dt^2} = k(y-x-c), \tag{8}$$

where k is a positive constant.

Alternatively, the driver of car B might try to match his speed to that of car A, leading to the differential equation

$$\frac{d^2x}{dt^2} = -k'\left(\frac{dx}{dt}-\frac{dy}{dt}\right). \tag{9}$$

More complex models of traffic flow take into account the delay introduced by the reaction time of the driver.

Ex. 5. *A learning model.*

Let p_n be the probability that a student, in a 'learning environment', knows how to integrate on day n, and q_n the probability that he does not. Whether or not he knows how to integrate on day $n+1$ depends, in a simple learning model, on p_n and q_n:

$$p_{n+1} = ap_n+bq_n$$

and

$$q_{n+1} = rp_n+sq_n.$$

However, since $p_n+q_n = p_{n+1}+q_{n+1} = 1$, it follows that both $a+r$ and $b+s$ must be equal to unity. The learning model therefore leads to the pair of difference equations

$$\left.\begin{aligned} p_{n+1} &= ap_n+bq_n \\ q_{n+1} &= (1-a)p_n+(1-b)q_n \end{aligned}\right\} \tag{10}$$

where $0 \leqslant a \leqslant 1$ and $0 \leqslant b \leqslant 1$.

9. Arbitrary constants, initial and boundary conditions

Suppose that

$$y = c_1 e^{-x}+c_2 e^{-2x}, \tag{1}$$

where c_1 and c_2 are constants that may take on any numerical values. We call these *arbitrary constants*. Differentiating (1), we have

$$\frac{dy}{dx} = -c_1 e^{-x}-2c_2 e^{-2x} \tag{2}$$

and

$$\frac{d^2y}{dx^2} = c_1 e^{-x}+4c_2 e^{-2x}. \tag{3}$$

Adding (1) and (2),

$$y+\frac{dy}{dx} = -c_2 e^{-2x},$$

and adding (2) and (3),

$$\frac{dy}{dx}+\frac{d^2y}{dx^2} = 2c_2 e^{-2x}$$

$$= -2\left(y+\frac{dy}{dx}\right).$$

Thus
$$\frac{d^2y}{dx^2}+3\frac{dy}{dx}+2y = 0. \tag{4}$$

We have therefore found a second-order differential equation by eliminating the arbitrary constants c_1 and c_2 from (1) and its first two derivatives. This differential equation is satisfied by (1) for *all* values of c_1 and c_2. In a roundabout way we have therefore shown that the second-order differential equation (4) has a general solution containing two arbitrary constants.

In general, it can be shown† that the general solution of an nth-order differential equation of any degree contains n arbitrary constants; this solution is called the *complete primitive* of the differential equation. Any solution derived from it by giving the constants particular values is called a *particular integral*. In some types of non-linear equations it happens that a solution can be found which cannot be derived from the complete primitive in this way; such a solution is called a *singular solution*.

It can also be shown that the general solution of an nth-order difference equation contains n arbitrary constants. It follows from this that whenever a practical problem leads to a differential or difference equation, additional information must be provided which enables one to determine the arbitrary constants. For example, in writing down the solutions to the population growth problems of § 2, it was assumed that we were given the population at time $t = 0$ in § 3 (1), or the size of the first generation in § 3 (2).

In order to solve a problem in particle dynamics where we have a second-order differential equation like § 6 (1), we would need

† See, for example, Ince, *Ordinary Differential Equations* (Longmans); other references on differential equations are Piaggio, *Differential Equations* (Bell); Forsyth, *Treatise on Differential Equations* (Macmillan).

two additional pieces of information. For instance, we could be given the position and velocity of the particle at $t = 0$. Alternatively, we might only be interested in the motion of the particle for $0 \leqslant t \leqslant T$, in which case we might be given the position (or velocity) of the particle at $t = 0$ and $t = T$.

When the required additional information is specified only at the starting value of the independent variable, the problem is called an *initial value problem*. When we are interested in the solution over a definite range of the independent variable, and additional information is specified at both ends of the range, we have a *boundary value problem*. In either case, the values of the arbitrary constants are found by substituting the initial or boundary conditions into the general solution of the differential or difference equation.

For example, to find the solution of (4) with initial conditions $y = 1$ and $dy/dx = 0$ at $x = 0$, we first put $y = 1$ and $x = 0$ in (1), giving

$$1 = c_1+c_2. \qquad (5)$$

We then substitute $dy/dx = 0$ and $x = 0$ in (2), giving

$$0 = -c_1-2c_2. \qquad (6)$$

Solving (5) and (6) gives $c_1 = 2$ and $c_2 = -1$.
The required solution is therefore

$$y = 2e^{-x}-e^{-2x}.$$

Similarly, the solution of (4) with boundary conditions $y = 0$ at $x = 0$ and $y = 1$ at $x = a$, is found to be

$$y = (e^{-x}-e^{-2x})/(e^{-a}-e^{-2a}).$$

10. Simultaneous equations and phase diagrams

A set of m ordinary differential equations involving m dependent variables in terms of a single independent variable is called a system of m simultaneous differential equations. We have seen examples of two simultaneous differential equations in Exs. 2, 3, and 4 of § 7.

Let us look more closely at the pair of equations § 7 (3) for two competing species x and y,

$$\frac{dx}{dt} = (a-bx-cy)x \qquad (1)$$

and $$\frac{dy}{dt} = (e-fx-gy)y. \tag{2}$$

If the solution to these equations could be found, it would normally be expressed as two ordinary equations; one for x in terms of t and the other for y in terms of t. Alternatively, the independent variable t could be eliminated from these equations, leading to a single expression involving only the dependent variables x and y. This expression defines a curve in the xy-plane.

The xy-plane is known as the *phase plane* and the curve traced by the solution in the phase plane is called the *phase trajectory*. The shape of the phase trajectory very often gives useful practical information about the solution to the simultaneous equations. Sometimes this information can be obtained without even formally solving the differential equations.

For instance, looking at (1) and (2) we see that, apart from the trivial case $x = y = 0$,

$$\frac{dx}{dt} = 0 \quad \text{when } y = \frac{a}{c}-\frac{bx}{c}, \tag{3}$$

and $$\frac{dy}{dt} = 0 \quad \text{when } y = \frac{e}{g}-\frac{fx}{g}. \tag{4}$$

Equations (3) and (4) define the lines AB and CD in the phase diagram Fig. 1 (a). (We have assumed here that $a/c > e/g$ and $a/b < e/f$.) These two lines divide the positive quadrant of the phase plane into four regions. In region I we have $y > (e/g)-(fx/g)$ and $y < (a/c)-(bx/c)$. It follows from (1) and (2) that, in region I, $dx/dt > 0$ and $dy/dt < 0$, that is, the phase trajectory moves in the positive x-direction but negative y-direction in region I. These directions are indicated by the arrows in each of the regions in Fig. 1 (b).

Now suppose that the initial values of the two populations, x_0 and y_0, define the point P in Fig. 1 (b). The phase trajectory from P must move in the directions dictated by the arrows in region IV and hence must trace a curve terminating in the point O at the intersection of lines AB and CD. This can be seen to be true for any starting-point P in the phase plane. We can

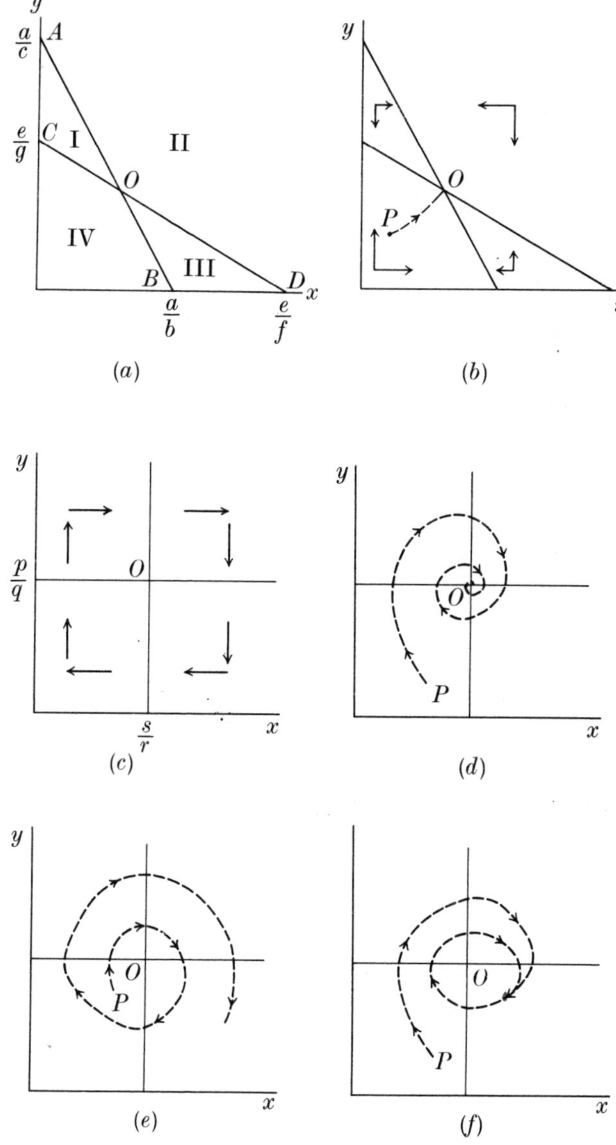

y
$\frac{a}{c}$
A
$\frac{e}{g}$
C
I
II
O
IV
III
B
D
x
$\frac{a}{b}$
$\frac{e}{f}$
(a)
y
O
P
x
(b)
y
$\frac{p}{q}$
O
$\frac{s}{r}$
x
(c)
y
O
P
x
(d)
y
O
P
x
(e)
O
P
x
(f)

FIG. 1.

therefore conclude that the coordinates of O define the equilibrium values of the two populations, and that no matter what the initial values of the populations may be, the populations will always tend to this equilibrium state. A very different result is obtained for the case $a/c < e/g$ and $a/b > e/f$ (cf. Ex. 6 at the end of this chapter).

It is not always possible to draw a definite conclusion from a rough sketch of the behaviour of the solution in the phase plane. For example, the phase diagram for the predator–prey model of § 7 (2) is shown in Fig. 1 (c). Here the solution could behave in one of three very different ways:

(i) The phase trajectory could spiral inwards from an initial point to equilibrium at O, as in Fig. 1 (d). This represents damped oscillations in the two populations.

(ii) The phase trajectory could trace a spiral that winds outwards away from O, as in Fig. 1 (e). This represents unstable oscillations of ever-increasing amplitude.

(iii) The phase trajectory could trace a curve that ends in a closed loop or *limit cycle*, as in Fig. 1 (f). This represents continuing undamped cyclic changes in the two populations.

To distinguish between these three cases, one would have to examine the solution of the differential equations in the neighbourhood of O. (See Ex. 14 at the end of Chapter III.)

In the general case of m simultaneous differential equations, elimination of the independent variable leads to a trajectory in an m-dimensional 'phase space'.

EXAMPLES ON CHAPTER I

1. Find the root between 0 and $\frac{1}{2}\pi$, of the equation $\tan x = 2/x$,

   (i) by the method of halving the interval

   (ii) by Newton's algorithm with an initial guess $x_0 = 0{\cdot}5$.

[Answer is $x = 1{\cdot}0769$.]

2. By applying Newton's method to the equation $x^2 - c = 0$, show that an algorithm for finding the square root of c is

$$x_{r+1} = \frac{1}{2}\left(x_r + \frac{c}{x_r}\right).$$

Show that this converges to $\sqrt{c}$ if the initial guess $x_0 > 0$, and to $-\sqrt{c}$ if the initial guess $x_0 < 0$.

3. Use Newton's algorithm with initial guess $\alpha_0 = 0$ to show that $f(\alpha) = \alpha^3+3\alpha^2+6\alpha+6$ has a root at $\alpha = -1{\cdot}60$, working to two decimal places (cf. § 13 Ex. 6).

4. Given two polynomials

$$P(x) = a_0 x^n + a_1 x^{n-1} \ldots + a_{n-1} x + a_n$$

and

$$Q(x) = b_0 x^n + b_1 x^{n-1} \ldots + b_{n-1} x + b_n,$$

write down an algorithm for finding the coefficients of the polynomial $S(x) = P(x)Q(x)$.

5. Given a function $f(x)$, the derivative df/dx at $x = x_0$ can be estimated numerically using the formula $\{f(x_0+h)-f(x_0-h)\}/(2h)$, where h is a small number. Use this formula to compute the derivative of x^5 at $x = 2$, using the following values of h: 0·1, 0·01, 0·001,..., 0·000000001. Comment on the influence of rounding errors.

6. Draw a phase diagram for the competing species problem of §7 Ex. 3 for the case $a/c < e/g$ and $a/b > e/f$. Hence show that, depending on the initial populations, one or other of the species will eventually become extinct.

7. Draw a phase diagram for the glucose–insulin model of § 7 Ex. 4 for the case $I = 0$. Hence show that if the constants a, b, r, and s are all positive, an initial imbalance in either the glucose or insulin level in the bloodstream will eventually be corrected; that is, the phase trajectory will tend to the equilibrium state $u = v = 0$.

II

ORDINARY LINEAR DIFFERENTIAL EQUATIONS WITH CONSTANT COEFFICIENTS

11. Introductory

It was remarked in Chapter I that many important problems of applied mathematics lead to ordinary linear differential equations with constant coefficients. Since explicit solutions of these are easily obtained, such problems form the core of the theory and are studied first in detail.

In this chapter we give first the classical methods of finding the general solution containing n arbitrary constants of the equation of order n. Then the case of simultaneous differential equations is discussed, and finally the determination of the arbitrary constants in initial and boundary value problems.

The chapter concludes with a brief account of the Laplace transformation method of solving linear differential equations with constant coefficients and given initial conditions. This is a completely alternative approach to that of §§ 12–15 and may be omitted if desired. It is perhaps simpler to learn and teach than the classical methods, but at the same time these are so much part of the common language of mathematics that it is impossible to omit them, and in the sequel they will usually be used. The advantages of the Laplace transformation method increase with the complexity of the problem, and it will occasionally be used to solve relatively complicated problems, in particular, transient problems involving several simultaneous differential equations.

12. The operator D

In the study of linear differential equations with constant coefficients it is convenient to use the symbol D for the operation of differentiation, so that

$$Dy = \frac{dy}{dx}. \tag{1}$$

Again, we write D^2y for $D(Dy)$, so that

$$D^2y = D(Dy) = D\left(\frac{dy}{dx}\right) = \frac{d^2y}{dx^2}, \tag{2}$$

and, similarly, when n is a positive integer,

$$D^ny = D(D^{n-1}y) = \frac{d^ny}{dx^n}. \tag{3}$$

It follows that when m and n are positive integers

$$D^m(D^ny) = D^{m+n}y, \tag{4}$$

so that D operating on y obeys the index law.

Also, if a and b are constants,

$$D(ay) = \frac{d}{dx}(ay) = a\frac{dy}{dx} = a\,Dy, \tag{5}$$

$$D(ay+bz) = a\frac{dy}{dx}+b\frac{dz}{dx} = a\,Dy+b\,Dz, \tag{6}$$

so that the orders of multiplication by a constant and the operation D may be interchanged.

Thus, provided we are considering only the operations of differentiation and multiplication by constants,† the operator D may be treated like an ordinary number. For example

$$\begin{aligned}\frac{d^2y}{dx^2}+2b\frac{dy}{dx}+b^2y &= (D^2+2bD+b^2)y\\ &= D(D+b)y+b(D+b)y\\ &= (D+b)^2y.\end{aligned}$$

In the same way, the so-called general differential expression of order n, namely

$$\frac{d^ny}{dx^n}+a_1\frac{d^{n-1}y}{dx^{n-1}}+a_2\frac{d^{n-2}y}{dx^{n-2}}+\ldots+a_n y, \tag{7}$$

where $a_1,\ldots, a_n$ are constants, may be written

$$(D^n+a_1 D^{n-1}+\ldots+a_{n-1} D+a_n)y, \tag{8}$$

and for shortness this will usually be written

$$f(D)y,$$

where $$f(D) \equiv D^n+a_1 D^{n-1}+\ldots+a_n. \tag{9}$$

† But if variables enter this is not the case, for example $D(xy) = xDy+y$.

Now suppose that $\alpha_1,\ldots,\alpha_n$ are the roots (not necessarily all real or different) of the equation in α

$$f(\alpha) \equiv \alpha^n+a_1\alpha^{n-1}+\ldots+a_n = 0. \tag{10}$$

Then $f(\alpha)$ can be factorized in the form

$$f(\alpha) \equiv (\alpha-\alpha_1)(\alpha-\alpha_2)\ldots(\alpha-\alpha_n), \tag{11}$$

and the general differential expression (7) or (8) may be put in the form

$$f(D)y \equiv (D-\alpha_1)(D-\alpha_2)\ldots(D-\alpha_n)y. \tag{12}$$

That this is equivalent to the form (7) follows on multiplying out and using the results (4) and (5).

As remarked above, the roots α_1, $\alpha_2,\ldots,\alpha_n$ need not all be different: if α_1 is an r-ple root, α_2 an s-ple root, and α_p an m-ple root, (12) may be written

$$f(D)y \equiv (D-\alpha_1)^r(D-\alpha_2)^s\ldots(D-\alpha_p)^m y. \tag{13}$$

The general homogeneous linear differential equation of order n may now be written

$$f(D)y = 0, \tag{14}$$

where $f(D)$ is defined by (7) and (9); the equation (10) is called the *auxiliary equation* for this differential equation, and the first step in the solution of the differential equation is to find the roots $\alpha_1,\ldots,\alpha_n$ of (10), that is, in effect, to express the operator $f(D)$ in the form (13).

As remarked earlier, the simple results given above apply only to the operations of multiplication by constants and differentiation. If variables enter, the ordinary laws of differentiation have to be used, for example

$$D(xy) = xDy+y,$$

$$D^2(xy) = xD^2y+2Dy,$$

etc., and the simplicity is lost. General results are obtained by using Leibnitz's theorem, viz.

$$D^n(yz) = yD^nz+{}^nC_1\,DyD^{n-1}z+\ldots+zD^ny. \tag{15}$$

We now prove three simple theorems which are of great importance in the sequel. These are all proved only for the case in which the function $f(D)$ is a polynomial in D.

THEOREM 1. *If $f(D)$ is a polynomial in D and k is a constant, then*

$$f(D)e^{kx} = f(k)e^{kx} \tag{16}$$

and

$$f(D)\left\{\frac{1}{f(k)}e^{kx}\right\} = e^{kx}. \tag{17}$$

For

$$De^{kx} = ke^{kx},$$
$$D^2e^{kx} = k^2e^{kx},$$

etc. Therefore

$$\begin{aligned} f(D)e^{kx} &= (D^n+a_1 D^{n-1}+...+a_n)e^{kx} \\ &= (k^n+a_1 k^{n-1}+...+a_n)e^{kx} \\ &= f(k)e^{kx}. \end{aligned}$$

(17) is, of course, equivalent to (16) and is stated here for reference.

THEOREM 2. *If $f(D^2)$ is a polynomial in D^2, then*

$$f(D^2)\,\frac{\sin}{\cos}\,\omega x = f(-\omega^2)\,\frac{\sin}{\cos}\,\omega x. \tag{18}$$

For

$$D^2 \sin \omega x = -\omega^2 \sin \omega x,$$
$$D^2 \cos \omega x = -\omega^2 \cos \omega x,$$

etc., and adding results of this type as in the proof of Theorem 1 gives (18).

THEOREM 3. *If $f(D)$ is a polynominal in D, and k is a constant, then*

$$f(D)\{e^{kx}y\} = e^{kx}f(D+k)y \tag{19}$$

and

$$f(D+k)y = e^{-kx}f(D)\{e^{kx}y\}. \tag{20}$$

It follows from Leibnitz's theorem (15) that

$$\begin{aligned} D^n(e^{kx}y) &= e^{kx}D^ny+{}^nC_1 D(e^{kx})D^{n-1}y+...+yD^n(e^{kx}) \\ &= e^{kx}\{D^ny+{}^nC_1 kD^{n-1}y+...+k^ny\} \\ &= e^{kx}\{(D^n+{}^nC_1 kD^{n-1}+{}^nC_2 k^2D^{n-2}+...+k^n)y\} \\ &= e^{kx}\{(D+k)^ny\}. \end{aligned}$$

Adding results of this type as in the proof of Theorem 1 we get

$$f(D)\{e^{kx}y\} = e^{kx}\{f(D+k)y\},$$

as required. The alternative form (20) which allows $f(D+k)y$ to be expressed in terms of $f(D)$ operating on $e^{kx}y$ is also often useful.

13. The homogeneous ordinary linear differential equation of order n

As in § 12 this may be written

$$f(D)y \equiv (D^n+a_1 D^{n-1}+...+a_n)y = 0. \tag{1}$$

Also as in § 12, we call the equation

$$f(\alpha) \equiv \alpha^n+a_1\alpha^{n-1}+...+a_n = 0 \tag{2}$$

the auxiliary equation and find its roots. If it has an r-ple root α_1, an s-ple root α_2, etc., (1) may be written

$$(D-\alpha_1)^r(D-\alpha_2)^s...(D-\alpha_p)^m y = 0. \tag{3}$$

We consider first the simple case of the first-order equation

$$\frac{dy}{dx}-\alpha_1 y = 0,$$

or

$$(D-\alpha_1)y = 0. \tag{4}$$

Obviously†

$$y = c_1 e^{\alpha_1 x}, \tag{5}$$

where c_1 is any constant, satisfies this, and since it contains one arbitrary constant is the general solution.

Next suppose that the roots $\alpha_1,..., \alpha_n$ of the auxiliary equation $f(\alpha) = 0$ are all different, so that (3) takes the form

$$(D-\alpha_1)(D-\alpha_2)...(D-\alpha_n)y = 0. \tag{6}$$

Clearly if y satisfies

$$(D-\alpha_n)y = 0 \tag{7}$$

it also satisfies (6). By (5) the solution of (7) is

$$c_n e^{\alpha_n x}. \tag{8}$$

In the same way, since the order of the operations $(D-\alpha_1),..., (D-\alpha_n)$ in (6) may be interchanged in any way,

$$c_1 e^{\alpha_1 x},\ c_2 e^{\alpha_2 x},...,\ c_n e^{\alpha_n x}, \tag{9}$$

where $c_1, c_2,..., c_n$ are any constants, are all independent solutions

† Alternatively, integrating with respect to x gives

$$\int \frac{1}{y}\frac{dy}{dx}\,dx = \int \alpha_1\,dx,$$

or

$$\ln y = \alpha_1 x+\text{constant}.$$

of (6). And, by the fundamental property of a linear differential equation derived in § 4, the sum of these solutions, namely

$$c_1 e^{\alpha_1 x}+c_2 e^{\alpha_2 x}+\ldots+c_n e^{\alpha_n x}, \tag{10}$$

is also a solution of (6); since it contains n arbitrary constants it is the general solution.

Ex. 1. $$(D^2+3D+2)y = 0.$$

The auxiliary equation is

$$\alpha^2+3\alpha+2 = 0,$$

and its roots are -1 and -2. Thus the solution is

$$y = Ae^{-x}+Be^{-2x},$$

where A and B are arbitrary constants.

Ex. 2. $$(D^2+2D+5)y = 0.$$

The auxiliary equation is

$$\alpha^2+2\alpha+5 = 0,$$

and its roots are $-1\pm 2i$.

Thus the solution is

$$y = Ae^{-x+2ix}+Be^{-x-2ix}, \tag{11}$$

where A and B are arbitrary constants. This form of solution is not the most useful one and it is usually better to express it in real form. Using the result

$$e^{\pm ix} = \cos x \pm i\sin x,$$

(11) becomes

$$(A+B)e^{-x}\cos 2x+i(A-B)e^{-x}\sin 2x$$

and, since A and B are arbitrary, $A+B$ and $i(A-B)$ are also two arbitrary constants which may be taken as the fundamental constants in the general solution. Thus the solution may be written

$$y = Ce^{-x}\cos 2x+Ge^{-x}\sin 2x. \tag{12}$$

Alternatively, it may be put in the form

$$y = Ee^{-x}\cos(2x+F), \tag{13}$$

where E and F are arbitrary constants.

From the point of view of pure mathematics, that is, of finding a solution of the equation with two arbitrary constants, (11), (12), and (13) are all equally good answers. In the application to practical problems the subsequent work may be simplified by choosing the most suitable form to begin with, so it is desirable to bear all three types in mind and not to write down solutions always in one standard form. For examples of this see § 16 (11) and Exs. 13 and 14 on Chapter II.

Ex. 3.
$$(D^2-1)y = 0.$$

The roots of the auxiliary equation are ± 1, so the solution is
$$y = Ae^{x}+Be^{-x},$$
or, alternatively,
$$y = C\cosh x+E\sinh x.$$

Ex. 4.
$$(D^4-1)y = 0.$$

The auxiliary equation is $\alpha^4-1 = 0$, and its roots are ± 1, $\pm i$. The solution is
$$y = A\cosh x+B\sinh x+C\cos x+E\sin x,$$
or one of the other forms discussed above.

Ex. 5.
$$(D^4+4\omega^4)y = 0.$$

The auxiliary equation has roots $\omega(\pm 1\pm i)$, and the solution is
$$y = e^{\omega x}(A\cos\omega x+B\sin\omega x)+e^{-\omega x}(C\cos\omega x+E\sin\omega x),$$
or, alternatively,
$$y = A_1 e^{\omega x}\cos(\omega x+\theta_1)+B_1 e^{-\omega x}\cos(\omega x+\theta_2).$$

Ex. 6.
$$(D^3+3D^2+6D+6)y = 0.$$

The auxiliary equation is
$$\alpha^3+3\alpha^2+6\alpha+6 = 0,$$
and, like many equations that arise in practice, does not have simple roots and must be solved numerically. To two places of decimals the roots are
$$-1{\cdot}60 \quad \text{and} \quad -0{\cdot}70\pm 1{\cdot}81i.$$

Thus the solution is
$$y = Ae^{-1\cdot 60x}+Be^{-0\cdot 70x}\cos(1{\cdot}81x+C).$$

Passing now to the general case (3) in which the auxiliary equation has repeated roots, it appears that any solution of
$$(D-\alpha_1)^r y = 0 \tag{14}$$
will satisfy (3). Now, by § 12 (20), this can be written in the form
$$e^{\alpha_1 x}D^r\{e^{-\alpha_1 x}y\} = 0. \tag{15}$$

Thus $e^{-\alpha_1 x}y$ must be a function whose rth differential coefficient is zero, that is, a polynomial of degree $r-1$, so that

$$y = e^{\alpha_1 x}(A_0+A_1 x+...+A_{r-1}x^{r-1}), \tag{16}$$

and since this contains r arbitrary constants it is the general solution of (14).

Adding results of this type, it follows that the general solution of (3), containing $r+s+...+m = n$ arbitrary constants, is

$$y = (A_0+A_1 x+...+A_{r-1}x^{r-1})e^{\alpha_1 x}+...+$$
$$+(K_0+K_1 x+...+K_{m-1}x^{m-1})e^{\alpha_p x}. \tag{17}$$

Ex. 7.
$$(D-1)^3(D-2)y = 0.$$
The solution is $\quad y = (A+Bx+Cx^2)e^x+Ee^{2x}.$

Ex. 8. The general second-order equation
$$(D^2+2bD+c)y = 0.$$
The roots of the auxiliary equation are $-b\pm\sqrt{(b^2-c)}$. The solution is
$$e^{-bx}\{Ae^{x\sqrt{(b^2-c)}}+Be^{-x\sqrt{(b^2-c)}}\} \quad \text{if } b^2 > c,$$
$$(A+Bx)e^{-bx} \quad \text{if } b^2 = c,$$
$$Ae^{-bx}\cos\{x\sqrt{(c-b^2)}+B\} \quad \text{if } b^2 < c.$$

Finally, we note an alternative approach to the theory which will often be useful. Suppose we have to solve the equation

$$f(D)y = 0 \tag{18}$$

and we seek a solution in which y is proportional to $e^{\alpha x}$, say $y = Ae^{\alpha x}$. Substituting in (18) gives, using § 12, Theorem 1,

$$Ae^{\alpha x}f(\alpha) = 0.$$

Thus α must be a root of $f(\alpha) = 0$, which is just the auxiliary equation (2). If the roots of this equation are all different, combining all solutions of this type gives the general solution (10).

If the roots are not all different, suppose that α is an r-ple root, so that the equation is of the form

$$(D-\alpha)^r F(D)y = 0. \tag{19}$$

We now seek a solution of the form $y = Y(x)e^{\alpha x}$, where $Y(x)$ is a function of x instead of a constant as before. If this is to satisfy (19) we must have

$$(D-\alpha)^r F(D)\{Y(x)e^{\alpha x}\} = 0,$$

or, by § 12, Theorem 3,

$$e^{\alpha x}F(D+\alpha)D^r Y(x) = 0. \tag{20}$$

(20) will be satisfied if we choose $Y(x)$ so that $D^r Y(x) = 0$, that is

$$Y(x) = A_0+A_1 x+...+A_{r-1} x^{r-1},$$

where $A_0,..., A_{r-1}$ are constants.

Thus we have the solution

$$(A_0+A_1 x+...+A_{r-1} x^{r-1})e^{\alpha x}$$

as in (16).

14. The inhomogeneous ordinary linear differential equation of order n

The general form of this is

$$f(D)y = \phi(x), \tag{1}$$

where $f(D)$ is a polynomial in D as in §§ 12, 13, and $\phi(x)$ is any function of x.

The first step in the solution is to find, by the methods of § 13, the general solution $y_1(x)$ containing n arbitrary constants of the homogeneous equation corresponding to (1), namely,

$$f(D)y = 0. \tag{2}$$

$y_1(x)$ is called the *Complementary Function* of (1).

Next we find a function of x, say $y_2(x)$, which satisfies (1); methods for doing this will be discussed below. This function $y_2(x)$ is called a *Particular Integral* of (1).

The sum of the complementary function and a particular integral, viz.

$$y_1(x)+y_2(x) \tag{3}$$

is the general solution of (1) containing n arbitrary constants. To prove this we notice that it contains n arbitrary constants since $y_1(x)$ does, also it satisfies (1) since, substituting,

$$f(D)\{y_1(x)+y_2(x)\} = f(D)y_1(x)+f(D)y_2(x) = \phi(x),$$

since $y_1(x)$ satisfies (2) and $y_2(x)$ satisfies (1).

It must be emphasized that any function whatever which satisfies (1) can be taken as a particular integral; many different particular integrals could be written down, but these differ by terms already included in the complementary function. Also if $\phi(x)$ is the sum of several functions, say,

$$\phi(x) = \phi_1(x)+...+\phi_r(x),$$

a particular integral of (1) is the sum of particular integrals of

$$f(D)y = \phi_r(x),$$

for $r = 1,..., n$.

We now have to study in detail methods for finding a particular integral of (1) when $\phi(x)$ has one of the common forms: a constant, a polynomial, a trigonometrical or exponential function, etc. There are many methods of doing this, but here we shall adopt the very simple one of seeking a particular integral similar in form to $\phi(x)$. Thus if $\phi(x)$ is a polynomial in x, we shall seek a polynomial for the particular integral; if $\phi(x)$ is e^{ax}, we shall seek a particular integral with e^{ax} as a factor, and so on.

(i) $\phi(x)$ *a constant* c

If the differential equation is

$$(D^n+a_1 D^{n-1}+...+a_n)y = c, \tag{4}$$

where $a_n \neq 0$, the constant

$$c/a_n \tag{5}$$

satisfies it and so is a particular integral.

If $f(D)$ has a factor D^r so that the differential equation is

$$(D^n+a_1 D^{n-1}+...+a_{n-r} D^r)y = c, \tag{6}$$

and we choose y so that

$$a_{n-r} D^r y = c \tag{7}$$

it will also satisfy (6). To satisfy (7), y has to be a function which, when differentiated r times, gives the constant c/a_{n-r}; the simplest such function is

$$y = \frac{cx^r}{a_{n-r}\, r!}. \tag{8}$$

Ex. 1. $$(D^2+3D+2)y = 1.$$

By (5), $\frac{1}{2}$ is a particular integral.

The complementary function is

$$Ae^{-2x}+Be^{-x}.$$

Thus the general solution is

$$\tfrac{1}{2}+Ae^{-2x}+Be^{-x}.$$

Ex. 2. $$D^2(D+1)y = 1.$$

The particular integral is to be a function which when differentiated twice gives 1, that is, $\frac{1}{2}x^2$.

The complementary function is

$$A+Bx+Ce^{-x}.$$

Thus the general solution is

$$A+Bx+\tfrac{1}{2}x^2+Ce^{-x}.$$

(ii) *$\phi(x)$ a polynomial $P_k(x)$ of degree k in x*

We seek a particular integral in the form of a polynomial. If the differential equation is

$$(D^n+a_1 D^{n-1}+...+a_n)y = P_k(x), \tag{9}$$

with $a_n \neq 0$, we assume for the particular integral a polynomial of degree k with unknown coefficients. The coefficients are found by substituting this in the left-hand side of (9) and comparing coefficients with $P_k(x)$.

If $a_n = ... = a_{n-r+1} = 0$, the differential equation has the form

$$(D^n+a_1 D^{n-1}+...+a_{n-r} D^r)y = P_k(x), \tag{10}$$

and we must assume a polynomial of degree $(k+r)$ for the particular integral, so that, when this is substituted in (10), there will be powers of x up to the kth on the left-hand side.

Ex. 3. $$(D^2+D+1)y = x+x^2+x^3. \tag{11}$$

We seek a polynomial particular integral

$$y = Ax^3+Bx^2+Cx+E.$$

If this is to satisfy (11) we must have

$$(6Ax+2B)+(3Ax^2+2Bx+C)+(Ax^3+Bx^2+Cx+E) \equiv x+x^2+x^3.$$

Equating coefficients of x^3, x^2, x, and the constant terms on the two sides gives

$$A = 1,$$
$$3A+B = 1,$$
$$6A+2B+C = 1,$$
$$2B+C+E = 0.$$

Solving, successively, gives $A = 1$, $B = -2$, $C = -1$, $E = 5$, and the required particular integral is

$$x^3-2x^2-x+5.$$

Adding the complementary function gives the general solution

$$x^3-2x^2-x+5+Fe^{-\frac{1}{2}x}\cos(\tfrac{1}{2}x\sqrt{3}+G).$$

Ex. 4. $$D^2(D+1)y = 1+x^3. \tag{12}$$

We must assume a polynomial of the fifth degree for y so that terms in x^3 will occur after it has been differentiated twice. Assume

$$y = Ax^5+Bx^4+Cx^3+Ex^2+Fx+G.$$

If this is to satisfy (12) we must have

$$(60Ax^2+24Bx+6C)+(20Ax^3+12Bx^2+6Cx+2E) \equiv 1+x^3.$$

Equating coefficients we find

$$A = \tfrac{1}{20}, \qquad B = -\tfrac{1}{4}, \qquad C = 1, \qquad E = -\tfrac{5}{2}.$$

There is no restriction on F and G, so we may take them to be zero. Thus a particular integral is

$$\tfrac{1}{20}x^5-\tfrac{1}{4}x^4+x^3-\tfrac{5}{2}x^2.$$

The complementary function is

$$Hx+J+Ke^{-x},$$

where H, J, K are arbitrary constants. If we had not taken $F = G = 0$ above, the terms so obtained would have been of the same type as $Hx+J$ in the complementary function.

The general solution is

$$\tfrac{1}{20}x^5-\tfrac{1}{4}x^4+x^3-\tfrac{5}{2}x^2+Hx+J+Ke^{-x}.$$

(iii) *$\phi(x)$ the exponential e^{ax} where $f(a) \neq 0$*

We require a particular integral of

$$f(D)y = e^{ax}. \tag{13}$$

As remarked earlier, we seek for a particular integral a solution of (13) of the same type as the right-hand side of (13), i.e. we seek a solution

$$y = Ye^{ax}, \tag{14}$$

where Y is a constant. Substituting in (13) we require

$$Yf(D)e^{ax} = e^{ax},$$

or, using § 12, Theorem 1,

$$Yf(a)e^{ax} = e^{ax}.$$

Thus $$Y = \frac{1}{f(a)},$$

and the required particular integral is

$$\frac{1}{f(a)}\,e^{ax}. \tag{15}$$

This argument breaks down if $f(a) = 0$: this case is discussed in (vi) below.

Ex. 5. $$(D^2+3D+2)y = \cosh 3x = \tfrac{1}{2}e^{3x}+\tfrac{1}{2}e^{-3x}.$$

Using (15), a particular integral is
$$\tfrac{1}{40}e^{3x}+\tfrac{1}{4}e^{-3x}.$$

The general solution is
$$\tfrac{1}{40}e^{3x}+\tfrac{1}{4}e^{-3x}+Ae^{-x}+Be^{-2x}.$$

(iv) *$\phi(x)$ the sine or cosine of ωx, provided $f(i\omega) \neq 0$*

Suppose that Y is a particular integral of
$$f(D)y = e^{i\omega x}, \tag{16}$$
found as in (iii): this will be a complex function of x, say $Y_1(x)+iY_2(x)$. Substituting this in the left-hand side of (16), and using the result
$$e^{i\omega x} = \cos\omega x+i\sin\omega x, \tag{17}$$
gives $$f(D)Y_1(x)+if(D)Y_2(x) = \cos\omega x+i\sin\omega x.$$
Equating real and imaginary parts on the two sides we see that the real and imaginary parts of Y, respectively, are particular integrals of
$$f(D)y = \cos\omega x, \tag{18}$$
and
$$f(D)y = \sin\omega x. \tag{19}$$

Thus to find a particular integral either of (18) or (19) we find the particular integral
$$\frac{1}{f(i\omega)}e^{i\omega x} \tag{20}$$
of (16) as in (15), and pick out its real part if a particular integral of (18) is required, or its imaginary part for (19).

In the reduction it is usually best to use the modulus–argument notation for a complex number. If
$$z = x+iy \tag{21}$$
is a complex number, x is called the real part of z, written $\mathbf{R}(z)$; y is called the imaginary part of z, written $\mathbf{I}(z)$; and $z^* = x-iy$ is called the conjugate of z. The modulus of z, $|z|$, is defined by
$$|z| = \surd(x^2+y^2), \tag{22}$$
and the argument of z, $\arg z$, is the angle θ defined by any two of
$$\sin\theta = \frac{y}{|z|}, \qquad \cos\theta = \frac{x}{|z|}, \qquad \tan\theta = \frac{y}{x}. \tag{23}$$

Since $y = |z|\sin\theta$ and $x = |z|\cos\theta$, the use of this notation enables us to write the complex number z in the form

$$z = x+iy = |z|(\cos\theta+i\sin\theta) = |z|e^{i\theta}. \tag{24}$$

The specification of the angle θ in (23) needs some care; it is equivalent to the statement that†

$$\left.\begin{aligned} \arg z &= \tan^{-1}(y/x) & \text{if } x > 0,\\ \arg z &= \pi+\tan^{-1}(y/x) & \text{if } x < 0 \end{aligned}\right\}, \tag{25}$$

where $\tan^{-1}(y/x)$ is the angle between $-\frac{1}{2}\pi$ and $\frac{1}{2}\pi$ whose tangent is y/x.

Ex. 6. $$(D^3+4D^2+5D+2)y = \cos\omega x. \tag{26}$$

We find the particular integral of

$$(D^3+4D^2+5D+2)y = e^{i\omega x}.$$

By (15) this is

$$\frac{1}{(i\omega)^3+4(i\omega)^2+5i\omega+2}e^{i\omega x} = \frac{1}{(2-4\omega^2)+i(5\omega-\omega^3)}e^{i\omega x}. \tag{27}$$

Now let Z and θ be the modulus and argument of the complex number

$$(2-4\omega^2)+i(5\omega-\omega^3), \tag{28}$$

these may be written down by (22) and (25). Then (28) may be written

$$(2-4\omega^2)+i(5\omega-\omega^3) = Ze^{i\theta}.$$

Using this, (27) becomes

$$\frac{1}{Z}e^{i(\omega x-\theta)}.$$

For the particular integral of (26) we need the real part of this, namely

$$\frac{1}{Z}\cos(\omega x-\theta). \tag{29}$$

The form (29) of the particular integral is probably the most useful for applications. A form involving $\cos\omega x$ and $\sin\omega x$ may be obtained by putting in it the values of Z and θ obtained from (28); alternatively this form may be obtained from (27) by multiplying it above and below by the conjugate of the denominator; this gives

$$\frac{[(2-4\omega^2)-i(5\omega-\omega^3)][\cos\omega x+i\sin\omega x]}{(2-4\omega^2)^2+(5\omega-\omega^3)^2}.$$

† The student is warned against the common practice of writing $\tan^{-1}(y/x)$ in place of $\arg z$ in general formulae in which the sign of x is not known—this may lead to errors. The value of θ defined above is ambiguous by a multiple of 2π. The angle θ such that $-\pi < \theta \leqslant \pi$ is usually used, and is called the principal value of the argument.

The real part of this is

$$\frac{(2-4\omega^2)\cos\omega x+(5\omega-\omega^3)\sin\omega x}{(2-4\omega^2)^2+(5\omega-\omega^3)^2}.$$

(v) *$\phi(x)$ of the form $e^{ax}\cos\omega x$ or $e^{ax}\sin\omega x$, provided $f(a+i\omega)\neq 0$*

In this case we proceed as in (iv), finding a particular integral of

$$f(D)y = e^{(a+i\omega)x} \tag{30}$$

and taking its real or imaginary part as required.

Ex. 7. $$(D^2+D+1)y = e^{ax}\sin\omega x. \tag{31}$$

As in (iii) the particular integral of

$$(D^2+D+1)y = e^{(a+i\omega)x}$$

is

$$\frac{1}{(a+i\omega)^2+(a+i\omega)+1}e^{(a+i\omega)x}$$

$$=\frac{1}{(a^2+a+1-\omega^2)+i(\omega+2a\omega)}e^{(a+i\omega)x}$$

$$=\frac{1}{Z}e^{(a+i\omega)x-i\theta}, \tag{32}$$

where Z and θ are the modulus and argument of

$$(a^2+a+1-\omega^2)+i(\omega+2a\omega).$$

Taking the imaginary part of (32) gives as a particular integral of (31)

$$\frac{1}{Z}e^{ax}\sin(\omega x-\theta).$$

Adding the complementary function, the complete solution is

$$\frac{1}{Z}e^{ax}\sin(\omega x-\theta)+Ae^{-\frac{1}{2}x}\cos(\tfrac{1}{2}x\sqrt{3}+B).$$

(vi) *$\phi(x)$ of the form e^{ax} with $f(a)=0$. This includes the case in which $\phi(x)$ is $\sin\omega x$ or $\cos\omega x$ with $f(i\omega)=0$*

Since $f(a)=0$, $f(D)$ must have $D-a$ as a factor so that the differential equation will be of the form

$$(D-a)^r F(D)y = e^{ax}, \tag{33}$$

where $F(a)\neq 0$.

For a particular integral we seek a solution of this of the form

$$y = Y(x)e^{ax} \tag{34}$$

in which the quantity $Y(x)$ multiplying e^{ax} is a function of x instead of a constant as in (14). If this is to satisfy (33) we must have

$$(D-a)^r F(D)\{Y(x)e^{ax}\} = e^{ax},$$

or, by § 12, Theorem 3,

$$e^{ax}D^rF(D+a)Y(x) = e^{ax}.$$

That is, $Y(x)$ is to be a function such that

$$D^rF(D+a)Y(x) = 1, \tag{35}$$

and the finding of such a function was discussed in (i).

Ex. 8. $$(D-a)^2(D-b)y = e^{ax}. \tag{36}$$

Assuming a particular integral

$$y = Y(x)e^{ax}$$

and substituting in (36) gives

$$(D-a)^2(D-b)\{Y(x)e^{ax}\} = e^{ax}.$$

Using § 12, Theorem 3, this becomes

$$D^2(D+a-b)Y(x) = 1,$$

and this is satisfied by $$Y(x) = \frac{x^2}{2(a-b)}.$$

Thus the required particular integral is

$$\frac{x^2}{2(a-b)}e^{ax}.$$

Adding the complementary function, the general solution of (36) is found to be

$$\left\{A+Bx+\frac{x^2}{2(a-b)}\right\}e^{ax}+Ce^{bx}.$$

Ex. 9. $$(D^2+n^2)y = \sin(nx+\beta). \tag{37}$$

We seek a particular integral of

$$(D^2+n^2)y = e^{i(nx+\beta)} \tag{38}$$

of the form $$y = Y(x)e^{inx}.$$

Substituting in (38) gives

$$(D+in)(D-in)\{Y(x)e^{inx}\} = e^{i(nx+\beta)},$$

or, using § 12, Theorem 3,

$$D(D+2in)Y(x) = e^{i\beta}.$$

This is satisfied by $$Y(x) = \frac{x}{2in}e^{i\beta},$$

so the required particular integral of (38) is

$$\frac{x}{2in}e^{i(nx+\beta)}.$$

The imaginary part of this, which is a particular integral of (37), is

$$-\frac{x}{2n}\cos(nx+\beta).$$

Adding the complementary function, the general solution of (37) is

$$A\cos nx+B\sin nx-\frac{x}{2n}\cos(nx+\beta).$$

(vii) $\phi(x)$ *of the form* $e^{ax}P_k(x)$, *where* $P_k(x)$ *is a polynomial in* x

The differential equation is

$$f(D)y = e^{ax}P_k(x). \tag{39}$$

For a particular integral we seek a solution of (39) of the form

$$y = Y(x)e^{ax}.$$

Substituting in (39) and using § 12, Theorem 3, as before, gives the equation for $Y(x)$

$$f(D+a)Y(x) = P_k(x). \tag{40}$$

The finding of $Y(x)$ from (40) was discussed in (ii).

(viii) $\phi(x)$ *any function of* x

We consider first the first-order equation

$$(D-\alpha_1)y = \phi(x), \tag{41}$$

and seek a particular integral

$$y = Y(x)e^{\alpha_1 x}. \tag{42}$$

Substituting (42) in (41) and using § 12, Theorem 3 gives

$$e^{\alpha_1 x}DY(x) = \phi(x).$$

That is

$$Y(x) = \int^{x} e^{-\alpha_1\xi}\phi(\xi)\,d\xi, \tag{43}$$

and the required particular integral of (41) is

$$e^{\alpha_1 x}\int^{x} e^{-\alpha_1\xi}\phi(\xi)\,d\xi. \tag{44}$$

In the same way, a particular integral of

$$(D-\alpha_1)^2y = \phi(x) \tag{45}$$

is found to be the repeated integral

$$e^{\alpha_1 x}\int^{x} d\eta \int^{\eta} e^{-\alpha_1\xi}\phi(\xi)\,d\xi. \tag{46}$$

Next consider the equation

$$(D-\alpha_1)(D-\alpha_2)y = \phi(x). \tag{47}$$

As before, we seek a particular integral of type

$$y = Y_1(x)e^{\alpha_1 x}. \tag{48}$$

Substituting in (47) and using § 12, Theorem 3, as before, gives

$$D(D+\alpha_1-\alpha_2)Y_1(x) = e^{-\alpha_1 x}\phi(x),$$

or
$$(D+\alpha_1-\alpha_2)Y_1(x) = \int^x e^{-\alpha_1\xi}\phi(\xi)\,d\xi. \tag{49}$$

(49) is of the same type as (41), so that, by (44),

$$Y_1(x) = e^{(\alpha_2-\alpha_1)x}\int^x e^{(\alpha_1-\alpha_2)\eta}\,d\eta\int^\eta e^{-\alpha_1\xi}\phi(\xi)\,d\xi. \tag{50}$$

From (48) and (50) the required particular integral of (47) is

$$e^{\alpha_2 x}\int^x e^{(\alpha_1-\alpha_2)\eta}\,d\eta\int^\eta e^{-\alpha_1\xi}\phi(\xi)\,d\xi \tag{51}$$

$$= \frac{e^{\alpha_1 x}}{\alpha_1-\alpha_2}\int^x e^{-\alpha_1\xi}\phi(\xi)\,d\xi - \frac{e^{\alpha_2 x}}{\alpha_1-\alpha_2}\int^x e^{-\alpha_2\xi}\phi(\xi)\,d\xi, \tag{52}$$

on integrating by parts.

For the equation
$$(D^2+n^2)y = \phi(x), \tag{53}$$

the particular integral (52) becomes

$$\frac{1}{n}\int^x \sin n(x-\xi)\phi(\xi)\,d\xi. \tag{54}$$

The method leading to (52) may be extended step by step, and it is found that the particular integral of

$$(D-\alpha_1)(D-\alpha_2)...(D-\alpha_n)y = \phi(x) \tag{55}$$

is
$$\sum_{r=1}^{n}\frac{1}{\beta_r}e^{\alpha_r x}\int^x e^{-\alpha_r\xi}\phi(\xi)\,d\xi, \tag{56}$$

where
$$\beta_r = (\alpha_r-\alpha_1)...(\alpha_r-\alpha_{r-1})(\alpha_r-\alpha_{r+1})...(\alpha_r-\alpha_n), \tag{57}$$

provided, of course, that $\alpha_1,..., \alpha_n$ are all different.

These results allow particular integrals to be written down for any form of $\phi(x)$. For the simple explicit functions discussed earlier this method is slower than those already given, but the general formulae are occasionally useful.

15. Simultaneous ordinary linear differential equations with constant coefficients

The solution of these follows from the theory of §§ 12–14, but with minor complications. To illustrate, we discuss the pair of homogeneous equations for u and v in terms of x

$$Du-a_{11}u-a_{12}v = 0 \tag{1}$$

$$Dv-a_{21}u-a_{22}v = 0. \tag{2}$$

We can eliminate v by differentiating (1)

$$D^2u-a_{11}Du-a_{12}Dv = 0$$

and substituting for Dv from (2)

$$D^2u-a_{11}Du-a_{12}(a_{21}u+a_{22}v) = 0. \qquad (3)$$

But from (1)

$$a_{12}v = Du-a_{11}u \qquad (4)$$

so that (3) becomes

$$D^2u-(a_{11}+a_{22})Du+(a_{11}a_{22}-a_{12}a_{21})u = 0$$

which is a second-order homogeneous equation in u. Its solution is

$$u = c_1e^{\alpha_1 x}+c_2e^{\alpha_2 x} \qquad (5)$$

where α_1 and α_2 are the roots of the auxiliary equation

$$\alpha^2-(a_{11}+a_{22})\alpha+(a_{11}a_{22}-a_{12}a_{21}) = 0. \qquad (6)$$

To find v, substitute (5) in (4), whence

$$v = \{(\alpha_1-a_{11})c_1e^{\alpha_1 x}+(\alpha_2-a_{11})c_2e^{\alpha_2 x}\}/a_{12}.$$

This same approach can be extended to more than two equations as well as to inhomogeneous equations with some function of x on the right-hand side of (1) or (2). (See Ex. 1 below.) However, for large sets of simultaneous equations, elimination can be complicated and tedious. The matrix methods described in § 109 can then be used.

We notice that a pair of first-order equations leads, on elimination, to a second-order equation in one of the unknowns. The solution therefore contains two arbitrary constants. In general a set of n first-order equations leads to an order n equation in one of the unknowns and hence to n arbitrary constants in the general solution. Similarly, a set of n second-order equations will have $2n$ arbitrary constants in its general solution. It is important to bear this rule in mind, for in more complicated examples spurious arbitrary constants can easily be introduced. For instance, if we had substituted the solution (5) for u in (2) instead of (4), we would have obtained

$$Dv-a_{22}v = a_{21}(c_1e^{\alpha_1 x}+c_2e^{\alpha_2 x})$$

and in solving this we would have introduced an extra arbitrary constant in the complementary function for v. It would then have been necessary to substitute both u and v in (1) to evaluate this constant.

Occasionally, in special cases, there are less (but never more) arbitrary constants than the preceding discussion suggests.

Ex. 1.
$$\left.\begin{aligned}(L_1 D+R_1)u+MDv &= 1\\ MDu+(L_2 D+R_2)v &= 0\end{aligned}\right\}, \tag{7}$$
with $L_1 L_2 = M^2$. Eliminating Du from these equations gives
$$MR_1 u+(M^2-L_1 L_2)Dv-L_1 R_2 v = M,$$
but $M^2 = L_1 L_2$, therefore
$$MR_1 u-L_1 R_2 v = M. \tag{8}$$
Differentiating (8) $\quad MR_1 Du-L_1 R_2 Dv = 0$

and substituting for MDu from (7)
$$(L_1 R_2+L_2 R_1)Dv+R_1 R_2 v = 0.$$
The solution of this is $\quad v = Ae^{-\alpha x} \quad (9)$

where $\quad \alpha = R_1 R_2/(L_1 R_2+L_2 R_1)$.

Substituting (9) in (8) gives
$$u = \frac{1}{R_1}+\left(\frac{L_1 R_2}{MR_1}\right)Ae^{-\alpha x}.$$

This example describes the case of 'perfect coupling' in a transformer. The fact that the solution contains only one arbitrary constant A is a direct consequence of the relationship $L_1 L_2 = M^2$. If $L_1 L_2 \neq M^2$, elimination of u leads to a second-order equation for v and hence to two arbitrary constants.

Just as one of the variables in a pair of simultaneous first-order equations can be eliminated to give a second-order equation in the remaining variable, so a single second-order equation can be reduced to a pair of simultaneous first-order equations by the introduction of a 'dummy' variable. This is a standard technique in the numerical solution of ordinary differential equations of order two or more (cf. § 116).

Ex. 2.

The equation for the undamped harmonic oscillator is
$$(D^2+\omega^2)y = 0. \tag{10}$$

Introduce a dummy variable v such that

$$Dy = v. \tag{11}$$

Then (10) becomes
$$Dv+\omega^2 y = 0, \tag{12}$$
and (11) and (12) are equivalent to the second-order equation (10).

The solution of (10) has the general form

$$y = A\sin\omega x+B\cos\omega x. \tag{13}$$

It follows that
$$v = Dy = \omega A\cos\omega x-\omega B\sin\omega x, \tag{14}$$
and eliminating x from (13) and (14) gives

$$v^2+\omega^2 y^2 = \omega^2(A^2+B^2), \tag{15}$$

which is the equation of an ellipse in the vy-phase plane. The phase trajectories for the undamped harmonic oscillator are therefore a set of concentric ellipses, each ellipse corresponding to particular values of the constants A and B.

16. Problems leading to ordinary linear differential equations

Problems on dynamics and electric circuit theory will be given in subsequent chapters; here we give a few in other fields.

Ex. 1. *The problem of § 8, Ex. 1.*

Writing D for d/dt, the equations to be solved are

$$(D+k)n = 0, \tag{1}$$

$$(D+k_1)n_1-kn = 0. \tag{2}$$

The solution of (1) is
$$n = Ae^{-kt}, \tag{3}$$
where A is an arbitrary constant. Putting (3) in (2) gives

$$(D+k_1)n_1 = kAe^{-kt},$$

of which the solution is

$$n_1 = \frac{kA}{k_1-k}e^{-kt}+Be^{-k_1 t}, \tag{4}$$

provided $k_1 \neq k$. The constants A and B can be determined from the initial conditions. Suppose that at $t = 0$ there were N atoms of A and none of B. That is, $n = N$, $n_1 = 0$, when $t = 0$. Substituting these values in (3) and (4) gives $A = N$, $B = -kN/(k_1-k)$.

Ex. 2. *The problem of § 8, Ex. 2; the linear problem with initial temperature T_1.*

We have to solve
$$M\frac{dT}{dt}+HT = HT_0 \tag{5}$$
with $T = T_1$ when $t = 0$.

The general solution of (5) is

$$T = T_0 + Ae^{-Ht/M},$$

and to make $T = T_1$ when $t = 0$ we must have

$$A = T_1 - T_0.$$

Ex. 3. *Steady flow of heat in a uniform rod.*

We take the x-axis in the direction of the rod, and suppose it to be so thin that its temperature T at x is constant across its cross-section. If a and K are the area and thermal conductivity of the rod, it is known from the theory of conduction of heat that the rate of flow of heat along it at the point x is

$$-Ka\frac{dT}{dx} \tag{6}$$

per unit time. Also we assume that the rod loses heat from its surface to its surroundings at the rate HT per unit time per unit surface area at any point.

Now consider the element of the rod between x and $x+\delta x$. Heat flows into this across the face x at the rate (6); it flows out across the face $x+\delta x$ at the rate

$$-Ka\frac{dT}{dx} - Ka\frac{d^2T}{dx^2}\delta x, \tag{7}$$

neglecting terms in δx^2; and if p is the perimeter of the rod, heat is lost from the surface at the rate

$$HpT\,\delta x. \tag{8}$$

Since the temperature in the rod is steady, the amount of heat flowing into the element must be equal to the amount flowing out, that is, by (6), (7), and (8),

$$\frac{d^2T}{dx^2} - \frac{Hp}{Ka}T = 0. \tag{9}$$

The differential equation (9) has to be solved with given boundary conditions. Suppose, for example, that the end $x = 0$ of the rod is at temperature T_1 and the end $x = l$ at T_2. The general solution of (9) may be written

$$T = A\sinh\mu x + B\sinh\mu(l-x),$$

where

$$\mu^2 = Hp/Ka, \tag{10}$$

and A and B are arbitrary constants. The conditions at the ends give

$$T_1 = B\sinh\mu l, \qquad T_2 = A\sinh\mu l,$$

and we get finally

$$T = \frac{T_1\sinh\mu(l-x) + T_2\sinh\mu x}{\sinh\mu l}. \tag{11}$$

Ex. 4. *Heat is generated within a slab $0 < x < l$ of solid at the rate $a+bT$ per unit time per unit volume, where a and b are constants and T is the temperature.*

We consider the case in which there is no flow of heat over the plane

$x = 0$, and flow at a rate H times the temperature over the plane $x = l$. That is, as in (6)

$$\frac{dT}{dx} = 0, \qquad x = 0, \tag{12}$$

$$-K\frac{dT}{dx} = HT, \qquad x = l. \tag{13}$$

We calculate the steady temperature under these conditions. The differential equation for T is

$$\frac{d^2T}{dx^2}+\frac{b}{K}T = -\frac{a}{K}; \tag{14}$$

this may be obtained by an argument similar to that of Ex. 3 or by putting $\partial T/\partial t = 0$ in § 119 (21).

The general solution of (14) is

$$T = A\sin\omega x+B\cos\omega x-\frac{a}{b}, \tag{15}$$

where $$\omega^2 = b/K.$$

The boundary conditions (12) and (13) give

$$A = 0,$$

$$\omega KB\sin\omega l = HB\cos\omega l-Ha/b.$$

Thus, finally, $$T = \frac{Ha\cos\omega x}{b(H\cos\omega l-\omega K\sin\omega l)}-\frac{a}{b}. \tag{16}$$

If b is small, ω is small and the denominator in (16) is positive. But the denominator decreases as b increases, and is zero when ω is the smallest root of

$$K\omega\tan\omega l = H. \tag{17}$$

Thus if b, H, l, etc., are connected by this critical relation the steady temperature becomes infinite. Physically this means that heat is being produced more rapidly than it can be conducted away through the solid. An effect of this sort always appears in questions involving explosions or spontaneous combustion, but in fact heat is usually generated at rates such as ae^{kT} or $ae^{-k/T}$ which lead to more difficult non-linear equations.

Ex. 5. *Parallel flow in a heat interchanger.*

Suppose that two fluids are flowing steadily in the direction of the x-axis on either side of a thin partition. The fluids are supposed to be so well stirred that their temperatures are constant in any plane $x =$ constant. Let M_1 be the mass of the first fluid in contact with unit area of the partition, c_1 its specific heat, u_1 its velocity, and T_1 its temperature at x; let M_2, c_2, u_2, T_2 be the corresponding quantities for the second fluid.

The partition is supposed to be such that the rate of flow of heat across it at any point, in the direction from the first fluid to the second, is

$$b(T_1-T_2) \tag{18}$$

per unit time per unit area, where b is a constant.

To find the differential equations satisfied by T_1 and T_2 consider the first fluid in the region between x and $x+\delta x$. Heat is carried into this region at the rate

$$M_1 u_1 c_1 T_1, \tag{19}$$

per unit time, per unit width normal to the direction of flow, and is carried out of it over the plane $x+\delta x$ at the rate

$$M_1 u_1 c_1 T_1 + M_1 u_1 c_1 \frac{dT_1}{dx}\,\delta x. \tag{20}$$

Also heat flows through the partition at the rate

$$b(T_1 - T_2)\,\delta x. \tag{21}$$

Since the flow of heat is assumed to be steady, the rate of flow of heat into the region must be equal to the rate of flow out, that is, by (19), (20), (21)

$$M_1 u_1 c_1 \frac{dT_1}{dx}\delta x + b(T_1 - T_2)\,\delta x = 0.$$

Therefore

$$k_1 \frac{dT_1}{dx} + T_1 - T_2 = 0, \tag{22}$$

where

$$k_1 = M_1 u_1 c_1/b.$$

In the same way, considering the second fluid we should have

$$k_2 \frac{dT_2}{dx} - (T_1 - T_2) = 0, \tag{23}$$

where

$$k_2 = M_2 u_2 c_2/b.$$

Adding (22) and (23) gives

$$k_1 \frac{dT_1}{dx} + k_2 \frac{dT_2}{dx} = 0. \tag{24}$$

Any two of (22)–(24) may be taken as the differential equations of the problem. Suppose that the 'hot fluid' enters at temperature T and the 'cold fluid' at zero. Then we have to solve (22) and (23) with

$$T_1 = T, \qquad T_2 = 0 \qquad \text{when } x = 0. \tag{25}$$

(24) and (25) give immediately

$$k_1 T_1 + k_2 T_2 = k_1 T, \tag{26}$$

and substituting (26) in (22) gives

$$\frac{dT_1}{dx} + \frac{k_1 + k_2}{k_1 k_2} T_1 = \frac{T}{k_2}. \tag{27}$$

The solution of this with $T_1 = T$ when $x = 0$ is

$$T_1 = \frac{T}{k_1 + k_2}\{k_1 + k_2 e^{-(k_1+k_2)x/k_1k_2}\}, \tag{28}$$

and T_2 follows from (26). The temperatures of both fluids tend to $k_1 T/(k_1 + k_2)$ for large values of x.

17. Heaviside's unit function and Dirac's delta function

Heaviside's unit function $H(x)$ is defined by

$$\left.\begin{aligned} H(x) &= 0, \quad x \leqslant 0 \\ H(x) &= 1, \quad x > 0 \end{aligned}\right\}. \tag{1}$$

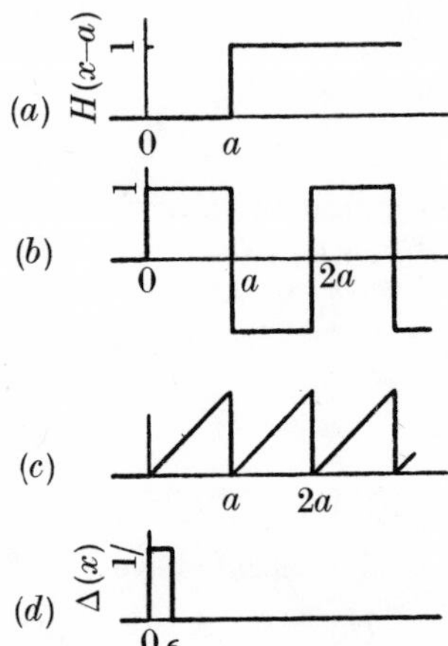

FIG. 2

It has an ordinary discontinuity at the point $x = 0$ and was defined to facilitate the representation of functions which have such discontinuities. The graph of $H(x-a)$ is shown in Fig. 2 (*a*).

For example, the function

$$H(x)-H(x-a)$$

is unity in $0 < x \leqslant a$, and zero for $x \leqslant 0$ and $x > a$.

The function

$$H(x)\sin x+H(x-\pi)\sin(x-\pi)$$

has the value $\sin x$ for $0 \leqslant x \leqslant \pi$ and is zero for $x < 0$ and $x > \pi$.

The 'square wave' of Fig. 2 (*b*) is represented by

$$H(x)+2\sum_{r=1}^{\infty}(-1)^r H(x-ra), \tag{2}$$

and the saw-tooth wave of Fig. 2 (*c*) by

$$x-\sum_{r=1}^{\infty} H(x-ra). \tag{3}$$

Periodic functions of forms such as these are of considerable importance in modern technical practice.

Since $H(x)$ has only an ordinary discontinuity it is integrable and may be used in formulae involving definite integrals—the results again may often be expressed simply in terms of unit functions. Thus

$$\left.\begin{aligned} \int_0^x H(\xi-a)\,d\xi &= 0, \qquad & x < a \\ &= (x-a), \qquad & x > a \end{aligned}\right\},$$

which may be written

$$\int_0^x H(\xi-a)\,d\xi = (x-a)H(x-a). \tag{4}$$

Similarly

$$\int_0^x (\xi-a)^n H(\xi-a)\,d\xi = \frac{1}{n+1}(x-a)^{n+1}H(x-a). \tag{5}$$

Particular integrals of differential equations of type

$$f(D)y = \phi(x),$$

where $\phi(x)$ contains Heaviside functions, may be written down from the integral formulae of § 14 (viii). For example, a particular integral of

$$(D^2+n^2)y = H(x-a)$$

is, by § 14 (54),

$$\frac{1}{n}\int_0^x \sin n(x-\xi)H(\xi-a)\,d\xi = \frac{1}{n^2}\{1-\cos n(x-a)\}H(x-a). \tag{6}$$

The particular integral (6) is available for all values of x. Using the ordinary method we would have to treat the regions $x < a$ and $x > a$ separately.

In all the above formulae, only integration of $H(x)$ has been in question and offers no difficulty. But care is needed in formulae involving differentiation: the differential coefficient of $H(x)$ is zero except at $x = 0$, where it is not defined. It may be regarded as being the function $\delta(x)$ defined below.

The Dirac delta function. This is not a mathematical function at all in the usual sense of the term. Its use is to represent symbolically an ideally concentrated quantity (such as a concentrated load on a beam, or an impulsive force in dynamics) in the same way that Heaviside's unit function was used to represent a discontinuous quantity.

We define the delta function $\delta(x)$ as the limit as $\epsilon \to 0$ of the function† $\Delta(x)$ shown in Fig. 2 (*d*) defined by

$$\left.\begin{aligned} \Delta(x) &= 0, & x &\leqslant 0 \\ \Delta(x) &= 1/\epsilon, & 0 < x &\leqslant \epsilon \\ \Delta(x) &= 0, & x &> \epsilon \end{aligned}\right\}, \tag{7}$$

or, in terms of the unit function,

$$\delta(x) = \lim_{\epsilon \to 0} \frac{H(x) - H(x-\epsilon)}{\epsilon}. \tag{8}$$

Thus $\delta(x)$ is very large in a vanishingly small region to the right of $x = 0$ and is zero elsewhere. Also from (7)

$$\int_{-\infty}^{\infty} \Delta(x)\,dx = \frac{1}{\epsilon}\int_{0}^{\epsilon} dx = 1,$$

for all ϵ however small. As $\epsilon \to 0$ this becomes

$$\int_{-\infty}^{\infty} \delta(x)\,dx = 1. \tag{9}$$

In the same way

$$\left.\begin{aligned} \int_{-\infty}^{x} \Delta(\xi - a)\,d\xi &= 0, & x &\leqslant a \\ &= 1, & x &> a+\epsilon \end{aligned}\right\},$$

so that, in the limit as $\epsilon \to 0$,

$$\int_{-\infty}^{x} \delta(\xi - a)\,d\xi = H(x-a). \tag{10}$$

Also, if $f(x)$ is continuous in the range $a \leqslant x \leqslant a+\epsilon$,

$$\begin{aligned} \int_{-\infty}^{\infty} \Delta(x-a)f(x)\,dx &= \frac{1}{\epsilon}\int_{a}^{a+\epsilon} f(x)\,dx \\ &= f(a+\theta\epsilon), \quad 0 < \theta < 1, \end{aligned} \tag{11}$$

by the first mean value theorem for integrals. In the limit as $\epsilon \to 0$, (11) becomes

$$\int_{-\infty}^{\infty} f(x)\delta(x-a)\,dx = f(a). \tag{12}$$

† There are many other functions which possess the properties of $\delta(x)$ in the limit as a parameter tends to zero. For example, the continuous function $(1/\epsilon\pi^{\frac{1}{2}})\exp[-x^2/\epsilon^2]$ has often been used.

Since the δ function is defined by a limiting process, all operations on it except the very simplest will involve the interchange of order of limiting processes and be difficult to justify rigorously. This justification can be supplied by comparatively advanced mathematics, but here we shall simply use the function as a convenient short notation and regard results obtained by its use as suggested rather than proved. Many interesting results may be obtained by treating $\delta(x)$ as an ordinary function with the properties (9), (10), and (12), and these can usually be verified by more conventional analysis. For example, a particular integral of

$$(D^2+n^2)y = \delta(x-a) \tag{13}$$

is, by § 14 (54)

$$\left.\begin{aligned}\frac{1}{n}\int\limits_0^x \sin n(x-\xi)\delta(\xi-a)\,d\xi &= 0, & x<a\\ &= \frac{1}{n}\sin n(x-a), & x>a\end{aligned}\right\}.$$

This may be written

$$\frac{1}{n}\sin n(x-a)H(x-a). \tag{14}$$

An important use of the δ function is in the passage from a continuously varying quantity to a discrete or concentrated one. Suppose, for example, that formulae have been derived for the behaviour of a beam with a continuously varying load $w(x)$ per unit length. A concentrated load W at the point $x = a$ may be treated by putting

$$w(x) = W\,\delta(x-a)$$

in the formulae.

18. The Laplace transformation method

This is a method for the solution of an ordinary linear differential equation with constant coefficients (or of systems of such equations) with given values of the function and its derivatives at $x = 0$.

We define the *Laplace transform* $\bar{y}$ of a function y of x as

$$\bar{y} = \int\limits_0^\infty e^{-px}y\,dx, \tag{1}$$

where p is supposed to be a real positive number sufficiently large to make the integral (1) convergent. $\bar{y}$ is a function of p, and to emphasize this we shall sometimes write it as $\bar{y}(p)$.†

We begin by calculating the Laplace transforms of the common functions that will be needed.

If

$$y = 1, \qquad \bar{y} = \int\limits_0^\infty e^{-px}\,dx = \frac{1}{p}; \tag{2}$$

$$y = H(x-a), \quad \bar{y} = \int\limits_a^\infty e^{-px}\,dx = \frac{1}{p}\,e^{-ap}; \tag{3}$$

$$y = \delta(x-a), \quad \bar{y} = \int\limits_0^\infty e^{-px}\,\delta(x-a)\,dx = e^{-ap}; \tag{4}$$

$$y = e^{\alpha x}, \qquad \bar{y} = \int\limits_0^\infty e^{-(p-\alpha)x}\,dx = \frac{1}{p-\alpha}. \tag{5}$$

(5), in which α may be real or complex, may be used to give the transforms of $\cos\omega x$, etc. Thus if

$$y = \cos\omega x = \tfrac{1}{2}e^{i\omega x}+\tfrac{1}{2}e^{-i\omega x},$$

$$\bar{y} = \frac{1}{2(p-i\omega)}+\frac{1}{2(p+i\omega)} = \frac{p}{p^2+\omega^2}. \tag{6}$$

Proceeding in this way, or quoting the results as known definite integrals, we can construct a table of Laplace transforms containing all those needed for the solution of differential equations of the types discussed in this chapter.

This table can be extended by a simple theorem.

THEOREM 1. *If $\bar{y}(p)$ is the Laplace transform of y, then $\bar{y}(p+a)$ is the Laplace transform of $e^{-ax}y$.*

This follows immediately since

$$\int\limits_0^\infty e^{-px}e^{-ax}y\,dx = \int\limits_0^\infty e^{-(p+a)x}y\,dx = \bar{y}(p+a).$$

† The letter s is often used instead of p in various applications of the Laplace transform: $\bar{y} = \bar{y}(s)$.

y	$\bar{y}$	
1	$1/p$	(7)
x^n	$\dfrac{n!}{p^{n+1}}, \quad n = 0, 1, 2,...$	(8)
$H(x-a)$	$\dfrac{1}{p} e^{-ap}$	(9)
$\delta(x-a)$	e^{-ap}	(10)
$e^{\alpha x}$	$\dfrac{1}{p-\alpha}$	(11)
$x^n e^{\alpha x}$	$\dfrac{n!}{(p-\alpha)^{n+1}}, \quad n = 0, 1, 2,...$	(12)
$\sin \omega x$	$\dfrac{\omega}{p^2+\omega^2}$	(13)
$\cos \omega x$	$\dfrac{p}{p^2+\omega^2}$	(14)
$\sinh \alpha x$	$\dfrac{\alpha}{p^2-\alpha^2}$	(15)
$\cosh \alpha x$	$\dfrac{p}{p^2-\alpha^2}$	(16)
$\dfrac{x}{2\omega} \sin \omega x$	$\dfrac{p}{(p^2+\omega^2)^2}$	(17)
$\dfrac{1}{2\omega^3} (\sin \omega x - \omega x \cos \omega x)$	$\dfrac{1}{(p^2+\omega^2)^2}$	(18)

As an example, (12) above follows immediately from (8) by this theorem. Also from (13) and (14), respectively, it follows that

$$\frac{\omega}{(p+\alpha)^2+\omega^2} \text{ is the transform of } e^{-\alpha x} \sin \omega x, \tag{19}$$

and
$$\frac{p+\alpha}{(p+\alpha)^2+\omega^2} \text{ is the transform of } e^{-\alpha x} \cos \omega x. \tag{20}$$

Next we need a theorem on the Laplace transforms of the derivatives of a function.

THEOREM 2. *If $y, Dy,..., D^{n-1}y$ are continuous functions of x, and $y_0, y_1,..., y_{n-1}$ are their values when $x = 0$, then*

$p\bar{y}-y_0$	*is the transform of*	$Dy,$	(21)
$p^2\bar{y}-py_0-y_1$	,, ,,	$D^2y,$	(22)
$p^3\bar{y}-p^2y_0-py_1-y_2$	,, ,,	$D^3y,$	(23)
$p^n\bar{y}-p^{n-1}y_0-p^{n-2}y_1...-y_{n-1}$	,, ,,	$D^ny.$	(24)

These results follow immediately by integration by parts. Thus the Laplace transform of Dy is

$$\int\limits_0^\infty e^{-px}\frac{dy}{dx}\,dx = \left[ye^{-px}\right]_0^\infty + p\int\limits_0^\infty e^{-px}y\,dx = p\bar{y}-y_0,$$

since y_0 is the value of y when $x = 0$. Again

$$\int\limits_0^\infty e^{-px}\frac{d^2y}{dx^2}\,dx = \left[e^{-px}\frac{dy}{dx}\right]_0^\infty + p\int\limits_0^\infty e^{-px}\frac{dy}{dx}\,dx = p^2\bar{y}-py_0-y_1.$$

(23) and (24) follow in the same way.

Suppose, now, that we wish to find the solution of

$$(D^n+a_1D^{n-1}+...+a_n)y = \phi(x) \tag{25}$$

which has the values $y_0, y_1,..., y_{n-1}$ of $y, Dy,..., D^{n-1}y$ when $x = 0$. We take the Laplace transform of both sides of (25), using (21) to (24) in the left-hand side, and writing down the Laplace transform of $\phi(x)$ from the table. We thus obtain

$$(p^n+a_1p^{n-1}+...+a_n)\bar{y} = \bar{\phi}+a_{n-1}y_0+a_{n-2}(py_0+y_1)+...+$$
$$+(p^{n-1}y_0+p^{n-2}y_1+...+y_{n-1}). \tag{26}$$

The equation (26) is called the *subsidiary equation* corresponding to the differential equation (25) and its initial conditions. With a little practice, particularly for first- and second-order equations, subsidiary equations can be written down immediately.

(26) gives $\bar{y}$, the Laplace transform of the solution y. If $\phi(x)$ is one of the common functions appearing in the table, $\bar{y}$ is a quotient of polynomials in p, the degree of whose numerator is less than that of its denominator. Thus if the roots of the equation

$$p^n+a_1p^{n-1}+...+a_n = 0 \tag{27}$$

are known, $\bar{y}$ can be expressed in partial fractions. When this has been done, y can be found from $\bar{y}$ by looking up each fraction in the table of transforms (using Theorem 1 if any of the fractions have general quadratic denominators). The result found in this way is certainly the unique solution of the problem since there is a theorem (Lerch's theorem) which states that if two continuous functions have the same Laplace transform they must be identical.

The equation (27) is the auxiliary equation § 12 (10) in the present notation: naturally, whatever method of solving the differential equation is used this equation will appear.

It will be noticed that the whole of the algebra of the solution consists of the expressing of $\bar{y}$ in partial fractions: if the roots of (27) are all different this may be done by the formula†

$$\frac{f(p)}{g(p)} = \sum_{r=1}^{n} \frac{f(\alpha_r)}{g'(\alpha_r)(p-\alpha_r)} \tag{28}$$

$$= \sum_{r=1}^{n} \frac{1}{p-\alpha_r}\left[\frac{(p-\alpha_r)f(p)}{g(p)}\right]_{p=\alpha_r}, \tag{29}$$

where $\alpha_1,..., \alpha_n$ are the roots of

$$g(p) = 0, \tag{30}$$

provided these are all different (but not necessarily real).

Systems of simultaneous linear differential equations with constant coefficients may be treated in the same way.

Ex. 1. $$(D^2+n^2)y = \sin\omega x, \quad \omega \neq n,$$

with $y = y_0$, $Dy = y_1$, when $x = 0$.

The subsidiary equation is, by (13) and (22),

$$(p^2+n^2)\bar{y} = \frac{\omega}{p^2+\omega^2}+py_0+y_1.$$

Thus $$\bar{y} = \frac{\omega}{\omega^2-n^2}\left\{\frac{1}{p^2+n^2}-\frac{1}{p^2+\omega^2}\right\}+\frac{py_0+y_1}{p^2+n^2}.$$

Therefore, by (13) and (14),

$$y = \frac{1}{n(\omega^2-n^2)}\{\omega\sin nx-n\sin\omega x\}+y_0\cos nx+\frac{y_1}{n}\sin nx.$$

Ex. 2. $$(D^2+2\kappa D+n^2)y = 0$$

with $y = y_0$, $Dy = y_1$, when $x = 0$.

The subsidiary equation is

$$(p^2+2\kappa p+n^2)\bar{y} = py_0+y_1+2\kappa y_0.$$

Thus $$\bar{y} = \frac{(p+\kappa)y_0}{(p+\kappa)^2+(n^2-\kappa^2)}+\frac{y_1+\kappa y_0}{(p+\kappa)^2+(n^2-\kappa^2)}.$$

† Gibson, *Treatise on the Calculus* (1906), § 120.

Therefore, by (13), (14), and Theorem 1,

$$y = e^{-\kappa x}\left\{y_0 \cos x(n^2-\kappa^2)^{\frac{1}{2}} + \frac{(y_1+\kappa y_0)}{(n^2-\kappa^2)^{\frac{1}{2}}} \sin x(n^2-\kappa^2)^{\frac{1}{2}}\right\},$$

provided $n^2 > \kappa^2$.

Ex. 3.
$$\left.\begin{aligned}(D+1)y+Dz &= 0\\ (D-1)y+2Dz &= e^{-x}\end{aligned}\right\},$$

with $y = y_0$, $z = 0$, when $x = 0$.

The subsidiary equations are

$$(p+1)\bar{y}+p\bar{z} = y_0,$$

$$(p-1)\bar{y}+2p\bar{z} = \frac{1}{p+1}+y_0.$$

Solving, for example, for $\bar{y}$, we get

$$\bar{y} = \frac{y_0}{p+3}-\frac{1}{(p+1)(p+3)}$$

$$= \frac{y_0}{p+3}-\frac{1}{2(p+1)}+\frac{1}{2(p+3)}.$$

Therefore
$$y = (y_0+\tfrac{1}{2})e^{-3x}-\tfrac{1}{2}e^{-x}.$$

EXAMPLES ON CHAPTER II

1. Solve the following differential equations

(i) $(D^2+2D+4)y = 8.$
(ii) $D^3(D^2+n^2)y = 1.$
(iii) $(D^3+4D^2+5D+2)y = 4+2x+2x^2.$
(iv) $D^2(D^2-a^2)y = x-x^2.$
(v) $(D^3+3D^2+3D+2)y = 1+e^{-x}.$
(vi) $(D^3+2D^2+D+1)y = e^{2x}.$
(vii) $(D^2-D-2)y = \sin \omega x.$
(viii) $(D^3+D^2-D-1)y = \cos 2x.$
(ix) $(D^2+n^2)y = e^{-\alpha x}\sin \beta x.$
(x) $(D+1)y = e^{-x}\cos x.$
(xi) $(D+1)^2y = 1-e^{-x}.$
(xii) $(D^2-4)y = \cosh 2x.$
(xiii) $(D+1)(D^2+n^2)y = \sin nx.$
(xiv) $(D^2+2D+2)y = e^{-x}\sin x.$
(xv) $(D+1)(D+2)y = (1+x)e^{-x}.$
(xvi) $(D^2+4)y = x\sin x.$
(xvii) $(D^2+n^2)y = x-2(x-a)H(x-a).$
(xviii) $(D-1)u-2v = e^{-x};\ -2u+(D-1)v = 1.$
(xix) $Du-v = 0;\ Dv-w = 0;\ Dw-u = 0.$
(xx) $(D^2-4D)u-(D-1)v = e^{4x};\ (D+6)u+(D^2-D)v = 0.$

The solutions are as follows:

(i) $Ae^{-x}\cos(x\sqrt{3}+B)+2.$

(ii) $A+Bx+Cx^2+x^3/6n^2+E\cos(nx+F).$

(iii) $x^2-4x+8+(A+Bx)e^{-x}+Ce^{-2x}.$

(iv) $\dfrac{1}{12a^2}x^4-\dfrac{1}{6a^2}x^3+\dfrac{1}{a^4}x^2+A+Bx+Ce^{ax}+Ee^{-ax}.$

(v) $Ae^{-2x}+Be^{-\frac{1}{2}x}\cos(\frac{1}{2}x\sqrt{3}+C)+\frac{1}{2}+e^{-x}.$

(vi) $Ae^{-1\cdot 75x}+Be^{-0\cdot 12x}\cos(0\cdot 74x+C)+\frac{1}{19}e^{2x}.$

(vii) $Ae^{2x}+Be^{-x}-\dfrac{(\omega^2+2)\sin\omega x-\omega\cos\omega x}{\omega^4+5\omega^2+4}.$

(viii) $Ae^{x}+(B+Cx)e^{-x}-(\cos 2x+2\sin 2x)/25.$

(ix) $A\cos(nx+B)+\dfrac{(\alpha^2+n^2-\beta^2)\sin\beta x+2\alpha\beta\cos\beta x}{(\alpha^2+n^2-\beta^2)^2+4\alpha^2\beta^2}.$

(x) $Ae^{-x}+e^{-x}\sin x.$

(xi) $1+(A+Bx-\frac{1}{2}x^2)e^{-x}.$

(xii) $A\cosh 2x+(B+\frac{1}{4}x)\sinh 2x.$

(xiii) $Ae^{-x}+\left(B-\dfrac{x}{2(1+n^2)}\right)\sin nx+\left(C-\dfrac{x}{2n(1+n^2)}\right)\cos nx.$

(xiv) $(A-\frac{1}{2}x)e^{-x}\cos x+Be^{-x}\sin x.$

(xv) $(A+\frac{1}{2}x^2)e^{-x}+Be^{-2x}.$

(xvi) $A\sin 2x+B\cos 2x+\frac{1}{3}x\sin x-\frac{2}{9}\cos x.$

(xvii) $A\sin nx+B\cos nx+\dfrac{x}{n^2}-\dfrac{2}{n^3}\{n(x-a)-\sin n(x-a)\}H(x-a).$

(xviii) $u=-\frac{2}{3}-e^{-x}(2A+\frac{1}{4}-\frac{1}{2}x)+2Be^{3x},$
$v=\frac{1}{3}+2(A-\frac{1}{4}x)e^{-x}+2Be^{3x}.$

(xix) $u=Ae^{x}-\frac{1}{2}(B+C\sqrt{3})e^{-\frac{1}{2}x}\cos\frac{1}{2}x\sqrt{3}-\frac{1}{2}(C-B\sqrt{3})e^{-\frac{1}{2}x}\sin\frac{1}{2}x\sqrt{3},$
$v=Ae^{x}+(B\cos\frac{1}{2}x\sqrt{3}+C\sin\frac{1}{2}x\sqrt{3})e^{-\frac{1}{2}x},$
$w=Ae^{x}-\frac{1}{2}(B-C\sqrt{3})e^{-\frac{1}{2}x}\cos\frac{1}{2}x\sqrt{3}-\frac{1}{2}(C+B\sqrt{3})e^{-\frac{1}{2}x}\sin\frac{1}{2}x\sqrt{3}.$

(xx) $u=2Ae^{-x}+2Ce^{2x}+6Ee^{3x}+\frac{2}{5}e^{4x},$
$v=-5Ae^{-x}-7Be^{x}-8Ce^{2x}-9Ee^{3x}-\frac{1}{3}e^{4x}.$

2. Solve $x^2D^2y+4xDy+2y=x$. Equations such as this in which all terms are of the type x^rD^ry are known as 'homogeneous' or 'Euler' equations and can be reduced to linear equations, and thus solved, by the substitution $x=e^t$. The solution is $y=Ax^{-1}+Bx^{-2}+x/6$.

3. Solve the following with initial values y_0, y_1,... of y, Dy,..., etc.

(i) $(D+1)(D+2)y=e^{-x}.$

(ii) $(D^2+n^2)y=\sin nx.$

(iii) $(D+1)u-(5D+7)v=1;\ u+(D-1)v=0.$

The solutions are

(i) $(y_1+2y_0+x-1)e^{-x}-(y_1+y_0-1)e^{-2x}.$

(ii) $\left(\dfrac{1}{2n^2}+\dfrac{y_1}{n}\right)\sin nx+\left(y_0-\dfrac{x}{2n}\right)\cos nx.$

(iii) $u=-\frac{1}{6}+(12v_0-3u_0+\frac{3}{2})e^{-2x}+(4u_0-12v_0-\frac{4}{3})e^{-3x},$
$v=-\frac{1}{6}+(4v_0-u_0+\frac{1}{2})e^{-2x}+(u_0-3v_0-\frac{1}{3})e^{-3x}.$

4. The concentrations of two fission by-products in a nuclear reactor are x and y. Of these, y is called a 'poison' because it inhibits the fission process. When the reactor has been running at constant power for some time, the concentrations x and y reach equilibrium levels x_0 and y_0. If the reactor is then shut down, show that the 'poison' level y reaches its maximum value after a time

$$\left(\frac{1}{\lambda_2-\lambda_1}\right)\ln\left\{\frac{\lambda_2}{\lambda_1}\left(1-\frac{\lambda_2-\lambda_1}{\lambda_1}\frac{y_0}{x_0}\right)\right\}.$$

The relevant differential equations after shut-down are

$$\frac{dx}{dt} = -\lambda_1 x \quad \text{and} \quad \frac{dy}{dt} = \lambda_1 x-\lambda_2 y.$$

5. (i) Show that if $\bar{y}$ is the Laplace transform of y, $\bar{y}/p$ is the transform of

$$\int\limits_0^x y(\xi)\,d\xi.$$

(ii) Show that if $\bar{y}(p)$ is the Laplace transform of $y(x)$, $\bar{y}(p/\omega)$ is the transform of $\omega y(\omega x)$.

(iii) Show that if $\bar{y}$ and $\bar{z}$ are the Laplace transforms of y and z, $\bar{y}\bar{z}$ is the Laplace transform of

$$\int\limits_0^x y(x-\xi)z(\xi)\,d\xi = \int\limits_0^x y(\xi)z(x-\xi)\,d\xi.$$

The proof depends on a change of variable in the double integrals.

(iv) Using (iii) and § 18 (28), deduce § 14 (56).

6. A tank contains volume V of water, initially at zero temperature. Water is run off from it at a constant rate of volume v per second, and replaced at the same rate by water at temperature T_1. Show that the temperature at time t is $$T_1(1-e^{-vt/V}).$$

7. Two tanks A, B, each of volume V, contain water at time $t = 0$. For $t > 0$, volume v of solution containing mass m of solute flows into A per second; mixture flows from A to B at the same rate; and mixture flows away from B at the same rate. Show that the mass of solute in B at any time is $$(mV/v)(1-e^{-vt/V})-mte^{-vt/V}.$$

8. A substance A changes into a substance B at a rate α times the amount of A present; B changes into C at a rate β times the amount of B present. If initially only A is present and its amount is a, show that the amount of C at time t is

$$a+a(\beta e^{-\alpha t}-\alpha e^{-\beta t})/(\alpha-\beta).$$

9. Refer to the glucose–insulin model of § 7, Ex. 4. If $u = v = 0$ at $t = 0$ and if $I = \alpha e^{-\alpha t}$, show that in the limit as $\alpha \to \infty$,

$$u = \{(\beta_1+s)e^{\beta_1 t}+(\beta_2+s)e^{\beta_2 t}\}/(\beta_1-\beta_2),$$

where β_1 and β_2 are the roots of $\beta^2+(a+s)\beta+(rb+sa) = 0$. What is the physical interpretation of I as $\alpha \to \infty$? (Cf. § 17.)

10. y_1, y_2, y_3,... are the number of atoms present at time t of the elements of a radioactive series in which the rth element decays into the $(r+1)$th at a rate λ_r times the number of atoms y_r present. If $y_1 = N$, $y_2 = y_3 = ... = 0$ when $t = 0$, show (preferably by using § 18 (29)) that at time t

$$y_r = N\lambda_1\lambda_2...\lambda_{r-1}\sum_{s=1}^{r}\frac{1}{\beta_s}e^{-\lambda_s t},$$

where $$\beta_s = (\lambda_1-\lambda_s)...(\lambda_{s-1}-\lambda_s)(\lambda_{s+1}-\lambda_s)...(\lambda_r-\lambda_s).$$

11. In an epidemic the rate at which healthy people become infected is α times their number, the rates of recovery and death of sick people are, respectively, β and γ times their number. If initially there are N healthy people and no sick ones, show that the number of deaths up to time t is

$$\alpha\gamma N(c-d+de^{ct}-ce^{dt})/\{cd(c-d)\},$$

where c and d are the roots of $(z+\alpha)(z+\beta+\gamma)-\alpha\beta = 0$.

12. Mass M_1 of a perfect conductor of specific heat c_1, initially at temperature T, is placed at $t = 0$ in a calorimeter containing mass M_2 of water, of specific heat unity, at zero temperature. Heat is exchanged between the mass M_1 and the water at a rate k_1 times their temperature difference, and heat is lost by the water at a rate k_2 times its temperature. If T_1 and T_2 are the temperatures of the mass M_1 and the water, show that

$$(D+a)T_1-aT_2 = 0, \qquad (D+b+c)T_2-bT_1 = 0,$$

where $b = k_1/M_2$, $c = k_2/M_2$, $a = k_1/M_1 c_1$. If λ_1 and λ_2 are the roots of $\alpha^2+(a+b+c)\alpha+ac = 0$, show that these are both real and negative, and that

$$T_2 = bT(e^{\lambda_1 t}-e^{\lambda_2 t})/(\lambda_1-\lambda_2).$$

13. If the end $x = 0$ of a uniform rod of length l, which loses heat from its surface at a rate H times its temperature, is held at temperature T_1 and there is no loss of heat at the end $x = l$, show that the temperature at the point x is $T_1\cosh\mu(l-x)\operatorname{sech}\mu l$, in the notation of § 16, Ex. 3. Deduce that the ratio of the heat removed by a long thin cooling fin of thickness d on a surface to that which would be removed from the area at its base if it were not present is $(2/\mu d)\tanh\mu l$.

14. If a thin wire is heated by electric current the differential equation for its temperature T is

$$\frac{d^2T}{dx^2}+b^2T = -k,$$

where k is a positive constant depending on the current, and $b^2 = \alpha k-\mu^2$, where α is the temperature coefficient of resistance and μ^2 is defined in § 16 (10). If the ends $x = 0$ and $x = 2l$ of the wire are at zero temperature, show that

$$T = \frac{k}{b^2}\left\{\frac{\cos b(l-x)}{\cos bl}-1\right\}$$

if $b^2 > 0$, and find the corresponding solutions for $b^2 = 0$ and $b^2 < 0$.

15. In a counterflow heat exchanger the only changes from the conditions of § 16, Ex. 5, are that the partition extends from $x = 0$ to $x = a$, and that while the first fluid is admitted at $x = 0$ at temperature T and flows in the direction of x increasing as before, the second fluid is admitted at $x = a$ at zero temperature and flows with velocity u_2 in the direction of x decreasing. Show that, with the notation of § 16, Ex. 5, and writing $\alpha = (k_2 - k_1)/k_1 k_2$,

$$T_1 = T\{k_1 e^{-\alpha a} - k_2 e^{-\alpha x}\}\{k_1 e^{-\alpha a} - k_2\}^{-1}.$$

16. (i) Defining D^{-1} as the operation of indefinite integration, show that $D^r D^{-r} y = y$ but that this is not true of $D^{-r} D^r y$.

(ii) Let $f(D)$ be the polynomial $D^r(a_0 D^n + a_1 D^{n-1} + ... + a_n)$, where $a_n \neq 0$, and let $(b_0 + b_1 D + ... + b_k D^k)$ and $R_k(D)$ be the quotient and remainder obtained by expanding $(a_0 D^n + ... + a_n)^{-1}$ in ascending powers of D as if D were an ordinary algebraic variable. Show that if $P_k(x)$ is a polynomial in x of degree k

$$f(D)\{D^{-r}(b_0 + b_1 D + ... + b_k D^k) P_k(x)\} = P_k(x).$$

(iii) If $f(D)$ and $P_k(x)$ are defined in (ii), show that a particular integral of $f(D)y = P_k(x)$ can be found by writing this symbolically as

$$\frac{1}{f(D)} P_k(x),$$

expanding $1/f(D)$ in ascending powers of D, and operating on $P_k(x)$ with this series. Find particular integrals of Ex. 1, (i)–(iv) in this way.

17. The conventional method of finding a particular integral of $f(D)y = e^{ax} P_k(x)$, where $P_k(x)$ is a polynomial in x, is to write this symbolically in the form $\{1/f(D)\} e^{ax} P_k(x)$ and to proceed as if § 12, Theorem 3 held for this expression. If $f(D)$ has the form $(D-a)^r \phi(D)$ where $\phi(a) \neq 0$ this procedure gives

$$\frac{1}{(D-a)^r \phi(D)} \{e^{ax} P_k(x)\} = e^{ax} \left\{\frac{1}{D^r \phi(D+a)} P_k(x)\right\},$$

and the latter is evaluated as in Ex. 16.

Verify, by using Ex. 16 and § 12, Theorem 3, that the result obtained by this formal procedure is in fact a particular integral. Discuss the special cases $k = 0$ and $r = 0$ independently. Solve Ex. 1, (v)–(xvi) in this way.

18. Deduce § 14 (56) formally by expanding $1/f(D)$ in partial fractions by the formula § 18 (29). Verify, by operating on it with $f(D)$, that the result so obtained is a particular integral.

19. Refer to the 'car-following' models of § 8, Ex. 4. Let the position of car A be $y(t) = Vt + R \sin \omega t$, where V, R, and ω are constants. Show that the particular integral of § 8 (8) is then

$$x(t) = Vt - c + \frac{1}{k - \omega^2} R \sin \omega t,$$

and the particular integral of § 8 (9) is

$$x(t) = Vt + \frac{k'}{\sqrt{\{(k')^2+\omega^2\}}} R\sin(\omega t+\alpha),$$

where $$\tan\alpha = -\omega/k'.$$

Hence deduce that in the model of § 8 (8), small fluctuations in the velocity of car A can lead to large fluctuations in the velocity of car B, but in the model of § 8 (9), small fluctuations in the velocity of the front car give rise to even smaller fluctuations in the velocity of the following car.

III

DIFFERENTIAL EQUATIONS OF THE FIRST ORDER

19. Introductory

In Chapter I the importance of the fact that the ordinary linear differential equation of any order with constant coefficients was easy to solve was stressed, but the fact that many of the equations arising in applied mathematics are non-linear was noted. Equations of the first order occupy an important position because a number of non-linear equations, as well as the general linear equation of this order, are easily soluble.

In this chapter the commonest types are considered briefly and a few examples from fields other than dynamics are studied.

The general equation of the first order and degree may be written

$$f(x,y)\frac{dy}{dx}+g(x,y) = 0, \tag{1}$$

and its solution will contain one arbitrary constant. We shall usually write C for an arbitrary constant when it occurs.

20. Equations in which the variables are separable

If $f(x,y)$ and $g(x,y)$ in the general equation § 19 (1) are both of the form $\phi(x)\psi(y)$, the equation may be written

$$F(y)\frac{dy}{dx} = G(x). \tag{1}$$

Integrating this gives the solution

$$\int F(y)\,dy = \int G(x)\,dx+C. \tag{2}$$

Ex. $$x\tan y\frac{dy}{dx}-1 = 0.$$

This may be written $$\tan y\frac{dy}{dx} = \frac{1}{x}.$$

Integrating we have $$\int \tan y\,dy = \int\frac{dx}{x}.$$

That is, $$-\ln\cos y = \ln x+C.$$

Or $$x\cos y = A.$$

21. Problems leading to first-order equations with the variables separable

Many of the problems of particle dynamics are of this type. Here we give some other examples.

Ex. 1. *The non-linear problem of § 8, Ex. 2.*

The differential equation to be solved is

$$M\frac{dT}{dt} = -H'(T^4-T_0^4).$$

The solution is

$$\begin{aligned}-\frac{H't}{M}+C &= \int\frac{dT}{T^4-T_0^4}\\ &= \frac{1}{2T_0^2}\int\left\{\frac{1}{2T_0(T-T_0)}-\frac{1}{2T_0(T+T_0)}-\frac{1}{T^2+T_0^2}\right\}dT\\ &= \frac{1}{4T_0^3}\ln\frac{T-T_0}{T+T_0}-\frac{1}{2T_0^3}\tan^{-1}\frac{T}{T_0}.\end{aligned}$$

Ex. 2. *The law of mass action.*

This states that, if in a chemical reaction n_1 molecules of a substance A, n_2 molecules of a substance B, n_3 of C, and so on, combine to form any number of resultants according to the formula

$$n_1A+n_2B+n_3C+... = \text{any number of resultants,} \tag{1}$$

then the rate of the reaction, that is the rate of increase of the amount transformed, is proportional to

$$(a-x)^{n_1}(b-y)^{n_2}(c-z)^{n_3}..., \tag{2}$$

where a, b, c,... are the amounts of A, B, C,... initially present, and x, y, z,... are the amounts of A, B, C,... transformed up to time t. The quantities x, y, z,... are connected by (1), thus, for example,

$$x/n_1 = y/n_2 = z/n_3 = ... \tag{3}$$

For the *second-order reaction*

$$A+B \rightarrow \text{any resultants,}$$

(2) gives
$$\frac{dx}{dt} = k(a-x)(b-x),$$

where k is a known constant. Therefore

$$\int\frac{dx}{(a-x)(b-x)} = kt+C.$$

The solution of this for which $x = 0$ when $t = 0$ is

$$\frac{1}{b-a}\ln\frac{a(b-x)}{b(a-x)} = kt. \tag{4}$$

For the *third-order reaction*

$$2A+B \to \text{any resultants},$$

(2) becomes
$$\frac{dx}{dt} = k'(a-2x)^2(b-x).$$

The solution of this for which $x = 0$ when $t = 0$ is

$$\frac{1}{(a-2b)^2}\left\{\frac{2x(2b-a)}{a(a-2x)}+\ln\frac{b(a-2x)}{a(b-x)}\right\} = k't.$$

Finally it should be remarked that a unimolecular reaction leads to a linear equation with constant coefficients, and that a chain of such reactions leads to a chain of such equations similar to those of § 8, Ex. 1.

Ex. 3. *Ionization and recombination.*

If a gas is ionized so that the number n of electrons per unit volume is equal to the number of positive ions, the rate at which electrons and positive ions recombine to form neutral molecules is

$$\alpha n^2,$$

where α is a constant called the coefficient of recombination.

Suppose that ions are produced at a constant rate I per unit volume for $t > 0$ in an initially un-ionized gas. Then n satisfies

$$\frac{dn}{dt} = I-\alpha n^2.$$

The solution of this is

$$t = \int \frac{dn}{I-\alpha n^2}+C.$$

That is
$$t = \frac{1}{(\alpha I)^{\frac{1}{2}}}\tanh^{-1}n\left(\frac{\alpha}{I}\right)^{\frac{1}{2}}+C,$$

and since $n = 0$ when $t = 0$ gives $C = 0$, we get finally

$$n = \left(\frac{I}{\alpha}\right)^{\frac{1}{2}}\tanh t(\alpha I)^{\frac{1}{2}}.$$

Ex. 4. *Flow of liquid in an open channel.*

If liquid is flowing along a channel in the direction of the x-axis, its depth h at x satisfies the differential equation

$$\frac{dh}{dx} = i\frac{h^3-H^3}{h^3-\alpha H^3}, \tag{5}$$

where i is the (small and constant) slope of the channel, α is a constant, and H is a constant depending on the quantity of water flowing (it is the depth of a rectangular channel which gives the same flow). All the cases $h \gtrless H$ and $\alpha \gtrless 1$ may occur, so there are many different possibilities.

Putting $h = mH$, the solution of (5) is

$$\frac{ix}{H} = \int \frac{m^3-\alpha}{m^3-1}\,dm+C,$$

$$= m-(1-\alpha)\phi(m)+C,$$

where

$$\phi(m) = -\int \frac{dm}{m^3-1} = \frac{1}{6}\ln\frac{m^2+m+1}{(m-1)^2}+\frac{1}{\sqrt{3}}\tan^{-1}\frac{2m+1}{\sqrt{3}}. \tag{6}$$

The function $\phi(m)$ is called the 'backwater function'.

22. The first-order linear equation

The most general first-order linear equation is

$$\frac{dy}{dx}+Py = Q, \tag{1}$$

where P and Q are functions of x.

In the homogeneous case, $Q = 0$, this becomes the separable equation

$$\frac{1}{y}\frac{dy}{dx} = -P, \tag{2}$$

the solution of which is

$$\ln y = -\int P\,dx+C,$$

or

$$y = Ae^{-\int P\,dx}, \tag{3}$$

where A is a constant.

The equation (1) may now be solved by a general method, 'variation of parameters', which allows the solution of an inhomogeneous linear differential equation to be deduced from the solution of the corresponding homogeneous equation.

We seek a solution of (1) of the form

$$y = ze^{-\int P\,dx}, \tag{4}$$

where z is a function of x. This form is suggested by (3), the constant A being replaced by the function z. Substituting (4) in (1) gives

$$\frac{dz}{dx}e^{-\int P\,dx}-Pze^{-\int P\,dx}+Pze^{-\int P\,dx} = Q.$$

Thus

$$\frac{dz}{dx} = Qe^{\int P\,dx},$$

and

$$z = \int Qe^{\int P\,dx}\,dx+C. \tag{5}$$

Hence, finally, the solution of (1) is

$$y = e^{-\int P\,dx}\left\{\int Qe^{\int P\,dx}\,dx+C\right\}, \tag{6}$$

where C is an arbitrary constant.

Another method of solving this equation will be given in § 26. The way in which (6) generalizes the results of Chapter II for the case in which P is a constant, say α, is worth noting explicitly; αx is replaced by $\int P\,dx$.

Ex. $$x\frac{dy}{dx}+y = xe^{-x}.$$

Here, in the above notation, $P = 1/x$ and

$$\int P\,dx = \ln x.$$

Thus (6) gives

$$y = e^{-\ln x}\left\{\int e^{-x+\ln x}\,dx+C\right\}$$

$$= \frac{1}{x}\left\{\int xe^{-x}\,dx+C\right\}$$

$$= \frac{C}{x}-e^{-x}\left(1+\frac{1}{x}\right).$$

23. Equations reducible to the linear form

It is often possible to reduce an equation to the linear form by a suitable substitution. The following are two important types.

(i) *The equation*

$$\frac{dy}{dx}+Pe^{y} = Q, \tag{1}$$

where P and Q are functions of x only. Putting

$$z = e^{-y}, \qquad \frac{dz}{dx} = -e^{-y}\frac{dy}{dx},$$

the equation (1) becomes

$$\frac{dz}{dx}+Qz = P, \tag{2}$$

which is linear in z. Equations containing a term in e^{y} are of considerable importance in chemical kinetics and similar problems.

(ii) *Bernoulli's equation*,

$$\frac{dy}{dx}+Py = Qy^n, \tag{3}$$

where P and Q are functions of x only.

Put
$$z = y^{1-n},$$
$$\frac{dz}{dx} = \frac{1-n}{y^n}\frac{dy}{dx},$$

and (1) becomes
$$\frac{dz}{dx}+(1-n)Pz = (1-n)Q, \tag{4}$$

which is a linear equation in z.

Riccati's equation,
$$\frac{dy}{dx} = Py^2+Qy+R, \tag{5}$$

where P, Q, and R are functions of x only, which is a little more general than (3) and of considerable importance in dynamics, is an example of an equation with no simple method of solution (that is, in the general case).

Making the substitution

$$y = -\frac{1}{Pz}\frac{dz}{dx}$$

it transforms into

$$\frac{d^2z}{dx^2}-\left(Q+\frac{1}{P}\frac{dP}{dx}\right)\frac{dz}{dx}+PRz = 0, \tag{6}$$

which is a linear second-order equation in z with variable coefficients.

24. Problems leading to first-order linear equations

Ex. 1. *The problem of* § *8, Ex. 3, with* $M = M_0$ *and* $T = T_0$ *when* $t = 0$.
From § 8 (5) we have
$$M = mt+M_0, \tag{1}$$

and substituting this in § 8 (6) gives the linear equation for T

$$\frac{dT}{dt}+\frac{m+H}{mt+M_0}T = \frac{mT_1+HT_2}{mt+M_0}. \tag{2}$$

Using § 22 (6), and writing $\mu = (m+H)/m$, $t_0 = M_0/m$, the solution of (2) is seen to be

$$T = (t+t_0)^{-\mu}\left\{\int \frac{mT_1+HT_2}{m(t+t_0)}(t+t_0)^{\mu}\,dt+C\right\}$$

$$= \frac{mT_1+HT_2}{m\mu}+C(t+t_0)^{-\mu}. \tag{3}$$

The requirement that $T = T_0$ when $t = 0$ gives

$$T_0 = \frac{mT_1+HT_2}{m\mu}+Ct_0^{-\mu},$$

and substituting this value of C in (3) gives finally

$$T = \frac{mT_1+HT_2}{m\mu}+\left\{\frac{m\mu T_0-mT_1-HT_2}{m\mu}\right\}\left(1+\frac{t}{t_0}\right)^{-\mu}.$$

Ex. 2. *Fungus spores are destroyed by heat at a rate proportional to the product of the number present and an exponential function of the temperature T.*

That is, if n is the number of spores per unit volume at time t, n satisfies the differential equation

$$\frac{dn}{dt} = -kne^{cT}, \tag{4}$$

where k and c are constants, and the temperature T is a given function of the time. By § 22 (6) the solution of (4) is

$$n = A\exp\left\{-k\int e^{cT}\,dt\right\}. \tag{5}$$

If n_0 is the number of spores per unit volume when $t = 0$, this becomes

$$n = n_0\exp\left\{-k\int\limits_0^t e^{cT}\,dt\right\}. \tag{6}$$

Ex. 3. *The equation of radiative transfer.*

Suppose radiation is being propagated in the direction of the x-axis in a medium which absorbs the radiation at a rate ρkI per unit volume and also emits radiation at the rate ρkB per unit volume, where ρ, k, and B are known functions of x, and I is the intensity of the radiation at x. Then I satisfies the equation

$$\frac{dI}{dx} = -k\rho(I-B).$$

If I_0 is the intensity at $x = 0$, the solution of this is

$$I = \exp\left[-\int\limits_0^x \rho k\,dx\right]\left\{\int\limits_0^x k\rho B\exp\left[\int\limits_0^x \rho k\,dx\right]dx+I_0\right\}.$$

25. Homogeneous equations

These are of the form

$$\frac{dy}{dx} = f\left(\frac{y}{x}\right). \tag{1}$$

It follows that by the substitution

$$y = vx \tag{2}$$

the function $f(y/x)$ is reduced to a function of v only.

Also from (2) $$\frac{dy}{dx} = v + x\frac{dv}{dx}, \tag{3}$$

so that (1) becomes $$x\frac{dv}{dx} = f(v) - v. \tag{4}$$

(4) is an equation with the variables separable, and its solution is

$$\int \frac{dv}{f(v)-v} = \ln x + C. \tag{5}$$

Ex. 1. $$\frac{dy}{dx} = \frac{x^2+y^2}{xy}.$$

Making the substitutions (2) and (3), this becomes

$$x\frac{dv}{dx} + v = \frac{1+v^2}{v},$$

or $$x\frac{dv}{dx} = \frac{1}{v}.$$

Thus the solution is $$\tfrac{1}{2}v^2 = \ln x + C,$$

or $$\frac{y^2}{2x^2} = \ln x + C.$$

Ex. 2. $$\frac{dy}{dx} + \frac{x-3y+2}{3x-y+6} = 0.$$

This equation is not homogeneous as it stands, but if the variables are changed to Y and X defined by

$$X - 3Y = x - 3y + 2,$$

$$3X - Y = 3x - y + 6,$$

it becomes $$\frac{dY}{dX} + \frac{X-3Y}{3X-Y} = 0,$$

which is homogeneous. Substituting $Y = vX$, this gives

$$X\frac{dv}{dX} + \frac{1-v^2}{3-v} = 0.$$

The solution of this is $$\frac{(v+1)^2X}{v-1} = C,$$

or $$\frac{(X+Y)^2}{(Y-X)} = C,$$

or, finally, reverting to the original notation,

$$\frac{(x+y+2)^2}{y-x-2} = C.$$

26. Exact equations

The equation $$f(x,y)\frac{dy}{dx}+g(x,y) = 0 \tag{1}$$

is called an exact equation if the left-hand side, as it stands, is the differential coefficient of a function $\phi(x,y)$ of x and y. That is, if (1) has the form

$$\frac{d}{dx}\phi(x,y) = 0, \tag{2}$$

so that its solution is $$\phi(x,y) = C. \tag{3}$$

For example, the equation

$$x\frac{dy}{dx}+y = 0.$$

is exact, and its solution is

$$xy = C.$$

In the notation of differentials the definition requires that

$$f(x,y)\,dy+g(x,y)\,dx$$

is to be an exact differential $d\phi$.

The condition that (1) be exact is

$$\frac{\partial f}{\partial x} = \frac{\partial g}{\partial y}, \tag{4}$$

and this condition is both necessary and sufficient, that is, if the equation is exact (4) must hold; and if (4) holds, (1) must be exact. We prove these statements in order. First suppose that (1) is exact so that it can be expressed in the form (2); this may be written

$$\frac{\partial\phi}{\partial y}\frac{dy}{dx}+\frac{\partial\phi}{\partial x} = 0. \tag{5}$$

Comparing (1) and (5) we must have

$$f(x,y) = k\frac{\partial\phi}{\partial y}, \qquad g(x,y) = k\frac{\partial\phi}{\partial x}, \tag{6}$$

where k is a constant. Then from (6)

$$\frac{\partial f}{\partial x} = k\frac{\partial^2\phi}{\partial x\partial y} = k\frac{\partial^2\phi}{\partial y\partial x} = \frac{\partial g}{\partial y},$$

so that (4) is satisfied as required.

Next we prove that if (4) is satisfied we can find a function ϕ such that (1) can be put in the form (2). Write

$$G(x,y) = \int g(x,y)\,dx, \tag{7}$$

the integral being evaluated as if y were constant. Then

$$\frac{\partial G(x,y)}{\partial x} = g(x,y),$$

$$\frac{\partial^2 G(x,y)}{\partial y\partial x} = \frac{\partial g(x,y)}{\partial y} = \frac{\partial f(x,y)}{\partial x}, \tag{8}$$

using (4). (8) may be written

$$\frac{\partial}{\partial x}\left\{\frac{\partial G(x,y)}{\partial y} - f(x,y)\right\} = 0,$$

so that

$$\frac{\partial G(x,y)}{\partial y} - f(x,y) = \psi(y), \tag{9}$$

where $\psi(y)$ is a function of y only.

We now choose for the required function $\phi(x,y)$,

$$\phi(x,y) = G(x,y) - \int \psi(y)\,dy, \tag{10}$$

and show that this has the required properties. We have

$$\frac{\partial\phi(x,y)}{\partial x} = \frac{\partial G(x,y)}{\partial x} = g(x,y),$$

$$\frac{\partial\phi(x,y)}{\partial y} = \frac{\partial G(x,y)}{\partial y} - \psi(y) = f(x,y),$$

by (9), and thus (1) is put in the form (2). This process may be used to find $\phi(x,y)$ if this is not immediately obvious.

There are many equations which are not exact as they stand but can be made so by multiplying by a suitable function of x and y. Such a function is called an *integrating factor*. For example, the equation

$$x\frac{dy}{dx} - y = 0, \tag{11}$$

is not exact, but if multiplied by x^{-2} it becomes

$$\frac{1}{x}\frac{dy}{dx} - \frac{y}{x^2} = 0, \tag{12}$$

F

which is exact and can be written

$$\frac{d}{dx}\left(\frac{y}{x}\right) = 0,$$

so that the solution of (11) is

$$y = Cx.$$

One method of finding an integrating factor is to use the condition (4). We illustrate this by considering the important case of the linear equation

$$\frac{dy}{dx}+Py = Q, \tag{13}$$

where P and Q are functions of x. Suppose ψ is an integrating factor of (13), then

$$\psi\frac{dy}{dx}+\psi(Py-Q) = 0$$

must be exact. By (4) this requires

$$\frac{\partial\psi}{\partial x} = \frac{\partial}{\partial y}\{\psi Py-\psi Q\}$$

$$= (Py-Q)\frac{\partial\psi}{\partial y}+\psi P. \tag{14}$$

Any function which satisfies (14) will be suitable, in particular a function $\psi(x)$ of x only chosen so that

$$\frac{d\psi}{dx}-\psi P = 0. \tag{15}$$

A solution of (15) is

$$\psi = e^{\int P\,dx} \tag{16}$$

so that this is an integrating factor of (13). Multiplying (13) by it gives

$$\left\{\frac{dy}{dx}+Py\right\}e^{\int P\,dx} = Qe^{\int P\,dx},$$

or

$$\frac{d}{dx}\{ye^{\int P\,dx}\} = Qe^{\int P\,dx},$$

so that, integrating,

$$ye^{\int P\,dx} = \int Qe^{\int P\,dx}\,dx+C, \tag{17}$$

which is the solution already found in § 22.

27. Equations of the first order but of higher degree than the first

The equation of the nth degree will be

$$\phi_0(x,y)\left(\frac{dy}{dx}\right)^n+\phi_1(x,y)\left(\frac{dy}{dx}\right)^{n-1}+...+\phi_n(x,y) = 0. \tag{1}$$

This may be factorized into the form

$$\left\{f_1(x,y)\frac{dy}{dx}+g_1(x,y)\right\}...\left\{f_n(x,y)\frac{dy}{dx}+g_n(x,y)\right\} = 0. \tag{2}$$

Therefore if y satisfies any one of the n equations

$$\left.\begin{array}{r} f_1(x,y)\dfrac{dy}{dx}+g_1(x,y) = 0 \\ \cdot\quad\cdot\quad\cdot\quad\cdot\quad\cdot\quad\cdot\quad\cdot\quad\cdot \\ f_n(x,y)\dfrac{dy}{dx}+g_n(x,y) = 0 \end{array}\right\} \tag{3}$$

it will satisfy (2). The solution of (1) consists of the collection of all the solutions of (3). Suppose $\psi_1(x,y,c_1) = 0$, ..., $\psi_n(x,y,c_n) = 0$ are the solutions of these found by the methods of the previous sections, where $c_1,..., c_n$ are arbitrary constants, then any of these satisfies (1) and its general solution may, if desired, be written

$$\psi_1(x,y,c)\psi_2(x,y,c)...\psi_n(x,y,c) = 0, \tag{4}$$

where c is an arbitrary constant. One arbitrary constant c is sufficient in (4), since, as it varies, the individual solutions $\psi_1(x,y,c)$, etc., run through all possible values.

As an example consider the equation

$$\left(\frac{dy}{dx}\right)^2+3\left(\frac{dy}{dx}\right)+2 = 0, \tag{5}$$

that is, factorizing, $$\left(\frac{dy}{dx}+2\right)\left(\frac{dy}{dx}+1\right) = 0.$$

The solution of $$\frac{dy}{dx}+2 = 0$$

is the family of straight lines

$$y+2x = c_1, \tag{6}$$

and that of $$\frac{dy}{dx}+1 = 0$$

is the family of straight lines

$$y+x = c_2. \tag{7}$$

(5) is satisfied by all the lines (6) and (7): its general solution may be written

$$(y+2x+c)(y+x+c) = 0.$$

A phenomenon which appears in certain equations of this type is that they may have additional solutions which are not comprised in the general solution. Such solutions are called singular solutions. For example, for the equation

$$\left(\frac{dy}{dx}\right)^2 - y = 0, \tag{8}$$

or

$$\left(\frac{dy}{dx} - y^{\frac{1}{2}}\right)\left(\frac{dy}{dx} + y^{\frac{1}{2}}\right) = 0,$$

the general solution, found as above, is

$$\{2y^{\frac{1}{2}} - (x+c)\}\{2y^{\frac{1}{2}} + (x+c)\} = 0,$$

or

$$4y = (x+c)^2, \tag{9}$$

which is a family of parabolas. But (8) is also satisfied by

$$y = 0, \tag{10}$$

which is not included in the general solution (9) and is a singular solution. The line (10) is in fact the envelope of the parabolas (9).

EXAMPLES ON CHAPTER III

1. Solve the following differential equations:

(i) $\dfrac{dy}{dx} = 1 + y^2$.

(ii) $\dfrac{dy}{dx} - (1 + e^y)\sin x = 0$.

(iii) $x\dfrac{dy}{dx} - y = x^3$.

(iv) $(1 + x^2)\dfrac{dy}{dx} + 4xy = 1$.

(v) $\dfrac{dy}{dx} + y = xy^3$.

(vi) $(x^3 - y^3)\dfrac{dy}{dx} + 3x^2y = 0$.

(vii) $\dfrac{dy}{dx} - \dfrac{x+y}{3x-y-1} = 0$.

(viii) $(\sin x + x\cos y)\dfrac{dy}{dx} + (\sin y + y\cos x) = 0$.

(ix) $3xy^2\dfrac{dy}{dx} + 2y^3 + 3x = 0$.

The solutions are

(i) $x = \tan^{-1}y + C$. (ii) $\cos x - \ln(1 + e^{-y}) = C$.

(iii) $y = \frac{1}{2}x^3 + Cx$. (iv) $y = (x + \frac{1}{3}x^3 + C)/(1 + x^2)^2$.

(v) $y = \{\frac{1}{2} + x + Ce^{2x}\}^{-\frac{1}{2}}$. (vi) $4x^3y - y^4 = C$.

(vii) $\ln(y - x + \frac{1}{2}) + (4x - 1)/(2y - 2x + 1) = C$.

(viii) $x\sin y + y\sin x = C$. (ix) $x^3 + x^2y^3 = C$.

2. (i) Show that in the second-order reaction $2A \to$ any resultants (cf. § 21) if x is the amount of A transformed in time t, and a is the amount of A originally present,

$$\frac{x}{a(a-x)} = kt.$$

(ii) Show that if a, b, c are the amounts of A, B, and C originally present, and x is the amount transformed in the third-order reaction $A+B+C \to$ any resultants,

$$\frac{1}{(b-a)(c-a)}\ln\frac{a}{a-x}+\frac{1}{(a-b)(c-b)}\ln\frac{b}{b-x}+\frac{1}{(a-c)(b-c)}\ln\frac{c}{c-x} = kt.$$

3. The differential equation for a reversible second-order reaction in which the second substance is not present initially is

$$\frac{dx}{dt} = k_1(a-x)^2-k_2 x^2.$$

Putting $y = x/(a-x)$ show that

$$\frac{1}{2a\sqrt{(k_1 k_2)}}\ln\frac{(a-x)\sqrt{k_1}+x\sqrt{k_2}}{(a-x)\sqrt{k_1}-x\sqrt{k_2}} = t.$$

4. Molten metal in a cylindrical container is cooling by conduction of heat radially outwards. At the instant of solidification the volume of the metal solidifying is reduced by a fraction α. If h is the height of the free surface of the liquid when the radius of the surface at which solidification is taking place is r, show that

$$\frac{dh}{dr} = \frac{2\alpha h}{r}.$$

If $h = h_0$ when $r = r_0$ show that the shape of the upper surface when the metal is all solid is $h = h_0(r/r_0)^{2\alpha}$.

5. A substance decomposes according to the unimolecular law

$$\frac{dw}{dt} = -ke^{-E/RT}w,$$

where w is the amount of the substance present, k, E, and R are constants, and T is the absolute temperature. At time $t = 0$, $w = w_0$ and $T = T_0$. Heat is given off during the reaction at a rate proportional to the reaction velocity (no heat being lost to the surroundings) so that $dT/dt = -q\,dw/dt$, where q is a constant. Show that the temperature at any time is given by

$$kt = \int\limits_{T_0}^{T} \frac{e^{E/RT}\,dT}{T_0+qw_0-T}.$$

6. If there are initially N electrons per unit volume and they disappear by recombination (cf. § 21, Ex. 3) show that the number present at time t is $N/(1+N\alpha t)$.

7. The density ρ in the earth's atmosphere is assumed to vary with height h according to the law $\rho = \rho_0 \exp(-h/H)$, where ρ_0 and H are constants. Radiation is incident on it vertically from outside, its intensity at infinity being I_0. If the radiation is absorbed at the rate $k\rho I$ per unit volume, where I is its intensity [cf. § 24, Ex. 3], show that

$$I = I_0 \exp\{-Hk\rho_0 e^{-h/H}\}.$$

Ions are produced by the absorbed radiation at a rate β times the rate of absorption. Show that the rate of ion production is

$$\beta I_0 k\rho_0 \exp\{-h/H - Hk\rho_0 e^{-h/H}\}.$$

Show that this has a maximum at the height $H \ln(Hk\rho_0)$. This is Chapman's theory of the formation of the ionosphere.

8. Show that if h_1 and h_2 are the depths of a stream at distances x_1 and x_2 from an origin,

$$i(x_1 - x_2) = (h_1 - h_2) - (1-\alpha)H\{\phi(h_1/H) - \phi(h_2/H)\},$$

in the notation of § 21, Ex. 4.

9. In a sterilizing process the temperature T is raised linearly from zero to T_0 in time t_0, and subsequently decreases exponentially according to the law $T = T_0 \exp[-\alpha(t-t_0)]$. If n_0 is the number of spores initially present (cf. § 24 (4)), show that the number n_1 at time $t_1 > t_0$ is given by

$$\ln(n_0/n_1) = (kt_0/cT_0)\{e^{cT_0} - 1\} + k\{\mathrm{Ei}(cT_0) - \mathrm{Ei}(cT_0 e^{-\alpha(t_1-t_0)})\}/\alpha,$$

where $\mathrm{Ei}(x)$ denotes the tabulated function

$$\mathrm{Ei}(x) = \int_{-\infty}^{x} \eta^{-1} e^{\eta}\, d\eta.$$

10. The path of a ray in a spherically symmetrical refracting medium is determined by Snell's law, $\mu r \sin i =$ constant, where μ is a given function of r, and i is the angle between the path and the radius vector to the origin. Show that if $\mu \to 1$ as $r \to \infty$, the equation of the path of a ray which for large values of r is parallel to the axis of polar coordinates and distant p from it is

$$\theta = p \int_{r}^{\infty} \frac{dr}{r(\mu^2 r^2 - p^2)^{\frac{1}{2}}}.$$

11. Ions are produced for $t > 0$ in a medium at the rate kt per unit time per unit volume, and they disappear by recombination (cf. § 21, Ex. 3) at the rate αn^2. Making the substitution $n = (1/\alpha z)\, dz/dt$, show that z must satisfy

$$\frac{d^2 z}{dt^2} - \alpha k t z = 0.$$

For the solution of this see § 98. The problem corresponds to production of ions in the ionosphere near sunrise.

12. The bottoms of two equal tanks are on the same level and are connected by a pipe which is such that the rate of flow of water is proportional to $\sqrt{(h_1-h_2)}$, where h_1 and h_2 are the heights of water in the two tanks. If water is pumped into the first tank at a constant rate, show that h_1 and h_2 satisfy

$$\frac{dh_1}{dt} = k-k'(h_1-h_2)^{\frac{1}{2}}, \qquad \frac{dh_2}{dt} = k'(h_1-h_2)^{\frac{1}{2}},$$

where k and k' are constants.

If both tanks are empty at $t = 0$, show that

$$\frac{k}{2k'^2}\ln\left\{\frac{k-2k'(kt-2h_2)^{\frac{1}{2}}}{k}\right\}+\frac{1}{k'}(kt-2h_2)^{\frac{1}{2}}+t = 0.$$

13. Solve the bacterial growth problem of § 7, Ex. 1 for the following feed rates: (i) $ds/dt = c$; (ii) $ds/dt = c(1+\cos\omega t)$; (iii) $ds/dt = kx^2$, where c, ω, and k are constants.

14. Refer to the predator–prey model of § 7, Ex. 2. If equilibrium between the two species is possible, show that this will occur when $x = s/r$ and $y = p/q$. Let $u = x-s/r$ and $v = y-p/q$ be the deviations of the two populations from their equilibrium values. Show that near equilibrium, neglecting second-order terms such as the product uv, the differential equation for the phase trajectory is

$$\frac{dv}{du} = -\frac{u}{v}\left(\frac{pr^2}{sq^2}\right).$$

Hence deduce that near equilibrium the phase trajectories are ellipses, and so the two populations will oscillate indefinitely.

IV

DYNAMICAL PROBLEMS LEADING TO ORDINARY LINEAR DIFFERENTIAL EQUATIONS

28. Introductory

In this chapter we shall consider problems which lead to linear differential equations. These will usually be 'vibration problems' on the motion of several masses constrained by springs, or the rotation of a number of wheels on a shaft. As remarked in § 6, quite complicated problems of these types can be solved, provided they are idealized in such a way that the equations of motion become linear.

Newton's second law for the motion of a (constant) mass m along the x-axis is

$$m\frac{d^2x}{dt^2} = \text{force}, \tag{1}$$

and, as in § 6, we assume that the force is a sum of terms depending on position, velocity, and time: i.e.

$$m\ddot{x} = f(x)+g(\dot{x})+h(t), \tag{2}$$

where, as is usual in dynamics, a dot is used to denote differentiation with respect to the time.

If the equation (2) is to be linear, $f(x)$ must be proportional to x, say $f(x) = -\lambda x$, and $g(\dot{x})$ must be proportional to $\dot{x}$, say $g(\dot{x}) = -k\dot{x}$.

Restoring force proportional to displacement is provided by the strain of an ideally elastic body or a perfect spring; we use the term 'stiffness' for the constant of proportionality λ, so that the restoring force exerted is the stiffness times the relative displacement of the ends.

Resistance to motion proportional to velocity is provided by the shearing of an ideal viscous fluid.

If $f(x)$ and $g(\dot{x})$ have these forms, (2) becomes

$$m\ddot{x} = -\lambda x-k\dot{x}+h(t), \tag{3}$$

which is an inhomogeneous second-order linear differential equation.

We also get a linear equation if the resistance to motion is due to the 'coulomb' or 'solid' friction between two solids which slide on one another. In this case there is a constant force μR opposite to the direction of motion, R being the normal reaction at the contact, and μ the coefficient of dynamic friction.

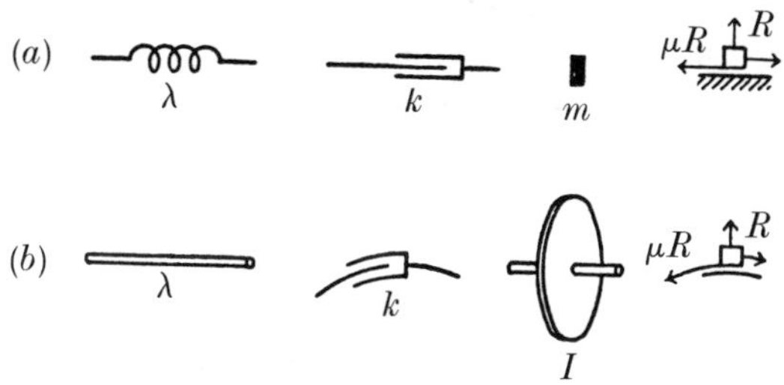

FIG. 3.

In this case, however, different equations of motion must be used for the two directions of motion since the sign of the constant force μR must be changed. This was not necessary in the case of resistance to motion $-k\dot{x}$, since $(-k\dot{x})$ changes sign automatically with $\dot{x}$ and thus is always in the opposite direction to it.

The fundamental elements considered above may be represented diagrammatically as in Fig. 3 (*a*). The first represents an ideal spring of stiffness λ, the second a dash-pot containing ideal viscous liquid giving resistance to motion k times the relative velocity of sliding, the third a mass m, and the fourth coulomb friction μ times the normal reaction R and in a direction opposite to an assumed direction of sliding. Complicated systems can be built up by combining these elements, and in all cases the equations of motion will be linear.

Similar equations arise for the angular motion of wheels on elastic shafts. Here the angular displacement θ of a marked line on a wheel from a fixed reference direction is the dependent variable corresponding to x, and the equation of motion of a wheel of moment of inertia I is

$$I\ddot{\theta} = \text{torque}. \tag{4}$$

An ideally elastic shaft of 'stiffness' λ provides a restoring torque λ times the relative angular displacement of its ends, and resistance to motion $k\dot{\theta}$ is provided by shearing of an ideal viscous liquid. Thus there are elements for rotational motion as in Fig. 3 (*b*) similar to those of Fig. 3 (*a*) for linear motion; also there will be an analogy between linear and rotational motion in the sense that certain systems will have the same differential equations, except for a change of notation, θ for x, I for m, etc. In Chapter V this analogy will be developed and extended to include electrical systems. In this chapter problems will be stated for linear motion, but the equivalent diagrams of types Fig. 3 (*a*) and (*b*) for linear and rotational motion will both be given.

Finally it should be stated that all the forces acting on the mass are supposed to be specified, and that unless gravity is mentioned explicitly (e.g. by saying that the mass moves vertically) it is supposed not to be effective, for example, the motion might be regarded as taking place on a smooth horizontal plane. There is no loss of generality in omitting gravity forces in these *linear* problems: if they are included there will be a position of static equilibrium calculable from the weights of the masses and the linear forces, and the equations of motion about this equilibrium position will be the same as those we discuss.

29. The damped harmonic oscillator: free vibrations

We consider the motion of a particle of mass m along the x-axis, Fig. 4; the particle is supposed to be constrained by a spring of stiffness λ, the displacement x of the particle being measured from the position in which the spring is unstrained;† there is resistance to motion k times the velocity, and the particle is acted on by an external force $F(t)$.

The equation of motion, § 28 (3) is

$$m\ddot{x} = -\lambda x - k\dot{x} + F(t), \tag{1}$$

† If the spring hangs vertically so that gravity acts on the mass m we may *either* include mg in the force $F(t)$, taking the origin in the unstrained position, cf. § 30 (i), *or* we may measure x from the position of static equilibrium (a deflexion of g/n^2), in which case the same equations result except that $F(t)$ is to be the external force other than gravity.

or
$$(D^2+2\kappa D+n^2)x = \frac{1}{m}F(t), \tag{2}$$
where, here and subsequently, we use the notation
$$\kappa = \frac{k}{2m}, \qquad n^2 = \frac{\lambda}{m}, \tag{3}$$
and write D for d/dt.

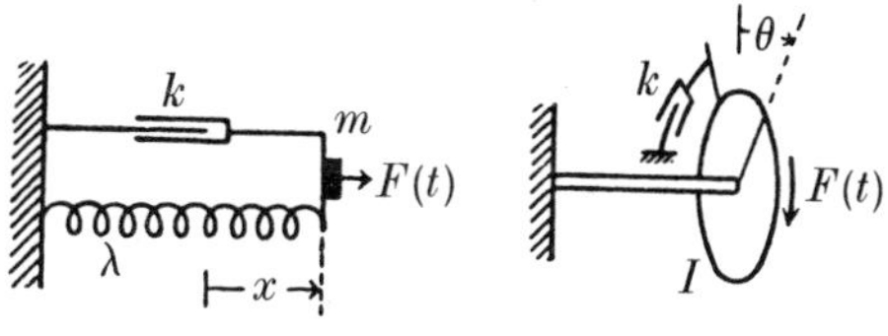

Fig. 4.

In this section we shall study the case $F(t) = 0$ in which there is no external force; this gives the 'free oscillations' of the system when disturbed in any way. This solution will also be needed in discussing the general equation (2) of which it is the complementary function.

Putting $F(t) = 0$ in (2) we have to solve
$$(D^2+2\kappa D+n^2)x = 0. \tag{4}$$

The auxiliary equation § 13 (2) is
$$\alpha^2+2\kappa\alpha+n^2 = 0,$$
and its roots are
$$\alpha = -\kappa\pm\sqrt{(\kappa^2-n^2)}. \tag{5}$$

Thus the solution takes three distinct forms according as $\kappa^2 \gtreqless n^2$.

(i) *The case* $\kappa < n$

Writing
$$n' = \sqrt{(n^2-\kappa^2)}, \tag{6}$$
the values (5) of α become $-\kappa\pm in'$ and the general solution of (4) is
$$x = Ae^{-\kappa t}\sin(n't+B), \tag{7}$$
where A and B are arbitrary constants determined by the way in which the motion was started when $t = 0$.

In the case $\kappa = 0$, no resistance to motion, the solution is simply
$$A\sin(nt+B), \tag{8}$$

a harmonic oscillation of frequency $n/2\pi$: this will be referred to as the *undamped natural frequency.*

The effect of resistance to motion κ in (7) is to make the solution the product of $e^{-\kappa t}$ with a sinusoidal term of frequency $n'/2\pi$. The motion is called *damped harmonic motion.* $n'/2\pi$ will be called the *damped natural frequency*, and κ the *damping coefficient*. The motion given by (7) is not strictly periodic since each oscillation is different from the preceding one, but it does possess certain periodic properties: for example $x = 0$ when

$$t = \frac{r\pi}{n'} - \frac{B}{n'} \quad (r = 1, 2,...), \tag{9}$$

that is, at a series of instants separated by the half-period π/n'. Also the velocity has the same property, for

$$\dot{x} = Ae^{-\kappa t}\{n'\cos(n't+B) - \kappa\sin(n't+B)\}, \tag{10}$$

and so $\dot{x} = 0$, when

$$t = \frac{r\pi}{n'} + \frac{1}{n'}\tan^{-1}\frac{n'}{\kappa} - \frac{B}{n'} \quad (r = 1, 2,...). \tag{11}$$

Comparing (9) and (11) it appears that the particle always takes time

$$\frac{1}{n'}\tan^{-1}\frac{n'}{\kappa}$$

to swing from its equilibrium position $x = 0$ to a point at which it is at rest, and time

$$\frac{1}{n'}\left\{\pi - \tan^{-1}\frac{n'}{\kappa}\right\}$$

to swing back from rest to $x = 0$. It thus moves back more slowly than it moves out. This may be seen from Fig. 5, in which the curves $e^{-\kappa t}$, $\sin n't$, and $e^{-\kappa t}\sin n't$ are shown.

Substituting the times (11) in (7) gives the displacements of the particle at successive instants when it is at rest. These are seen to be in geometric progression with common ratio

$$-e^{-\pi\kappa/n'}. \tag{12}$$

The amplitudes of successive swings on the same side of the origin diminish by the factor

$$e^{-2\pi\kappa/n'},$$

and the logarithm of this, namely,

$$\delta = 2\pi\kappa/n', \tag{13}$$

is called the logarithmic decrement.† This quantity, which is the ratio of the damping coefficient κ to the damped natural frequency $n'/2\pi$, appears also in connexion with resonance. The quantity in electric circuit theory analogous to $n'/2\kappa$ is called the 'Q' of a circuit.

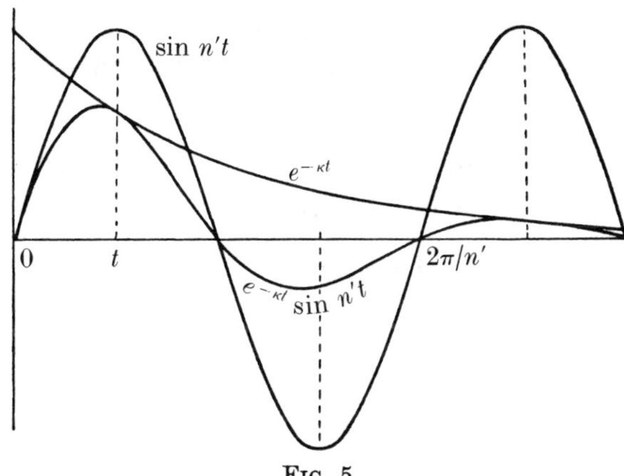

FIG. 5.

Finally it should be remarked that if κ/n is small, the damped natural frequency

$$\frac{1}{2\pi}n' = \frac{1}{2\pi}\sqrt{(n^2-\kappa^2)} = \frac{n}{2\pi}\left(1-\frac{\kappa^2}{2n^2}+...\right) \tag{14}$$

differs very little from the undamped natural frequency $n/2\pi$. Thus a small amount of resistance to motion has a much more important effect on the amplitude than on the frequency.

(ii) *The case* $\kappa > n$

Here the general solution of (4) is

$$x = Ae^{-\{\kappa-\sqrt{(\kappa^2-n^2)}\}t}+Be^{-\{\kappa+\sqrt{(\kappa^2-n^2)}\}t}. \tag{15}$$

The velocity $\dot{x}$ can vanish once at most. The motion is not oscillatory and consists either of a single swing or of a creep back

† Several different definitions are in use. Some authors use half this quantity which corresponds to comparing successive swings on *opposite* sides of the origin. Sometimes, also, the quantity δ in (13) is called the decrement.

to the origin, according to the circumstances of projection. This case $\kappa > n$ is referred to as 'dead beat'.

(iii) *The case of 'critical damping'*, $\kappa = n$

Here the auxiliary equation to (4) has equal roots n and the general solution is

$$(A+Bt)e^{-nt}. \tag{16}$$

As in case (ii) the motion consists of at most a single swing and is not oscillatory.

Finally we consider the effect of increasing the damping coefficient of a system with undamped natural frequency n. If $\kappa < n$ we have from (7)

$$|x| < |A|e^{-\kappa t};$$

thus as κ is increased towards n, the motion, while remaining oscillatory, dies away like

$$e^{-\kappa t}. \tag{17}$$

When $\kappa = n$, the motion ceases to be oscillatory and dies away like

$$te^{-nt}, \tag{18}$$

by (16). Finally, when $\kappa > n$ the motion given by (15) dies away like

$$e^{-\{\kappa-\sqrt{(\kappa^2-n^2)}\}t}. \tag{19}$$

Since both (17) for any value of $\kappa < n$, and (19) for any value of $\kappa > n$, are larger than (18) for sufficiently large values of t, it follows that, in order to make the motion die away as rapidly as possible we must give κ the critical value n. If this value is exceeded the motion does not die away so rapidly. Thus in recording instruments which have to take up a reading as quickly as possible, critical damping is often aimed at.

Ex. *The particle is set in motion with velocity V at $t = 0$ from its equilibrium position $x = 0$: to find the motion.*

The conditions $x = 0$, $\dot{x} = V$, when $t = 0$ require that in (7)

$$A \sin B = 0,$$

$$A\{n' \cos B - \kappa \sin B\} = V.$$

Thus $B = 0$, $A = V/n'$ and the solution is

$$x = \frac{V}{n'} e^{-\kappa t} \sin n't, \quad \text{if } \kappa < n.$$

Similarly from (16) and (15) we find

$$x = Vte^{-nt}, \quad \text{if } \kappa = n,$$

$$x = \frac{V}{\sqrt{(\kappa^2-n^2)}} e^{-\kappa t} \sinh t\sqrt{(\kappa^2-n^2)}, \quad \text{if } \kappa > n.$$

30. The harmonic oscillator with external force applied to it

In this section we consider the harmonic oscillator with applied forces of various types; the most important case, namely harmonic applied force, will be treated at length in § 31. Resistance to motion proportional to velocity is not considered, but it could be added in all cases.

(i) *Constant applied force*

The effect of such a force, as remarked in §§ 28, 29, is to give oscillation about a displaced position of equilibrium.

Ex. 1. *A particle of mass m is hung vertically by a spring of stiffness mn^2. At $t = 0$, when the spring is unstrained, the particle is released.*

Taking the x-axis vertically downwards, with the origin at the position in which the spring is unstrained, the equation of motion of the particle, § 28 (1), is

$$m\ddot{x} = -mn^2x+mg,$$

or

$$\ddot{x}+n^2x = g. \tag{1}$$

The general solution of this is

$$x = A\sin nt+B\cos nt+\frac{g}{n^2}. \tag{2}$$

The constants A and B in (2) have to be determined from the conditions $x = \dot{x} = 0$, when $t = 0$. These give

$$B+\frac{g}{n^2} = 0,$$

$$A = 0.$$

Therefore the solution is

$$x = \frac{g}{n^2}(1-\cos nt),$$

an oscillation about $x = g/n^2$, which is the equilibrium position of the mass when hanging from the spring.

(ii) *The applied force any function $F(t)$ of the time*

In this case the equation of motion, § 29 (2), is

$$(D^2+n^2)x = \frac{1}{m}F(t). \tag{3}$$

By § 14 (54) the general solution of (3) is

$$x = A\sin nt+B\cos nt+\frac{1}{mn}\int\limits_0^t \sin n(t-\xi)F(\xi)\,d\xi. \tag{4}$$

Ex. 2. *The particle is set in motion at $t = 0$ from rest in its equilibrium position by a constant force F_0 which acts for time T and then ceases.*

In this case $F(\xi)$ in (4) is F_0 for $0 < t < T$, and zero for $t > T$. We have to determine A and B in (4) to give $x = 0$ and $\dot{x} = 0$ when $t = 0$. These require

$$B = 0,$$

$$nA + \left[\frac{F_0}{m}\int_0^t \cos n(t-\xi)\,d\xi\right]_{t=0} = 0.$$

Thus $A = B = 0$, and the solution is

$$x = \frac{F_0}{mn}\int_0^t \sin n(t-\xi)\,d\xi = \frac{F_0}{mn^2}(1-\cos nt) \quad (0 < t < T),$$

$$x = \frac{F_0}{mn}\int_0^T \sin n(t-\xi)\,d\xi$$

$$= \frac{F_0}{mn^2}\{\cos n(t-T)-\cos nt\} = \frac{2F_0}{mn^2}\sin n(t-\tfrac{1}{2}T)\sin\tfrac{1}{2}nT \quad (t > T).$$

Thus the residual effect after the force has ceased is an oscillation of amplitude $(2F_0/mn^2)\sin\frac{1}{2}nT$. This, of course, could have been found by studying the motions for $0 < t < T$ and for $t > T$ separately.

(iii) *Solid or coulomb friction*

In this case, as remarked in § 28, the direction of motion must be known before the differential equation is written down, and the frictional force, μ times the normal reaction, is in the direction opposite to the direction of motion. This differential equation is valid only for motion in this direction.

It should be remarked that the assumption that kinetic friction is equal to a constant times the normal reaction is a very crude first approximation which is used because it gives linear equations of motion.

Actually, for most rubbing substances the frictional force varies with the velocity of sliding according to a complicated law such as that of Fig. 6 (a) in which for low velocities of sliding the frictional force is less than the static value. A second approximation which leads to interesting results is to assume a constant coefficient of kinetic friction μ which is less than the static value μ'; cf. Ex. 4 below.

Ex. 3. *A particle of mass m rests on a horizontal plane, the coefficient of friction being μ. It is attached to a fixed point by an elastic spring of stiffness mn^2. The particle is displaced a distance a from its equilibrium position and then released.*

Initially we have $x = a$, $\dot{x} = 0$, when $t = 0$. Also, when released, the particle will start to move backwards and so the frictional force will act forwards. The equation of motion for this part of the motion is then

$$m\ddot{x} = -mn^2x+\mu mg,$$

or

$$\ddot{x}+n^2x = \mu g. \tag{5}$$

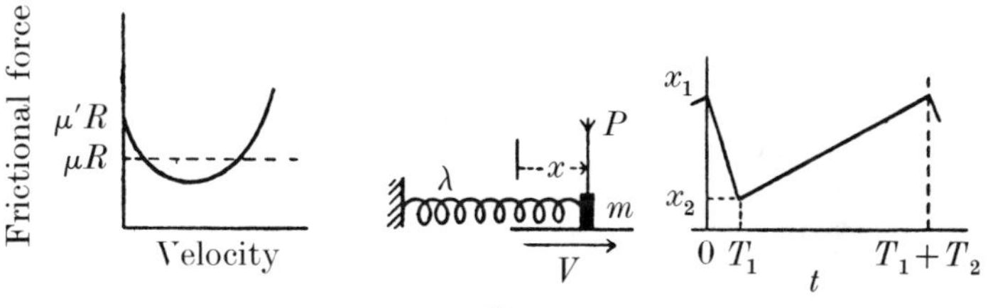

(a) (b) (c)

Fig. 6.

The general solution of this is

$$x = A\sin nt+B\cos nt+\frac{\mu g}{n^2}.$$

The conditions $x = a$, $\dot{x} = 0$, when $t = 0$ give $A = 0$, $B = a-\mu g/n^2$, and so the solution is

$$x = \frac{\mu g}{n^2}+\left(a-\frac{\mu g}{n^2}\right)\cos nt, \tag{6}$$

$$\dot{x} = -n\left(a-\frac{\mu g}{n^2}\right)\sin nt. \tag{7}$$

The equation of motion (5) and the solutions (6) and (7) hold so long as $\dot{x}$ is negative. This is the case until $t = \pi/n$, at which time the particle comes to rest, its displacement then being

$$x = -a+\frac{2\mu g}{n^2},$$

that is, a distance $(a-2\mu g/n^2)$ on the opposite side of its equilibrium position.

The same argument will show that it will next come to rest at time $2\pi/n$ at a distance $(a-4\mu g/n^2)$ from the equilibrium position, and so on.

Thus the particle oscillates with the period $2\pi/n$ which it would have in the absence of friction, but the amplitude diminishes by a constant amount $2\mu g/n^2$ in each half-swing. This continues until the particle comes to rest so near the equilibrium position that the restoring force is less than the frictional force. Motion then ceases.

If this happens after r half-swings we must have

$$2r-1 < \frac{n^2a}{\mu g} < 2r+1,$$

that is, r must be the first integer greater than $\tfrac{1}{2}(n^2a/\mu g)-\tfrac{1}{2}$.

Ex. 4. *Chattering in a system in which kinetic friction is less than static friction.*

This type of problem has many important technical applications. The motion is periodic but is quite different from the harmonic oscillations treated hitherto and is related to the 'relaxation oscillations' discussed in § 59.

As a definite problem suppose that a mass m is pressed by force P against a plane which moves with velocity V, motion of the mass with the plane being resisted by a spring of stiffness mn^2 [Fig. 6 (b)]. The coefficients of static and dynamic friction between the mass and the plane are μ' and μ, with $\mu' > \mu$.

We suppose the mass to be moving with the plane in the direction of the x-axis; this will continue until its displacement from the position in which the spring is unstrained is

$$x_1 = \frac{\mu' P}{mn^2}. \tag{8}$$

At this instant, which we shall take as the origin of time, $t = 0$, it will commence to slip and the frictional force on it will be dynamic friction acting forwards. That is, its equation of motion is

$$m\ddot{x} = -mn^2x+\mu P, \tag{9}$$

to be solved with $\dot{x} = V$, $x = \mu' P/mn^2$, when $t = 0$. The solution of (9) with these initial conditions is

$$x = \frac{\mu P}{mn^2}+\frac{V}{n}\sin nt+\frac{(\mu'-\mu)P}{mn^2}\cos nt, \tag{10}$$

$$\dot{x} = V\cos nt-\frac{(\mu'-\mu)P}{mn}\sin nt. \tag{11}$$

This motion continues until the velocity of the mass relative to the plane vanishes, and then sliding ceases. This happens at time T_1 given by the smallest (non-zero) root of

$$V\cos nt-\frac{(\mu'-\mu)P}{mn}\sin nt = V,$$

i.e. of

$$\tan\tfrac{1}{2}nt = -\frac{(\mu'-\mu)P}{mnV}. \tag{12}$$

This gives

$$T_1 = \frac{2\pi}{n}-\frac{2}{n}\tan^{-1}\frac{(\mu'-\mu)P}{mnV}. \tag{13}$$

At this time its displacement x_2 is, by (10),

$$x_2 = \frac{(2\mu-\mu')P}{mn^2}. \tag{14}$$

When the particle has come to rest relative to the plane it stays moving with it until its displacement is x_1 given by (8), when the process repeats itself. The time taken in moving from x_2 to x_1 is

$$T_2 = \frac{x_1-x_2}{V} = \frac{2(\mu'-\mu)P}{mn^2V}, \tag{15}$$

and the period of the whole process is T_1+T_2 given by (13) and (15). The curve of displacement against time for the motion is sketched in Fig. 6 (c).

31. The damped harmonic oscillator: forced oscillations

We now consider the system of § 29, namely a mass m with restoring force mn^2 times its displacement x, and resistance to motion $2m\kappa$ times its velocity $\dot{x}$, acted on by an external force $F_0\sin(\omega t+\beta)$.

Putting $F(t) = F_0\sin(\omega t+\beta)$ in § 29 (2), the equation of motion is

$$(D^2+2\kappa D+n^2)x = \frac{F_0}{m}\sin(\omega t+\beta). \tag{1}$$

The complementary function of (1) has been discussed in § 29; we now have to find a particular integral. Proceeding as in § 14 (iv), we find a particular integral of

$$(D^2+2\kappa D+n^2)x = \frac{F_0}{m}e^{i(\omega t+\beta)} \tag{2}$$

and take its imaginary part. We seek a particular integral of (2) of the form $x = Xe^{i\omega t}$: substituting this in (2) gives

$$(n^2-\omega^2+2\kappa i\omega)X = \frac{F_0}{m}e^{i\beta},$$

and the required particular integral of (2) becomes

$$x = \frac{F_0}{m(n^2-\omega^2+2\kappa i\omega)}e^{i(\omega t+\beta)}$$

$$= \frac{F_0}{m\{(n^2-\omega^2)^2+4\kappa^2\omega^2\}^{\frac{1}{2}}}e^{i(\omega t+\beta-\phi)}, \tag{3}$$

where $$\phi = \arg\{(n^2-\omega^2)+2\kappa i\omega\}. \tag{4}$$

Taking the imaginary part of (3), the particular integral of (1) is

$$x = \frac{F_0}{m\{(n^2-\omega^2)^2+4\kappa^2\omega^2\}^{\frac{1}{2}}}\sin(\omega t+\beta-\phi), \tag{5}$$

where ϕ is defined in (4).

Adding the complementary function [§ 29 (7) if $\kappa < n$, or the corresponding expression of § 29 if $\kappa \geqslant n$] gives the general solution of (1)

$$x = Ae^{-\kappa t}\sin(n't+B)+ \\ +\frac{F_0}{m\{(n^2-\omega^2)^2+4\kappa^2\omega^2\}^{\frac{1}{2}}}\sin(\omega t+\beta-\phi), \quad (6)$$

where $n' = \sqrt{(n^2-\kappa^2)}$, and A and B are constants to be determined from the initial conditions. $n'/2\pi$ is the damped natural frequency. The first term of (6) dies away exponentially as the time increases and is termed the *transient* part of the solution; for large values of the time, only the second term of (6), which is the particular integral (5) of (1), remains. This is called the *steady state solution* or *forced oscillation*; it has the same frequency $\omega/2\pi$ as the applied force and a lag in phase of ϕ behind it. In the time during which the transient is not negligible the oscillation builds up from its initial value to the final steady state value: the smaller the damping coefficient, the longer the time taken in this building-up process.

We proceed to discuss in detail the way in which the amplitude and phase of the forced oscillation (5) vary as ω varies between 0 and ∞. The velocity of the mass in the forced oscillation (5) is

$$\dot{x} = \frac{F_0\,\omega}{m\{(n^2-\omega^2)^2+4\kappa^2\omega^2\}^{\frac{1}{2}}}\sin\{\omega t+\beta-(\phi-\tfrac{1}{2}\pi)\}, \quad (7)$$

and we discuss also the variation of the amplitude and phase lag $(\phi-\frac{1}{2}\pi)$ of this.

Treating the phase lag first, we need the value of ϕ from (4). If $\omega < n$, $n^2-\omega^2$ is positive, and

$$\phi = \tan^{-1}\frac{2\kappa\omega}{n^2-\omega^2} \quad (\omega < n). \quad (8)$$

If $\omega > n$, $n^2-\omega^2$ is negative, and [cf. § 14 (25)]

$$\phi = \pi-\tan^{-1}\frac{2\kappa\omega}{\omega^2-n^2} \quad (\omega > n). \quad (9)$$

It appears from (8) and (9) that as $\omega \to 0$, $\phi \to 0$, that is the phase lag ϕ of the displacement is small for small ω; as $\omega \to n$,

$\phi \to \frac{1}{2}\pi$, a phase lag of $\frac{1}{2}\pi$ when $\omega/2\pi$ is equal to the undamped natural frequency; as $\omega \to \infty$, $\phi \to \pi$, a phase lag of π for high frequencies. Since by (7) the phase lag of the velocity is $\phi - \frac{1}{2}\pi$ it follows that this is $-\frac{1}{2}\pi$ for very low frequencies, $\frac{1}{2}\pi$ for very high frequencies, and zero at the undamped natural frequency. Graphs of ϕ and $\phi - \frac{1}{2}\pi$ are shown in Fig. 7 (*a*) and (*b*) respectively, for the case $\kappa = n/10$.

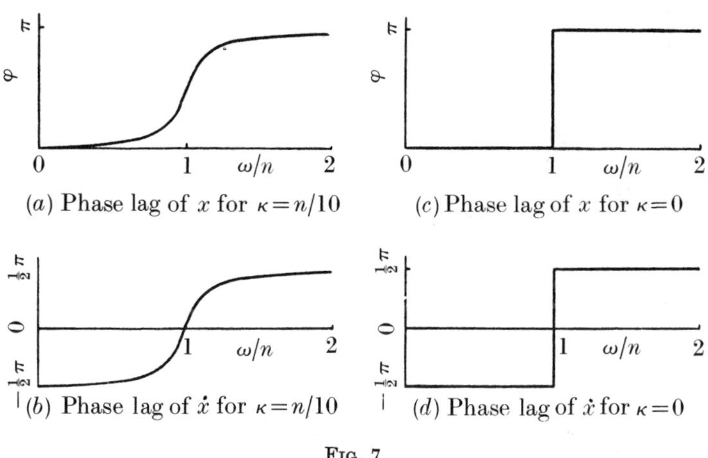

FIG. 7.

Next we have to consider the variation with ω of the amplitude of the forced oscillation, and we study first the amplitude

$$A_v = \frac{F_0}{2\kappa m\{1+(n^2-\omega^2)^2/4\kappa^2\omega^2\}^{\frac{1}{2}}} \tag{10}$$

of the velocity $\dot{x}$ given by (7). Clearly $A_v = 0$ when $\omega = 0$, and $A_v \to 0$ as $\omega \to \infty$. Also, when $\omega = n$, the denominator of (10) has its least value, so that A_v has a maximum of $F_0/2\kappa m$ when $\omega = n$, that is when the frequency is equal to the undamped natural frequency. The curve of A_v against ω is shown in Fig. 8 (*a*) for $\kappa = n/10$. Clearly, the smaller the value of the damping coefficient κ, the larger the value of the maximum. Thus when the frequency of the applied force is equal to the undamped natural frequency of the system, vibrations of large amplitude can be set up, particularly if the damping coefficient is small. This phenomenon is called *resonance*.

It is of importance to have some information about the sharpness of the peak of the curve of A_v against ω. To specify this we determine the values of ω at which A_v has a value $(1/s)$th of its maximum. By (10) this is the case when

$$\frac{1}{\{1+(n^2-\omega^2)^2/4\kappa^2\omega^2\}^{\frac{1}{2}}} = \frac{1}{s}, \tag{11}$$

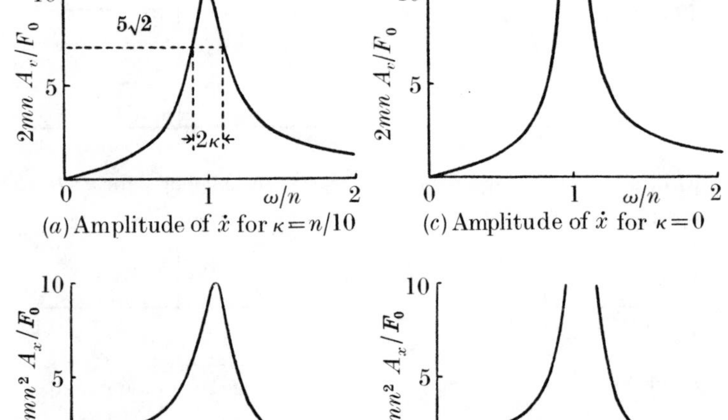

(a) Amplitude of $\dot{x}$ for $\kappa = n/10$

(c) Amplitude of $\dot{x}$ for $\kappa = 0$

(b) Amplitude of x for $\kappa = n/10$

(d) Amplitude of x for $\kappa = 0$

FIG. 8.

or, squaring, when

$$(n^2-\omega^2)^2 = 4\kappa^2\omega^2(s^2-1),$$

that is, when

$$\omega^2 \pm 2\kappa\omega(s^2-1)^{\frac{1}{2}} - n^2 = 0. \tag{12}$$

The roots of this quadratic are

$$\omega = \mp\kappa(s^2-1)^{\frac{1}{2}} + \{\kappa^2(s^2-1)+n^2\}^{\frac{1}{2}}, \tag{13}$$

where the positive sign has been chosen for the second square root since we are only interested in positive values of ω.

These two values of ω differ by $2\kappa\sqrt{(s^2-1)}$, so this is the width of the peak at the point where the amplitude is $(1/s)$th of its maximum value. In particular, the width of the curve is 2κ at $1/\sqrt{2} = 0{\cdot}7071$ of the maximum amplitude, and the values (13) of ω for this value of s are $(\kappa^2+n^2)^{\frac{1}{2}}\pm\kappa$. In Fig. 9 curves of A_v against ω for various values of κ/n are shown in order to illustrate the effect of decreasing κ on the sharpness of the curve.

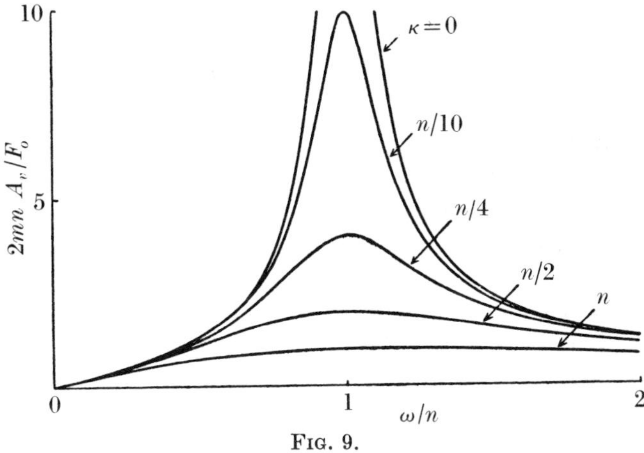

Fig. 9.

The amplitude of the velocity was considered in detail above because it is the more important in the analogous problem in electric circuit theory (cf. § 43) and also because it leads to the simple exact result (13). We now consider the amplitude A_x of the displacement x, which by (3) is

$$A_x = \frac{F_0}{m\{(n^2-\omega^2)^2+4\kappa^2\omega^2\}^{\frac{1}{2}}}. \tag{14}$$

When $\omega = 0$, $A_x = F_0/mn^2$, and as $\omega \to \infty$, $A_x \to 0$. Also

$$\frac{dA_x}{d\omega} = \frac{2\omega F_0\{n^2-2\kappa^2-\omega^2\}}{m\{(n^2-\omega^2)^2+4\kappa^2\omega^2\}^{\frac{3}{2}}}. \tag{15}$$

It follows that A_x has a minimum at $\omega = 0$ and a maximum at $\omega = \sqrt{(n^2-2\kappa^2)}$, provided $n^2 > 2\kappa^2$. If $n^2 < 2\kappa^2$ the curve decreases steadily and there is no maximum. The maximum value of A_x is

$$\frac{F_0}{2\kappa m(n^2-\kappa^2)^{\frac{1}{2}}}, \tag{16}$$

and it should be noticed that the frequency $(n^2-2\kappa^2)^{\frac{1}{2}}/2\pi$ at which it is attained does not coincide with either the undamped or the damped natural frequency. The curve of A_x against ω is shown in Fig. 8 (b).

As before we specify the sharpness of the peak by seeking the values of ω at which A_x has $(1/s)$th of its maximum value (16). These are given by

$$\omega = \{(n^2-2\kappa^2)\pm 2\kappa(s^2-1)^{\frac{1}{2}}(n^2-\kappa^2)^{\frac{1}{2}}\}^{\frac{1}{2}}. \tag{17}$$

If we assume that κ/n is small and use the binomial theorem in (17), we get approximately, neglecting terms in κ^2/n^2,

$$\omega = n \pm \kappa\surd(s^2-1), \tag{18}$$

which gives the same value as (13) for the width of the peak, but (18) is approximate whereas (13) was exact.

Finally, we treat the case $\kappa = 0$ *ab initio* in order to show clearly the connexion between the results for this case and those obtained earlier. This connexion is of considerable importance, since the study of more complicated systems involving several masses is not particularly difficult if damping is neglected, but becomes extremely complicated if it is included. It is therefore desirable to be able to obtain a qualitative idea of the behaviour of such a system from the solution for the case of no damping.

If $\kappa = 0$, the differential equation (1) becomes

$$(D^2+n^2)x = \frac{F_0}{m}\sin(\omega t+\beta), \tag{19}$$

and its particular integral, found as before, is

$$x = \frac{F_0}{m(n^2-\omega^2)}\sin(\omega t+\beta), \tag{20}$$

provided $\omega \neq n$.

If $\omega < n$, x is thus exactly in phase with the applied force. As $\omega \to n$ the amplitude of x tends to infinity. When $\omega > n$, x becomes negative and may be written

$$x = \frac{F_0}{m(\omega^2-n^2)}\sin(\omega t+\beta-\pi),$$

which corresponds to an oscillation of amplitude $F_0/m(\omega^2-n^2)$ with a phase lag of π behind the applied force. Thus (20) may be written

$$x = \frac{F_0}{m|n^2-\omega^2|}\sin(\omega t+\beta-\phi), \tag{21}$$

where the phase lag

$$\phi = 0 \quad (\omega < n)$$

$$\phi = \pi \quad (\omega > n)$$

changes discontinuously by π as ω passes through n. Also the amplitude tends to infinity as ω tends to n. These curves are shown in Figs. 7 (*c*) and 8 (*d*), respectively, and are obvious limiting cases of those for finite damping as $\kappa \to 0$.

In the same way the velocity is

$$\dot{x} = \frac{\omega F_0}{m|n^2-\omega^2|} \sin(\omega t+\beta+\tfrac{1}{2}\pi-\phi).$$

Its phase lag and amplitude are shown in Figs. 7 (*d*) and 8 (*c*).

The general solution of (19) is

$$A \sin(nt+B)+\frac{F_0}{m(n^2-\omega^2)} \sin(\omega t+\beta), \tag{22}$$

where A and B are arbitrary constants to be found from the initial conditions. It should be remarked that in this ideal case the complementary function $A\sin(nt+B)$ does not die away for large values of the time.

It remains to consider what happens in the case of no damping when ω has the resonance frequency n, so that (19) becomes

$$(D^2+n^2)x = \frac{F_0}{m} \sin(nt+\beta). \tag{23}$$

The particular integral of this has been found in § 14, Ex. 9, to be

$$-\frac{F_0}{2mn} t \cos(nt+\beta). \tag{24}$$

It thus corresponds to an oscillation of steadily increasing amplitude. If a system with negligible damping is set in motion by a harmonic force of the resonance frequency, the amplitude of its oscillations will increase steadily until they are so great that the assumed equations of motion cease to hold.

32. The harmonic oscillator: forced oscillations caused by motion of the support

The differential equations discussed in detail in § 31 arise again here but in a slightly different manner.

Suppose a particle of mass m is attached to a point S by a spring of stiffness mn^2, and that there is resistance to motion $2m\kappa$ times the velocity of the particle. Let the point S be given a prescribed motion $\xi(t)$ in the direction of the spring. If the origin of x is taken so that the spring is unstrained when $x = 0$

and $\xi = 0$, the restoring force on the particle for any values of x and ξ will be

$$mn^2\{x-\xi(t)\},$$

and the equation of motion of the particle will be

$$m\ddot{x} = -mn^2\{x-\xi(t)\}-2m\kappa\dot{x},$$

or
$$\ddot{x}+2\kappa\dot{x}+n^2x = n^2\xi(t), \tag{1}$$

which is the same as § 29 (2) except for the change of notation.

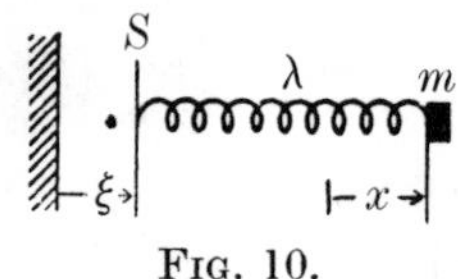

Fig. 10.

If the point of support is given a harmonic motion

$$\xi = a\sin(\omega t+\beta),$$

(1) becomes
$$\ddot{x}+2\kappa\dot{x}+n^2x = n^2a\sin(\omega t+\beta), \tag{2}$$

which is the same as § 31 (1) with n^2a in place of F_0/m.

Ex. *A machine of mass M is supported on springs of total stiffness $(M+m)n^2$, and its vibration is damped by resistance $2(M+m)\kappa$ times its velocity. It carries a mass m which executes a vertical simple harmonic motion $\xi = a\sin(\omega t+\beta)$ relative to the mass M. Find the steady periodic motion of the bed.*

Let x be the displacement of the mass M measured upwards from the position in which the springs are unstrained, so that $(x+\xi)$ is the position of the mass m relative to the same origin. Let $f(\xi)$ be the force exerted by the machine on the mass m, then the equation of motion of the mass m is

$$m(\ddot{x}+\ddot{\xi}) = -mg+f(\xi), \tag{3}$$

and the equation of motion of the mass M is

$$M\ddot{x} = -(M+m)n^2x-2(M+m)\kappa\dot{x}-f(\xi)-Mg. \tag{4}$$

Adding (3) and (4) gives

$$\ddot{x}+2\kappa\dot{x}+n^2x = -g-\frac{m}{M+m}\ddot{\xi}$$

$$= -g+\frac{ma\omega^2}{M+m}\sin(\omega t+\beta). \tag{5}$$

Using § 31 (4) and (5), the particular integral of (5) is

$$-\frac{g}{n^2}+\frac{ma\omega^2}{(M+m)\{(n^2-\omega^2)^2+4\kappa^2\omega^2\}^{\frac{1}{2}}}\sin(\omega t+\beta-\phi),$$

corresponding to a forced oscillation about the equilibrium position $-g/n^2$ of the machine.

33. Systems of several masses

When several masses are involved there is a great increase in algebraical complexity. We discuss in detail a case involving two masses to show the new ideas involved. In this section we shall assume that there is no resistance to motion; the same problem, with resistance included, will be studied in § 34.

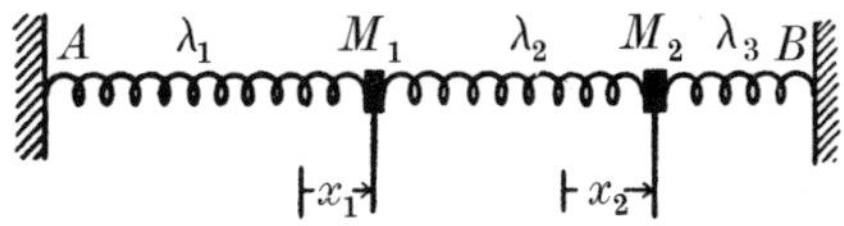

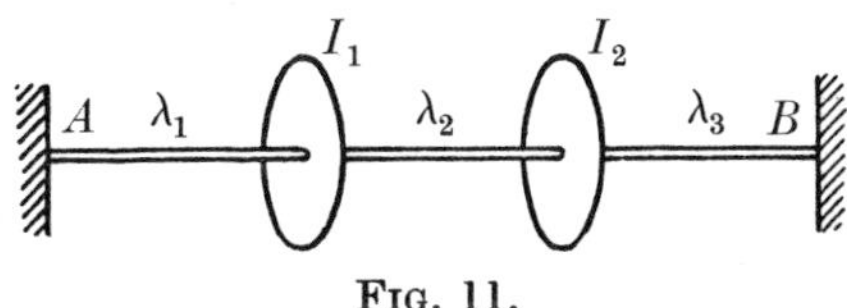

FIG. 11.

Suppose that two masses M_1 and M_2 are connected as shown in Fig. 11 by springs of stiffnesses λ_1, λ_2, and λ_3 to fixed points A and B, and that they oscillate in the straight line AB. Let x_1 and x_2 be the displacements of M_1 and M_2 from their equilibrium positions. Then, if there are no external forces, the equations of motion of M_1 and M_2 are

$$M_1\ddot{x}_1 = \lambda_2(x_2-x_1)-\lambda_1 x_1, \tag{1}$$

$$M_2\ddot{x}_2 = -\lambda_2(x_2-x_1)-\lambda_3 x_2. \tag{2}$$

Writing

$$n_1^2 = \frac{\lambda_1}{M_1}, \qquad n_2^2 = \frac{\lambda_2}{M_2}, \qquad n_{12}^2 = \frac{\lambda_2}{M_1}, \qquad n_{23}^2 = \frac{\lambda_3}{M_2}, \tag{3}$$

these become

$$(D^2+n_1^2+n_{12}^2)x_1-n_{12}^2 x_2 = 0, \tag{4}$$

$$-n_2^2 x_1+(D^2+n_2^2+n_{23}^2)x_2 = 0. \tag{5}$$

(4) and (5) are a pair of simultaneous linear differential equations for x_1 and x_2. We could solve them by the methods of § 15, but it is more usual in problems of this type to use the

equivalent method indicated at the end of § 13 and seek a solution in which x_1 and x_2 are proportional to $e^{\alpha t}$. Thus we assume†

$$x_1 = X_1 e^{\alpha t}, \qquad x_2 = X_2 e^{\alpha t}, \tag{6}$$

where X_1 and X_2 are independent of t. Substituting (6) in (4) and (5) gives

$$(\alpha^2+n_1^2+n_{12}^2)X_1 - n_{12}^2 X_2 = 0, \tag{7}$$

$$-n_2^2 X_1 + (\alpha^2+n_2^2+n_{23}^2)X_2 = 0. \tag{8}$$

These may be written

$$\frac{X_1}{X_2} = \frac{n_{12}^2}{\alpha^2+n_1^2+n_{12}^2} = \frac{\alpha^2+n_2^2+n_{23}^2}{n_2^2}. \tag{9}$$

The second of equations (9) gives the equation

$$(\alpha^2+n_2^2+n_{23}^2)(\alpha^2+n_1^2+n_{12}^2) - n_2^2 n_{12}^2 = 0 \tag{10}$$

for α, and corresponding to each root of this, the first of equations (9) gives the associated value of the ratio $X_1 : X_2$. (9) is the auxiliary equation of the differential equations (4) and (5) for x_1 and x_2. It will be called the *frequency equation*, since it will appear that its roots determine the natural frequencies of the system.

For simplicity we now restrict ourselves to the case of equal masses and springs, so that

$$M_1 = M_2, \qquad \lambda_1 = \lambda_2 = \lambda_3, \qquad n_1 = n_2 = n_{12} = n_{23},$$

and (9) and (10) become

$$\frac{X_1}{X_2} = \frac{n_1^2}{\alpha^2+2n_1^2} = \frac{\alpha^2+2n_1^2}{n_1^2}, \tag{11}$$

and

$$\alpha^4+4\alpha^2 n_1^2+3n_1^4 = 0. \tag{12}$$

The roots of (12) are $\alpha^2 = -n_1^2$ and $\alpha^2 = -3n_1^2$.

The root $\alpha^2 = -n_1^2$, $\alpha = \pm in_1$, gives by (11)

$$\frac{X_1}{X_2} = 1. \tag{13}$$

† It will appear in (13) that α is pure imaginary so that we might have started by assuming solutions proportional to e^{int}, and this is often done. The form (6) is used here, partly because of the correspondence with the work of Chapter II, and partly because it is a little better when there is resistance to motion as in § 34.

Thus the solution of type (6) is

$$x_1 = Ce^{in_1t}+De^{-in_1t},$$

$$x_2 = Ce^{in_1t}+De^{-in_1t},$$

where C and D are any constants. These may be written

$$\left.\begin{aligned} x_1 &= A_1\sin(n_1t+\beta_1) \\ x_2 &= A_1\sin(n_1t+\beta_1) \end{aligned}\right\}, \tag{14}$$

where A_1 and β_1 are any constants.

Thus in this solution, x_1 and x_2 execute harmonic oscillations of the same amplitude and phase, and with period $2\pi/n_1$.

The other root $\alpha^2 = -3n_1^2$, $\alpha = \pm in_1\sqrt{3}$, of (12) gives from (11)

$$\frac{X_1}{X_2} = -1. \tag{15}$$

For this root the solutions (6) take the form

$$\left.\begin{aligned} x_1 &= A_2\sin(n_1t\sqrt{3}+\beta_2) \\ x_2 &= A_2\sin(n_1t\sqrt{3}+\beta_2-\pi) \end{aligned}\right\}, \tag{16}$$

where A_2 and β_2 are arbitrary constants. Thus x_1 and x_2 execute harmonic vibrations of the same amplitude, of period $2\pi/n_1\sqrt{3}$, and 180° out of phase.

These two solutions (14) and (16) are called the two *normal modes of oscillation* of the system, and their frequencies $n_1/2\pi$ and $n_1\sqrt{3}/2\pi$ are called the *natural frequencies* of the system. In the 'lower mode', that is the mode of lower natural frequency, Fig. 12 (*a*), the displacements of the particles are equal and in phase; in the higher mode, Fig. 12 (*b*), they are equal and out of phase by 180°.

The general solution of (4) and (5) consists of a combination of the two normal modes (14) and (16) with arbitrary amplitudes and phases, that is,

$$\left.\begin{aligned} x_1 &= A_1\sin(n_1t+\beta_1)+A_2\sin(n_1t\sqrt{3}+\beta_2) \\ x_2 &= A_1\sin(n_1t+\beta_1)+A_2\sin(n_1t\sqrt{3}+\beta_2-\pi) \end{aligned}\right\}. \tag{17}$$

This contains four arbitrary constants as it should (cf. § 15), and these can be determined from the initial displacements and velocities of the two masses. For initial value problems of this sort it is better to use the Laplace transformation method as in § 40; for most purposes a knowledge of the normal modes and natural frequencies is sufficient.

The same type of result holds in general: if there had been r masses, the equation for α^2 would have had r roots, corresponding to r natural frequencies, each of these gives a normal

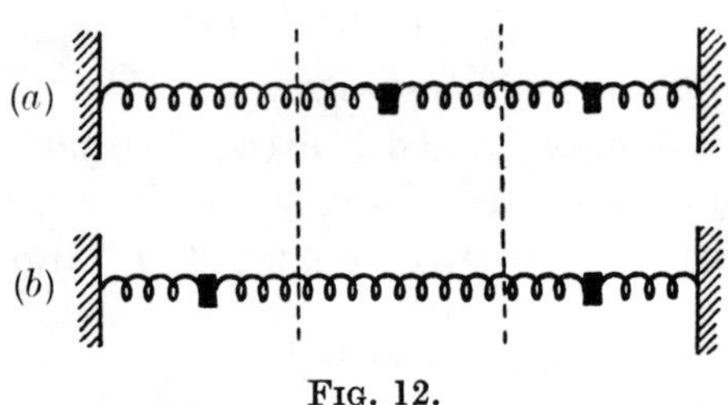

FIG. 12.

mode in which the relative amplitudes and phases of the masses are known.

Next we consider forced oscillations of the system of Fig. 11 due to a force $F_0 \sin \omega t$ applied to the mass M_1.

The differential equations (4) and (5) are replaced by

$$(D^2+n_1^2+n_{12}^2)x_1-n_{12}^2 x_2 = (F_0/M_1)\sin \omega t, \tag{18}$$

$$-n_2^2 x_1+(D^2+n_2^2+n_{23}^2)x_2 = 0. \tag{19}$$

We now require a particular integral of these, and as usual replace $\sin \omega t$ in (18) by $e^{i\omega t}$, giving

$$(D^2+n_1^2+n_{12}^2)x_1-n_{12}^2 x_2 = (F_0/M_1)e^{i\omega t}, \tag{20}$$

find a particular integral of (19) and (20), and take its imaginary part. To find this particular integral we seek solutions of (19) and (20) of the form

$$x_1 = X_1 e^{i\omega t}, \qquad x_2 = X_2 e^{i\omega t}. \tag{21}$$

Substituting (21) in (19) and (20) gives

$$-n_2^2 X_1+(n_2^2+n_{23}^2-\omega^2)X_2 = 0,$$

$$(n_1^2+n_{12}^2-\omega^2)X_1-n_{12}^2 X_2 = F_0/M_1.$$

Solving we find

$$X_1 = \frac{F_0(n_2^2+n_{23}^2-\omega^2)}{M_1\{(n_1^2+n_{12}^2-\omega^2)(n_2^2+n_{23}^2-\omega^2)-n_2^2\,n_{12}^2\}}, \tag{22}$$

$$X_2 = \frac{n_2^2\,F_0}{M_1\{(n_1^2+n_{12}^2-\omega^2)(n_2^2+n_{23}^2-\omega^2)-n_2^2\,n_{12}^2\}}. \tag{23}$$

X_1 and X_2 are both real, so, using these values in (21) and taking the imaginary part, the required particular integrals of (18) and (19) are found to be

$$x_1 = X_1 \sin \omega t, \tag{24}$$

$$x_2 = X_2 \sin \omega t. \tag{25}$$

To study the behaviour of these in greater detail we consider again the special case $M_1 = M_2$, $\lambda_1 = \lambda_2 = \lambda_3$ discussed earlier. In this case X_1 and X_2 become

$$X_1 = \frac{F_0(2n_1^2-\omega^2)}{M_1(3n_1^2-\omega^2)(n_1^2-\omega^2)}, \tag{26}$$

$$X_2 = \frac{n_1^2\,F_0}{M_1(3n_1^2-\omega^2)(n_1^2-\omega^2)}. \tag{27}$$

The behaviour of X_1 and X_2, given by (26) and (27), as functions of ω is shown in Fig. 13 (a) and (b). They tend to infinity as ω tends to either of the values n_1 or $n_1\sqrt{3}$ corresponding to natural frequencies of the system, and they change sign on passing through these points. As in § 31 (21) we regard x_1 and x_2 as oscillations of amplitudes $|X_1|$ and $|X_2|$ respectively, and when X_1 or X_2 is negative we express this as a phase lag of π of x_1 or x_2 behind the applied force. The amplitudes of x_1 and x_2 are shown in Fig. 13 (c) and (d) and their phases in Fig. 13 (e) and (f). These may be compared with the curves of Fig. 15 for the same system when resistance to motion is taken into account.

The case of harmonic force applied to one of the masses only has been considered above. Clearly results of the same general type will be obtained if harmonic force is applied to both the masses, or if, as in § 32, one of the supports is given a harmonic oscillation.

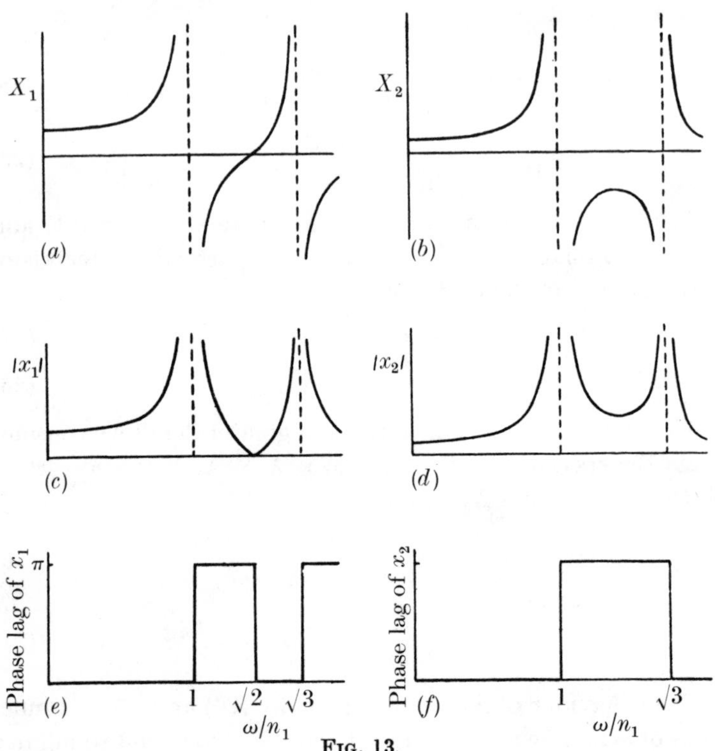

FIG. 13.

34. Systems of several masses with resistance proportional to velocity

To illustrate the effect of resistance to motion we consider again in detail the system of § 33, but now include resistance terms. These resistance terms may arise in two ways: there may be resistance proportional to the absolute velocities $\dot{x}_1$ and $\dot{x}_2$ of the masses, and there may in addition be resistance proportional to the relative velocity $(\dot{x}_2-\dot{x}_1)$ of the masses. These are indicated by the appropriate symbols in Fig. 14.

The equations of motion of this system when subject to no external forces are

$$M_1\ddot{x}_1 = \lambda_2(x_2-x_1)-\lambda_1 x_1-k_1\dot{x}_1+k_{12}(\dot{x}_2-\dot{x}_1), \tag{1}$$

$$M_2\ddot{x}_2 = -\lambda_2(x_2-x_1)-\lambda_3 x_2-k_2\dot{x}_2-k_{12}(\dot{x}_2-\dot{x}_1). \tag{2}$$

Using the abbreviations § 33 (3) and in addition

$$2\kappa_1 = \frac{k_1}{M_1}, \qquad 2\kappa_2 = \frac{k_2}{M_2}, \qquad 2\kappa_{12} = \frac{k_{12}}{M_1}, \qquad 2\kappa_{21} = \frac{k_{12}}{M_2}, \tag{3}$$

these become

$$\{D^2+2(\kappa_1+\kappa_{12})D+n_1^2+n_{12}^2\}x_1-(2\kappa_{12}D+n_{12}^2)x_2 = 0, \tag{4}$$

$$-(2\kappa_{21}D+n_2^2)x_1+\{D^2+2(\kappa_2+\kappa_{21})D+n_2^2+n_{23}^2\}x_2 = 0. \tag{5}$$

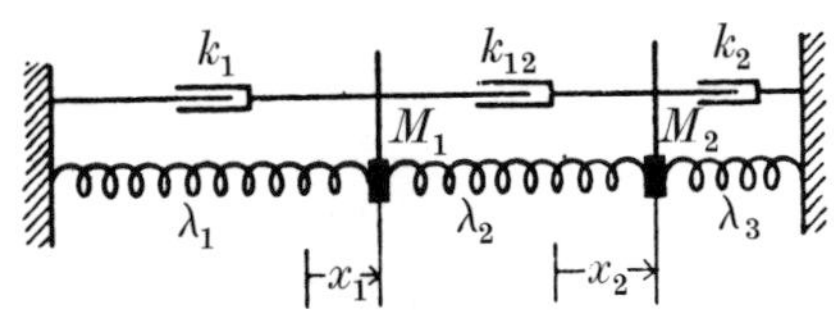

Fig. 14.

As before, we seek solutions of (4) and (5) of the form

$$x_1 = X_1 e^{\alpha t}, \qquad x_2 = X_2 e^{\alpha t}, \tag{6}$$

where X_1 and X_2 are independent of t. Substituting from (6) in (4) and (5) gives the equations for X_1 and X_2, namely,

$$\{\alpha^2+2(\kappa_1+\kappa_{12})\alpha+n_1^2+n_{12}^2\}X_1-(2\kappa_{12}\alpha+n_{12}^2)X_2 = 0, \tag{7}$$

$$-(2\kappa_{21}\alpha+n_2^2)X_1+\{\alpha^2+2(\kappa_2+\kappa_{21})\alpha+n_2^2+n_{23}^2\}X_2 = 0. \tag{8}$$

To simplify the algebra we now restrict ourselves to the special case considered in detail in § 33, namely, $M_1 = M_2$, $\lambda_1 = \lambda_2 = \lambda_3$, so that $n_2^2 = n_{12}^2 = n_{23}^2 = n_1^2$, and in addition we take $k_{12} = 0$ so that $\kappa_{12} = \kappa_{21} = 0$, and $k_1 = k_2$ so that $\kappa_2 = \kappa_1$. The equations (7) and (8) then become

$$(\alpha^2+2\kappa_1\alpha+2n_1^2)X_1-n_1^2X_2 = 0, \tag{9}$$

$$-n_1^2X_1+(\alpha^2+2\kappa_1\alpha+2n_1^2)X_2 = 0. \tag{10}$$

These require

$$\frac{X_1}{X_2} = \frac{\alpha^2+2\kappa_1\alpha+2n_1^2}{n_1^2} = \frac{n_1^2}{\alpha^2+2\kappa_1\alpha+2n_1^2}. \tag{11}$$

The second of equations (11) gives the frequency equation

$$(\alpha^2+2\kappa_1\alpha+2n_1^2)^2-n_1^4 = 0,$$

or

$$(\alpha^2+2\kappa_1\alpha+n_1^2)(\alpha^2+2\kappa_1\alpha+3n_1^2) = 0. \tag{12}$$

The roots of this are

$$\alpha = -\kappa_1 \pm i\surd(n_1^2-\kappa_1^2), \tag{13}$$

and

$$\alpha = -\kappa_1 \pm i\surd(3n_1^2-\kappa_1^2), \tag{14}$$

provided that $n_1 > \kappa_1$.

The roots (13) correspond to an oscillation with time factor

$$e^{-\kappa_1 t \pm it\surd(n_1^2-\kappa_1^2)}$$

and in which, using (13) in the first of equations (11),

$$\frac{X_1}{X_2} = 1.$$

This gives the normal mode

$$\left.\begin{aligned} x_1 &= A_1 e^{-\kappa_1 t}\sin\{t\surd(n_1^2-\kappa_1^2)+\theta_1\} \\ x_2 &= A_1 e^{-\kappa_1 t}\sin\{t\surd(n_1^2-\kappa_1^2)+\theta_1\} \end{aligned}\right\}, \tag{15}$$

where A_1 and θ_1 are arbitrary constants. This differs only from the corresponding result § 33 (14) for the undamped case by the occurrence of the damping factor $e^{-\kappa_1 t}$ and the change in the natural frequency from $n_1/2\pi$ to $(n_1^2-\kappa_1^2)^{\frac{1}{2}}/2\pi$.

In the same way, using (14) in the first of equations (11) gives $X_1/X_2 = -1$, and (14) leads to the normal mode

$$\left.\begin{aligned} x_1 &= A_2 e^{-\kappa_1 t}\sin\{t\surd(3n_1^2-\kappa_1^2)+\theta_2\} \\ x_2 &= A_2 e^{-\kappa_1 t}\sin\{t\surd(3n_1^2-\kappa_1^2)+\theta_2-\pi\} \end{aligned}\right\}; \tag{16}$$

cf. § 33 (16). It should be remarked that the fact that the ratios X_1/X_2 are the same as those of the undamped case of § 33 is accidental: usually they become complex.

Considering now forced oscillations, suppose that, as in § 33, a force $F_0 \sin \omega t$ is applied to the first of the masses. We require a particular integral of the equations

$$(D^2+2\kappa_1 D+2n_1^2)x_1 - n_1^2 x_2 = (F_0/M_1)e^{i\omega t}, \tag{17}$$

$$-n_1^2 x_1 + (D^2+2\kappa_1 D+2n_1^2)x_2 = 0, \tag{18}$$

and we assume this to be of the form

$$x_1 = X_1 e^{i\omega t}, \qquad x_2 = X_2 e^{i\omega t}. \tag{19}$$

Substituting (19) in (17) and (18) gives

$$(2n_1^2-\omega^2+2\kappa_1 i\omega)X_1 - n_1^2 X_2 = F_0/M_1,$$

$$-n_1^2 X_1 + (2n_1^2-\omega^2+2\kappa_1 i\omega)X_2 = 0.$$

Solving for X_1 gives

$$X_1 = \frac{F_0(2n_1^2-\omega^2+2\kappa_1 i\omega)}{M_1(3n_1^2-\omega^2+2\kappa_1 i\omega)(n_1^2-\omega^2+2\kappa_1 i\omega)} = A_1 e^{-i\phi}, \quad (20)$$

where

$$A_1 = \frac{F_0}{M_1}\left\{\frac{(2n_1^2-\omega^2)^2+4\kappa_1^2\omega^2}{[(3n_1^2-\omega^2)^2+4\kappa_1^2\omega^2][(n_1^2-\omega^2)^2+4\kappa_1^2\omega^2]}\right\}^{\frac{1}{2}}, \quad (21)$$

$$\phi = \arg(3n_1^2-\omega^2+2\kappa_1 i\omega)+\arg(n_1^2-\omega^2+2\kappa_1 i\omega)-$$
$$-\arg(2n_1^2-\omega^2+2\kappa_1 i\omega). \quad (22)$$

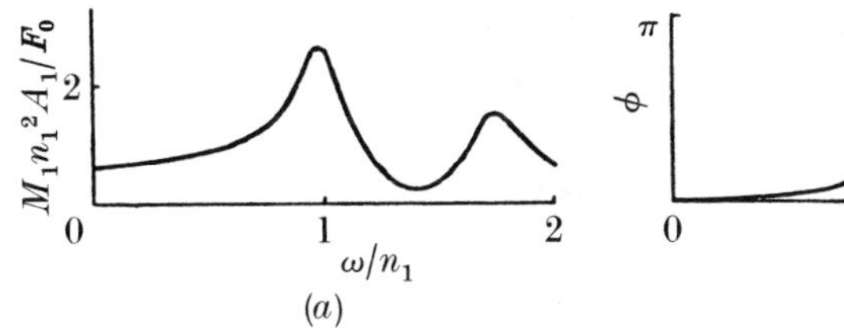

Fig. 15.

Using (20) in (19) and taking the imaginary part gives for the forced oscillation of x_1

$$x_1 = A_1 \sin(\omega t-\phi), \quad (23)$$

a vibration of frequency $\omega/2\pi$, amplitude A_1, and phase lag ϕ behind the applied force.

We now have to discuss the variation of A_1 and ϕ with ω as was done in § 31, and to compare the results with those of § 33.

The discussion of ϕ is easy: it consists of three terms of type § 31 (4), Fig. 7 (*a*), and is shown in Fig. 15 (*b*) for the case $\kappa_1 = n_1/10$.

The amplitude A_1 contains the product of two terms of type § 31 (14) and thus may be expected to have two maxima near the values n_1 and $n_1\sqrt{3}$ of ω. The curve is too complicated for general discussion: its graph for the case $\kappa_1 = n_1/10$ is shown in Fig. 15 (*a*).

The correspondence between the case $\kappa = 0$ of Fig. 13 (*c*) and (*e*) with the case of small damping, Fig. 15 (*a*) and (*b*), shows clearly on comparing these figures.

35. Systems of several masses: variation of the natural frequencies with the number of masses

In § 33 the natural frequencies of two equal masses constrained by three equal springs as in Fig. 11 have been studied: in this section we consider the effect of adding more masses connected

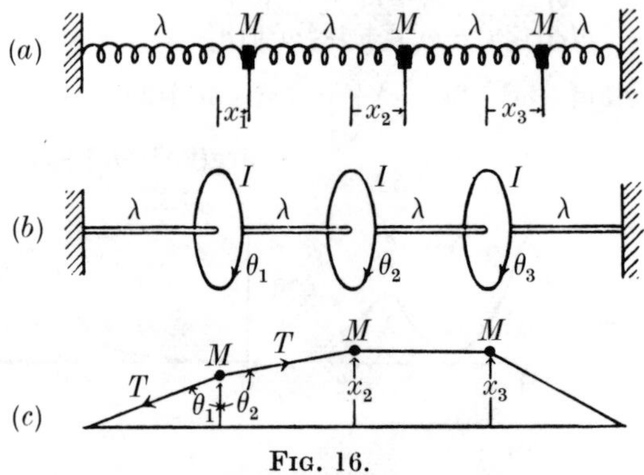

FIG. 16.

in the same way. Resistance to motion will not be included—if it is added, a change corresponding to that from the results of § 33 to those of § 34 appears.

The system to be considered here is that of Fig. 16 (*a*), longitudinal vibrations of three equal masses M constrained by equal springs of stiffness λ. The corresponding system for torsional oscillation of a shaft is shown in Fig. 16 (*b*). In order to introduce another interesting system, and because the normal oscillations can be represented more picturesquely for it, we shall state the problem in terms of the system of Fig. 16 (*c*), *small transverse oscillations of three equal masses attached at distances l, $2l$, $3l$ along a light elastic string of length $4l$ stretched to tension T*.

Let x_1, x_2, x_3 be the displacements of the masses; these are assumed to be so small that they do not affect the tension of the string. The restoring force on the first mass whose displacement is x_1 is

$$T\cos\theta_1+T\cos\theta_2 = T\frac{x_1}{l}-T\frac{x_2-x_1}{l}, \tag{1}$$

approximately. Those on the second and third masses can be written down in the same way.

Thus the equations of motion of the three masses are

$$M\ddot{x}_1 = -\frac{Tx_1}{l}+\frac{T(x_2-x_1)}{l},$$

$$M\ddot{x}_2 = -\frac{T(x_2-x_1)}{l}+\frac{T(x_3-x_2)}{l},$$

$$M\ddot{x}_3 = -\frac{T(x_3-x_2)}{l}-\frac{Tx_3}{l}.$$

Writing $n^2 = T/Ml$ these become

$$(D^2+2n^2)x_1-n^2x_2 = 0, \tag{2}$$

$$-n^2x_1+(D^2+2n^2)x_2-n^2x_3 = 0, \tag{3}$$

$$-n^2x_2+(D^2+2n^2)x_3 = 0. \tag{4}$$

These are identical with the equations which would be written down as in § 33 for the system of Fig. 16 (*a*) with $n^2 = \lambda/M$, or for the system of Fig. 16 (*b*) with $n^2 = \lambda/I$. In the problem of Fig. 16 (*c*) the linear equations (2) to (4) are only an approximation for small oscillations—the restoring forces in fact contain in addition terms involving the squares and higher powers of the displacements which we neglect. The equations of the problems of Fig. 16 (*a*) and (*b*) are accurately linear if we assume the springs and shafts to be perfectly elastic.

To solve (2) to (4) we assume a solution

$$x_1 = X_1 e^{\alpha t}, \qquad x_2 = X_2 e^{\alpha t}, \qquad x_3 = X_3 e^{\alpha t}, \tag{5}$$

and substitute in (2) to (4) which give

$$(\alpha^2+2n^2)X_1-n^2X_2 = 0, \tag{6}$$

$$-n^2X_1+(\alpha^2+2n^2)X_2-n^2X_3 = 0, \tag{7}$$

$$-n^2X_2+(\alpha^2+2n^2)X_3 = 0. \tag{8}$$

(6)–(8) are three homogeneous linear equations for X_1, X_2, X_3, and, in order that they may have a solution, α must be a root of

$$\begin{vmatrix} \alpha^2+2n^2 & -n^2 & 0 \\ -n^2 & \alpha^2+2n^2 & -n^2 \\ 0 & -n^2 & \alpha^2+2n^2 \end{vmatrix} = 0, \tag{9}$$

which is the frequency equation. Expanding the determinant this becomes

$$(\alpha^2+2n^2)\{(\alpha^2+2n^2)^2-2n^4\} = 0. \tag{10}$$

The roots of (10) are

$$\alpha^2 = -2n^2, \quad \alpha^2 = -(2+\surd 2)n^2, \quad \alpha^2 = -(2-\surd 2)n^2, \tag{11}$$

corresponding to the natural frequencies of

$$[n\surd(2-\surd 2)]/2\pi, \quad [n\surd 2]/2\pi, \quad [n\surd(2+\surd 2)]/2\pi. \tag{12}$$

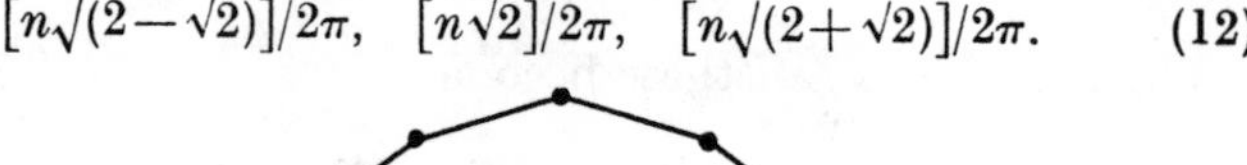

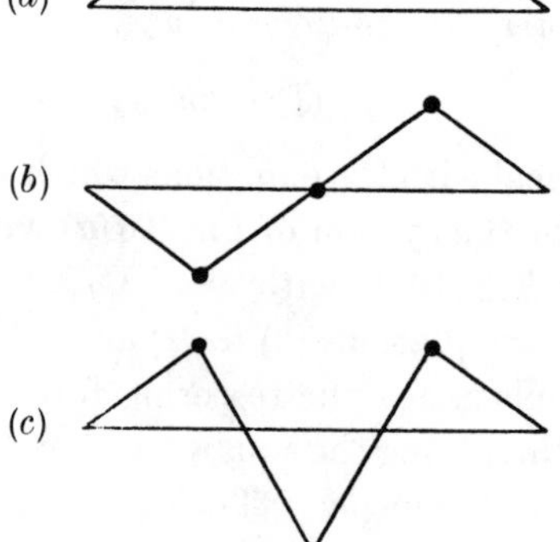

FIG. 17.

To find the normal mode of oscillation corresponding to each natural frequency we insert the value of α in any two of (6)–(8) and solve for the ratio $X_1:X_2:X_3$.

Thus if

$$\alpha^2 = -(2-\surd 2)n^2, \qquad X_1:X_2:X_3 = 1:\surd 2:1, \tag{13}$$

if
$$\alpha^2 = -2n^2, \qquad X_1:X_2:X_3 = 1:0:-1, \tag{14}$$

if
$$\alpha^2 = -(2+\surd 2)n^2, \qquad X_1:X_2:X_3 = 1:-\surd 2:1. \tag{15}$$

The displacements of the particles in the three normal modes are shown in Fig. 17: (*a*), corresponding to (13), is the lowest mode; while (*c*), corresponding to (15), is the highest.

The method of solving (6)–(8) used above is the general one applicable to any number of equations; the simpler method used in §§ 33, 34 for the case of two equations is equivalent to it.

If a fourth mass is added in the chain of Fig. 16, the frequency equation corresponding to (9) becomes

$$\begin{vmatrix} \alpha^2+2n^2 & -n^2 & 0 & 0 \\ -n^2 & \alpha^2+2n^2 & -n^2 & 0 \\ 0 & -n^2 & \alpha^2+2n^2 & -n^2 \\ 0 & 0 & -n^2 & \alpha^2+2n^2 \end{vmatrix} = 0, \qquad (16)$$

which has roots $\alpha^2 = -0{\cdot}382n^2, \alpha^2 = -1{\cdot}382n^2, \alpha^2 = -2{\cdot}618n^2$, $\alpha^2 = -3{\cdot}618n^2$, corresponding to natural frequencies

$$0{\cdot}618n/2\pi, \quad 1{\cdot}176n/2\pi, \quad 1{\cdot}618n/2\pi, \quad 1{\cdot}902n/2\pi. \qquad (17)$$

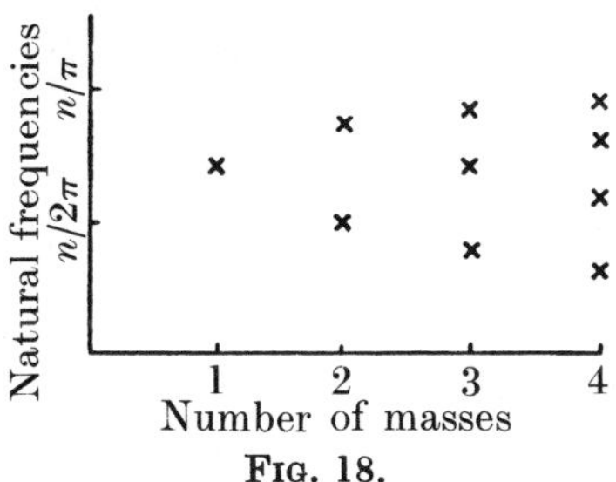

FIG. 18.

The natural frequencies for the same system with two masses have been found in § 33 to be $n/2\pi$ and $(n\sqrt{3})/2\pi$, while the natural frequency for the system with one mass only is the same as that for the mass attached to a spring of twice the stiffness, namely, $(n\sqrt{2})/2\pi$.

These results are shown by crosses in Fig. 18: it appears that as more masses are added to the system the highest natural frequency increases and tends to n/π, and the lowest decreases and tends to zero. The individual natural frequencies steadily approach closer together. The general result for any number of masses is given in Ex. 15 at the end of the chapter.

36. Systems of several masses, variation of the natural frequencies with the masses: vibration dampers

Although the general frequency equation for any values of the masses and stiffnesses of the springs was given in § 33, only the simple case of equal masses and stiffnesses was studied in order to have simple numerical results for discussion. The question of how the natural frequencies vary when the masses are varied naturally arises, and in this section we discuss another system from this point of view.

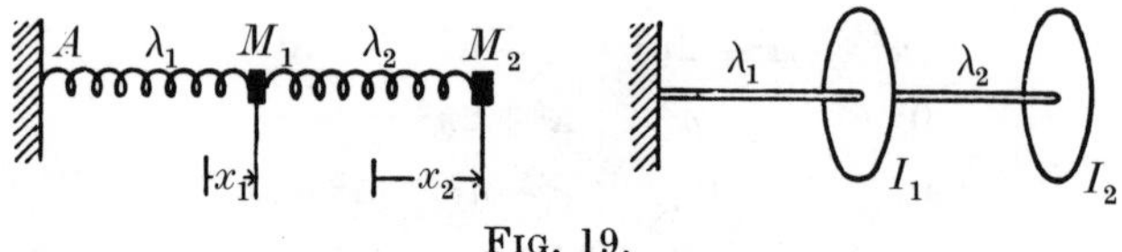

FIG. 19.

The problem considered here is that of longitudinal vibrations of two masses M_1 and M_2, connected by a spring of stiffness λ_2; M_1 is attached to a fixed point by a spring of stiffness λ_1. As before, resistance to motion is neglected: it may be included as in § 34.

Let x_1 and x_2 be the displacements of the masses from their equilibrium positions. Then, if there are no external forces, the equations of motion are

$$M_1\ddot{x}_1 = -\lambda_1 x_1+\lambda_2(x_2-x_1), \tag{1}$$

$$M_2\ddot{x}_2 = -\lambda_2(x_2-x_1). \tag{2}$$

Writing

$$n_1^2 = \frac{\lambda_1}{M_1}, \qquad n_2^2 = \frac{\lambda_2}{M_2}, \qquad n_{12}^2 = \frac{\lambda_2}{M_1}, \tag{3}$$

these become

$$(D^2+n_1^2+n_{12}^2)x_1-n_{12}^2x_2 = 0, \tag{4}$$

$$-n_2^2x_1+(D^2+n_2^2)x_2 = 0. \tag{5}$$

As usual, we seek solutions

$$x_1 = X_1e^{\alpha t}, \qquad x_2 = X_2e^{\alpha t}. \tag{6}$$

Substituting in (4) and (5) gives

$$(\alpha^2+n_1^2+n_{12}^2)X_1-n_{12}^2X_2 = 0,$$

$$-n_2^2X_1+(\alpha^2+n_2^2)X_2 = 0.$$

That is

$$\frac{X_1}{X_2} = \frac{n_{12}^2}{\alpha^2+n_1^2+n_{12}^2} = \frac{\alpha^2+n_2^2}{n_2^2}. \tag{7}$$

The second of equations (7) is the frequency equation

$$\alpha^4+\alpha^2(n_1^2+n_2^2+n_{12}^2)+n_1^2n_2^2 = 0. \tag{8}$$

We propose to discuss the way in which the roots of this vary with M_2, and, in order to have only one variable parameter we suppose that the stiffnesses of the springs are equal. Writing $M_2 = M_1/k$ we have

$$n_2^2 = kn_1^2, \qquad n_{12}^2 = n_1^2, \tag{9}$$

and (8) becomes

$$\left(\frac{\alpha}{n_1}\right)^4 + \left(\frac{\alpha}{n_1}\right)^2(2+k)+k = 0. \tag{10}$$

The roots of this are

$$\left(\frac{\alpha}{n_1}\right)^2 = -(1+\tfrac{1}{2}k)\pm(1+\tfrac{1}{4}k^2)^{\frac{1}{2}}. \tag{11}$$

Corresponding to each of these roots the first of equations (7) gives the ratio X_1/X_2.

If $k = 1$, *both masses equal*, the roots are

$$\alpha = \pm 0{\cdot}618in_1, \quad \text{with } X_1/X_2 = 0{\cdot}618,$$

and
$$\alpha = \pm 1{\cdot}618in_1, \quad \text{with } X_1/X_2 = -1{\cdot}618.$$

If $k = 2$, *the second mass half the first*, the roots are

$$\alpha = \pm 0{\cdot}766in_1, \quad \text{with } X_1/X_2 = 0{\cdot}707,$$

and
$$\alpha = \pm 1{\cdot}848in_1, \quad \text{with } X_1/X_2 = -0{\cdot}707.$$

If $k = 0{\cdot}5$, *the second mass double the first*, the roots are

$$\alpha = \pm 0{\cdot}468in_1, \quad \text{with } X_1/X_2 = 0{\cdot}562,$$

and
$$\alpha = \pm 1{\cdot}510in_1, \quad \text{with } X_1/X_2 = -3{\cdot}562.$$

As $k \to 0$, *i.e.* $M_2 \to \infty$, the roots are, neglecting k^2,

$$\alpha = \pm in_1(1+\tfrac{1}{8}k)\surd 2, \quad \text{with } X_1/X_2 = -2/k, \tag{12}$$

and
$$\alpha = \pm in_1\surd(\tfrac{1}{2}k), \qquad \text{with } X_1/X_2 = \tfrac{1}{2}(1+\tfrac{1}{4}k). \tag{13}$$

In the limiting case $M_2 \to \infty$ its position becomes fixed and the system becomes that of a single mass attached to two fixed points by equal springs. The motion in this case has a single natural frequency given by (12). (13) shows that the second natural frequency of the system of two masses tends to zero as $M_2 \to \infty$.

As $k \to \infty$, *i.e.* $M_2 \to 0$, writing (11) in the form

$$\left(\frac{\alpha}{n_1}\right)^2 = -\frac{k}{2}\left\{1+\frac{2}{k}\mp\left(1+\frac{2}{k^2}+\ldots\right)\right\}, \tag{14}$$

it appears that, neglecting terms in $1/k^2$, (14) has roots

$$\alpha = \pm in_1\left(1-\frac{1}{2k}\right), \quad \text{with } X_1/X_2 = 1-1/k, \tag{15}$$

and
$$\alpha = \pm in_1\left(1+\frac{1}{2k}\right)\sqrt{k}, \quad \text{with } X_1/X_2 = -1/k. \tag{16}$$

In the limiting case $M_2 = 0$ the second mass is not present and the system has the single natural frequency (15): the second natural frequency, given by (16), tends to infinity as $M_2 \to 0$.

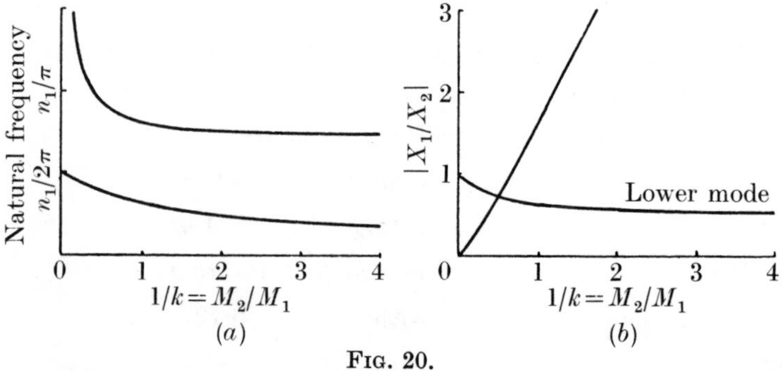

FIG. 20.

These results are plotted in Fig. 20. The variation of the natural frequencies with the values of the second mass is shown in Fig. 20 (*a*): the effect of a small second mass is to introduce a higher second natural frequency; as the second mass is increased both natural frequencies decrease. The ratio of the amplitudes of the first and second masses is shown in Fig. 20 (*b*). For the vibration of lower frequency this decreases steadily from 1 to $\frac{1}{2}$ as the second mass is increased: for the higher mode it increases steadily.

Results of this type have important practical applications: most practical mechanical systems inevitably have natural frequencies; also they are usually designed to operate over a fixed range of frequency. If any of the natural frequencies fall in the range of operating frequencies resonance will occur and vibrations of large amplitude will be set up. It appears from Fig. 20 that the natural frequencies of a system may be altered by the addition of extra components: for example the addition of a suitable second mass M_2 to a single mass will remove the natural frequencies of the system from the range between $n_1/2\pi$ and $n_1/4\pi$. The main object of so-called 'vibration dampers' is to remove all natural frequencies of a system from a specified range in this way; a secondary object is to damp some of the vibrations. A simple example, related to the system of Fig. 19, is that of Fig. 21.

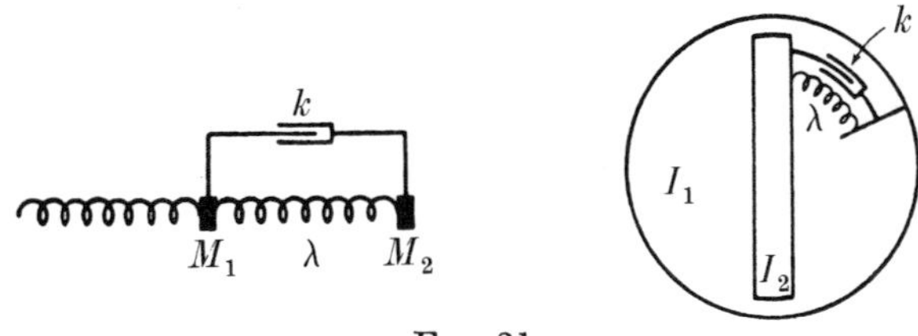

FIG. 21.

The mass M_1 is a portion of a mechanical system and it is required to damp its vibration; a second, suitably chosen, mass M_2 shifts the natural frequencies to a safe region, and a dash-pot between M_1 and M_2 giving resistance to motion proportional to relative velocity damps the vibration. The corresponding rotation system for two wheels I_1 and I_2 on the same shaft is widely used.

37. Geared systems

If a number of shafts carrying wheels are connected by gears, or a number of masses by levers, a small additional complication occurs. We discuss it here for the rather more interesting case of gearing.

Suppose a gear of moment of inertia I_1 and radius a is in mesh with a gear of moment of inertia I_2 and radius b, [Fig. 22 (a)]. I_1 is connected to a wheel of moment of inertia I by a shaft of stiffness λ_1, and I_2 to a wheel of moment of inertia I_3 by a shaft of stiffness λ_2. Let θ and θ_1 be the angular displacements of the wheels I and I_1 from fixed reference positions, and θ_2 and θ_3 those of the wheels I_2 and I_3, as shown in Fig. 22 (a). It is convenient to measure θ and θ_1 in the same direction, opposite to that of θ_2 and θ_3, since the wheels I_1 and I_2 rotate in opposite directions.

The relationship imposed by the gears can be seen from Fig. 22 (*b*). Firstly, since the circumferences of the two wheels must travel the same distance

$$a\theta_1 = b\theta_2. \tag{1}$$

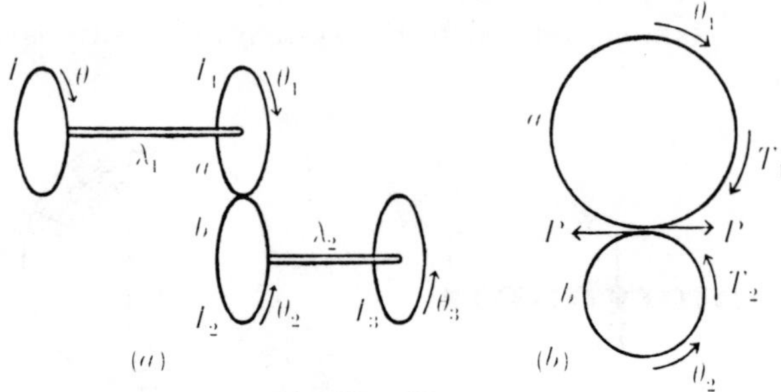

FIG. 22.

Secondly, if T_1 and T_2 are the torques on the two gear-wheels, measured in the directions of θ_1 and θ_2 increasing, respectively, and if P is the reaction at the point of contact of the wheels,

$$T_2 = Pb, \qquad T_1 = -Pa,$$

and therefore
$$aT_2+bT_1 = 0. \tag{2}$$

The equations of motion of the wheels are

$$I\ddot{\theta} = \lambda_1(\theta_1-\theta), \tag{3}$$

$$I_1\ddot{\theta}_1 = -\lambda_1(\theta_1-\theta)+T_1, \tag{4}$$

$$I_2\ddot{\theta}_2 = T_2+\lambda_2(\theta_3-\theta_2), \tag{5}$$

$$I_3\ddot{\theta}_3 = -\lambda_2(\theta_3-\theta_2). \tag{6}$$

Eliminating T_1 and T_2 from (4) and (5) by (2), and expressing θ_2 in terms of θ_1 by (1), we obtain the following three equations for θ, θ_1, and θ_3:

$$(D^2+n^2)\theta-n^2\theta_1 = 0, \tag{7}$$

$$-n_{12}^2\theta+\{(r^2+\rho)D^2+n_{12}^2+n_2^2r^2\}\theta_1-rn_2^2\theta_3 = 0, \tag{8}$$

$$-rn_3^2\theta_1+(D^2+n_3^2)\theta_3 = 0, \tag{9}$$

where

$$r = \frac{a}{b}, \quad \rho = \frac{I_1}{I_2}, \quad n^2 = \frac{\lambda_1}{I}, \quad n_{12}^2 = \frac{\lambda_1}{I_2}, \quad n_2^2 = \frac{\lambda_2}{I_2}, \quad n_3^2 = \frac{\lambda_2}{I_3}. \tag{10}$$

If as usual we seek a solution of (7)–(9) of the form

$$\theta_1 = \Theta_1 e^{\alpha t}, \qquad \theta_2 = \Theta_2 e^{\alpha t}, \qquad \theta_3 = \Theta_3 e^{\alpha t}, \tag{11}$$

the frequency equation for α is

$$\begin{vmatrix} \alpha^2+n^2 & -n^2 & 0 \\ -n_{12}^2 & (r^2+\rho)\alpha^2+n_{12}^2+n_2^2 r^2 & -rn_2^2 \\ 0 & -rn_3^2 & \alpha^2+n_3^2 \end{vmatrix} = 0. \tag{12}$$

On expanding, (12) becomes

$$\alpha^2\{(r^2+\rho)\alpha^4+[(r^2+\rho)(n^2+n_3^2)+n_{12}^2+r^2n_2^2]\alpha^2+$$
$$+[(r^2+\rho)n^2n_3^2+r^2n^2n_2^2+n_3^2 n_{12}^2]\} = 0. \tag{13}$$

There is thus a zero root in α^2 as well as a pair of roots giving rise to a pair of natural frequencies with their associated normal modes. A factor α^2 in the auxiliary equation gives rise to a term of type

$$A+Bt$$

in the solution, corresponding to steady rotation of the system.

38. Mechanical models illustrating the rheological behaviour of common substances

In § 28 we described three simple elements, namely perfectly elastic solid, perfectly viscous liquid, and coulomb friction, from which idealized mechanical systems such as those of §§ 29–37 could be built up.

The same elements may be combined to give a useful approximation to the behaviour of many common substances ranging from metals and rubber to flour dough. All substances exhibit to a greater or less extent phenomena such as creep, elastic hysteresis, plastic flow, etc. We give below a number of typical examples.

(i) *Rubber-like substances*

These may be represented by the combination of a spring giving restoring force λ times the displacement and a dash-pot giving resistance k times the velocity. λ and k are regarded as constants of the rubber itself. The extension of a piece of rubber when constant stress S_0 is applied to it is thus given by the equation

$$k\dot{x}+\lambda x = S_0, \tag{1}$$

or

$$\left(D+\frac{\lambda}{k}\right)x = \frac{S_0}{k}. \tag{2}$$

If the stress S_0 is applied at $t = 0$ when $x = 0$, the solution of (2) is

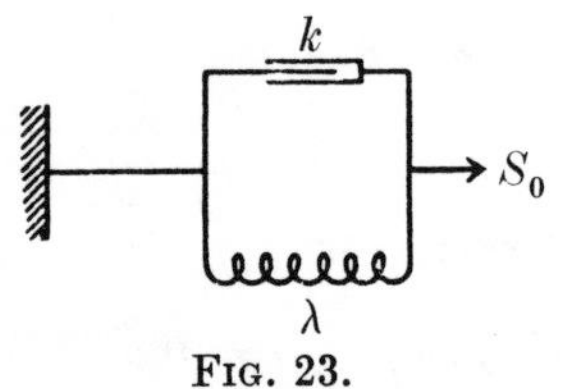

FIG. 23.

$$x = \frac{S_0}{\lambda}(1-e^{-\lambda t/k}). \tag{3}$$

That is, the substance does not take up the deflexion S_0/λ instantaneously as a perfectly elastic solid does, but moves out to it according to the law (3).

If a mass M is attached to a piece of rubber its equation of motion will be that of § 29: by measurements of two quantities such as the natural frequency and damping coefficient of the motion the constants λ and k of the rubber may be determined.

Next we illustrate the phenomenon of elastic after effect which occurs for such substances. Instead of applying a constant stress suddenly we apply a cyclically varying stress: the simplest case, which we consider here, is that of a stress S which increases linearly for time T and then decreases again to zero in time T: this is given by

$$\left.\begin{aligned} S &= S_0 t, & 0 < t < T \\ S &= S_0(2T-t), & T < t < 2T \\ S &= 0, & t > 2T \end{aligned}\right\}, \tag{4}$$

where S_0 is a constant.

For $0 < t < T$, the differential equation is

$$\left(D+\frac{\lambda}{k}\right)x = \frac{S_0}{k}t, \tag{5}$$

with $x = 0$, when $t = 0$.

The solution of this is

$$x = \frac{S_0 t}{\lambda}-\frac{S_0 k}{\lambda^2}(1-e^{-\lambda t/k}). \tag{6}$$

When $t = T$, this gives

$$x = \frac{S_0 T}{\lambda} - \frac{S_0 k}{\lambda^2}(1 - e^{-\lambda T/k}). \tag{7}$$

When $T < t < 2T$, the differential equation is

$$\left(D + \frac{\lambda}{k}\right)x = \frac{S_0}{k}(2T - t), \tag{8}$$

to be solved with the value (7) of x when $t = T$.

The solution is

$$x = \frac{S_0}{\lambda}\left(2T - t + \frac{k}{\lambda}\right) + \frac{S_0 k}{\lambda^2}(1 - 2e^{\lambda T/k})e^{-\lambda t/k}. \tag{9}$$

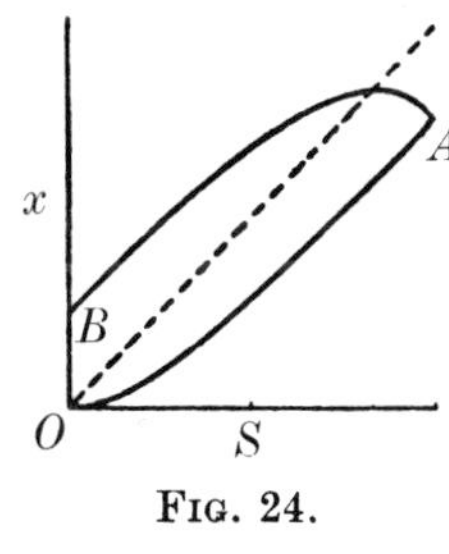

FIG. 24.

When $t = 2T$, so that the stress has returned to zero, the strain x is

$$x = \frac{kS_0}{\lambda^2}(1 - e^{\lambda T/k})^2 e^{-2\lambda T/k}. \tag{10}$$

Finally, when $t > 2T$, the differential equation is

$$\left(D + \frac{\lambda}{k}\right)x = 0, \tag{11}$$

with initial value (10) of x. This has solution

$$x = \frac{kS_0}{\lambda^2}(1 - e^{\lambda T/k})^2 e^{-\lambda t/k}. \tag{12}$$

If we plot the displacement x against the stress S as in Fig. 24 we get the curve OAB, OA being the portion $0 < t < T$, and AB that for $T < t < 2T$. OB is the residual deflexion when the stress is zero.

(ii) *Substances exhibiting plastic flow*

To include the phenomena of plastic flow an element providing constant friction F has to be introduced.

The simplest system is that of Fig. 25 (a) which corresponds roughly to the behaviour of metals. This is elastic until the stress S is equal to F (the yield point) which is the greatest stress the substance can resist.

The system of Fig. 25 (b) (a Bingham solid) is again elastic until S reaches F: for values of $S > F$ it flows at a rate which increases with S. If x is the displacement in this system

$$x = \frac{S}{\lambda} \qquad (S < F), \tag{13}$$

$$x = \frac{F}{\lambda}+\frac{1}{k}(S-F)t \quad (S > F). \tag{14}$$

(a) F λ S

(b) 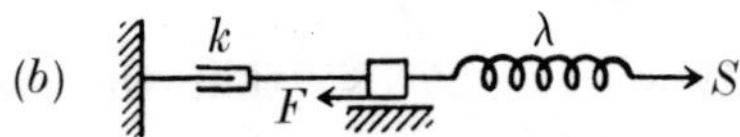

(c) 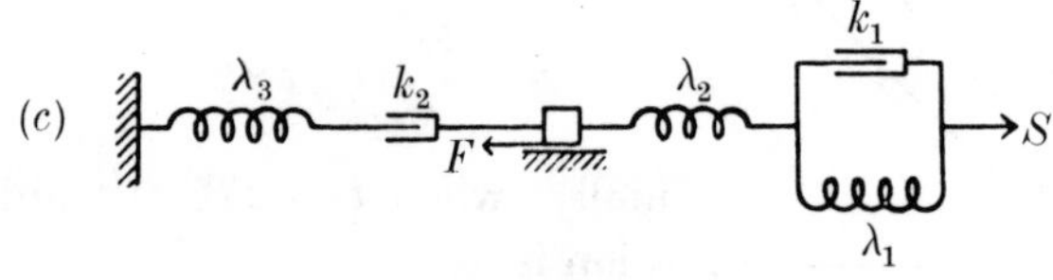

FIG. 25.

Finally Fig. 25 (c) is a model which has been proposed for flour dough. The displacement x is given by

$$x = \frac{S}{\lambda_1}(1-e^{-\lambda_1 t/k_1})+\frac{S}{\lambda_2} \qquad (S < F), \tag{15}$$

$$x = \frac{S}{\lambda_1}(1-e^{-\lambda_1 t/k_1})+\frac{S}{\lambda_2}+\frac{(S-F)t}{k_2}+\frac{S-F}{\lambda_3} \quad (S > F). \tag{16}$$

39. Impulsive motion

We shall speak of an 'impulsive force' or 'a blow of impulse P' meaning a very great force applied for a time τ so short that the motion of its point of application during this time is negligible while the force is so large that its integral over this time is finite —this will be called the impulse P of the blow.

Thus if $F(t)$ is an impulsive force applied at $t = 0$, we have

$$\int\limits_0^\tau F(t)\,dt = P. \tag{1}$$

Suppose now that a particle of mass m is struck by a blow of impulse P at $t = 0$. The equation of motion is

$$m\ddot{x} = F(t). \tag{2}$$

Integrating (2) over the small time τ during which the blow operates, we find

$$m[\dot{x}]_0^\tau = \int\limits_0^\tau F(t)\,dt = P,$$

or

$$mu_f - mu_i = P, \tag{3}$$

where u_i and u_f are the velocities of the particle before and after the blow. Thus the change in momentum of the particle in the direction of the blow is equal to the impulse of the blow.

If the particle is attached to a spring with viscous damping [§ 29 (1)] the same result holds: (2) is replaced by

$$m\ddot{x} + k\dot{x} + \lambda x = F(t).$$

Integrating as before gives

$$m[\dot{x}]_0^\tau + k[x]_0^\tau + \lambda \int\limits_0^\tau x\,dt = P. \tag{4}$$

Now we have postulated above that τ shall be negligibly small, and that the change of x in this time shall be negligible, so the second and third terms of the left-hand side of (4) are negligible and we regain (3).

Thus setting a particle of mass m in motion by a blow of impulse P is equivalent to giving it an initial velocity P/m.

The same result applies to systems of masses such as those considered in this chapter; since the motion of any particle is negligible during the blow, it can exert no influence on its neighbours. If the particles are in contact or rigidly connected this is not the case.

Using the δ-function notation, a blow of impulse P at $t = 0$ may be regarded as a force $P\,\delta(t)$, and the above results obtained.

Ex. *Motion of a damped harmonic oscillator maintained by impulses.*

We consider the oscillator of § 29, Fig. 4, with no applied forces. Choosing the origin of time so that $x = 0$ when $t = 0$, the solution § 29 (7)

$$x = Ae^{-\kappa t}\sin n't \tag{5}$$

represents an oscillation which dies away as discussed in § 29.

Now suppose that each time the mass passes through the position $x = 0$, that is, when

$$t = r\pi/n' \quad (r = 1, 2,...), \tag{6}$$

the particle is given a blow P just sufficient to raise the magnitude of its velocity to the value it had when $t = 0$. This is roughly the action of the escapement of a clock. If this is done, all half-swings of the mass will be the same, and the motion is maintained indefinitely. To calculate P we have from (5)

$$\dot{x} = Ae^{-\kappa t}\{n' \cos n't - \kappa \sin n't\}, \tag{7}$$

and so

$$\dot{x} = n'A, \quad \text{when } t = 0,$$

$$\dot{x} = -n'Ae^{-\kappa\pi/n'}, \quad \text{when } t = \pi/n'.$$

Thus the magnitude of the velocity when $t = \pi/n'$ is less than that when $t = 0$ by an amount

$$n'A(1-e^{-\kappa\pi/n'}), \tag{8}$$

and a blow of impulse

$$P = mn'A(1-e^{-\kappa\pi/n'}) \tag{9}$$

is required.

If we regard P as a given quantity, (9) gives A, and the motion is

$$x = \frac{P}{mn'(1-e^{-\kappa\pi/n'})} e^{-\kappa t} \sin n't, \tag{10}$$

for $0 < t < \pi/n'$, all subsequent half-swings being the same.

40. Initial value problems: use of the Laplace transformation

In the preceding sections the general solutions of a number of problems have been found: in initial value problems sufficient information is always given in the initial conditions to find the arbitrary constants in these solutions. In this section we shall solve some typical initial value problems by the Laplace transformation method—it is for such problems that the method was designed and its advantage over the older methods increases as the complexity of the problem increases.

Ex. 1. *Force $F_0 \sin \omega t$ is applied at $t = 0$ to the damped harmonic oscillator of §§ 29, 31, the initial displacement and velocity of the mass being zero.*

We have to solve

$$m(D^2+2\kappa D+n^2)x = F_0 \sin \omega t \tag{1}$$

with $x = Dx = 0$, when $t = 0$.

The subsidiary equation, § 18 (26), is

$$m(p^2+2\kappa p+n^2)\bar{x} = \frac{F_0 \omega}{p^2+\omega^2}.$$

Thus, considering the case $n^2 > \kappa^2$, and writing $n'^2 = n^2-\kappa^2$,

$$m\bar{x} = \frac{F_0\,\omega}{(p+\kappa-in')(p+\kappa+in')(p-i\omega)(p+i\omega)}$$

$$= \left\{\frac{F_0\,\omega}{2in'(p+\kappa-in')\{\omega^2+(\kappa-in')^2\}}+\text{conjugate}\right\}+$$

$$+\left\{\frac{F_0\,\omega}{2i\omega(p-i\omega)(n^2-\omega^2+2\kappa i\omega)}+\text{conjugate}\right\}, \quad (2)$$

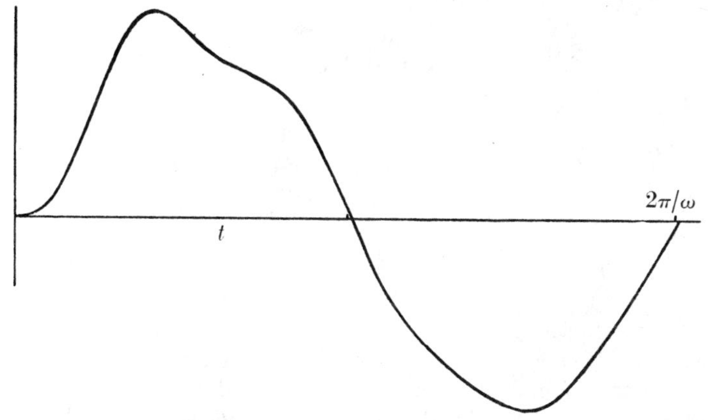

Fig. 26.

on putting $\bar{x}$ in partial fractions by § 18 (29). Writing 'conjugate' implies a similar term except that that the sign of i is changed. From (2) and § 18 (11)

$$mx = \left\{\frac{F_0\,\omega}{2in'(\omega^2+\kappa^2-n'^2-2in'\kappa)}e^{-\kappa t+in't}+\text{conjugate}\right\}+$$

$$+\left\{\frac{F_0\,\omega}{2i\omega(n^2-\omega^2+2\kappa i\omega)}e^{i\omega t}+\text{conjugate}\right\}$$

$$= \frac{F_0\,\omega}{n'Z_1}e^{-\kappa t}\sin(n't-\theta)+\frac{F_0}{\{(n^2-\omega^2)^2+4\kappa^2\omega^2\}^{\frac{1}{2}}}\sin(\omega t-\phi), \quad (3)$$

where

$$\phi = \arg(n^2-\omega^2+2\kappa i\omega), \quad (4)$$

and Z_1 and θ are the modulus and argument of

$$\omega^2+\kappa^2-n'^2-2n'\kappa i. \quad (5)$$

The second term of (3) is the forced oscillation § 31 (5), and the first the transient part of § 31 (6) with the arbitrary constants determined to make $x = \dot{x} = 0$ when $t = 0$.

The function (3) is graphed in Fig. 26 for the case $n = 4\omega$, $\kappa = n/10$ and shows how the starting transient dies out, leaving the forced oscillation.

Ex. 2. *The system of § 33, Fig. 11, with equal masses and springs. The first mass is set in motion with velocity V at $t = 0$ when the masses are at rest in their equilibrium positions.*

We have to solve

$$(D^2+2n^2)x_1-n^2x_2 = 0,$$
$$-n^2x_1+(D^2+2n^2)x_2 = 0,$$

where $n^2 = \lambda/M$, with $x_1 = x_2 = \dot{x}_2 = 0$, $\dot{x}_1 = V$, when $t = 0$.

The subsidiary equations are

$$(p^2+2n^2)\bar{x}_1-n^2\bar{x}_2 = V,$$
$$-n^2\bar{x}_1+(p^2+2n^2)\bar{x}_2 = 0.$$

Solving for $\bar{x}_1$ we get

$$\bar{x}_1 = \frac{(p^2+2n^2)V}{(p^2+3n^2)(p^2+n^2)} = \frac{V}{2(p^2+n^2)}+\frac{V}{2(p^2+3n^2)}.$$

Therefore $$x_1 = \frac{V}{2n}\sin nt+\frac{V}{2n\sqrt{3}}\sin nt\sqrt{3}.$$

EXAMPLES ON CHAPTER IV

1. The displacement x in an undamped harmonic oscillation $x = a\cos nt$ may be written in the form $x = \mathbf{R}(ae^{int})$ and so interpreted as the real part of the complex number $z = ae^{int}$ which describes a circle in the z-plane with constant speed. Show that the velocity $\dot{x}$ may be represented in the same way as $\mathbf{R}(nae^{i(nt+\frac{1}{2}\pi)})$. Show that damped harmonic motion $x = ae^{-\kappa t}\cos n't$ may be treated in the same way, but the complex number z now describes an equiangular spiral.

2. For forced oscillations of the damped harmonic oscillator of § 31 show that the phase lag of the velocity is $\pm\tan^{-1}(s^2-1)^{\frac{1}{2}}$ at the points at which the amplitude of the velocity has $(1/s)$th of its maximum value. In particular it is $\pm\frac{1}{4}\pi$ if $s = \sqrt{2}$.

Show also that the slope of the curve of the phase-lag ϕ against ω is $1/\kappa$ when $\omega = n$.

3. Force $F_0\sin\omega t$ is applied to the damped harmonic oscillator of § 29 with critical damping, $\kappa = n$. Show that the displacement in the forced oscillation is

$$\frac{F_0}{m(n^2+\omega^2)}\sin(\omega t-2\phi),$$

where $\phi = \tan^{-1}(\omega/n)$.

4. In the damped harmonic motion $x = Ae^{-\kappa t}\sin n't$ of § 29 the potential energy V of the system is $\frac{1}{2}\lambda x^2$ and the kinetic energy T is $\frac{1}{2}m\dot{x}^2$. Show that the total energy $T+V$ is given by

$$T+V = \tfrac{1}{2}mA^2e^{-2\kappa t}\{n^2-\kappa^2\cos 2n't-\kappa n'\sin 2n't\},$$

and that the average value of this quantity over the cycle $2r\pi/n'$ to $2(r+1)\pi/n'$ is

$$(mA^2n'^3/8\pi\kappa)\{e^{-4\kappa r\pi/n'}-e^{-4\kappa(r+1)\pi/n'}\}.$$

Find the loss in total energy in the same cycle, and show that if Q is defined as 2π times the ratio of the average total energy over a cycle to the energy loss in the cycle,

$$Q = n'/2\kappa = \pi/\delta,$$

where δ is defined in § 29 (13). This provides a physical interpretation for the important quantity Q mentioned in § 29.

5. A mass m attached to a spring of stiffness mn^2 is set in motion from rest at $t = 0$ in its equilibrium position by a force which is $F_0 \sin \omega t$ for $0 < t < \pi/\omega$, and zero for $t > \pi/\omega$. Show that its displacement x is

$$x = F_0\{\omega \sin nt - n \sin \omega t\}/\{mn(\omega^2 - n^2)\} \qquad (0 < t < \pi/\omega).$$

$$x = \{2\omega F_0/mn(\omega^2 - n^2)\} \sin n(t - \pi/2\omega) \cos(n\pi/2\omega) \quad (t > \pi/\omega).$$

6. A mass m is constrained to move in a straight line and is attached to a fixed point in the line by a spring of stiffness mn^2. Its motion is resisted by coulomb friction μR. The mass can be set in any position by an adjusting screw whose end touches it. At $t = 0$, when the mass is at rest and the spring compressed by an amount x_1, the screw is turned so that its end moves away from the mass with constant velocity v. Show that the end of the screw and the mass will lose contact until a time t given by the smallest root of

$$(x_1 - \mu R/mn^2)(1 - \cos nt) = vt.$$

This implies that if the screw is released with constant velocity a jerky motion always results. Show that if the inertia of the screw is taken into account it is possible to have a smooth motion.

7. A mass hangs at rest in its equilibrium position at the end of a spring whose unstretched length is a and whose stretched length is b. At $t = 0$ the point of support is given a downwards motion $c \sin \omega t$. Show that the length of the spring at time t is

$$b - \frac{cn\omega}{n^2 - \omega^2} \sin nt + \frac{c\omega^2}{n^2 - \omega^2} \sin \omega t,$$

where $n/2\pi$ is the natural frequency of oscillation of the mass, and $n \neq \omega$.

8. In the system of Fig. 19, $M_2 = 2M_1/3$, $\lambda_1 = \lambda_2$; show that the natural frequencies are $(n_1\sqrt{3})/2\pi$ and $n_1/(2\pi\sqrt{2})$, where $n_1^2 = \lambda_1/M_1$, and find the normal modes of oscillation.

9. In the system of Fig. 19 the masses M and the stiffnesses λ of the springs are equal, and in addition a third mass M is connected by an equal spring λ. Writing $n^2 = \lambda/M$, show that the natural frequencies are

$$0{\cdot}445n/2\pi, \quad 1{\cdot}247n/2\pi, \quad 1{\cdot}802n/2\pi.$$

10. Show that if in the system of Fig. 19 the stiffnesses of the springs are equal, the squares of the natural frequencies are rational if

$$M_1 : M_2 = r : s,$$

where r and s are integers such that $r^2 + 4s^2$ is a perfect square. Show that such integers can be found by splitting up any odd perfect square into a sum of integers differing by unity, e.g. $49 = 24+25$ gives $24^2+7^2 = 25^2$, corresponding to $M_1:M_2 = 7:12$.

11. Three wheels A, B, and C are of moment of inertia I. A and B, and B and C, are connected by shafts of stiffness λ. Show that the natural frequencies of the system are $n/2\pi$ and $(n\surd 3)/2\pi$, where $n^2 = \lambda/I$.

12. If in the system of Fig. 14, $M_1 = M_2$, $\lambda_1 = \lambda_2 = \lambda_3$, $k_1 = k_2 = 0$, show that the normal modes of oscillation are

$$x_1 = x_2 = A_1 \sin(nt+\beta_1),$$
$$x_1 = -x_2 = A_2 e^{-2\kappa t}\sin\{t\surd(3n^2-4\kappa^2)+\beta_2\},$$

where $n^2 = \lambda_1/M_1$, and $2\kappa = k_{12}/M_1$.

13. If in the system of Fig. 14, $M_1 = M_2$, $\lambda_1 = \lambda_2 = \lambda_3$, $k_2 = k_{12} = 0$, show that the frequency equation is

$$\alpha^4+2\kappa\alpha^3+4n^2\alpha^2+4\kappa\alpha n^2+3n^4 = 0,$$

where $n^2 = \lambda_1/M_1$ and $2\kappa = k_1/M_1$. Show that if κ is so small that κ^2 is negligible, the natural frequencies are still $n/2\pi$ and $(n\surd 3)/2\pi$, but the oscillations both have a damping factor $\exp(-\frac{1}{2}\kappa t)$.

14. n particles, each of mass m, are attached at equal distances along a string of length $(n+1)l$ which is stretched to tension T and whose ends are fixed. If the particles execute small transverse oscillations and x_r is the displacement of the rth particle, show that

$$\ddot{x}_r = c^2(x_{r+1}-2x_r+x_{r-1}) \quad (r = 1,..., n),$$

with $x_0 = x_{n+1} = 0$, and $c^2 = T/ml$.

15. Show that the system of equations of Ex. 14 is satisfied by

$$x_r = (Ae^{r\beta}+Be^{-r\beta})e^{\alpha t},$$

where β is a root of $\qquad \cosh\beta = 1+(\alpha^2/2c^2)$.

Show that the conditions $x_0 = x_{n+1} = 0$ require $\beta = s\pi i/(n+1)$, and thus that the natural frequencies of the system are

$$c\{2-2\cos s\pi/(n+1)\}^{\frac{1}{2}}/2\pi \quad (s = 1, 2,..., n).$$

16. A wheel of moment of inertia I is connected to a gear-box by a shaft of stiffness λ. The gear-box gives a step up of r, and is connected to a wheel of moment of inertia I_1 by a shaft of stiffness λ_1. Show that if the moments of inertia of the gears in the gear-box are negligible, the natural frequency of the system is $\omega/2\pi$, where

$$\omega^2 = \frac{\lambda\lambda_1(I_1 r^2+I)}{II_1(\lambda+r^2\lambda_1)}.$$

17. If the system of Ex. 16 is at rest with the shafts unstrained, and the wheel I is set in motion at $t = 0$ with angular velocity Ω, show that its subsequent angular velocity is

$$\Omega(I+I_1 r^2\cos\omega t)/(I_1 r^2+I),$$

where ω is defined in Ex. 16.

18. The motion of a mass M acted on by a force $F\sin\omega t$ and constrained by supports of rubber in shear is given by

$$(MD^2+KD+S)x = F\sin\omega t,$$

where K and S depend on the nature of the rubber. Find the amplitude of the forced oscillation and show that, neglecting terms in K^2, its maximum value is approximately $F/2\pi K\Omega$, where Ω, the resonance frequency, is approximately $(S/M)^{\frac{1}{2}}/2\pi$.

19. Show that if a varying stress $S(t)$ is applied to a rubber-like substance [cf. § 38 (i)] for $t > 0$ with zero initial strain, the strain x at time t is given by

$$kx = e^{-\lambda t/k}\int\limits_0^t e^{-\lambda\xi/k}S(\xi)\,d\xi.$$

If $S(\xi) = \sin\omega\xi$ show that, writing $\alpha = \lambda/k$, $\phi = \tan^{-1}(k\omega/\lambda)$,

$$x = \frac{\omega}{k(\alpha^2+\omega^2)}e^{-\alpha t}+\frac{1}{k(\alpha^2+\omega^2)^{\frac{1}{2}}}\sin(\omega t-\phi).$$

20. In the system of Fig. 11 the masses and springs are equal. At $t = 0$, when the masses are at rest and the springs unstrained, a point of support is given the motion $a\sin\omega t$, starting in the direction towards the masses. Show that, if ω is not equal to either natural frequency, the displacement of the nearer mass is, writing $n^2 = \lambda_1/M_1$,

$$\frac{na\omega}{2(\omega^2-n^2)}\sin nt+\frac{na\omega}{2(\omega^2-3n^2)\sqrt{3}}\sin nt\sqrt{3}+\frac{an^2(2n^2-\omega^2)}{(n^2-\omega^2)(3n^2-\omega^2)}\sin\omega t.$$

21. A mass $M/3$ is connected by a light string of length l to a mass M which is connected to a fixed point by an equal string. At $t = 0$, when the two masses are hanging vertically and at rest, the mass M is given a small horizontal blow of impulse P; show that its subsequent displacement is

$$P\{3\sin nt\sqrt{2}+\sqrt{3}\sin nt\sqrt{(2/3)}\}/(4nM\sqrt{2}),$$

where $n^2 = g/l$.

22. In the geared system of Fig. 22 (a), $I = I_1 = I_2 = I_3$, $\lambda_1 = \lambda_2$, and $a = b$. For $t < 0$ the wheels I and I_1 are at rest, the gears are not in mesh, and the wheels I_1 and I_2 are rotating with constant angular velocity ω. At $t = 0$ the gears are forced into mesh. Show that the angular velocity of I_3 at time t is $\frac{1}{2}\omega(1+\cos nt)$, where $n^2 = \lambda_1/I$.

V
ELECTRIC CIRCUIT THEORY

41. Introductory

In this chapter we shall consider the elementary theory of electric circuits with 'lumped' or concentrated properties. This theory is so closely related to that of the mechanical systems of Chapter IV that they can best be studied simultaneously.

We regard electric circuits as being built up of 'elements' of three types: namely, inductance L, resistance R, and capacitance C. The current I at a point of a circuit in a certain direction is the rate at which positive charge passes that point in that direction. In the problems with which we shall be concerned, the current is usually caused by a 'voltage' V applied to two terminals, one of which we select, by convention, as the positive one, so that V is positive when this terminal is at the higher voltage and the current I is positive when flowing away from it.

These circuit elements are illustrated in Fig. 27. The information provided by the theory of electricity which we shall assume is as follows.

(i) The voltage drop across a resistance R is R times the current in it [Fig. 27 (a)],

$$RI = V. \tag{1}$$

(ii) The voltage drop across an inductance L is L times the rate of change of current in it [Fig. 27 (b)],

$$L\frac{dI}{dt} = V. \tag{2}$$

(iii) The voltage drop across a capacitance C is $(1/C)$ times the charge Q on it [Fig. 27 (c)],

$$V = \frac{Q}{C}. \tag{3}$$

Also, since the current I is the rate of flow of positive charge, we must have

$$I = \frac{dQ}{dt}, \tag{4}$$

and, with the conventions of sign introduced above, the current is positive when flowing towards the high-voltage side of the capacitance.

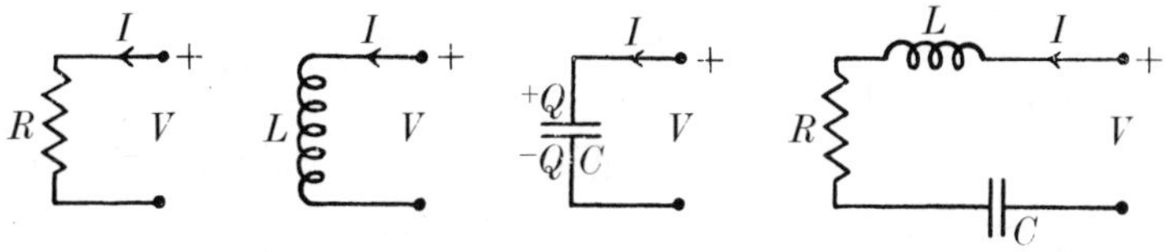

(a) $RI = V$ (b) $L\frac{dI}{dt} = V$ (c) $\frac{Q}{C} = V$ (d) $L\frac{dI}{dt}+RI+\frac{Q}{C} = V$

FIG. 27.

The equations (1)–(4) are practically all that is needed for the work of this chapter; they are linear, and it is because of this that it is possible to go so far in electric circuit theory. They are, of course, approximations, but they are very good approximations (except for iron-cored inductances), so results calculated by using them will be very near the truth. The more accurate equations are non-linear and will be discussed in § 57.

More complicated circuits can be regarded as being built up of the simple elements described above, but it is a little shorter to take as the fundamental unit the 'L, R, C circuit' consisting of inductance L, resistance R, and capacitance C in series [Fig. 27 (d)]. If voltage V is applied to such a circuit, the sum of the voltage drops over the inductance, resistance, and capacitance, given by (1), (2), and (3), respectively, must be equal to V, that is

$$L\frac{dI}{dt}+RI+\frac{Q}{C} = V. \tag{5}$$

(4) and (5) are the fundamental equations for this circuit; they are a pair of simultaneous ordinary linear differential equations for the two unknowns I and Q in terms of V, which is supposed to be a given function of the time. If we substitute from (4) in (5) we get the equation

$$L\frac{d^2Q}{dt^2}+R\frac{dQ}{dt}+\frac{Q}{C} = V \tag{6}$$

for Q. Writing $$\kappa = \frac{R}{2L}, \qquad n^2 = \frac{1}{LC}, \tag{7}$$

this becomes $$\frac{d^2Q}{dt^2}+2\kappa\frac{dQ}{dt}+n^2Q = \frac{V}{L}, \tag{8}$$

which is the same as § 29 (2), except that it has Q in place of x, and V/L in place of F/m. This is a special case of a far-reaching analogy between mechanical and electrical systems which will be discussed in §§ 42, 43. For the present we simply note that the similarity in form of (8) and § 29 (2) allows us to quote many of the results and much of the general discussion of §§ 29–32; some examples of this are given below.

Ex. 1. *Constant voltage E applied at $t = 0$ to the circuit of Fig. 27 (d) with the initial values $Q = 0$ and $I = 0$ when $t = 0$.*

We have to solve $$(D^2+2\kappa D+n^2)Q = \frac{E}{L}. \tag{9}$$

A particular integral of this is

$$\frac{E}{n^2L} = CE.$$

Adding the complementary function given by § 29 (7), the general solution is found to be

$$Q = ae^{-\kappa t}\sin n't+be^{-\kappa t}\cos n't+CE, \tag{10}$$

where a and b are arbitrary constants, $n' = \sqrt{(n^2-\kappa^2)}$, and we assume that this is real (the other cases follow precisely as in § 29). The initial conditions

$$Q = 0, \qquad I = \frac{dQ}{dt} = 0, \quad \text{when } t = 0$$

give the equations for a and b

$$b+CE = 0, \qquad n'a-\kappa b = 0.$$

Thus the final solution is

$$Q = CE\{1-e^{-\kappa t}\cos n't-(\kappa/n')e^{-\kappa t}\sin n't\},$$

and the current I is by (4)

$$I = (E/n'L)e^{-\kappa t}\sin n't.$$

Ex. 2. *Alternating voltage $E\sin(\omega t+\beta)$ applied to the circuit of Fig. 27 (d).*

The differential equation is now

$$(D^2+2\kappa D+n^2)Q = \frac{E}{L}\sin(\omega t+\beta),$$

and its general solution, obtained by quoting § 31 (6) with x replaced by Q and F_0/m by E/L, is

$$Q = Ae^{-\kappa t}\sin(n't+B)+\frac{E}{L\{(n^2-\omega^2)^2+4\kappa^2\omega^2\}}\sin(\omega t+\beta-\phi), \tag{11}$$

where $n' = \sqrt{(n^2-\kappa^2)}$ and ϕ is defined in § 31 (4).

Differentiating, we have by (4)

$$I = Ae^{-\kappa t}\{n'\cos(n't+B)-\kappa\sin(n't+B)\}+$$
$$+\frac{E\omega}{L\{(n^2-\omega^2)^2+4\kappa^2\omega^2\}}\sin\{\omega t+\beta-(\phi-\tfrac{1}{2}\pi)\}. \quad (12)$$

The first term of (12) is the 'transient' current which dies away with time; the second is the steady state alternating current. The variation of its amplitude and phase with frequency are shown in Figs. 8 (*a*) and (*c*) and Figs. 7 (*b*) and (*d*).

If the voltage is supposed to be switched on at time $t = 0$ the constants A and B can be determined from the given values of the charge Q and current I at this instant, either by direct substitution, or as in § 40, Ex. 1, where the complete solution of this problem is given.

42. Electrical networks

The principles of § 41 may be extended immediately to any network, however complicated. A network can be divided up into a number of simple 'branches' AB, BC,..., Fig. 28 (*a*), which are connected at 'junctions' or 'nodes' A, B,.... For definiteness, we suppose each branch to contain L, R, C in series, together, possibly, with a source of applied voltage.

First we assign by convention a positive direction for current in each branch. Suppose that in the branch AB this direction is from A to B, that I_1 is the current in this branch, and that V_1 is the external voltage applied at terminals in the branch AB, V_1 being reckoned positive if it would cause a current to flow in the direction from A to B. These conventions are shown in Fig. 28 (*b*).

Then, as in § 41, the voltage drop v_1 over AB is the sum of the voltage drops over the elements L_1, R_1, C_1, and the applied voltage V_1 in AB, that is

$$L_1\frac{dI_1}{dt}+R_1 I_1+\frac{Q_1}{C_1}-V_1 = v_1. \quad (1)$$

V_1 appears in the left-hand side of (1) with a negative sign, since, with the convention chosen above, it is a *rise* in voltage.

Now consider any closed circuit round several branches of Fig. 28 (*a*), e.g. the closed circuit AB, BC, CA. If v_1, v_2, v_3 are the voltage drops from A to B, B to C, and C to A, respectively, the algebraic sum of these must be zero. Adding the three corresponding equations of type (1), we get the result that *the algebraic sum of the voltage drops across the elements and the sources of applied voltage in a closed circuit is zero.* Clearly this is true for any closed circuit, and the statement is known as *Kirchhoff's first law.* Alternatively it may be stated in the form that *the algebraic sum of the voltage drops across the elements of a closed circuit is equal to the algebraic sum of the applied voltages in the closed circuit.*

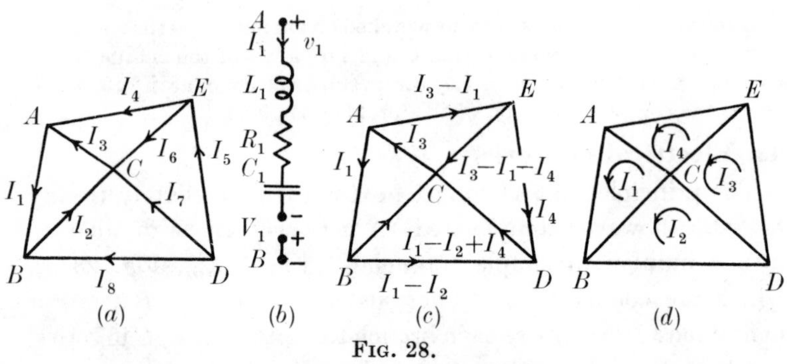

FIG. 28.

We also require *Kirchhoff's second law*, which states that *the algebraic sum of the currents flowing towards any junction is zero.* This is simply the expression of the physical fact that charge does not accumulate at the junctions.

By using Kirchhoff's two laws we can always write down sufficient equations to determine all the currents and charges in any network. There are various ways of shortening and systematizing this work which are commonly used by engineers in solving such problems. These will be indicated briefly later: for the present we solve a number of problems from first principles.

Ex. 1. *The circuit of Fig. 29 with applied voltage V.*

Let I_1, I_2, and I_3 be the currents in AB, BC, and BE in the directions of the arrows. Let Q_1, Q_2, and Q_3 be the charges on C_1, C_3, and C_2; then by § 41 (4)

$$I_1 = DQ_1, \qquad I_2 = DQ_2, \qquad I_3 = DQ_3. \tag{2}$$

Next, Kirchhoff's second law at the junction B gives

$$I_1 - I_2 - I_3 = 0. \tag{3}$$

Kirchhoff's first law for the closed circuit $ABEF$ gives

$$(L_1 D + R_1)I_1 + \frac{Q_1}{C_1} + \frac{Q_3}{C_2} = V, \tag{4}$$

and for the closed circuit $BCDE$ it gives

$$(L_2 D + R_2)I_2 + \frac{Q_2}{C_3} - \frac{Q_3}{C_2} = 0. \tag{5}$$

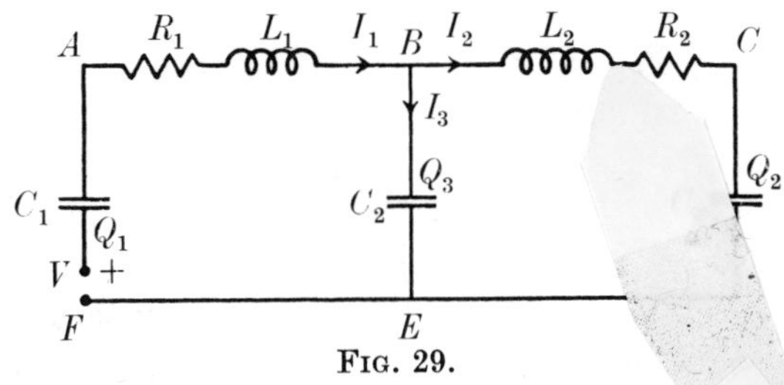

Fig. 29.

(2), (3), (4), and (5) are six equations for the s[illegible]knowns $I_1,\ldots, Q_3$. If we substitute for the I in terms of the Q by (2), equations (4), (5), and (3), respectively, become

$$\left(L_1 D^2 + R_1 D + \frac{1}{C_1}\right)Q_1 + \frac{Q_3}{C_2} = V, \tag{6}$$

$$\left(L_2 D^2 + R_2 D + \frac{1}{C_3}\right)Q_2 - \frac{Q_3}{C_2} = 0, \tag{7}$$

$$D(Q_1 - Q_2 - Q_3) = 0. \tag{8}$$

Integrating (8) gives

$$Q_1 - Q_2 - Q_3 = Q, \tag{9}$$

where Q is a constant to be determined from the known conditions at the instant when the voltage V was switched on. For simplicity we assume that $Q = 0$; this would be the case, for example, if the condensers were initially uncharged. We then have

$$Q_3 = Q_1 - Q_2, \tag{10}$$

and substituting this in (6) and (7) gives

$$\left(L_1 D^2 + R_1 D + \frac{1}{C_1} + \frac{1}{C_2}\right)Q_1 - \frac{Q_2}{C_2} = V, \tag{11}$$

$$-\frac{Q_1}{C_2} + \left(L_2 D^2 + R_2 D + \frac{1}{C_2} + \frac{1}{C_3}\right)Q_2 = 0. \tag{12}$$

Writing

$$\kappa_1 = \frac{R_1}{2L_1}, \quad n_1^2 = \frac{1}{L_1 C_1}, \quad n_{12}^2 = \frac{1}{L_1 C_2},$$
$$\kappa_2 = \frac{R_2}{2L_2}, \quad n_2^2 = \frac{1}{L_2 C_2}, \quad n_{23}^2 = \frac{1}{L_2 C_3}, \tag{13}$$

(11) and (12) become

$$(D^2+2\kappa_1 D+n_1^2+n_{12}^2)Q_1-n_{12}^2 Q_2 = V/L_1, \tag{14}$$
$$-n_2^2 Q_1+(D^2+2\kappa_2 D+n_2^2+n_{23}^2)Q_2 = 0. \tag{15}$$

These equations are exactly those for forced oscillations of the mechanical system of Fig. 14 with $k_{12} = 0$ and Q_1 and Q_2 replacing x_1 and x_2 (cf. § 34 (4) and (5)) and it follows that the whole discussion of free and forced oscillations of this system in § 34 can be taken over bodily: there will be two natural frequencies, each with its own damping factor and normal mode of oscillation of the currents and charges.

For the special case $L_1 = L_2$, $R_1 = R_2$, $C_1 = C_2 = C_3$, so that

$$n_1^2 = n_{12}^2 = n_2^2 = n_{23}^2, \qquad \kappa_1 = \kappa_2, \tag{16}$$

(14) and (15) become

$$(D^2+2\kappa_1 D+2n_1^2)Q_1-n_1^2 Q_2 = V/L_1, \tag{17}$$
$$-n_1^2 Q_1+(D^2+2\kappa_1 D+2n_1^2)Q_2 = 0. \tag{18}$$

Free oscillations for this case have been discussed in detail in § 34 (9) and (10), and forced oscillations in § 34 (17) and (18).

Ex. 2. *A system containing mutual inductance.*

In addition to inductance, resistance, and capacitance in the branches of a circuit there may be mutual inductance between some of the branches. If there is current I_r in the rth branch and current I_s in the sth branch, Fig. 30 (*a*), and mutual inductance M between them, there will be voltage drops

$$M\frac{dI_r}{dt} \quad \text{and} \quad M\frac{dI_s}{dt}, \tag{19}$$

respectively in the sth and rth branches. The mutual inductance M may be positive or negative according to the way in which the coils are wound.

As an example we write down the equations for the circuit of Fig. 30 (*b*) which has mutual inductance M between the inductances L_1 and L_2. Choosing currents I_1, I_2, and I_3 in the directions of the arrows, Kirchhoff's second law at the junction C gives

$$I_1-I_2-I_3 = 0. \tag{20}$$

The first law for the circuits $ABCDEA$ and $CGFDC$, respectively, gives

$$(L_1 D+R_1)I_1-MDI_2+L_3 DI_3 = V,$$
$$(L_2 D+R_2)I_2-L_3 DI_3-MDI_1 = 0.$$

Eliminating I_3 from these by (20) gives

$$\{(L_1+L_3)D+R_1\}I_1-(M+L_3)DI_2 = V, \tag{21}$$
$$-(M+L_3)DI_1+\{(L_2+L_3)D+R_2\}I_2 = 0. \tag{22}$$

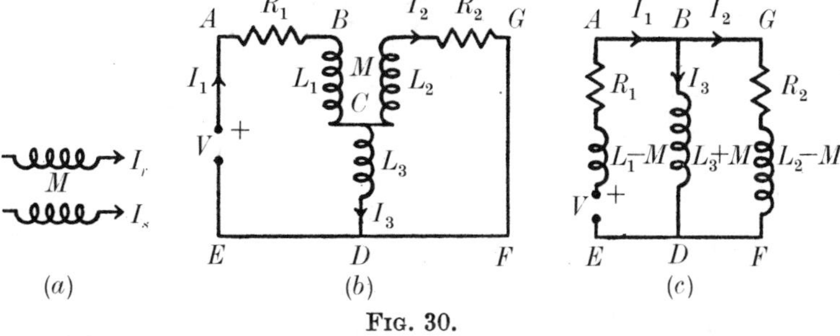

FIG. 30.

It should be noticed that the equations (21) and (22) are the same as those which would be found for the network of Fig. 30 (*c*) which contains self-inductances and resistances only. This device is much used in circuit theory.

Ex. 3. *The transformer.*

Two circuits L_1, R_1 and L_2, R_2 are coupled solely by mutual inductance M. Alternating voltage $E \sin \omega t$ is applied to the primary.

If I_1 and I_2 are the primary and secondary currents, Fig. 31, the circuit equations are

$$(L_1 D+R_1)I_1+MDI_2 = E \sin \omega t, \tag{23}$$
$$MDI_1+(L_2 D+R_2)I_2 = 0. \tag{24}$$

Suppose we wish to find the forced oscillations of the system. As usual we replace $\sin \omega t$ by $e^{i\omega t}$ and find a particular integral of the resulting system by assuming

$$I_1 = I_1' e^{i\omega t}, \qquad I_2 = I_2' e^{i\omega t}, \tag{25}$$

where I_1' and I_2' are constants. This gives

$$(R_1+L_1 i\omega)I_1'+Mi\omega I_2' = E,$$
$$Mi\omega I_1'+(R_2+L_2 i\omega)I_2' = 0.$$

Solving for I_2' we get

$$I_2' = -\frac{ME}{(L_1 R_2+R_1 L_2)+i\{(L_1 L_2-M^2)\omega-R_1 R_2/\omega\}}.$$

Thus the steady value of I_2 is

$$-\frac{ME}{Z}\sin(\omega t-\phi),$$

where $\tan\phi = [(L_1 L_2-M^2)\omega^2-R_1 R_2]/(L_1 R_2+R_1 L_2)\omega,$

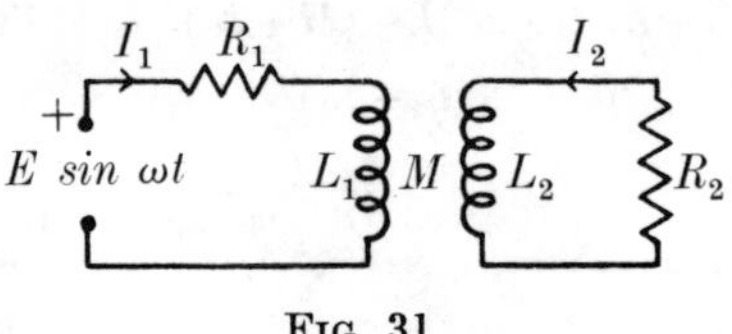

FIG. 31.

and $Z = \{(L_1 R_2+R_1 L_2)^2-2R_1 R_2\sigma+\sigma^2\omega^2+R_1^2 R_2^2/\omega^2\}^{\frac{1}{2}},$

where $\sigma = L_1 L_2-M^2$.

$Z\to\infty$ as $\omega\to 0$ or $\omega\to\infty$. It has a minimum when $\omega = (R_1 R_2/\sigma)^{\frac{1}{2}}$.

In the examples above we have written down the equations directly from Kirchhoff's two laws. There are two ways in which this procedure can be shortened.

(i) Instead of assuming an unknown current in each branch of the circuit and writing down algebraic equations connecting them by Kirchhoff's second law at the junctions, we may choose the unknown currents to satisfy the second law automatically. For example in Fig. 28 (c), if we assume unknown currents I_1 and I_2 in AB and BC, the current in BD must be I_1-I_2. Proceeding in this way, it appears that only four unknown currents appear in Fig. 28 (c) instead of the eight in Fig. 28 (a). We shall often use this procedure in future.

(ii) The same result is achieved if we assume a 'mesh-current' to be circulating round each of the four meshes ABC, BDC,... of Fig. 28 (d). In this case the current in BC, for example, is I_1-I_2, and Kirchhoff's second law is satisfied automatically at B and similarly at all the other junctions.

43. Mechanical analogies

In §§ 41, 42 it was seen that certain electrical circuits led to precisely the same differential equations as corresponding mechanical systems, so that the algebra of the solutions is the same for both. The obvious correspondence appears on comparing (5) and (4) of § 41, namely,

$$L\frac{dI}{dt}+RI+\frac{Q}{C} = V, \tag{1}$$

$$\frac{dQ}{dt} = I, \tag{2}$$

with the equations of motion § 29 (1) for a mass m attached to a spring of stiffness λ, with resistance to motion k times the velocity v, and acted on by force F. These are

$$m\frac{dv}{dt}+kv+\lambda x = F, \tag{3}$$

$$\frac{dx}{dt} = v. \tag{4}$$

These two sets of equations correspond precisely if we replace Q by x, I by v, L by m, $1/C$ by λ, V by F and R by k. This is the most common and useful form of analogy.

These considerations can be carried a good deal farther. First we notice that by integrating (2) with respect to the time from $t = 0$ to $t = t$ we get

$$Q = \overset{0}{Q}+\int\limits_0^t I\,dt, \tag{5}$$

where $\overset{0}{Q}$ is the value of Q when $t = 0$. For shortness we shall write (5) in the form

$$Q = \int I\,dt, \tag{6}$$

and other integrals of the same type below are to be understood in the same sense.

Using (6) in (1) gives

$$L\frac{dI}{dt}+RI+\frac{1}{C}\int I\,dt = V, \tag{7}$$

which is a form in which the fundamental equation for the L, R, C circuit is often written. (7) is called an integrodifferential equation since it contains both the integral and the differential coefficient of the unknown I.

In § 41 the equations

$$RI = V, \qquad L\frac{dI}{dt} = V, \qquad \frac{1}{C}Q = \frac{1}{C}\int I\,dt = V, \tag{8}$$

were given for determining the current in a resistance, inductance, or capacitance in terms of the voltage V applied to it.

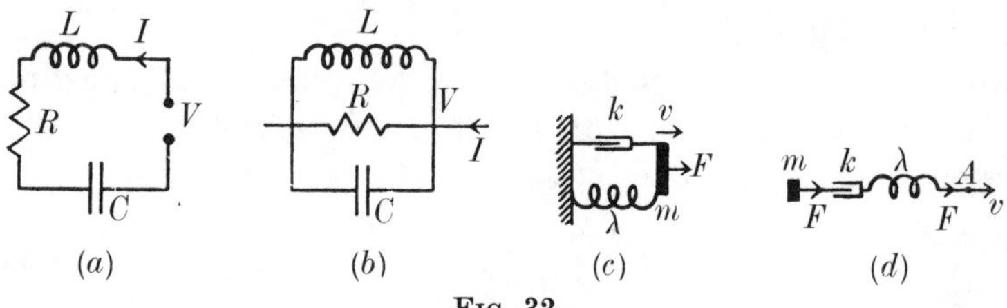

FIG. 32.

Now suppose we regard the *current* I in each of these elements as known and wish to find the voltage drop V in terms of it. The equations (8) may be rewritten from this point of view as†

$$\frac{1}{R}V = I, \qquad \frac{1}{L}\int V\,dt = I, \qquad C\frac{dV}{dt} = I, \tag{9}$$

where the second and third of (9) are obtained by integrating the second and differentiating the third of (8), respectively.

Next consider the 'L, R, C parallel circuit', Fig. 32 (b), in which the total current I into the combination of L, R, and C in parallel is regarded as given, and it is required to find the voltage drop across the elements. Adding the currents in the separate elements given by (9), we get

$$C\frac{dV}{dt} + \frac{1}{R}V + \frac{1}{L}\int V\,dt = I. \tag{10}$$

This is of the same form as (7) with V and I interchanged, L and C interchanged, and R replaced‡ by $1/R$. The circuit Fig. 32 (b) is called the 'dual' of Fig. 32 (a), and this duality can be extended to more complicated circuits.

† These equations may be used to develop the theory of electrical networks in the same way that we have used (8).

‡ It will be noticed that in (7) and (10) all of L, R, C and their reciprocals appear. It is usual to give these reciprocals names and to use them when they occur: thus $G = 1/R$ is the conductance, $S = 1/C$ the elastance, etc.

Similar relations apply to mechanical systems: writing x for displacement, $v = \dot{x}$ for velocity, we have

$$kv = F, \qquad m\frac{dv}{dt} = F, \qquad \lambda x = \lambda \int v\,dt = F, \tag{11}$$

for a force F applied respectively to a dash-pot which gives resistance to motion k times the velocity, to a mass m, and to a spring of stiffness λ. Combining these we have for the mechanical system, Fig. 32 (c),

$$m\frac{dv}{dt}+kv+\lambda\int v\,dt = F \tag{12}$$

as in (3) and (4). If, on the other hand, we regard the velocity as given and wish to determine the force across the elements, we write (11) in the form

$$\frac{1}{k}F = v, \qquad \frac{1}{m}\int F\,dt = v, \qquad \frac{1}{\lambda}\frac{dF}{dt} = v. \tag{13}$$

The equation of motion of the system of Fig. 32 (d), in which the point A is moved in a straight line with velocity v which is a known function of t, is

$$\frac{1}{\lambda}\frac{dF}{dt}+\frac{1}{k}F+\frac{1}{m}\int F\,dt = v, \tag{14}$$

which, again, is of the same form as (7), (10), and (12). The system of Fig. 32 (d) is the dual of the system of Fig. 32 (c), and all four systems of Fig. 32 lead to equations of the same type in symbols which correspond as follows:

Electrical		*Mechanical*	
L, R, C series Fig. 32 (a)	*Dual Fig. 32 (b)*	*Damped harmonic oscillator Fig. 32 (c)*	*Dual Fig. 32 (d)*
L	C	m	$1/\lambda$
R	$1/R$	k	$1/k$
C	L	$1/\lambda$	m
I	V	v	F
V	I	F	v

It is possible, by using the relations given above, to set up electrical analogues of dynamical systems. The correspondence between the circuit of Fig. 29 and a special case of the mechanical system of Fig. 14 has been noted in § 42.

In the same way the circuits of Fig. 33 (*a*), (*b*), and (*c*) will be found to be the analogues of the mechanical systems of Figs. 14, 19, and 16 (*a*), respectively, of the type in which inductance corresponds to mass, etc. For example, if the differential equations for Q_1 and Q_2 in Fig. 33 (*a*) are written down they will be

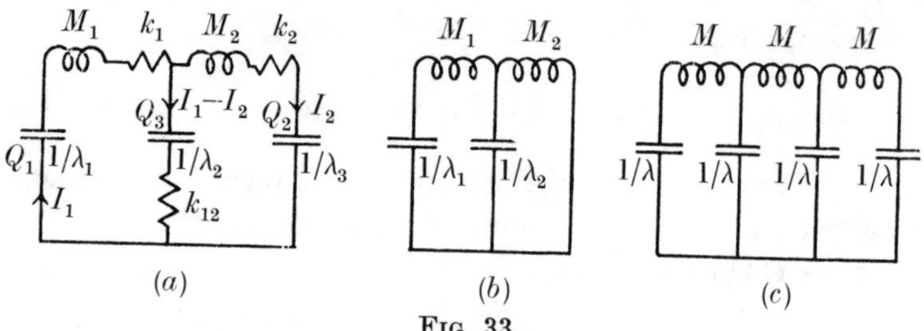

FIG. 33.

found to be identical with § 34 (4) and (5) with Q_1 and Q_2 replacing x_1 and x_2. In doing this the charge in any branch is taken to be the total amount of charge transported by the current in this branch, so that Q_3, the charge associated with I_1-I_2, is equal to Q_1-Q_2; cf. § 42 (10).

The procedure for setting up such analogous systems has been extensively studied and extended to systems containing gearing and also to acoustical systems.†

44. Steady state theory. Impedance

It has been shown in the last three sections that there is no fundamental difference between the equations to be solved in problems of electric circuit theory and of mechanical vibrations. But owing to the enormous complexity of many practical circuits, special techniques have been developed for their study which concentrate attention on the quantities of interest to the electrical engineer. This leads to two important changes in point of view.

Firstly, the electrical engineer is most often interested in the steady response of a circuit to an alternating voltage rather than in its transient behaviour. Thus he does not usually look for

† Cf. Olson, *Dynamical Analogies* (van Nostrand).

natural frequencies as such, but regards them as frequencies at which resonance occurs: this corresponds to the second of the two methods developed in Chapter IV. Similarly, the normal modes of oscillation corresponding to the natural frequencies are rarely calculated—for a complete solution of a transient problem

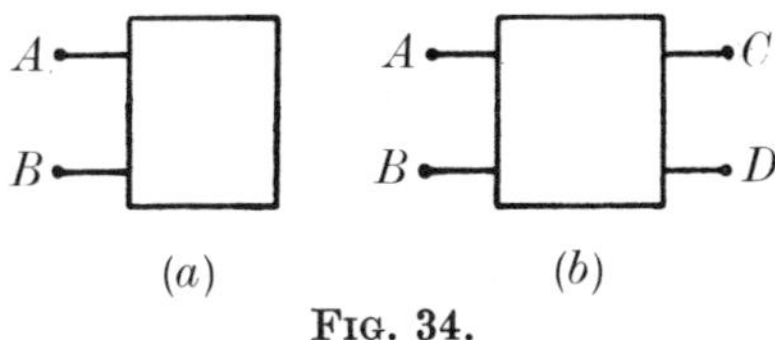

FIG. 34.

by the classical methods it is necessary to do this, but the labour can be avoided by the use of the Laplace transformation or operational methods which have, in fact, largely been developed for problems of this type.

Secondly, the engineer is very little interested in what goes on in many parts of the complicated circuits with which he deals. Instead, he regards them as boxes, with, for example, two terminals A, B, Fig. 34 (a), which is a *two-terminal network*, or with four terminals, Fig. 34 (b), which is a *four-terminal network*.

The commonest types of problem are then: (i) to find the steady state current flowing into the network of Fig. 34 (a) due to sinusoidal voltage applied to the terminals AB, or (ii) to find the steady state current in a known load connected to the terminals CD of Fig. 34 (b) when a sinusoidal voltage is applied to the terminals AB.

In a complicated circuit it would be a waste of labour to write down a complete set of equations for all the currents in all the branches in the boxes of Fig. 34 and this labour can often be avoided by the methods now to be discussed.

Consider first the L, R, C circuit, Fig. 35 (a), with sinusoidal voltage $E\cos(\omega t+\beta)$ applied to it, and suppose we require the steady state current in the circuit. The equations for the current I in the circuit are, by § 41 (5),

$$(LD+R)I+\frac{Q}{C} = E\cos(\omega t+\beta), \tag{1}$$

$$DQ = I. \tag{2}$$

To find the steady state current we replace $E\cos(\omega t+\beta)$ in (1) by $E'e^{i\omega t}$, where

$$E' = Ee^{i\beta}. \tag{3}$$

Then we seek a solution of the form

$$I = I'e^{i\omega t}, \qquad Q = Q'e^{i\omega t}, \tag{4}$$

and take its real part. Substituting (4) in (1) and (2) gives

$$(R+Li\omega)I'+\frac{Q'}{C} = E', \qquad i\omega Q' = I'.$$

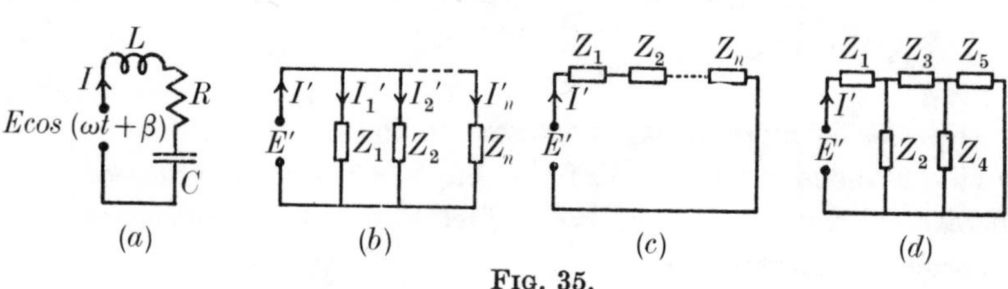

FIG. 35.

Therefore, eliminating Q',

$$\left\{R+i\left(L\omega-\frac{1}{C\omega}\right)\right\}I' = E'. \tag{5}$$

We call

$$z = R+i\left(L\omega-\frac{1}{C\omega}\right) \tag{6}$$

the *complex impedance* of the circuit. Its imaginary part

$$X = L\omega-\frac{1}{C\omega} \tag{7}$$

is called the *reactance* of the circuit, and its modulus

$$|z| = \left\{R^2+\left(L\omega-\frac{1}{C\omega}\right)^2\right\}^{\frac{1}{2}} = \{R^2+X^2\}^{\frac{1}{2}} \tag{8}$$

is called the *impedance* of the circuit. The angle

$$\theta = \tan^{-1}\frac{X}{R} \tag{9}$$

is called the phase angle.

With this notation, (5) gives

$$I' = \frac{E'}{z}. \tag{10}$$

Using (10) and (3) in (4), the steady state current is thus the real part of

$$\frac{E'}{z}e^{i\omega t} = \frac{E}{|z|}e^{i(\omega t+\beta-\theta)},$$

that is,

$$\frac{E}{|z|}\cos(\omega t+\beta-\theta), \tag{11}$$

where $|z|$ and θ are defined in (8) and (9). This is the result† which was quoted in § 41 (12), and the argument above is merely a repetition of that of § 31 in which this particular integral of the equations (1) and (2) was found. One minor change has been made, namely including the phase angle β of $E\cos(\omega t+\beta)$ in the complex quantity E' of (3), but the main point has been the expressing of the solution in terms of the complex impedance z which is now regarded as a quantity which can be written down immediately for this circuit.

For shortness, we shall call E' the *complex voltage* applied to the circuit and I' the *complex current* in it, it being understood that these terms refer to the steady state only, and that to get actual real voltage or current we multiply E' or I' by $e^{i\omega t}$ and take the real part. Thus (10) states that the complex current in an L, R, C circuit is obtained by dividing the complex voltage across it by the complex impedance z. This relation has the same form as Ohm's law for direct current with complex impedance z replacing resistance; thus we can write down formulae for the impedance of complicated networks by the same rules which allow us to write down direct current resistance.

First we define the complex impedance of any two-terminal network, Fig. 34 (a), as the complex voltage across its terminals divided by the complex current into them.

Next consider the circuit of Fig. 35 (b) with n impedances $z_1,\ldots, z_n$ in parallel and complex voltage E' applied to it. Let $I'_1,\ldots, I'_n$ be the complex currents in the impedances, then

$$I'_1 = \frac{E'}{z_1}, \quad \ldots, \quad I'_n = \frac{E'}{z_n}.$$

† Except that the voltage has been taken here to be $E\cos(\omega t+\beta)$ to conform with the usual engineering practice. It should be added, also, that engineers use the small letters e and i for the quantities denoted here by E' and I'.

Thus if I' is the total complex current into the network

$$I' = I'_1+...+I'_n = E'\left(\frac{1}{z_1}+...+\frac{1}{z_n}\right). \tag{12}$$

Thus the complex impedance z of the n impedances in parallel is given by

$$\frac{1}{z} = \frac{1}{z_1}+\frac{1}{z_2}+...+\frac{1}{z_n}. \tag{13}$$

Similarly the complex impedance z of a number of impedances $z_1,..., z_n$ in series, Fig. 35 (c), is

$$z = z_1+z_2+...+z_n. \tag{14}$$

For more complicated circuits the complex impedance may be written down by a combination of these rules. For example, for the 'ladder' network of Fig. 35 (d), it is

$$z_1+\cfrac{1}{\cfrac{1}{z_2}+\cfrac{1}{z_3+\cfrac{1}{(1/z_4)+(1/z_5)}}}. \tag{15}$$

The reciprocal $1/z$ of the complex impedance of a circuit is its complex *admittance*. Both quantities are equally useful in the theory: for example (13) may be better stated in the form that the admittance of n circuits in parallel is the sum of their admittances.

For four-terminal networks the position is a little more complicated. Consider the network of Fig. 34 (b) and suppose a load of complex impedance z_L is connected to the terminals CD. If I'_L is the complex current in this load caused by complex voltage E' applied at the terminals AB, the quantity

$$E'/I'_L$$

is called the *transfer impedance* between the pairs of terminals AB and CD: it can be calculated from a knowledge of the network and the load impedance. To do this it is usually necessary to use Kirchhoff's laws discussed in § 42. In the steady state with which we are now concerned the voltage drop over any element of complex impedance z carrying complex current I' will be $zI'e^{i\omega t}$ and the applied voltage will be $E'e^{i\omega t}$:

the time-factors $e^{i\omega t}$ cancel and we are left with equations connecting the complex currents and voltages. Kirchhoff's first law then becomes *the algebraic sum of the complex voltage drops over the elements of a closed circuit is equal to the algebraic sum of the complex applied voltages in the circuit.* The second law becomes *the algebraic sum of the complex currents at a junction is zero.*

45. Variation of impedance with frequency. Filter circuits

It was remarked in § 44 that most of the information required by engineers could be obtained by writing down the complex impedance of a circuit and studying its variation with frequency. In this section we discuss some simple networks from this point of view.

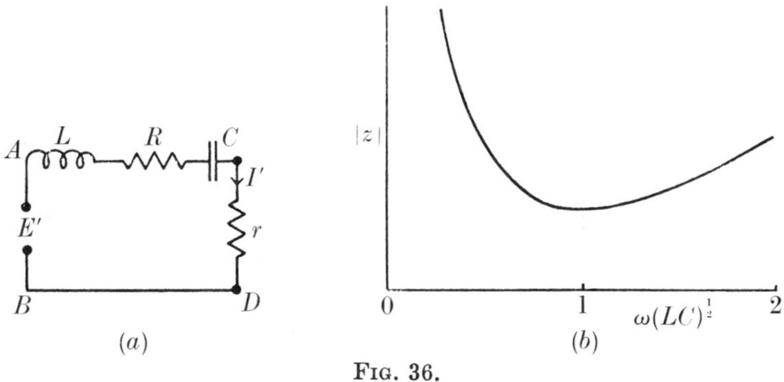

FIG. 36.

(i) *The L, R, C circuit in series with a load*

We suppose that complex voltage E' is applied to the terminals AB of Fig. 36 (a), and that the complex current I' in a terminating resistance r is to be calculated. By § 44 (10)

$$I' = \frac{E'}{z},$$

where
$$z = R+r+\left(L\omega-\frac{1}{C\omega}\right)i$$

is the complex impedance.

The impedance is

$$|z| = \left\{(R+r)^2+\left(L\omega-\frac{1}{C\omega}\right)^2\right\}^{\frac{1}{2}}, \tag{1}$$

and this has its least value $(R+r)$ when $\omega^2 = 1/LC$, that is, at the undamped natural frequency of the circuit. The variation of $|z|$ with ω is shown in Fig. 36 (b). When $\omega = (LC)^{-\frac{1}{2}}$, the impedance is least and the current greatest: thus this circuit tends to favour the passage of this frequency relative to others. The problem is essentially that of § 31 and Fig. 8 (a) except that here we have studied the impedance instead of the amplitude which is proportional to its reciprocal.

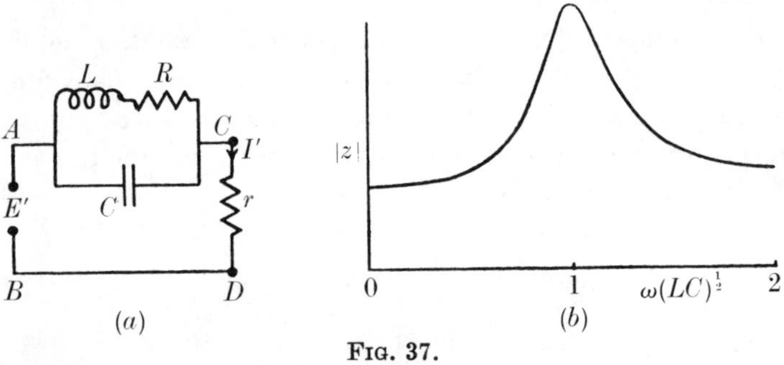

FIG. 37.

(ii) *The 'choke' circuit, Fig. 37 (a)*

Here we require the complex current I' in a load consisting of resistance r when complex voltage E' is applied at the terminals AB.

The complex impedance z of the system $ACDB$ is, by § 44 (13) and (14),

$$z = r + \frac{1}{C\omega i + 1/(R+L\omega i)}$$

$$= r + \frac{R+L\omega i}{(1-LC\omega^2)+RC\omega i}.$$

$$|z| = \left\{\frac{\{r(1-LC\omega^2)+R\}^2+(RCr\omega+L\omega)^2}{(1-LC\omega^2)^2+R^2C^2\omega^2}\right\}^{\frac{1}{2}}. \qquad (2)$$

Here, if R is small which is usually the case, the denominator of (2) is least for a value of ω near $(LC)^{-\frac{1}{2}}$. Thus the impedance has a maximum value near the undamped natural frequency of the L, C circuit—its variation is shown in Fig. 37 (b). Thus this circuit tends to discourage the passage of the frequency $(LC)^{-\frac{1}{2}}/2\pi$ and to favour higher or lower frequencies.

(iii) *The 'parallel-T' circuit*

The circuit is shown in Fig. 38 (a). Complex voltage E' is applied at the terminals AB, and we require the complex current I' in a load (which is taken to be a resistance r) connected across the terminals EF.

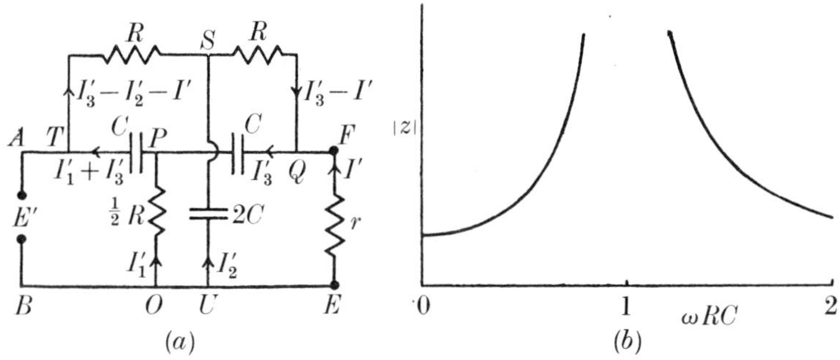

FIG. 38.

This is a more complicated problem than those discussed above because we require the current in a different branch to that in which the voltage is applied. We use Kirchhoff's laws for complex currents and voltages as in § 44. Let I_1', I_2', and I_3' be the complex currents in the branches OP, US, and QP of Fig. 38 (a), those in the other branches being chosen to satisfy Kirchhoff's second law automatically. Then Kirchhoff's first law for the complex voltage drops round the closed circuits $PABOP$, $OEFPO$, $USQFEU$, and $PTSQP$, respectively, give

$$-\tfrac{1}{2}RI_1'+\frac{i}{C\omega}(I_1'+I_3') = E', \tag{3}$$

$$rI'-\frac{i}{C\omega}I_3'-\tfrac{1}{2}RI_1' = 0, \tag{4}$$

$$-\frac{i}{2C\omega}I_2'+R(I_3'-I')-rI' = 0, \tag{5}$$

$$-\frac{i}{C\omega}(I_1'+I_3')+R(I_3'-I_2'-I')+R(I_3'-I')-\frac{i}{C\omega}I_3' = 0. \tag{6}$$

If we write
$$x = \omega RC, \qquad k = r/R, \tag{7}$$

and solve for I' we find

$$-\frac{RI'}{E'} = \begin{vmatrix} -\frac{1}{2} & 0 & -i/x \\ 0 & -i/2x & 1 \\ -i/x & -1 & 2(1-i/x) \end{vmatrix} \div \begin{vmatrix} 0 & (\frac{1}{2}-i/x) & 0 & -i/x \\ k & -\frac{1}{2} & 0 & -i/x \\ -(k+1) & 0 & -i/2x & .1 \\ -2 & -i/x & -1 & 2(1-i/x) \end{vmatrix}$$

$$= \frac{x^2-1}{(kx^2-2-k)-2ix(1+2k)}. \tag{8}$$

If we write
$$I' = E'/z,$$
where, in the notation of § 44, z is the transfer impedance between the branches AB and EF, we find from (8)
$$|z| = \frac{R\{(kx^2-2-k)^2+4x^2(1+2k)^2\}^{\frac{1}{2}}}{|x^2-1|}. \tag{9}$$

The variation of $|z|$ with $x = \omega RC$ is shown in Fig. 38 (*b*). The impedance is infinite at the frequency $1/2\pi RC$.

(iv) *Filter circuits*

In the above examples it has been seen that simple combinations of circuit elements have rather crude filtering properties. Thus the circuit of Fig. 36 (*a*) tends to 'stop' low and high frequencies and to 'pass' those near $(LC)^{-\frac{1}{2}}/2\pi$. The circuits of Figs. 37 (*a*) and 38 (*a*), on the other hand,

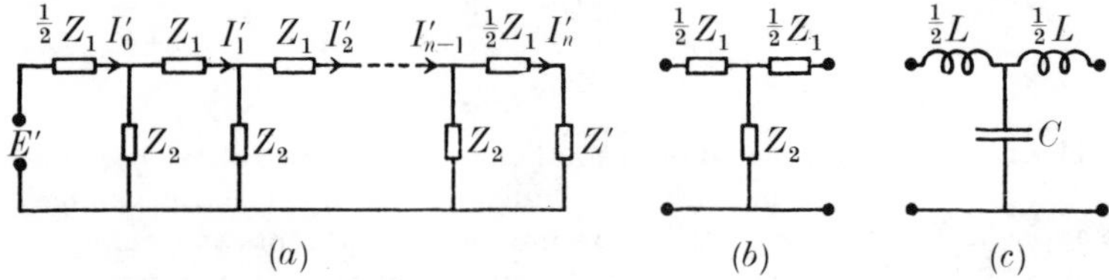

FIG. 39.

pass the low and high frequencies and stop, in part or wholly, intermediate frequencies. This behaviour can be sharpened by connecting together a number of these combinations. The simplest example of this is the 'ladder' network of Fig. 39 (*a*) which may be regarded as composed of n of the sections of Fig. 39 (*b*) connected in tandem and terminated by a complex impedance z'.

We suppose complex voltage E' to be applied to the circuit and we wish to find the complex current I'_n in the terminating impedance z'. Let I'_0, I'_1,..., I'_n be the mesh currents as shown in Fig. 39 (*a*). Then Kirchhoff's laws for the successive meshes give
$$\tfrac{1}{2}z_1 I'_0+z_2(I'_0-I'_1) = E', \tag{10}$$
$$z_1 I'_r+z_2(I'_r-I'_{r+1})-z_2(I'_{r-1}-I'_r) = 0 \quad (r = 1,..., n-1), \tag{11}$$
$$(\tfrac{1}{2}z_1+z')I'_n-z_2(I'_{n-1}-I'_n) = 0. \tag{12}$$

Equations (11) are a set of simultaneous linear difference equations with constant coefficients.† We seek a solution of the form
$$I'_r = Ae^{r\theta}, \tag{13}$$

† An alternative approach to ladder networks is to be found in Ex. 5 at the end of Chapter XIII, where the 'transmission matrix' is introduced.

where A and θ are independent of r. Substituting in (11) gives

$$(z_1+2z_2)e^{r\theta}-z_2\,e^{(r+1)\theta}-z_2\,e^{(r-1)\theta} = 0.$$

Or
$$\cosh\theta = 1+\frac{z_1}{2z_2}. \tag{14}$$

Since (14) gives two values of θ with opposite signs, we get the final result that all the equations (11) are satisfied by

$$I'_r = Ae^{r\theta}+Be^{-r\theta}, \tag{15}$$

where A and B are independent of r, and θ is given by (14). The constants A and B are determined by substituting (15) in (10) and (12) which give, using (14),

$$(A+B)\cosh\theta-(Ae^{\theta}+Be^{-\theta}) = E'/z_2$$
$$(\cosh\theta+z'/z_2)(Ae^{n\theta}+Be^{-n\theta})-(Ae^{(n-1)\theta}+Be^{-(n-1)\theta}) = 0.$$

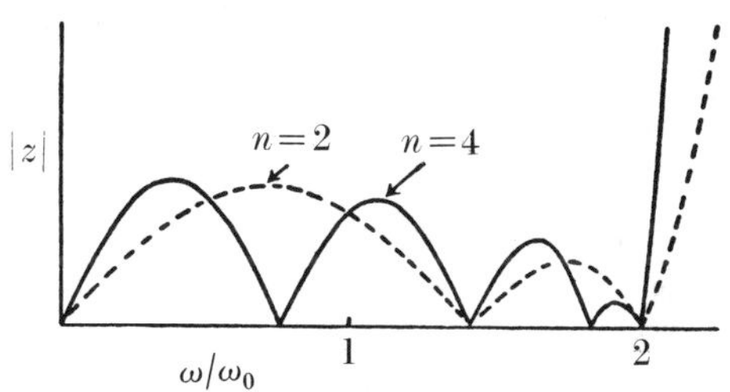

FIG. 40.

Solving for A and B and putting these values in (15) we get, after some reduction,

$$I'_r = \frac{E'[\sinh\theta\cosh(n-r)\theta+(z'/z_2)\sinh(n-r)\theta]}{z_2\sinh\theta[\sinh n\theta\sinh\theta+(z'/z_2)\cosh n\theta]}. \tag{16}$$

For a simple special case to study in detail, we consider I'_n in the case in which $z' = 0$ (the output terminals short-circuited) for the circuit of Fig. 39 (c) in which

$$z_1 = L\omega i, \qquad z_2 = -\frac{i}{C\omega}, \tag{17}$$

so that
$$\cosh\theta = 1-\tfrac{1}{2}LC\omega^2. \tag{18}$$

In this case (16) gives
$$I'_n = E'/z, \tag{19}$$

where z, the transfer impedance in which we are interested, is given by

$$\begin{aligned} z &= z_2\sinh\theta\sinh n\theta \\ &= -(i/C\omega)2^{n-1}(\cosh\theta-1)(\cosh\theta-\cos\pi/n)...\times \\ &\qquad\times(\cosh\theta-\cos(n-1)\pi/n)(\cosh\theta+1), \end{aligned} \tag{20}$$

where in (20) we have used the finite product for $\sinh\theta\,\sinh n\theta$. Using the value of (18) of $\cosh\theta$ in (20) we get finally

$$z = 2^{n-2}iL\omega(1-\cos\pi/n-\tfrac{1}{2}LC\omega^2)(1-\cos 2\pi/n-\tfrac{1}{2}LC\omega^2)...(2-\tfrac{1}{2}LC\omega^2). \tag{21}$$

In Fig. 40, $|z|$ is plotted against ω/ω_0, where $\omega_0 = (LC)^{-\frac{1}{2}}$, for $n = 2$ and $n = 4$.

It appears that $|z|$ is small if $\omega < 2\omega_0$, but $|z|$ increases rapidly as ω is increased beyond $2\omega_0$ and that this tendency increases as n, the number of sections, is increased. The filter is a 'low-pass' filter with 'cut-off' frequency ω_0/π.

46. Semiconductor circuits. The amplifier

We shall illustrate the properties of semiconductor circuits by analysing the simple amplifier circuit of Fig. 41 (*a*). This contains a device, known as a tunnel-diode, which is more often used in oscillator circuits (cf. § 47) than amplifier circuits. However, it has the advantage that it is relatively straightforward to describe and exhibits all the properties of semiconductors that are of interest to us.

The behaviour of the tunnel-diode† is defined by the current–voltage or 'characteristic' curve of Fig. 41 (*b*). The steady voltage V_s across the diode is chosen so as to 'bias' the tunnel-diode to operate around some point O on the negative slope portion BC of the characteristic curve. If I and V are the variations in current and voltage from the steady values at O,

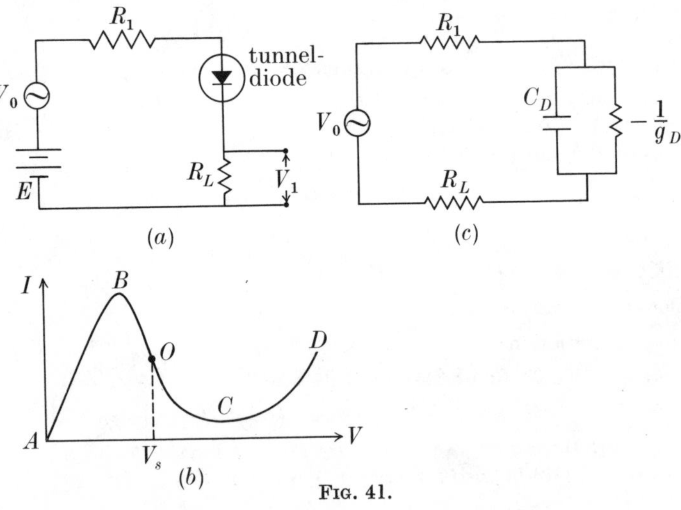

Fig. 41.

† See, for example, Chow, *Principles of Tunnel-Diode Circuits* (Wiley).

they are related by the shape of the characteristic curve in the neighbourhood of O. This will usually be non-linear, and in general we can write

$$I = f(V). \tag{1}$$

However, if the variations I and V are both small, we can approximate $f(V)$ by the tangent to the characteristic curve at O, and (1) can be written

$$I = -g_D V, \tag{2}$$

where $g_D = -dI/dV$ evaluated at O. Equation (2) shows that for small signals the tunnel-diode acts as a negative resistance device.

Fig. 41 (c) is the small signal 'equivalent circuit'† for Fig. 41 (a). In it, the tunnel-diode is represented as the negative resistance $-1/g_D$ in parallel with the capacitance C_D which represents the junction capacitance of the diode. Fig. 41 (c) contains only the standard linear components of an ordinary electric circuit, and hence can be analysed by the methods of §§ 41–5. The complex impedance of the circuit is

$$z = R_1+R_L+1/(i\omega C_D-g_D), \tag{3}$$

so that the complex current I' due to a complex voltage V_0' is

$$I' = V_0'/z. \tag{4}$$

The output voltage is $$V_1' = R_L I', \tag{5}$$

so that, by (3), (4), and (5),

$$\frac{V_1'}{V_0'} = \frac{R_L(i\omega C_D-g_D)}{1+(R_1+R_L)(i\omega C_D-g_D)}. \tag{6}$$

The voltage amplification is

$$A = \left|\frac{V_1'}{V_0'}\right| = \frac{R_L(\omega^2 C_D^2+g_D^2)}{[\omega^2 C_D^2+\{(R_1+R_L)(\omega^2 C_D^2+g_D^2)-g_D\}^2]^{\frac{1}{2}}}$$

and from the argument of the complex function (6) it follows that the output voltage lags the input voltage by

$$\theta = \tan^{-1}\left\{\frac{\omega C_D}{g_D-(R_1+R_L)(\omega^2 C_D^2+g_D^2)}\right\}.$$

† Notice that steady voltage sources are usually omitted from the equivalent circuit; it is assumed that these have been chosen so that the circuit is correctly 'biased'.

The same general approach holds for other semiconductor circuits. The operating points on a set of characteristic curves are determined, linear approximations are obtained for these curves in the neighbourhood of the operating points, and equivalent linear circuits are then developed and solved by standard methods. In the next section we consider the stability and non-linear behaviour of semiconductor circuits.

47. Stability. Oscillator circuits

In dealing with the solutions of linear differential equations in Chapter II, it was found that the solution of

$$(a_0 D^n + \ldots + a_n)y = 0 \tag{1}$$

was of type

$$A_1 e^{\alpha_1 t} + \ldots + A_n e^{\alpha_n t}, \tag{2}$$

where $A_1,\ldots, A_n$ were arbitrary constants, and $\alpha_1,\ldots,\alpha_n$ were the roots of the auxiliary equation

$$a_0 \alpha^n + \ldots + a_n = 0. \tag{3}$$

For the moment we suppose $\alpha_1,\ldots,\alpha_n$ to be all different. In most of the mechanical and electrical systems studied so far, all the roots of (3) have been either real and negative, or complex with negative real parts, corresponding to exponential decay or to damped oscillations. If, however, any of the roots has a positive real part, the solution (2) would contain a term of exponentially increasing amplitude, and the system is said to be *unstable*. Such effects arise most commonly in systems in which energy is supplied from an external source, such as transistor circuits, servomechanisms, and so on. It must always be remembered that the linear equations with which we deal are approximate, and only hold for a restricted range of the dependent variable; thus if we say that a system is unstable, this implies that the dependent variable will become so large that our linear equations will no longer be valid.

The criterion for stability is that *all* the roots of (3) should have negative real parts. For the quadratic

$$\alpha^2 + a_1 \alpha + a_2 = 0, \tag{4}$$

this requires that

$$a_1 > 0 \quad \text{and} \quad a_2 > 0. \tag{5}$$

In general, suppose that in (3) $a_0 > 0$ and that the left-hand side can be resolved into linear and quadratic factors in the form

$$a_0(\alpha+b_1)(\alpha+b_2)...(\alpha^2+c_1\alpha+d_1)(\alpha^2+c_2\alpha+d_2)... = 0. \tag{6}$$

For stability, all the quantities $b_1, b_2,...$ in (6) must be positive, and by (5), all the quantities $c_1, d_1, c_2, d_2,...$ must also be positive. Multiplying out the left-hand side of (6), it follows that if $a_0 > 0$, *it is a necessary condition for stability that the coefficients of all the powers of* α *in* (3) *be positive.*

This simple criterion is often useful, but it is not sufficient. The complete conditions for equation (3) with $a_0 > 0$ are that all the determinants

$$a_1,\quad \begin{vmatrix} a_1 & a_0 \\ a_3 & a_2 \end{vmatrix},\quad \begin{vmatrix} a_1 & a_0 & 0 \\ a_3 & a_2 & a_1 \\ a_5 & a_4 & a_3 \end{vmatrix},\ldots \quad \begin{vmatrix} a_1 & a_0 & 0 & . & . & . & . & 0 \\ a_3 & a_2 & a_1 & 0 & . & . & . & 0 \\ . & & & & & & & \\ . & & & & & & & \\ . & & & & & & & \\ . & & & & & & & \\ a_{2n-1} & a_{2n-2} & . & . & . & . & . & a_n \end{vmatrix} \tag{7}$$

should be positive, where, in writing them down, a_r is replaced by zero if $r > n$. If any of the determinants in (7) is negative, the system is unstable. This is known as Hurwitz's criterion and is related to Routh's rule.† It provides a compact statement, but is rather clumsy to use.

For the cubic $\qquad \alpha^3+a_1\alpha^2+a_2\alpha+a_3 = 0, \qquad (8)$

Hurwitz's criterion for stability gives

$$a_1 > 0, \quad a_1 a_2 - a_3 > 0, \quad a_3 > 0. \tag{9}$$

Criteria for stability are important in two ways:

(i) *Systems required to be stable.* In systems such as amplifiers and servomechanisms, oscillations of large amplitude may be initiated by a chance disturbance unless the system is stable. The criteria for stability set limits to some of the system parameters, as in (16) below.

† See J. L. Willems, *Stability Theory of Dynamical Systems* (Nelson, 1970), chapter 3.

(ii) *Systems required to be unstable.* In studying oscillation generators, the first step is to write down approximate linear equations for the system. These equations are required to be unstable so that the system will execute large oscillations. The amplitude of the large oscillations must then be determined from the accurate non-linear system equations. We proceed to discuss a tunnel-diode oscillator from this point of view.

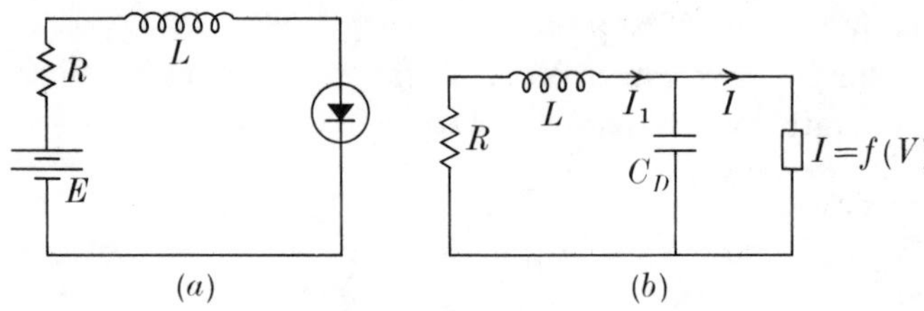

FIG. 42.

The circuit is as shown in Fig. 42 (*a*). The equivalent circuit is shown in Fig. 42 (*b*), where the tunnel-diode is represented by a non-linear element $I = f(V)$. The circuit equations are

$$L\frac{dI_1}{dt}+RI_1+V = 0 \tag{10}$$

and

$$I_1 = f(V)+C_D\frac{dV}{dt}. \tag{11}$$

For small I and V, $f(V) = -g_D V$ and (10) and (11) reduce to the second-order equation

$$LC_D\frac{d^2V}{dt^2}+(C_D R-Lg_D)\frac{dV}{dt}+(1-Rg_D)V = 0. \tag{12}$$

The auxiliary equation is

$$\alpha^2+(R/L-g_D/C_D)\alpha+(1-Rg_D)/LC_D = 0, \tag{13}$$

so that by (5) the conditions for the circuit to be *unstable* are

$$\left.\begin{aligned} R/L-g_D/C_D &< 0\\ \text{and}\qquad (1-Rg_D)/LC_D &< 0\end{aligned}\right\}. \tag{14}$$

If equations (14) are satisfied, the linear approximation will no longer be valid, and the actual amplitude and period of the oscillations must be determined from the non-linear equations

(10) and (11). The approximation $f(V) = AV+BV^3$ is sometimes used, in which case (10) and (11) reduce to

$$LC_D\frac{d^2V}{dt^2}+\frac{dV}{dt}\{C_D R+L(A+3BV^2)\}+V\{1+R(A+BV^2)\} = 0, \tag{15}$$

cf. Ex. 10 at the end of Chapter XIV.

For certain values of the circuit parameters, the oscillations will actually swing into the positive resistance region CD of Fig. 41 (b), back through the negative resistance region BC, into the positive region AB, and so on. The resulting oscillation will be similar to that of Fig. 49 (c) and is known as a 'relaxation oscillation'.

The circuit equations (10) and (11) can be applied to the amplifier circuit of Fig. 41 (c) if we put $L = 0$ and $R = R_1+R_L$. For small signals (10) and (11) then reduce to the first-order equation

$$(R_1+R_L)C_D\frac{dV}{dt}+\{1-g_D(R_1+R_L)\}V = V_0,$$

which shows that the amplifier circuit will be stable if and only if

$$g_D(R_1+R_L) < 1, \tag{16}$$

This limits the choice of components in Fig. 41 (a).

48. Impulsive motion

In § 39 the effect of an impulsive force or blow on a particle was studied: analogous problems arise in electric circuit theory, as might be expected from § 43. We define an impulsive voltage E_0 at $t = 0$ as a voltage $V(t)$ which is very large over the vanishingly small time-interval $0 < t < \tau$, zero outside this interval, and such that its time integral has the finite value E_0, i.e.

$$\int_{-\infty}^{\infty} V(t)\,dt = E_0. \tag{1}$$

In the notation of the δ-function, § 17, this voltage is $E_0\,\delta(t)$.

Ex. 1. *Impulsive voltage E_0 applied at $t = 0$ to an L, R, C circuit with initial current I_i and initial charge Q_i.*

The differential equations, § 41 (4) and (5) are

$$(LD+R)I+\frac{Q}{C} = V(t), \tag{2}$$

$$DQ = I. \tag{3}$$

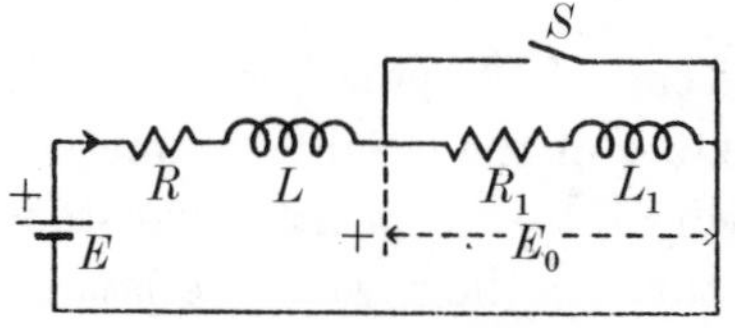

Fig. 43.

As in the analogous problem of § 39 we integrate these over the small time τ. I and Q must be finite, so that their integrals over this time will be of the order of τ. Omitting these small quantities (2) and (3) give

$$L[I]_{t=0}^{t=\tau} = E_0, \qquad [Q]_{t=0}^{t=\tau} = 0. \tag{4}$$

Writing I_f and Q_f for the values of I and Q when $t = \tau$, we get from (4)

$$I_f = I_i+E_0/L, \qquad Q_f = Q_i. \tag{5}$$

The current is thus changed instantaneously by an amount E_0/L [cf. § 39 (3)] by the impulsive voltage, and the new initial current (5) must be used in calculations on the subsequent behaviour of the circuit.

Ex. 2. *Steady current E/R is flowing on the circuit of Fig. 43 with the switch S closed, when at $t = 0$ the switch S is opened.*

In switching problems of this type there is an impulsive redistribution of currents between the inductances analogous to the redistribution of velocity when two masses collide. Suppose there is an impulsive voltage E_0 over L_1, R_1 caused by the closing of the switch. Then by (5) the currents in L_1, R_1, and L, R, respectively, change to

$$E_0/L_1 \quad \text{and} \quad E/R-E_0/L. \tag{6}$$

Now for $t > 0$ these must be equal, so

$$E_0 = \frac{ELL_1}{R(L+L_1)}, \tag{7}$$

and the initial value of the current in the circuit with the switch open is

$$\frac{EL}{R(L+L_1)}. \tag{8}$$

49. Transient problems. Use of the Laplace transformation

In transient problems differential equations such as those

written down earlier have to be solved with given values of the initial currents and charges in the system. The only difficulty is the amount of algebra involved, much of which can be avoided by the use of the Laplace transformation discussed in § 18.

We begin with the fundamental equations § 41 (5), (4) for an L, R, C circuit, namely,

$$L\frac{dI}{dt}+RI+\frac{Q}{C} = V, \tag{1}$$

$$\frac{dQ}{dt} = I. \tag{2}$$

Let $\bar{I}$, $\bar{Q}$, and $\bar{V}$ be the Laplace transforms of I, Q, and V, and let $\overset{0}{I}$ and $\overset{0}{Q}$ be the values of I and Q when $t = 0$. Then the subsidiary equations corresponding to (1) and (2), formed as in § 18, are

$$(Lp+R)\bar{I}+\frac{\bar{Q}}{C} = \bar{V}+L\overset{0}{I}, \tag{3}$$

$$p\bar{Q} = \bar{I}+\overset{0}{Q}. \tag{4}$$

(4) gives

$$\bar{Q} = \frac{\bar{I}}{p}+\frac{\overset{0}{Q}}{p}, \tag{5}$$

and substituting this in (3) gives

$$\left(Lp+R+\frac{1}{Cp}\right)\bar{I} = \bar{V}+L\overset{0}{I}-\frac{\overset{0}{Q}}{Cp}. \tag{6}$$

(6) will be called the *subsidiary equation for an L, R, C circuit*: it gives the Laplace transform of I, and is regarded as fundamental. If Q is needed, it is found from (5).

Ex. 1. As a simple example on the L, R, C circuit, suppose that the condenser has initial charge $\overset{0}{Q}$, and is discharged at $t = 0$ through L and R. Then in (6), $\bar{V} = 0$, $\overset{0}{I} = 0$, and (6) becomes

$$\bar{I} = -\frac{\overset{0}{Q}}{(LCp^2+RCp+1)} = -\frac{\overset{0}{Q}}{LC\{(p+\kappa)^2+n'^2\}}, \tag{7}$$

where

$$\kappa = R/2L \quad \text{and} \quad n'^2 = 1/LC-\kappa^2. \tag{8}$$

It follows from § 18, Theorem 1, and § 18 (13), (8), and (15), respectively, that

$$I = -\frac{\overset{0}{Q}}{n'LC}e^{-\kappa t}\sin n't, \qquad \text{if } n'^2 > 0, \tag{9}$$

$$I = -\frac{\overset{0}{Q}t}{LC}e^{-\kappa t}, \qquad \text{if } n'^2 = 0, \tag{10}$$

$$I = -\frac{\overset{0}{Q}}{kLC}e^{-\kappa t}\sinh kt, \qquad \text{if } n'^2 = -k^2 < 0. \tag{11}$$

The negative sign in these indicates that the current is flowing away from the high-voltage side of the condenser.

More complicated circuits are regarded as being built up of L, R, C, or simpler circuits, and sufficient equations can be written down to determine the Laplace transforms of all the currents in a circuit by adding equations of type (6) round closed circuits. When this is done, as in Kirchhoff's first law, only the transforms of applied voltages appear on the right-hand sides together with terms involving the initial conditions in which each inductance is given its own initial current and each condenser its own initial charge. The statement corresponding to Kirchhoff's second law is that the algebraic sum of the Laplace transforms of the currents flowing towards any junction is zero.

Ex. 2. *Voltage V applied at $t = 0$ to the circuit of Fig. 38(a), the condensers being initially uncharged.*

Choosing currents as shown in Fig. 38(a) (omitting the dashes, i.e. I_1 is the current in OP, etc.), Kirchhoff's second law is satisfied automatically. Then the subsidiary equations corresponding to Kirchhoff's first law for the circuits $PABOP$, $OEFPO$, $USQFEU$, $PTSQP$, respectively, are

$$\tfrac{1}{2}R\bar{I}_1 + \frac{1}{Cp}(\bar{I}_1 + \bar{I}_3) = -\bar{V}, \tag{12}$$

$$r\bar{I} + \frac{1}{Cp}\bar{I}_3 - \tfrac{1}{2}R\bar{I}_1 = 0, \tag{13}$$

$$\frac{1}{2Cp}\bar{I}_2 + R(\bar{I}_3 - \bar{I}) - r\bar{I} = 0, \tag{14}$$

$$\frac{1}{Cp}(\bar{I}_1 + \bar{I}_3) + R(\bar{I}_3 - \bar{I}_2 - \bar{I}) + R(\bar{I}_3 - \bar{I}) + \frac{1}{Cp}\bar{I}_3 = 0. \tag{15}$$

Writing $$\omega_0 = 1/RC, \quad k = r/R, \tag{16}$$

and solving for $\bar{I}$ we get

$$\bar{I} = -\frac{\bar{V}(p^2+\omega_0^2)}{R\{kp^2+2p\omega_0(1+2k)+(2+k)\omega_0^2\}}. \tag{17}$$

Suppose, for example, that

$$V = \sin\omega_0 t, \qquad \bar{V} = \frac{\omega_0}{p^2+\omega_0^2}; \tag{18}$$

then from (17)

$$\bar{I} = -\frac{\omega_0}{R\{kp^2+2p\omega_0(1+2k)+\omega_0^2(2+k)\}}. \tag{19}$$

Therefore $$I = \frac{\omega_0}{kR(\lambda_2-\lambda_1)}\{e^{\lambda_1 t}-e^{\lambda_2 t}\}, \tag{20}$$

where λ_1 and λ_2 are the roots (both real and negative) of

$$kp^2+2p\omega_0(1+2k)+\omega_0^2(2+k) = 0.$$

The current I thus dies away exponentially and has no component of frequency $\omega_0/2\pi$. In the steady-state treatment of § 45 it was found that this frequency was 'stopped' by the circuit: here we have verified this and found the transient caused by switching on the voltage.

It is instructive to compare the treatment of this section with that of § 45. If the bars in (12) to (17) are replaced by dashes, $\bar{V}$ is replaced by E' and p by $i\omega$, we get the equations § 45 (3) to (8). By reversing the process we may pass from § 45 (8) to equation (17) above which gives the complete transient solution. If the condensers are initially charged, there will be other terms on the right-hand sides of (12)–(15) and this simple correspondence no longer holds. There is a general correspondence between the Laplace transformation and the steady-state theory which is developed in works on the subject. As another example we notice that if $\overset{0}{I} = \overset{0}{Q} = 0$, (6) may be written

$$z(p)\bar{I} = \bar{V}, \tag{21}$$

where $$z(p) = Lp+R+\frac{1}{Cp} \tag{22}$$

is called the *generalized impedance* of the L, R, C circuit. The complex impedance, defined in § 44 (6), is just $z(i\omega)$. The rules given in § 44 (13), (14), (15) for writing down complex impedances also apply to generalized impedances.

Finally we remark that it can be shown that the Laplace transformation method always gives the correct result when applied to switching problems such as those of § 48 in which the currents or charges in a circuit redistribute themselves instantaneously at the instant when a switch is opened or closed.

Ex. 3. *The circuit of Fig. 31. At $t = 0$, when steady current E/R_1 is flowing in the primary from a battery of voltage E, and the secondary current is zero, the primary circuit is opened.*

We have $\overset{0}{I}_1 = E/R_1$, $\overset{0}{I}_2 = 0$, and the subsidiary equation for the secondary § 42 (24) gives

$$Mp\bar{I}_1+(L_2 p+R_2)\bar{I}_2 = ME/R_1. \tag{23}$$

Now $I_1 = 0$ for $t > 0$ and so

$$\bar{I}_1 = \int\limits_0^\infty e^{-pt}I_1\,dt = 0.$$

Using this in (23) gives

$$\bar{I}_2 = \frac{ME}{R_1(L_2 p+R_2)},$$

$$I_2 = \frac{ME}{R_1 L_2} e^{-R_2 t/L_2}. \tag{24}$$

The solution (24) shows that the secondary current changes discontinuously from zero to $ME/R_1 L_2$ at the instant the primary circuit is broken.

Ex. 4. *Impulsive voltage E_0 applied at $t = 0$ to an L, R, C circuit with zero initial current and charge.*

The voltage V in (1) is $E_0\,\delta(t)$, and using § 18 (10) the subsidiary equation is

$$\left(Lp+R+\frac{1}{Cp}\right)\bar{I} = E_0. \tag{25}$$

Thus, by (8),

$$\bar{I} = \frac{E_0\{(p+\kappa)-\kappa\}}{L\{(p+\kappa)^2+n'^2\}},$$

$$I = \frac{E_0}{L} e^{-\kappa t}\left\{\cos n't-\frac{\kappa}{n'}\sin n't\right\}. \tag{26}$$

50. Transfer functions and feedback control

In many problems of practical importance we have a linear differential equation with constant coefficients relating an 'output' or 'effect' V_2 to an 'input' or 'cause' V_1. The equation can be written most generally as

$$f(D)V_2 = g(D)V_1, \tag{1}$$

in which case V_2 can depend on the derivatives of V_1 as well as V_1 itself.

Taking Laplace transforms of (1) on the assumption that all initial conditions are zero,

$$f(p)\bar{V}_2 = g(p)\bar{V}_1$$

or

$$\frac{\bar{V}_2}{\bar{V}_1} = \frac{g(p)}{f(p)} = G(p). \tag{2}$$

The function $G(p)$† is called the *transfer function* for the system. It gives us certain information about the relationship between the input and output, irrespective of the precise nature of the input function.

Ex. 1. *The parallel-T circuit of Fig. 38 (a).*

The output current $\bar{I}$ is related to the input voltage $\bar{V}$ by § 49 (17). It follows that the transfer function for the circuit is

$$G(p) = \frac{\bar{I}}{\bar{V}} = \frac{-(p^2+\omega_0^2)}{R\{kp^2+2p\omega_0(1+2k)+(2+k)\omega_0^2\}}.$$

If the input function V_1 is the delta function $\delta(t)$, by § 18 (4) $\bar{V}_1 = 1$, and so by (2) above

$$\bar{V}_2 = G(p).$$

The transfer function can therefore be interpreted physically as the Laplace transform of the system response to a unit impulse input.

The transfer function has three important properties which explain why it is so widely used in linear systems theory and control theory.

Firstly, it is always possible to transform from the Laplace domain to the real-time domain by result (iii) of Ex. 5 at the end of Chapter II. For example, if $\bar{V}_2 = G(p)\bar{V}_1$ then by this

† Textbooks almost invariably use s instead of p as the transfer function variable; we use p only for the sake of consistency with § 18.

result

$$V_2(t) = \int_0^t G(t-\tau)V_1(\tau)\, d\tau. \tag{3}$$

Equation (3) expresses the output in terms of any input function $V_1(t)$.

Secondly, transfer functions of complex systems can easily be written down as the sum or product of the transfer functions of simpler subsystems. For instance, suppose the input V_1 to subsystem 1 with transfer function $G(p)$ produces an output V_2 which is fed directly into subsystem 2 with transfer function $K(p)$. The resulting output V_3 is, by (2),

$$\bar{V}_3 = K(p)\bar{V}_2,$$

but

$$\bar{V}_2 = G(p)\bar{V}_1,$$

therefore

$$\bar{V}_3 = G(p)K(p)\bar{V}_1, \tag{4}$$

and the transfer function of the total system is just the product $G(p)K(p)$.

Thirdly, the stability of a system can be inferred directly from its transfer function. Writing $G(p)$ as the ratio of two polynomials $g(p)/f(p)$, it follows from (1), (2), and the discussion of § 47 that the stability of the system is determined by the roots of the equation

$$f(p) = 0. \tag{5}$$

If the roots of (5) all have negative real parts, the system is stable.

An important example that illustrates the last two properties of the transfer function is the single-loop feedback system of Fig. 44 (a). Here the output V_2 of the system $G(p)$ is fed back into system $H(p)$ which produces a correction signal C that is subtracted from the input V_1, so that the actual input into $G(p)$ is the error signal $E = V_1 - C$.

Now

$$\bar{V}_2 = G(p)\bar{E} = G(p)(\bar{V}_1 - \bar{C}),$$

but

$$\bar{C} = H(p)\bar{V}_2$$

therefore

$$\bar{V}_2 = G(p)\bar{V}_1 - G(p)H(p)\bar{V}_2$$

or

$$\frac{\bar{V}_2}{\bar{V}_1} = \frac{G(p)}{1+H(p)G(p)}. \tag{6}$$

Feedback of this type is often used to stabilize an inherently unstable system. For instance, if $\alpha > 0$, the system with $G(p) = 1/(p-\alpha)$ is inherently unstable. However, if a feedback loop containing, for example,

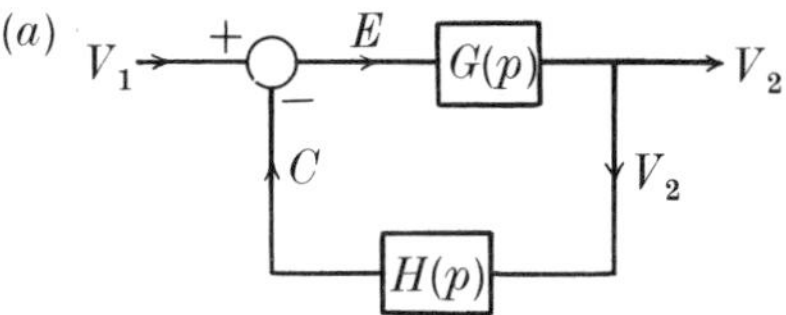

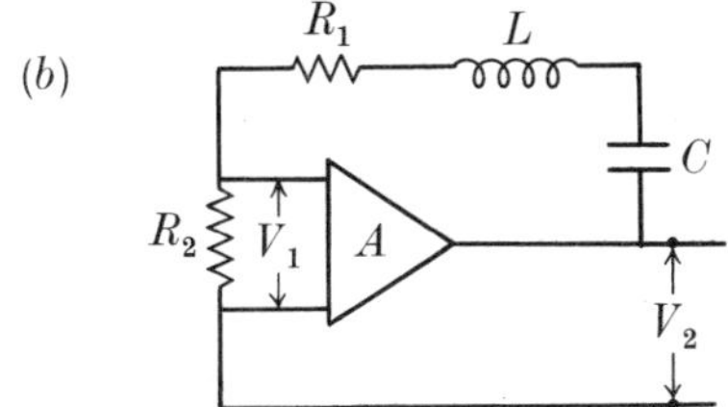

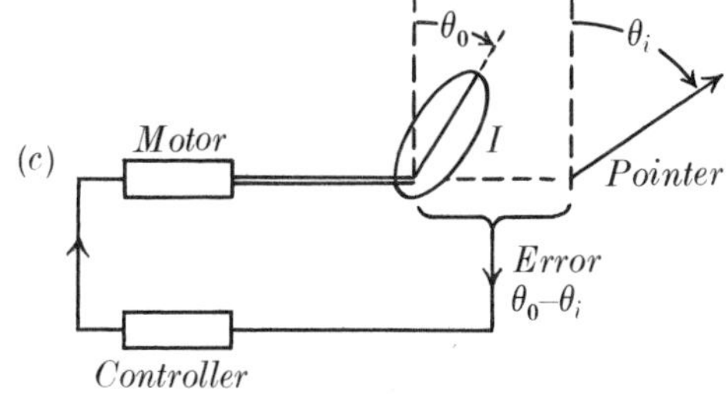

Fig. 44.

an amplifier with $H(p) = k$, is added to the system, by (6)

$$\frac{\bar{V}_2}{\bar{V}_1} = \frac{1/(p-\alpha)}{1+k/(p-\alpha)} = \frac{1}{p-\alpha+k}$$

which is stable for all $k > \alpha$.

Ex. 2. *A feedback amplifier.*

If the amplifier A in Fig. 44 (b) amplifies without phase lag or distortion, its transfer function will be $G(p) = k$, a constant. It can be shown that the transfer function for the feedback loop is

$$H(p) = R_2/\{R_1+R_2+pL+1/pC\}.$$

The transfer function for the feedback amplifier is therefore

$$\bar{V}_2/\bar{V}_1 = G(p)/\{1-H(p)G(p)\}\dagger$$
$$= \frac{k\{1+pC(R_1+R_2+pL)\}}{LCp^2+(R_1+R_2-kR_2)Cp+1}.$$

† There is a minus sign in the denominator because the feedback signal is added to the input, not subtracted as in (6).

By § 47 (5), the condition for stability is

$$R_1+R_2-kR_2 > 0$$

or

$$k < (R_1+R_2)/R_2.$$

51. Servomechanisms

The basic ideas involved may be understood from Fig. 44 (*c*). Suppose a heavy rotor of moment of inertia I is to be turned about an axis so that it always follows the motion of a light pointer. We call θ_i, the angular displacement of the pointer from a fixed direction, the *input displacement*: this is a prescribed function of the time. The angular displacement of the rotor from the same direction, θ_0, is the *output displacement*, and $\theta = \theta_0 - \theta_i$ is the *error*.

In order to make θ_0 follow θ_i, the error θ is fed into a controller which determines the torque T applied to the rotor I by the driving motor. The simplest case is that of 'proportional control' in which T is proportional to the error, so that

$$T = -\lambda(\theta_0 - \theta_i), \tag{1}$$

where λ is a constant. Then, if there is frictional resistance $k\dot{\theta}_0$ to the motion of the rotor, its equation of motion is

$$I\ddot{\theta}_0 = -k\dot{\theta}_0 - \lambda(\theta_0 - \theta_i),$$

or

$$(ID^2 + kD + \lambda)\theta_0 = \lambda\theta_i. \tag{2}$$

The way in which θ_0 follows θ_i may be studied by giving θ_i a simple prescribed motion, calculating θ_0, and comparing the two. Suppose, for example, that θ_i is given a steady rotation

$$\theta_i = \omega t, \tag{3}$$

starting at $t = 0$ when $\theta_0 = D\theta_0 = 0$. The solution of (2) with these initial values is

$$\theta_0 = \omega t - \frac{k\omega}{\lambda} + \frac{k\omega}{\lambda} e^{-\kappa t}\cos n't + \left(\frac{\kappa k\omega}{\lambda n'} - \frac{\omega}{n'}\right)e^{-\kappa t}\sin n't, \tag{4}$$

where $\kappa = k/2I, \qquad n'^2 = \lambda/I - \kappa^2.$

Comparing (3) and (4) it appears that the error $\theta_0-\theta_i$ contains an oscillation which dies away like $\exp(-\kappa t)$, and that for large values of the time θ_0 lags behind θ_i by $k\omega/\lambda$. If k is increased in order to make the oscillation die away more rapidly, the time lag is also increased. Thus this simple system cannot be made really efficient.

In the next stage of complexity terms depending on the successive derivatives and integrals of the error are included in T. Thus

$$T = -\lambda(\theta_0-\theta_i)-\lambda_1\int\limits_0^t(\theta_0-\theta_i)\,dt \tag{5}$$

has a term proportional to the error, together with a term proportional to the integral of the error. This gives the equation of motion

$$(ID^2+kD+\lambda)\theta_0+\lambda_1\int\limits_0^t\theta_0\,dt = \lambda\theta_i+\lambda_1\int\limits_0^t\theta_i\,dt, \tag{6}$$

or, differentiating,

$$(ID^3+kD^2+\lambda D+\lambda_1)\theta_0 = \lambda D\theta_i+\lambda_1\theta_i. \tag{7}$$

If, as before, we consider an input displacement $\theta_i = \omega t$, the particular integral of (7) is just ωt, and so we have eliminated the constant lag inherent in equation (4). It is, however, necessary to ensure the stability of the system described by (7). By § 47 (9) this requires

$$k\lambda > I\lambda_1, \tag{8}$$

which sets limits to the coefficients in (5).

Both the long-term error and the stability conditions for a servomechanism can be obtained directly from its transfer function. The stability conditions follow from § 50 (5), while the long-term error is obtained from the result

$$\lim_{t\to\infty} f(t) = \lim_{p\to 0}\, p\bar{f}(p) \tag{9}$$

if the first limit exists.

For example, the transfer function for the 'proportional control' system of equation (2) is

$$\bar{\theta}_0/\bar{\theta}_i = \lambda/(Ip^2+kp+\lambda).$$

The Laplace transform of the error is

$$\bar{\theta}_0-\bar{\theta}_i = -\bar{\theta}_i(Ip^2+kp)/(Ip^2+kp+\lambda). \tag{10}$$

If $\theta_i = \omega t$, then $\bar{\theta}_i = \omega/p^2$ and by (9) and (10) the long-term error is

$$\lim_{p\to 0} p(\bar{\theta}_0-\bar{\theta}_i) = \lim_{p\to 0} \{-\omega(Ip+k)/(Ip^2+kp+\lambda)\}$$
$$= -k\omega/\lambda$$

as before.

EXAMPLES ON CHAPTER V

1. A condenser of capacitance C is charged to voltage E_0 and discharged at $t = 0$ into a non-inductive resistance R. Show that the charge on the condenser at time t is

$$CE_0\, e^{-t/RC}.$$

2. A battery of voltage E_0 is applied at $t = 0$ to a circuit consisting of inductance L and capacitance C in series. Show that, if the initial charge and current are zero, the current in the circuit at time t is

$$(E_0/nL)\sin nt,$$

where $n = (LC)^{-\frac{1}{2}}$. Show also that if the circuit is opened at the instant π/n when the current changes sign, the condenser will then be charged to voltage $2E_0$.

3. A circuit consists of an inductive resistance L, R, a capacitance C, and a resistance $1/G$ in parallel (this corresponds to the effect of a leaky condenser). Show that its natural frequency of oscillation is

$$\frac{1}{2\pi}\left[\frac{1}{LC}-\frac{1}{4}\left(\frac{R}{L}-\frac{G}{C}\right)^2\right]^{\frac{1}{2}},$$

and that the oscillations have the damping factor $\exp[-(R/2L+G/2C)t]$.

4. A circuit consists of three branches in parallel. Two of them contain equal inductive resistances L, R, and the third a capacitance C. Show that the general solution contains an exponentially decaying term, together with a damped oscillation of frequency $(2n^2-\kappa^2)^{\frac{1}{2}}/2\pi$ and damping factor $\exp(-\kappa t)$, where $n^2 = 1/LC$, $\kappa = R/2L$. What is the mechanical analogue of the circuit?

5. A circuit consists of inductance L and capacitance C in series with a battery of voltage E_0. At $t = 0$, $I = I_0$, and the condenser is uncharged. Show that if I_1 is the current at the time when the voltage drop over the condenser attains the value $V < E_0$, then

$$E_0^2+(I_0/nC)^2 = (E_0-V)^2+(I_1/nC)^2,$$

where $n^2 = 1/LC$. Discuss the behaviour of such a circuit in which the condenser is automatically discharged when the voltage drop over it attains the value V.

6. The primary circuit of a transformer consists of inductance L_1 and capacitance C_1 in series, and the secondary consists of inductance L_2 and capacitance C_2 in series. There is mutual inductance M between L_1 and L_2. Find an equation for the natural frequencies, and show that in the special case $L_1 = L_2$, $C_1 = C_2$, these are

$$n(1+k)^{-\frac{1}{2}}/2\pi \quad \text{and} \quad n(1-k)^{-\frac{1}{2}}/2\pi,$$

where $n^2 = 1/L_1 C_1$ and $k = M/L_1$.

7. The primary of a transformer contains inductance L_1, resistance R_1, and capacitance C_1 in series, and the secondary L_2, R_2, C_2 in series. They are coupled by mutual inductance M. Alternating voltage $E'e^{i\omega t}$ is applied to the primary; show that the steady-state secondary current is

$$(E'/Z_0)e^{i(\omega t+\delta-\pi)},$$

where

$$Z_0 e^{i\delta} = \{R_1 X_2 + R_2 X_1 + i(M^2\omega^2 + R_1 R_2 - X_1 X_2)\}/M\omega,$$

and

$$X_1 = L_1\omega - 1/C_1\omega, \qquad X_2 = L_2\omega - 1/C_2\omega.$$

8. Show that the complex impedance of the circuit of Fig. 29 with $C_1 = C_2 = C_3$, $L_1 = L_2 = 0$, $R_1 = R_2$ is

$$\frac{4R_1 C_1\omega + i(R_1^2 C_1^2\omega^2 - 3)}{2C_1\omega + R_1 C_1^2\omega^2 i}.$$

9. The branches AB and CD of a circuit $ABCDA$ contain resistance R, and the branches BC and DA contain capacitance C. Show that if complex voltage E' is applied to the terminals A and C, the complex voltage drop between D and B is $E'\exp(2\gamma i)$, where $\gamma = \tan^{-1}(1/RC\omega)$. The magnitude of the voltage across D and B is equal to that of the applied voltage, but its phase may be varied at will by changing R or C.

10. Show that if complex voltage E' is applied to a three-stage ($n = 3$) filter circuit, Fig. 39 (a), with z_1 a capacitance C, z_2 an inductance L, and $z' = 0$, the complex output current I_3' is

$$I_3' = 4i\omega^7 E'\{L\omega_0^2(\omega^2-\omega_0^2)(3\omega^2-\omega_0^2)(4\omega^2-\omega_0^2)\}^{-1},$$

where $\omega_0^2 = 1/LC$. Show that the filter cuts off frequencies less than $\frac{1}{2}\omega_0$, i.e. it is a 'high-pass' filter.

11. A telephone line has resistance R_1 between posts and a leakage resistance R_2 at each post. If a battery of voltage E is connected to it half-way between two posts and the line is broken between the nth and $(n+1)$th posts [$z' \to \infty$ in § 45 (16)], show that the battery current is

$$\frac{E\sinh n\theta}{R_2\sinh\theta\cosh n\theta},$$

where $\cosh\theta = 1+(R_1/2R_2)$.

12. Show that $|I_n'|$, the current in the terminal impedance of the filter circuit of Fig. 39 (a), decreases exponentially as n is increased unless ω is such that θ defined in § 45 (14) is pure imaginary, and that the condition for this is that $\cosh\theta$ lie between ± 1.

For the circuit of Fig. 39 (a) in which z_1 consists of inductance L_1 and capacitance C_1 in series, while z_2 consists of inductance L_2, show that this criterion is satisfied if $(L_1 C_1 + 4L_2 C_1)^{-\frac{1}{2}} < \omega < (L_1 C_1)^{-\frac{1}{2}}$. The filter is a 'band-pass' filter and this is its pass band.

13. Regarding the applied stress as input and the resulting deflexion as output, show that the transfer function for the rheological model of § 38 (1) is $1/(kp+\lambda)$. Hence deduce that the deflexion due to an impulsive stress $S_0 \delta(t)$ is $x = S_0 e^{-\lambda t/k}$.

14. Refer to the 'car-following' models of § 8 Ex. 4. If the speed of the first car is regarded as input and the speed of the following car as output, show that the transfer functions for the two models are $k/(p^2+k)$ and $k'/(p+k')$ respectively. Hence deduce that only the second model is stable.

15. Suppose that the torque T applied to the rotor of the servomechanism in § 51 is

$$T = -\lambda(\theta_0 - \theta_i) - \lambda_2(\dot{\theta}_0 - \dot{\theta}_i).$$

Show that the transfer function for the servomechanism is then

$$(\lambda + \lambda_2 p)/\{Ip^2 + (k+\lambda_2)p + \lambda\},$$

and deduce that this is always stable, but does not eliminate the long-term lag of $k\omega/\lambda$ if $\theta_i = \omega t$.

16. In the circuit of Fig. 29 with $C_1 = C_2 = C_3$, $R_1 = R_2$, $L_1 = L_2 = 0$, the condenser C_1 is charged to voltage E_0 and discharged into the system at $t = 0$, the other two condensers being then uncharged. Show that the charge on the condenser C_1 at any time is

$$C_1 E_0\{2 + 3e^{-\alpha t} + e^{-3\alpha t}\}/6,$$

where $\alpha = 1/R_1 C_1$.

17. Voltage $E\sin(\omega t + \beta)$ is applied at $t = 0$ to an inductance L and a resistance R in series, the initial current in the inductance being zero. Show that the current at time t is

$$E\{\sin(\gamma - \beta)e^{-Rt/L} + \sin(\omega t + \beta - \gamma)\}(R^2 + L^2\omega^2)^{-\frac{1}{2}},$$

where $\gamma = \tan^{-1}(L\omega/R)$.

18. An inductive resistance L, R is carrying steady current E_0/R from a battery of voltage E_0, when at $t = 0$ an inductive resistance L_1, R_1 is switched into series with it. Show that the subsequent battery current is

$$\frac{E_0}{R+R_1} - \frac{E_0(RL_1 - LR_1)}{R(L+L_1)(R+R_1)} e^{-t(R+R_1)/(L+L_1)}.$$

VI

VECTORS

52. Vector algebra

In applied mathematics we are concerned with two different types of quantity: firstly, quantities such as temperature, mass, electric charge, etc., which are specified by their magnitude only and are called scalars, and secondly, quantities such as force, velocity, electric and magnetic field strength, etc., with which are associated both magnitude and a direction in space. It is for the handling of these latter that vector analysis has been developed. Its place in mathematical teaching is still a matter of argument; on the one hand it is possible to work exclusively in rectangular Cartesian coordinates or others of similar type—most of the classical English text-books do this and the student must be prepared to follow arguments set out in this way; on the other hand, vector methods do provide a conspicuous simplification in many problems of three-dimensional statics, dynamics, and electromagnetism, though, again, it should be said that they are not sufficiently powerful to handle many of the more complicated problems and that for these still other methods have been developed.

The most common practice, and that which will be adopted here, is to use vector analysis as a tool in subjects for which it is well adapted. Its main virtues from the present point of view are: (i) that it provides a useful physical picture so that it is better to *think* of quantities such as force, angular velocity, electric field, etc., as vectors even if manipulation is done with their components, and (ii) that the vector product arises naturally in fundamental formulae such as § 53 (11) of statics, § 53 (6) of dynamics, and § 60 (23) of electromagnetism, and that by the use of its elementary properties their discussion is greatly simplified.

In this chapter a brief account† of the fundamentals of vector

† For a complete treatment see Milne, *Vectorial Mechanics* (Methuen); Weatherburn, *Elementary Vector Analysis* and *Advanced Vector Analysis* (Bell).

algebra is given, and subsequently vector treatments will be used for problems particularly well adapted to them, while in other cases the vector equivalent of Cartesian formulae will be noted.

A vector **r** is defined as a number r associated with a definite direction in space.† The number r is called the magnitude of the

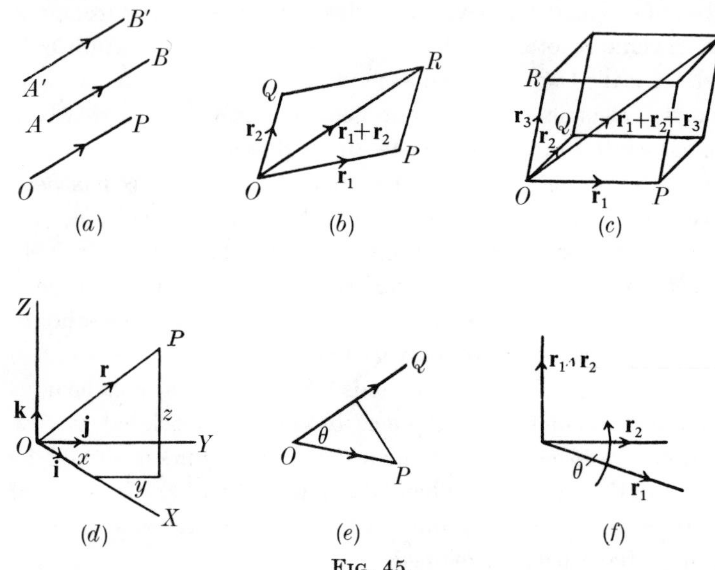

FIG. 45.

vector and is to be positive or zero. A vector may thus be represented by *any* line such as AB in Fig. 45 (a) which has length r and is such that the direction from A to B is parallel to the given direction and is in the same sense. Any other line equal and parallel to AB represents the vector **r** equally well, for example, $A'B'$, and, in particular, just one such line OP can be drawn from a chosen origin O.

(i) *The sum of two vectors*. If OP and OQ, drawn from an origin O, represent the vectors $\mathbf{r}_1$ and $\mathbf{r}_2$, then the sum $\mathbf{r}_1+\mathbf{r}_2$ is defined

† Clarendon type is usually used in print for vectors. In manuscript a bar or a wavy line above or below the letter is convenient; if written below it cannot be confused with the bar used to denote an average or a Laplace transform. $|\mathbf{r}|$ is sometimes written for the magnitude of **r**. Normally the space with which we are dealing is three-dimensional; the results, except those involving vector products, still hold in two dimensions.

as the vector represented by the diagonal OR of the parallelogram $OPRQ$; cf. Fig. 45 (*b*). The sum of three non-coplanar vectors $\mathbf{r}_1, \mathbf{r}_2, \mathbf{r}_3$, represented by OP, OQ, OR in Fig. 45 (*c*), will be represented by the diagonal OS of the parallelepiped with sides OP, OQ, OR.

Algebraic laws such as

$$\mathbf{r}_1+\mathbf{r}_2 = \mathbf{r}_2+\mathbf{r}_1, \tag{1}$$

$$\mathbf{r}_1+(\mathbf{r}_2+\mathbf{r}_3) = (\mathbf{r}_1+\mathbf{r}_2)+\mathbf{r}_3 = \mathbf{r}_1+\mathbf{r}_2+\mathbf{r}_3 \tag{2}$$

follow immediately from the geometry of Figs. 45 (*b*) and (*c*).

$(-\mathbf{r})$ is defined as a vector of the same magnitude r as $\mathbf{r}$, but oppositely directed. The difference $\mathbf{r}_1-\mathbf{r}_2$ is defined as $\mathbf{r}_1+(-\mathbf{r}_2)$.

The product $k\mathbf{r}$ of a number k with a vector $\mathbf{r}$ is defined as a vector whose magnitude is $|k|r$, and which is in the same direction as $\mathbf{r}$ if $k > 0$, and in the opposite direction to $\mathbf{r}$ if $k < 0$.

An immediate consequence of the parallelogram law of addition is that a vector represented by OS, Fig. 45 (*c*), may be expressed as a sum of vectors represented by OP, OQ, OR in three chosen (non-coplanar) directions. These vectors are called the components of the vector in the chosen directions. The most important case is that in which the directions are mutually perpendicular. Suppose that OX, OY, OZ are a right-handed system of rectangular axes, and that $\mathbf{i}$, $\mathbf{j}$, $\mathbf{k}$ are vectors of unit magnitude ('unit vectors') in these directions, Fig. 45 (*d*). Then if OP represents a vector $\mathbf{r}$ of magnitude r, and (x, y, z) are the Cartesian coordinates of P relative to OX, OY, OZ, we have

$$\mathbf{r} = x\mathbf{i}+y\mathbf{j}+z\mathbf{k}. \tag{3}$$

In future we shall always use right-handed rectangular axes, and for shortness will describe x, y, z as the 'components' of the vector $\mathbf{r}$; strictly, of course, the components, as defined above, are $x\mathbf{i}$, $y\mathbf{j}$, $z\mathbf{k}$. The magnitude r of the vector $\mathbf{r}$ is

$$(x^2+y^2+z^2)^{\frac{1}{2}}. \tag{4}$$

If x_1, y_1, z_1 are the components of another vector $\mathbf{r}_1$ relative to the same axes, the components of the sum $\mathbf{r}+\mathbf{r}_1$, are $x+x_1$, $y+y_1$, $z+z_1$.

(ii) *The product of two vectors*. The choice of a definition is rather arbitrary, since any expression which involves the product of the magnitudes of the vectors might be regarded as a product. Two of these expressions are chosen because of their usefulness: they are called the 'scalar' (or 'dot') product, and the 'vector' (or 'cross') product.

If two vectors $\mathbf{r}_1$ and $\mathbf{r}_2$ are represented by OP and OQ, Fig. 45 (*e*), the angle θ such that $0 \leqslant \theta \leqslant \pi$ between OP and OQ is called the angle between the vectors. The *scalar product* $\mathbf{r}_1.\mathbf{r}_2$ of the two vectors is then defined by

$$\mathbf{r}_1.\mathbf{r}_2 = r_1 r_2 \cos\theta. \tag{5}$$

As its name implies, it is a scalar or pure number. It follows immediately that if the two vectors are perpendicular their scalar product is zero. Also that

$$\mathbf{r}_1.\mathbf{r}_2 = \mathbf{r}_2.\mathbf{r}_1, \tag{6}$$

and that
$$\mathbf{r}_1.\mathbf{r}_1 = r_1^2. \tag{7}$$

Finally, it is easy to show geometrically that (cf. Ex. 1)

$$\mathbf{r}_1.(\mathbf{r}_2+\mathbf{r}_3) = \mathbf{r}_1.\mathbf{r}_2+\mathbf{r}_1.\mathbf{r}_3. \tag{8}$$

For the system of mutually perpendicular unit vectors $\mathbf{i}$, $\mathbf{j}$, $\mathbf{k}$ introduced above the scalar products are

$$\mathbf{i}.\mathbf{i} = \mathbf{j}.\mathbf{j} = \mathbf{k}.\mathbf{k} = 1, \tag{9}$$

$$\mathbf{i}.\mathbf{j} = \mathbf{j}.\mathbf{k} = \mathbf{k}.\mathbf{i} = 0. \tag{10}$$

If x_1, y_1, z_1 and x_2, y_2, z_2 are the components of the vectors $\mathbf{r}_1$ and $\mathbf{r}_2$ relative to the axes of Fig. 45 (*d*), we have by (3)

$$\begin{aligned}\mathbf{r}_1.\mathbf{r}_2 &= (x_1\mathbf{i}+y_1\mathbf{j}+z_1\mathbf{k})(x_2\mathbf{i}+y_2\mathbf{j}+z_2\mathbf{k})\\ &= x_1x_2\mathbf{i}.\mathbf{i}+y_1y_2\mathbf{j}.\mathbf{j}+z_1z_2\mathbf{k}.\mathbf{k}+y_1z_2\mathbf{j}.\mathbf{k}+z_1y_2\mathbf{k}.\mathbf{j}+\ldots\\ &= x_1x_2+y_1y_2+z_1z_2,\end{aligned} \tag{11}$$

using (9) and (10).

(iii) *The vector product* $\mathbf{r}_1 \wedge \mathbf{r}_2$ of the vectors $\mathbf{r}_1$ and $\mathbf{r}_2$ is defined as the vector whose direction is perpendicular to the plane of $\mathbf{r}_1$ and $\mathbf{r}_2$ and in the sense of the progression of a right-handed screw rotating from $\mathbf{r}_1$ towards $\mathbf{r}_2$, and whose magnitude is

$$r_1 r_2 \sin\theta, \tag{12}$$

where θ is the angle between the directions of $\mathbf{r}_1$ and $\mathbf{r}_2$ [Fig. 45 (*f*)].

It follows that† $$\mathbf{r}_2 \wedge \mathbf{r}_1 = -\mathbf{r}_1 \wedge \mathbf{r}_2. \tag{13}$$

For the unit vectors of Fig. 45 (*d*) the vector products are

$$\mathbf{i} \wedge \mathbf{i} = \mathbf{j} \wedge \mathbf{j} = \mathbf{k} \wedge \mathbf{k} = 0, \tag{14}$$

$$\mathbf{i} \wedge \mathbf{j} = \mathbf{k}, \qquad \mathbf{j} \wedge \mathbf{k} = \mathbf{i}, \qquad \mathbf{k} \wedge \mathbf{i} = \mathbf{j}. \tag{15}$$

It is easy to show that (cf. Ex. 2)

$$\mathbf{r}_1 \wedge (\mathbf{r}_2 + \mathbf{r}_3) = \mathbf{r}_1 \wedge \mathbf{r}_2 + \mathbf{r}_1 \wedge \mathbf{r}_3. \tag{16}$$

If $\mathbf{r}_1$ and $\mathbf{r}_2$ are vectors of components x_1, y_1, z_1 and x_2, y_2, z_2, we have by (3)

$$\begin{aligned} \mathbf{r}_1 \wedge \mathbf{r}_2 &= (x_1\mathbf{i} + y_1\mathbf{j} + z_1\mathbf{k}) \wedge (x_2\mathbf{i} + y_2\mathbf{j} + z_2\mathbf{k}) \\ &= (y_1 z_2 - z_1 y_2)\mathbf{i} + (z_1 x_2 - x_1 z_2)\mathbf{j} + (x_1 y_2 - y_1 x_2)\mathbf{k}, \end{aligned} \tag{17}$$

on multiplying out and using (14) and (15). This may be written in the form

$$\mathbf{r}_1 \wedge \mathbf{r}_2 = \begin{vmatrix} \mathbf{i} & \mathbf{j} & \mathbf{k} \\ x_1 & y_1 & z_1 \\ x_2 & y_2 & z_2 \end{vmatrix} \tag{18}$$

which is the easiest way of remembering the result.

The vector product of two vectors being a vector, the scalar and vector products of it with a third vector will be of importance. These are called scalar and vector triple products.

Using (11) and (18) it follows that the scalar triple product

$$\mathbf{r}_1 . (\mathbf{r}_2 \wedge \mathbf{r}_3) = \begin{vmatrix} x_1 & y_1 & z_1 \\ x_2 & y_2 & z_2 \\ x_3 & y_3 & z_3 \end{vmatrix} \tag{19}$$

and thus from the properties of the determinant in (19) that

$$\mathbf{r}_1 . (\mathbf{r}_2 \wedge \mathbf{r}_3) = \mathbf{r}_3 . (\mathbf{r}_1 \wedge \mathbf{r}_2) = \mathbf{r}_2 . (\mathbf{r}_3 \wedge \mathbf{r}_1). \tag{20}$$

† That is, the commutative law of multiplication does not hold. In developing any calculus of this sort, it is necessary to prove the laws of elementary algebra as in (1), (2), (6), (8), (16), etc.; this makes the early stages tedious, but it is essential since exceptions such as (13) do occur. The same point arises in connexion with the operator D; cf. § 12.

The vector triple product is by (18)

$$\mathbf{r}_1 \wedge (\mathbf{r}_2 \wedge \mathbf{r}_3) = \begin{vmatrix} \mathbf{i} & \mathbf{j} & \mathbf{k} \\ x_1 & y_1 & z_1 \\ y_2 z_3 - y_3 z_2 & z_2 x_3 - z_3 x_2 & x_2 y_3 - y_2 x_3 \end{vmatrix}$$

$$= \mathbf{i}\{x_2(x_3 x_1 + y_3 y_1 + z_3 z_1) - x_3(x_1 x_2 + y_1 y_2 + z_1 z_2)\} +$$
$$+ \mathbf{j}\{y_2(x_3 x_1 + y_3 y_1 + z_3 z_1) - y_3(x_1 x_2 + y_1 y_2 + z_1 z_2)\} +$$
$$+ \mathbf{k}\{z_2(x_3 x_1 + y_3 y_1 + z_3 z_1) - z_3(x_1 x_2 + y_1 y_2 + z_1 z_2)\}$$
$$= \mathbf{r}_2(\mathbf{r}_3 . \mathbf{r}_1) - \mathbf{r}_3(\mathbf{r}_1 . \mathbf{r}_2). \tag{21}$$

This result is of the greatest importance. The reason for its form may be seen from the following considerations: $\mathbf{r}_2 \wedge \mathbf{r}_3$ is perpendicular to the plane of $\mathbf{r}_2$ and $\mathbf{r}_3$, and since the vector product $\mathbf{r}_1 \wedge (\mathbf{r}_2 \wedge \mathbf{r}_3)$ is perpendicular to $\mathbf{r}_2 \wedge \mathbf{r}_3$ it must lie in the plane of $\mathbf{r}_2$ and $\mathbf{r}_3$, that is, there must be a relation of type

$$\mathbf{r}_1 \wedge (\mathbf{r}_2 \wedge \mathbf{r}_3) = \alpha \mathbf{r}_2 + \beta \mathbf{r}_3,$$

where α and β are scalars. The result† (21) gives the values of α and β.

53. The vector quantities of applied mathematics

The vector algebra developed in § 52 is really the pure mathematics of a type of generalized number. We now have to consider its relation to the mathematics of the physical quantities in which we are interested. The first stage in the development of any branch of applied mathematics consists essentially of establishing the vectorial character of many of the fundamental quantities which occur in it. This process needs considerable care, not only since this character is different for different quantities, but also because it does not follow without proof (and in fact is not always true) that if a quantity can be specified by a vector the combined effect of two such quantities is the same as the effect of the quantity specified by the sum of the vectors. In this section we sketch the foundations of statics and dynamics, very briefly, from this point of view.

Vectors as defined in § 52 only specify a magnitude and a

† Many direct proofs of (21) are available, cf. *Math. Gazette*, **23** (1939), 35; **33** (1949), 125, 212; Milne, loc. cit., § 31.

direction:† some physical quantities can be completely described in this way, for example a translation of a rigid body without rotation (i.e. in such a way that the displacements of all points of the body are equal and parallel); others are specified by a magnitude and a direction *at a point*, and for these a vector is required to specify the magnitude and direction, and another to specify the point.

(i) *Position.* The position of a point P relative to an origin O can be specified by a vector $\mathbf{r}$ of length and direction given by OP. As in Fig. 45 (a), OP is the one line which can be drawn from O which represents the vector $\mathbf{r}$: it will be called a *position vector*.

If the position vector of O relative to another origin O' is $\mathbf{r}'$, then the position vector of P relative to O' is $\mathbf{r}+\mathbf{r}'$.

(ii) *Velocity and acceleration.* To introduce velocity we have to define differentiation of a vector with respect to a scalar quantity, which in our case will be the time t. If the position vector of a moving point P relative to an origin O is $\mathbf{r}$ at time t and $\mathbf{r}+\delta\mathbf{r}$ at time $t+\delta t$, we define the velocity $\mathbf{v}$ or $\dot{\mathbf{r}}$ as

$$\mathbf{v} = \dot{\mathbf{r}} = \frac{d\mathbf{r}}{dt} = \lim_{\delta t\to 0}\frac{\delta\mathbf{r}}{\delta t}. \tag{1}$$

From its definition, cf. Fig. 46 (a), the velocity of the point P can be represented by a vector directed along the tangent to its path. The acceleration $\ddot{\mathbf{r}}$ is defined similarly as $d\mathbf{v}/dt$.

Rules for differentiating sums and products of vectors are proved in the same way as for scalars. Thus,

$$\frac{d}{dt}(\mathbf{r}_1+\mathbf{r}_2) = \dot{\mathbf{r}}_1+\dot{\mathbf{r}}_2, \tag{2}$$

$$\frac{d}{dt}(\mathbf{r}_1.\mathbf{r}_2) = \mathbf{r}_1.\dot{\mathbf{r}}_2+\dot{\mathbf{r}}_1.\mathbf{r}_2, \tag{3}$$

$$\frac{d}{dt}(\mathbf{r}_1\wedge\mathbf{r}_2) = \mathbf{r}_1\wedge\dot{\mathbf{r}}_2+\dot{\mathbf{r}}_1\wedge\mathbf{r}_2. \tag{4}$$

† They are often called *free vectors* to emphasize the fact that they are not restricted to any particular line and to distinguish them from *localized vectors* which have a definite line of action. Thus a couple in statics is a free vector, and a force a localized vector.

It follows from (2) that velocities combine according to the law of vector addition, that is, if $\dot{\mathbf{r}}_1$ is the velocity of a point P relative to an origin O, and $\dot{\mathbf{r}}_2$ is the velocity of O relative to another origin O', then $\dot{\mathbf{r}}_1+\dot{\mathbf{r}}_2$ is the velocity of P relative to O'.

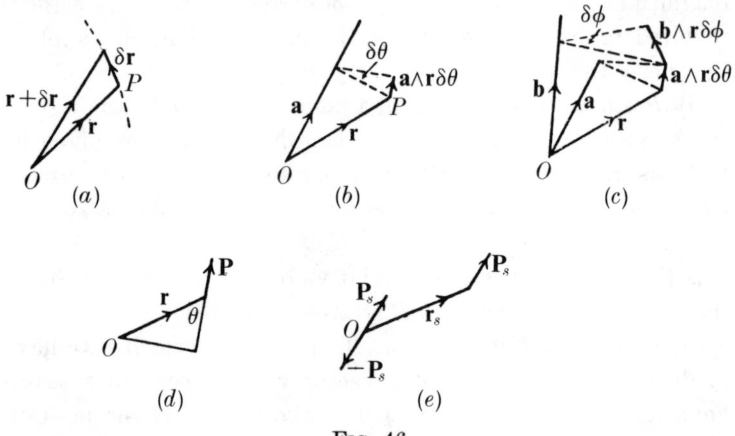

FIG. 46.

(iii) *Rotation.* Suppose a rigid body is rotated through a small angle $\delta\theta$ about an axis through the origin O whose direction is specified by a unit vector $\mathbf{a}$. Then the displacement of the point P whose position vector relative to O is $\mathbf{r}$ is given in magnitude and direction by

$$\mathbf{a}\wedge\mathbf{r}\,\delta\theta \tag{5}$$

[cf. Fig. 46 (*b*)] provided $\delta\theta$ is so small that $\delta\theta^2$ is negligible.

If the body is rotating about the axis, so that $\delta\theta$ is the angle turned through in time δt and

$$\dot{\theta} = \lim_{\delta t\to 0}\frac{\delta\theta}{\delta t},$$

then, taking the limit as $\delta t \to 0$, (5) gives for the velocity $\mathbf{v}$ of the point P

$$\mathbf{v} = \boldsymbol{\omega}\wedge\mathbf{r}, \tag{6}$$

where $\boldsymbol{\omega} = \mathbf{a}\dot{\theta}$. $\boldsymbol{\omega}$ is called the angular velocity of the body. Returning to (5), suppose that following the rotation $\delta\theta$ about $\mathbf{a}$, the whole system, including the axis $\mathbf{a}$, is given another small rotation $\delta\phi$ about an axis through O specified by the unit vector $\mathbf{b}$. The displacement of P due to the combined effect of these two small rotations is given by

$$\mathbf{a}\wedge\mathbf{r}\,\delta\theta+\mathbf{b}\wedge\mathbf{r}\,\delta\phi = (\mathbf{a}\,\delta\theta+\mathbf{b}\,\delta\phi)\wedge\mathbf{r}, \tag{7}$$

cf. Fig. 46 (*c*), and thus is the same as the displacement caused by a single rotation specified by the vector sum† $\mathbf{a}\,\delta\theta+\mathbf{b}\,\delta\phi$. Thus small rotations combine as vectors, and in the same way angular velocities combine as vectors.

The most general displacement of a rigid body can be specified by the displacement of a marked point on it from a fixed origin, together with a rotation of the body about an axis through the marked point (cf. Ex. 10). Thus its motion will be specified by the velocity $\mathbf{v}$ of the marked point and the angular velocity $\boldsymbol{\omega}$ of rotation about it.

(iv) *Force.* A force is specified by its magnitude, direction, and line of action. So far as magnitude and direction are concerned, it may be specified by a vector $\mathbf{P}$.

It is taken as an experimental fact (or, if preferred, as an axiom of statics) that a force may be regarded as applied at any point of its line of action (or that forces $\mathbf{P}$ and $-\mathbf{P}$ applied at any two points of this line annul each other). Thus a force is completely specified by $\mathbf{P}$ and the position vector $\mathbf{r}$ of any point of its line of action. We shall speak of 'a force $\mathbf{P}$ applied at a point $\mathbf{r}$'.

It is a second experimental fact (or an axiom of statics) that concurrent forces combine as vectors, that is, that forces $\mathbf{P}_1$ and $\mathbf{P}_2$ applied at a point O are equivalent to a force $\mathbf{P}_1+\mathbf{P}_2$ applied at O.

The work done by a force $\mathbf{P}$ in a small displacement $\delta\mathbf{r}$ of its point of application is

$$\mathbf{P}.\delta\mathbf{r}. \tag{8}$$

(v) *The equation of motion of a particle.* If the resultant of the

† The need for care is illustrated by the fact that this is not true of *finite* rotations. Such rotations can be *specified* by vectors, but the combined effect of two finite rotations is not equal to that of a single rotation corresponding to the sum of the vectors. In some treatments of vectors the parallelogram law of addition is included in the definition, and before any quantity is regarded as a vector quantity it has to be verified that such quantities combine according to the parallelogram law.

forces on a particle of constant mass m is $\mathbf{P}$, Newton's second law gives the equation of motion

$$\frac{d}{dt}(m\dot{\mathbf{r}}) = m\ddot{\mathbf{r}} = \mathbf{P}. \tag{9}$$

If the mass of the particle is not constant, Newton's law becomes

$$\frac{d}{dt}(m\dot{\mathbf{r}}) = \mathbf{P}, \tag{10}$$

but this is not applicable to all cases; cf. § 63.

(vi) *The moment of a force about a point.* The moment of a force $\mathbf{P}$ about a point O is defined as the vector whose magnitude is P times the perpendicular distance of O from the line of action of $\mathbf{P}$, and whose direction is normal to the plane of O and $\mathbf{P}$ and is related by the right-hand screw law to the direction of turning of $\mathbf{P}$ about O. If $\mathbf{r}$ is the position vector, relative to O, of any point on the line of action of $\mathbf{P}$, the moment is exactly [cf. Fig. 46 (d)]

$$\mathbf{r} \wedge \mathbf{P}. \tag{11}$$

(vii) *The general conditions of statical equilibrium.* Suppose a rigid body is acted on by forces $\mathbf{P}_1$, $\mathbf{P}_2$,..., $\mathbf{P}_n$ applied at the points $\mathbf{r}_1$, $\mathbf{r}_2$,..., $\mathbf{r}_n$, respectively. The general conditions for the body to be in equilibrium under this system of forces are

$$\sum_{s=1}^{n} \mathbf{P}_s = 0, \tag{12}$$

$$\sum_{s=1}^{n} \mathbf{r}_s \wedge \mathbf{P}_s = 0. \tag{13}$$

These may be established in either of two ways. Here the equations of motion of a rigid body, § 66 (8) and (9), will be established without reference to them, and (12) and (13) may be deduced from the condition that the body is to remain at rest under the application of these forces.

Alternatively, the system of forces is reduced in the following way. Corresponding to the force $\mathbf{P}_s$ at $\mathbf{r}_s$ we add two forces $\mathbf{P}_s$ and $-\mathbf{P}_s$ at the origin O: these have no effect on the system [cf. Fig. 46 (e)]. The force $\mathbf{P}_s$ at $\mathbf{r}_s$ and the force $-\mathbf{P}_s$ at O are defined to form a couple of moment $\mathbf{r}_s \wedge \mathbf{P}_s$ and the properties of this studied (cf. Ex. 11, 12). When this has been done for all forces, we are left with concurrent forces $\mathbf{P}_1,\ldots,\ \mathbf{P}_n$ at the origin, and (12) is the condition that the resultant of these vanish, while (13) is the condition that the resultant of the moments of the couples vanish.

(viii) *Linear momentum.* Let m be the mass of a typical particle of a rigid body (or any assemblage of particles), let $\mathbf{r}$ be its position vector relative to an origin and $\dot{\mathbf{r}}$ its velocity. Its linear momentum is defined as $m\dot{\mathbf{r}}$, and, using $\sum$ to denote a summation over all particles of the system, the linear momentum of the system is

$$\sum m\dot{\mathbf{r}}. \tag{14}$$

(ix) *Angular momentum.* If m is a typical particle of any assemblage of particles as in (viii), its linear momentum is $m\dot{\mathbf{r}}$, and the moment of this momentum about the origin O is as in (11)

$$m\mathbf{r}\wedge\dot{\mathbf{r}}.$$

This quantity, summed over all particles of the system, viz.

$$\sum m\mathbf{r}\wedge\dot{\mathbf{r}}, \tag{15}$$

is called the angular momentum of the system about the origin.

(x) *Vector fields.* In a vector field, at each point of space a vector is defined whose magnitude and direction are functions of the position of the point. Electric and magnetic fields are of this type.

(xi) *Vectors in electric circuit theory.* A complex number $z = x+iy = |z|e^{i\theta}$ may be represented by a vector in the (x, y)-plane of magnitude $|z|$ and in a direction inclined at θ to the x-axis. The sum of several complex numbers may be represented by the sum of the corresponding vectors. The product and quotient of two complex numbers z_1 and z_2 may be represented by vectors of magnitudes $|z_1|\times|z_2|$ and $|z_1|/|z_2|$, respectively, in the directions $\theta_1\pm\theta_2$, and so on; in particular, multiplying a complex number by i corresponds to rotating the vector which represents it through 90°.

In this way a representation of the currents and voltages in an electric circuit can be given which has been much used by engineers. Its most important application is to the steady state theory, cf. § 44, where it gives a representation of the complex currents and voltages, I' and E', and the complex impedances z in the various parts of the circuit, which shows the connexions between them very clearly: this may be used either as a diagram for illustrating and proving circuit properties or as a drawing-board method for calculating them numerically. The actual currents and voltages, I and E, are obtained by multiplying I' and E' by $e^{i\omega t}$ and taking the real part: since $e^{i\omega t}$ is represented by a vector of unit length which rotates steadily with angular velocity ω, the connexion between I and E is found by regarding the diagram connecting I' and E' as rotating steadily with angular velocity ω.

EXAMPLES ON CHAPTER VI

1. If OP and OQ represent the vectors $\mathbf{r}_1$ and $\mathbf{r}_2$, show that $\mathbf{r}_1.\mathbf{r}_2$ is equal to the product of the length of OP and the projection of OQ on it.

OPR is any triangle. Show that the sum of the projections of OP and PR on any line through O is equal to the projection of OR on it. Deduce § 52 (8).

2. Let OP, OQ, OR, OS represent the vectors $\mathbf{r}_2$, $\mathbf{r}_3$, $\mathbf{r}_2+\mathbf{r}_3$, and $\mathbf{r}_1$, respectively. Let OP', OQ', OR' be the projections of OP, OQ, and OR on the plane through O perpendicular to OS. Show that the vector products of $\mathbf{r}_1$ with $\mathbf{r}_2$, $\mathbf{r}_3$, and $\mathbf{r}_2+\mathbf{r}_3$ are represented by r_1 times the lines obtained by rotating OP', OQ', and OR' through 90°. Deduce § 52 (16).

3. If OP, OQ, OR represent $\mathbf{r}_1$, $\mathbf{r}_2$, $\mathbf{r}_3$, show that $\mathbf{r}_1.(\mathbf{r}_2\wedge\mathbf{r}_3)$ is equal in magnitude to the volume of a parallelepiped of sides OP, OQ, OR.

4. Derive § 52 (20) by expressing $\mathbf{r}_1$, $\mathbf{r}_2$, and $\mathbf{r}_3$ in terms of their components in the right-handed system of rectangular axes parallel to $\mathbf{r}_2$, $\mathbf{r}_2\wedge\mathbf{r}_3$, and $\mathbf{r}_2\wedge(\mathbf{r}_2\wedge\mathbf{r}_3)$.

5. Show that

(i) $$(\mathbf{r}_1\wedge\mathbf{r}_2).(\mathbf{r}_3\wedge\mathbf{r}_4) = \begin{vmatrix} \mathbf{r}_1.\mathbf{r}_3 & \mathbf{r}_1.\mathbf{r}_4 \\ \mathbf{r}_2.\mathbf{r}_3 & \mathbf{r}_2.\mathbf{r}_4 \end{vmatrix}.$$

(ii) $$(\mathbf{r}_1\wedge\mathbf{r}_2).(\mathbf{r}_1\wedge\mathbf{r}_2) = r_1^2 r_2^2-(\mathbf{r}_1.\mathbf{r}_2)^2.$$

(iii) $$(y_1 z_2-z_1 y_2)^2+(z_1 x_2-x_1 z_2)^2+(x_1 y_2-y_1 x_2)^2 = (x_1^2+y_1^2+z_1^2)(x_2^2+y_2^2+z_2^2)-(x_1 x_2+y_1 y_2+z_1 z_2)^2.$$

6. Show that

$$(\mathbf{a}\wedge\mathbf{b})\wedge(\mathbf{c}\wedge\mathbf{d}) = [\mathbf{a}.(\mathbf{c}\wedge\mathbf{d})]\mathbf{b}-[\mathbf{b}.(\mathbf{c}\wedge\mathbf{d})]\mathbf{a} = [\mathbf{a}.(\mathbf{b}\wedge\mathbf{d})]\mathbf{c}-[\mathbf{a}.(\mathbf{b}\wedge\mathbf{c})]\mathbf{d}.$$

Deduce that any vector $\mathbf{r}$ can be represented in terms of three non-coplanar vectors $\mathbf{a}$, $\mathbf{b}$, $\mathbf{c}$ by the formula

$$\mathbf{r} = \frac{[\mathbf{r}.(\mathbf{b}\wedge\mathbf{c})]\mathbf{a}+[\mathbf{r}.(\mathbf{c}\wedge\mathbf{a})]\mathbf{b}+[\mathbf{r}.(\mathbf{a}\wedge\mathbf{b})]\mathbf{c}}{\mathbf{a}.(\mathbf{b}\wedge\mathbf{c})}.$$

7. Show that the solution of the vector equation

$$\mathbf{a}x+\mathbf{b}y+\mathbf{c}z = \mathbf{d}$$

is
$$x = \frac{\mathbf{d}.(\mathbf{b}\wedge\mathbf{c})}{\mathbf{a}.(\mathbf{b}\wedge\mathbf{c})},$$

etc., unless the denominator vanishes. Deduce the usual rule for solving linear equations using determinants.

8. Show that the equation of a straight line may be written

$$\mathbf{r} = \mathbf{a}+t\mathbf{b},$$

where $\mathbf{r}$ is the position vector of any point on it. Show that the shortest distance between the above line and the line $\mathbf{r} = \mathbf{a}'+t\mathbf{b}'$ is

$$\frac{\mathbf{b}.(\mathbf{b}'\wedge(\mathbf{a}-\mathbf{a}'))}{|\mathbf{b}\wedge\mathbf{b}'|}.$$

9. Show that

(i) The equation of the plane through $\mathbf{r}_1$ with its normal in the direction of a unit vector $\mathbf{n}$ is

$$(\mathbf{r}-\mathbf{r}_1).\mathbf{n} = 0,$$

where $\mathbf{r}$ is the position vector of any point on the plane.

(ii) If the length of the perpendicular from the origin to the plane is p, the equation of the plane is $\mathbf{n}.\mathbf{r} = p$.

(iii) The equation of the plane through three points whose position vectors are $\mathbf{a}$, $\mathbf{b}$, $\mathbf{c}$, is

$$\mathbf{r}.(\mathbf{b}\wedge\mathbf{c}+\mathbf{c}\wedge\mathbf{a}+\mathbf{a}\wedge\mathbf{b}) = \mathbf{a}.(\mathbf{b}\wedge\mathbf{c}).$$

10. The point O of a rigid body is fixed. OA and OB are two marked lines in the body, and OA' and OB' are their positions after the body has been moved in any way about O. Show that the body can be brought from its first position to its second by a rotation about the line of intersection of the plane which is perpendicular to the plane AOA' and bisects the angle AOA' with the plane related in the same way to BOB'.

11. A force $\mathbf{P}$ applied at $\mathbf{r}_1$ and a force $-\mathbf{P}$ at $\mathbf{r}_2$ constitute a couple. The sum of the moments of the forces about O is $(\mathbf{r}_1-\mathbf{r}_2)\wedge\mathbf{P}$, which is independent of the position of O and is called the moment of the couple.

(i) Show that the above couple and a couple consisting of $\mathbf{Q}$ at $\mathbf{r}_1$ and $-\mathbf{Q}$ at $\mathbf{r}_2$, where $\mathbf{Q}$ is such that $(\mathbf{r}_1-\mathbf{r}_2)\wedge\mathbf{Q} = -(\mathbf{r}_1-\mathbf{r}_2)\wedge\mathbf{P}$, are in static equilibrium. [Add forces $-\mathbf{P}-\mathbf{Q}$ at $\mathbf{r}_1$, and $\mathbf{P}+\mathbf{Q}$ at $\mathbf{r}_2$.]

(ii) Show that the above couple and a couple consisting of $-\mathbf{P}$ at $\mathbf{r}_1+\mathbf{r}$ and $\mathbf{P}$ at $\mathbf{r}_2+\mathbf{r}$ are in static equilibrium.

(iii) Deduce that a couple is statically equivalent to any couple of the same moment.

12. Show that any two couples are equivalent to a couple whose moment is the sum of their moments.

13. Show that forces $\mathbf{P}_1$ at $\mathbf{r}_1$,..., $\mathbf{P}_n$ at $\mathbf{r}_n$ are equivalent to a force

$$\mathbf{R} = \sum \mathbf{P}$$

at the origin O, together with a couple of moment

$$\mathbf{G} = \sum \mathbf{r} \wedge \mathbf{P}.$$

Deduce the conditions of statical equilibrium, § 53 (12) and (13).

If the origin is moved from O to the point O' whose position vector relative to O is $\mathbf{s}$, show that the resultant force and couple are $\mathbf{R}$ and $\mathbf{G}-\mathbf{s}\wedge\mathbf{R}$.

14. Show that the system of forces in Ex. 13 is equivalent to a single force if and only if $\mathbf{R}.\mathbf{G} = 0$, and that its line of action relative to the origin O is

$$\frac{\mathbf{R}\wedge\mathbf{G}}{R^2}+t\mathbf{R}.$$

15. Show that if the point O' in Ex. 13 is chosen on the line

$$\frac{\mathbf{R}\wedge\mathbf{G}}{R^2}+t\mathbf{R},$$

the couple and force are parallel and the ratio of their magnitudes is

$$(\mathbf{R}.\mathbf{G})/R^2.$$

16. If forces $\mathbf{P}_1,\ldots, \mathbf{P}_n$ act on a rigid body at the points $\mathbf{r}_1,\ldots, \mathbf{r}_n$, respectively, and the body is given a small displacement (which must be consistent with any constraints on the body) consisting of a translation $\delta\mathbf{a}$ and a rotation $\delta\theta$ about an axis specified by the unit vector $\mathbf{n}$, show that the work done by the forces is

$$\sum_{s=1}^{n} \mathbf{P}_s.\delta\mathbf{a}+\sum_{s=1}^{n} \mathbf{n}.(\mathbf{r}_s\wedge\mathbf{P}_s)\,\delta\theta.$$

Deduce that this work is zero if the forces are in equilibrium, and conversely that if the work is zero for all possible small displacements the forces are in equilibrium.

17. Complex voltage E' of frequency $\omega/2\pi$ is applied to an L, R, C circuit. Draw vector diagrams showing the voltage drops over the inductance, resistance, and capacitance, and their relation to E'.

18. Draw vector diagrams representing the combinations of impedances in Fig. 35 (b), (c), (d).

19. The voltage drop V over portion of a circuit carrying steady state alternating current is the real part of $V'e^{i\omega t}$, and the current I in it is the real part of $I'e^{i\omega t}$. The average power P_{av} in this portion of the circuit is defined as the average of VI over a period. Show that if $V' = V_1+iV_2$, $I' = I_1+iI_2$, then

$$P_{av} = \tfrac{1}{2}(V_1 I_1+V_2 I_2).$$

This provides a meaning for the scalar product of the vectors V' and I'.

VII
PARTICLE DYNAMICS

54. Introductory

In §§ 6, 28 the motion of a particle whose position is specified by a single coordinate, say x, with any laws of force and resistance, was discussed and found to lead to the equation

$$m\ddot{x} = f(x, \dot{x}, t), \tag{1}$$

and frequently to the simpler equation

$$m\ddot{x} = f(x)+g(\dot{x})+h(t). \tag{2}$$

The important special case in which $f(x)$ and $g(\dot{x})$ are proportional to x and $\dot{x}$, respectively, so that (2) becomes linear, has been discussed in the preceding chapters.

We now return to the non-linear equations (1) and (2). There are no general methods for solving these, but two important special cases can be treated in detail, namely those in which the right-hand sides consist of functions of x or $\dot{x}$ only. These cases are considered in §§ 55, 56. Occasionally, exact solutions of more complicated equations can be obtained in the same way; for example, the equation

$$\ddot{x}+\dot{x}^2 f(x)+g(x) = 0 \tag{3}$$

is reduced to a first-order linear equation in v^2 by § 55 (3), but the discussion of such equations has usually been restricted to the study of cases in which the non-linear terms are small. These are treated by the method of successive approximations described in § 58. When the non-linear terms are not small, numerical or graphical integration is often resorted to.

The remainder of the chapter is devoted to a study of the motion of particles in two or more dimensions.

In all cases it is assumed without further statement that the mass of the particles involved is constant. The motion of a particle whose mass varies is considered in § 63.

55. The force a function of position only

In this case the equation of motion § 54 (2) becomes

$$m\ddot{x} = f(x). \tag{1}$$

The methods of this section and the next consist essentially of transforming the second-order equation § 54 (2) into a separable first-order equation in either the velocity $v = \dot{x}$ or in v^2. We have

$$\ddot{x} = \frac{d^2x}{dt^2} = \frac{dv}{dt} \tag{2}$$

$$= \frac{dv}{dx}\frac{dx}{dt} = v\frac{dv}{dx} = \frac{1}{2}\frac{d(v^2)}{dx}. \tag{3}$$

Using the form (3) of $\ddot{x}$ we may write (1) in the form

$$\tfrac{1}{2}m\frac{d(v^2)}{dx} = f(x), \tag{4}$$

and thus v^2 is determined in terms of x by a simple integration

$$\tfrac{1}{2}mv^2 = \int f(x)\,dx + C. \tag{5}$$

The arbitrary constant C in (5) is to be determined from the initial conditions. Alternatively, using these, a definite integral for v^2 may be written down immediately; suppose that, when $t = 0$, $x = a$ and $v = V$, then, integrating (4) with respect to x between the limits a and x, we get

$$\tfrac{1}{2}m[v^2]_a^x = \int_a^x f(x)\,dx,$$

or

$$\tfrac{1}{2}mv^2 - \tfrac{1}{2}mV^2 = \int_a^x f(x)\,dx. \tag{6}$$

A slightly different method of integrating (1) which is often used is as follows: multiply both sides of (1) by $\dot{x}$ which gives

$$m\dot{x}\ddot{x} = \dot{x}f(x), \tag{7}$$

and integrate with respect to the *time* which gives

$$\tfrac{1}{2}m\dot{x}^2 = \int f(x)\frac{dx}{dt}\,dt + C = \int f(x)\,dx + C, \tag{8}$$

as before, since

$$\frac{d}{dt}(\dot{x}^2) = 2\dot{x}\ddot{x}. \tag{9}$$

Equations (5) or (6) give the velocity as a function of position, and are referred to as the *first integral* of the equation of motion. Each of them is the energy equation for the motion and will be discussed from this point of view in § 73. For the present we merely remark that, for motion to be possible, v must be real and thus v^2 given by (6) must be positive.

To complete the solution we have from (6)

$$\frac{dx}{dt} = \pm\left\{V^2+\frac{2}{m}\int\limits_a^x f(x)\,dx\right\}^{\frac{1}{2}}. \tag{10}$$

The sign before the square root is determined by the circumstances of projection, that is, by the sign of dx/dt at the instant $t = 0$ when the particle was set in motion. Finally, integrating (10) between the corresponding limits of 0 to t in t, and a to x in x, gives

$$t = \pm\int\limits_a^x \left\{V^2+\frac{2}{m}\int\limits_a^x f(x)\,dx\right\}^{-\frac{1}{2}} dx. \tag{11}$$

Needless to say, these formulae should not be quoted; the whole process should be gone through in each special case, remembering only the initial step (4) or (7).

Ex. 1. *The simple pendulum.*

A particle of mass m is attached to a light rigid rod of length l, freely hinged at a fixed point O, and moves in a vertical plane. If θ is the inclination of the rod to the vertical, the equation of motion of m is [cf. § 61 (11)]

$$ml\ddot{\theta} = -mg\sin\theta, \tag{12}$$

or

$$\ddot{\theta}+n^2\sin\theta = 0, \tag{13}$$

where

$$n^2 = g/l. \tag{14}$$

For small oscillations $\sin\theta$ in (13) may be replaced by θ, and (13) becomes the linear equation

$$\ddot{\theta}+n^2\theta = 0. \tag{15}$$

Here we do not make this assumption, but solve (13) by the methods described earlier. As in (4) we write (13) in the form

$$\frac{1}{2}\frac{d}{d\theta}(\dot{\theta}^2) = -n^2\sin\theta,$$

so that, integrating,

$$\dot{\theta}^2 = 2n^2\cos\theta+C. \tag{16}$$

Suppose that the pendulum is released from rest at $\theta = \alpha$, that is, when $t = 0$, $\theta = \alpha$, $\dot{\theta} = 0$. Substituting these values in (16) gives $C = -2n^2\cos\alpha$, and thus (16) becomes

$$\dot{\theta}^2 = 2n^2(\cos\theta - \cos\alpha).$$

Therefore
$$\frac{d\theta}{dt} = -n(2\cos\theta - 2\cos\alpha)^{\frac{1}{2}}, \tag{17}$$

where the negative sign has been chosen for the square root in (17) since the particle begins to move backwards. Integrating (17) between the limits 0 and t in t, and α and θ in θ, gives

$$nt = -\int\limits_{\alpha}^{\theta} \frac{d\theta}{(2\cos\theta - 2\cos\alpha)^{\frac{1}{2}}}. \tag{18}$$

This integral cannot be expressed in terms of elementary functions, but, like many that arise in the present context, it is an elliptic integral.

The elliptic integrals $F(k,\phi)$ and $E(k,\phi)$ of the first and second kinds, respectively, are defined by

$$F(k,\phi) = \int\limits_0^{\phi} \frac{d\psi}{(1-k^2\sin^2\psi)^{\frac{1}{2}}} = \int\limits_0^{\sin\phi} \frac{dt}{(1-t^2)^{\frac{1}{2}}(1-k^2t^2)^{\frac{1}{2}}}, \tag{19}$$

$$E(k,\phi) = \int\limits_0^{\phi} (1-k^2\sin^2\psi)^{\frac{1}{2}}\,d\psi = \int\limits_0^{\sin\phi} \left\{\frac{1-k^2t^2}{1-t^2}\right\}^{\frac{1}{2}} dt, \tag{20}$$

and are tabulated functions.† Many integrals can be expressed in terms of them, for example integrals whose integrands are the reciprocals of the square roots of cubics or quartics.

To reduce (18) to one of these forms we write it as

$$2nt = \int\limits_{\theta}^{\alpha} \frac{d\theta}{(\sin^2\frac{1}{2}\alpha - \sin^2\frac{1}{2}\theta)^{\frac{1}{2}}}. \tag{21}$$

In this put
$$\sin\tfrac{1}{2}\theta = \sin\tfrac{1}{2}\alpha\sin\phi,$$

$$\frac{d\theta}{d\phi} = \frac{2\sin\frac{1}{2}\alpha\cos\phi}{\cos\frac{1}{2}\theta} = \frac{2\sin\frac{1}{2}\alpha\cos\phi}{(1-\sin^2\frac{1}{2}\alpha\sin^2\phi)^{\frac{1}{2}}},$$

† Cf. Jahnke, Emde, and Lösch, *Tables of Higher Functions* (McGraw-Hill). They also give many formulae for expressing integrals in terms of F and E.

and (21) becomes

$$nt = \int\limits_{\phi}^{\frac{1}{2}\pi} \frac{d\phi}{(1-\sin^2\tfrac{1}{2}\alpha\sin^2\phi)^{\frac{1}{2}}} \tag{22}$$

$$= F(\sin\tfrac{1}{2}\alpha, \tfrac{1}{2}\pi) - F(\sin\tfrac{1}{2}\alpha, \phi). \tag{23}$$

The period T of the oscillation is

$$T = \frac{4}{n} F(\sin\tfrac{1}{2}\alpha, \tfrac{1}{2}\pi). \tag{24}$$

While (23) and (24) are accurate solutions of the problem in terms of tabulated functions, they are not particularly informative in the sense that they do not show how solutions based on the approximate linear equation (15) go wrong for larger values of the amplitude. This may be seen by expanding the integrand of (22) by the binomial theorem. Suppose we wish to find the effect of the amplitude of the oscillation on the period: by (22) the period T is

$$T = \frac{4}{n}\int\limits_0^{\frac{1}{2}\pi} (1-\sin^2\tfrac{1}{2}\alpha\sin^2\phi)^{-\frac{1}{2}}\, d\phi$$

$$= \frac{4}{n}\int\limits_0^{\frac{1}{2}\pi} (1+\tfrac{1}{2}\sin^2\tfrac{1}{2}\alpha\sin^2\phi+\tfrac{3}{8}\sin^4\tfrac{1}{2}\alpha\sin^4\phi+\ldots)\, d\phi$$

$$= \frac{4}{n}\left(\frac{\pi}{2}+\frac{1}{2}\frac{\pi}{4}\sin^2\tfrac{1}{2}\alpha+\frac{3}{8}\frac{3\pi}{16}\sin^4\tfrac{1}{2}\alpha+\ldots\right). \tag{25}$$

Neglecting fourth and higher powers of α this gives

$$T = \frac{2\pi}{n}\left(1+\frac{\alpha^2}{16}\right), \tag{26}$$

thus the period increases with increasing amplitude.

Ex. 2. *The anharmonic oscillator.*

The linear or harmonic oscillator has equation of motion

$$\ddot{x} = -n^2 x. \tag{27}$$

In many important cases the law of force is not that of (27), but there is a change of restoring force with displacement which may be expressed by adding higher powers of x to the right-hand side of (27). The next power for which the restoring force is independent of the sign of x is the third, giving

$$\ddot{x} = -n^2 x - bx^3. \tag{28}$$

If b is positive in (28) the restoring force increases steadily above the linear value as the displacement increases; this is the case with a non-linear spring whose stiffness increases with increase of displacement. If b is negative the restoring force falls below the linear value: the case $b = -\frac{1}{6}n^2$, corresponding to retaining the first two terms in the expansion of $\sin\theta$, is often used as a second approximation to the equation of motion (13) of a pendulum.

If a quadratic term in x is included as in

$$\ddot{x} = -n^2x - ax^2, \tag{29}$$

or

$$\ddot{x} = -n^2x - ax^2 - bx^3, \tag{30}$$

the restoring force is not symmetrical about $x = 0$.

Exact solutions can easily be written down. Taking the law (30), suppose that when $t = 0$, $\dot{x} = V$, and $x = 0$. Then, as in (6),

$$\tfrac{1}{2}v^2 - \tfrac{1}{2}V^2 = -\tfrac{1}{2}n^2x^2 - \tfrac{1}{3}ax^3 - \tfrac{1}{4}bx^4, \tag{31}$$

$$v = \{V^2 - n^2x^2 - \tfrac{2}{3}ax^3 - \tfrac{1}{2}bx^4\}^{\frac{1}{2}},$$

and

$$t = \int\limits_0^x \{V^2 - n^2x^2 - \tfrac{2}{3}ax^3 - \tfrac{1}{2}bx^4\}^{-\frac{1}{2}}\, dx. \tag{32}$$

As remarked above, the integral (32) can be expressed in terms of the elliptic integral $F(k,\phi)$ defined in (19), but if higher powers of x than the third appear in the equation of motion (30), this cannot be done.

Ex. 3. *A rigid cone of semi-vertical angle α and mass m, moving with velocity V in the direction of its axis, impinges normally on a plastic substance which provides resistance to motion of the cone which may be represented by a uniform pressure F (the 'flow pressure') over the region of contact.*

Let x be the depth of penetration of the apex of the cone at time t after the instant of contact $t = 0$. Then at $t = 0$ we have $x = 0$, $\dot{x} = V$.

The force on an element of area δA of the surface of the cone is $-F\,\delta A$; the resolved part of this in the direction of the axis is $-F\,\delta A \sin\alpha$, which is just $-F$ times the projection of the area δA on the plane $x = 0$. Thus the total force on the cone in the direction of its axis is $-F$ times the area of the impression made by the cone in the plane $x = 0$, that is

$$-F\pi x^2 \tan^2\alpha.$$

Thus the equation of motion of the cone is

$$m\ddot{x} = -\pi F x^2 \tan^2\alpha. \tag{33}$$

Integrating as in (6) and using the initial conditions, the velocity v of the cone is found to be

$$\tfrac{1}{2}mv^2 = \tfrac{1}{2}mV^2 - \tfrac{1}{3}\pi F x^3 \tan^2\alpha. \tag{34}$$

The cone comes to rest when $v = 0$, that is, when the depth of penetration is

$$\left(\frac{3mV^2}{2\pi F \tan^2\alpha}\right)^{\frac{1}{3}}.$$

F can be determined by measuring the size of the impression. The time of penetration to any depth can be expressed as an elliptic integral.

Ex. 4. *Collision of equal spheres.*

Suppose a sphere A of mass m and radius a, moving with velocity V along the x-axis, collides with an equal sphere B at rest with its centre on the x-axis. We choose the origins of time and distance so that the spheres touch at $t = 0$, and the centres of the spheres A and B are at $x = 0$ and $x = 2a$ at that time.

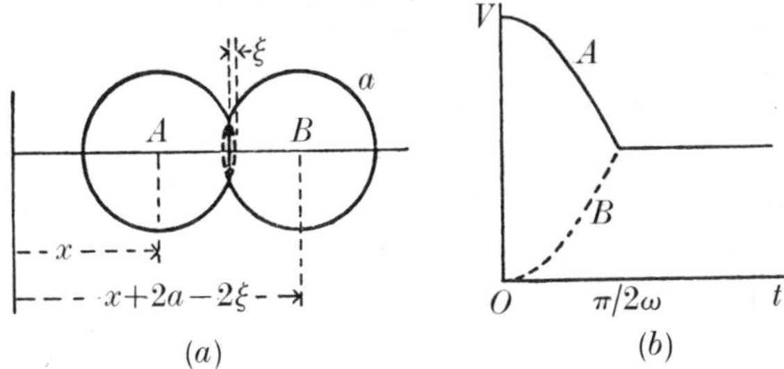

FIG. 47.

In the process of collision a small flat is squashed on either sphere, and, since we have assumed that the spheres are of the same size and material, this squashing will be the same for both and the surface of separation will be plane. Let ξ be the depth of the impression on either sphere at time t, then if x is the position of the centre of the sphere A at time t, that of the sphere B will be $x+2a-2\xi$.

Also, when $t = 0$, $x = 0$, $\xi = 0$, $\dot{x} = V$, and the velocity of the centre of the sphere B is zero, that is

$$\frac{d}{dt}(x+2a-2\xi) = 0,$$

i.e.
$$\dot{\xi} = \tfrac{1}{2}V, \quad \text{when } t = 0. \tag{35}$$

The spheres exert forces on each other across the area of contact: suppose the force is $f(\xi)$, a function of the depth of the impression which will be discussed later. The equations of motion of the spheres A and B are

$$m\ddot{x} = -f(\xi), \tag{36}$$

$$m(\ddot{x}-2\ddot{\xi}) = f(\xi). \tag{37}$$

Adding (36) and (37) gives

$$\ddot{x}-\ddot{\xi} = 0.$$

Integrating, and using $\dot{x} = V$, $\dot{\xi} = \tfrac{1}{2}V$, when $t = 0$, gives

$$\dot{x}-\dot{\xi} = \tfrac{1}{2}V. \tag{38}$$

This result could have been written down by the principle of conservation of momentum.

Subtracting (36) and (37) gives

$$m\ddot{\xi} = -f(\xi), \tag{39}$$

an equation for ξ of the type being studied in this section.

If the spheres are perfectly elastic, $f(\xi)$ can be calculated by the methods of the theory of elasticity, being the force necessary to squash a flat of depth ξ on a sphere, and it is found that $f(\xi) = k\xi^3$, where k depends on the radius of the sphere and its elastic properties.

Here we shall consider the case in which the spheres are perfectly plastic, that is, that they exert a constant pressure P (the flow pressure, characteristic of the material) over the area of contact while the spheres are approaching one another. If r is the radius of contact, so that, for small ξ, $r^2 = 2a\xi$ nearly, we have

$$f(\xi) = \pi r^2 P = 2\pi a\xi P,$$

and (39) becomes

$$\ddot{\xi} + \omega^2\xi = 0, \tag{40}$$

where

$$\omega^2 = 2\pi aP/m. \tag{41}$$

(40) is linear and its solution with $\xi = 0$, $\dot{\xi} = \frac{1}{2}V$, when $t = 0$, is

$$\xi = \frac{V}{2\omega}\sin\omega t. \tag{42}$$

The spheres stop approaching when $\dot{\xi} = 0$, that is, when $t = \pi/2\omega$ and the value of ξ then is $V/2\omega$.

The velocity of the sphere A is by (38) and (42)

$$\dot{x} = \tfrac{1}{2}V + \dot{\xi} = \tfrac{1}{2}V(1+\cos\omega t). \tag{43}$$

The velocity of the sphere B is

$$\frac{d}{dt}(x+2a-2\xi) = \dot{x}-2\dot{\xi} = \tfrac{1}{2}V-\dot{\xi} = \tfrac{1}{2}V(1-\cos\omega t). \tag{44}$$

The velocities of the spheres A and B, given by (43) and (44), are shown in Fig. 47 (b). For $t > \pi/2\omega$ the spheres both move with velocity $\frac{1}{2}V$.

56. Motion with resistance a function of the velocity

In this section we consider motion with differential equation

$$m\ddot{x} = f(\dot{x}). \tag{1}$$

The process of solution is very simple provided the integrals can be evaluated. As in § 55 (2) and (3), we write

$$v = \dot{x}$$

for the velocity. Then

$$\ddot{x} = \frac{dv}{dt}, \tag{2}$$

and

$$\ddot{x} = \frac{dv}{dx}\frac{dx}{dt} = v\frac{dv}{dx}. \tag{3}$$

The method of solution is now as follows:

(i) *Velocity in terms of distance.* Writing $\ddot{x}$ in (1) in the form (3) gives

$$mv\frac{dv}{dx} = f(v), \tag{4}$$

a separable first-order equation for v in terms of x.

(ii) *Velocity in terms of time.* Writing $\ddot{x}$ in (1) in the form (2) gives

$$m\frac{dv}{dt} = f(v), \tag{5}$$

a separable first-order equation for v in terms of t.

(iii) *Position in terms of time.* The simplest way of finding this is usually to eliminate v between the solutions of (4) and (5). Alternatively, the solution of (5) gives dx/dt as a function of t, and another integration gives x as a function of t.

The function $f(\dot{x})$ in (1) may contain a term independent of the velocity, that is, a constant force F. In this case we write $f(\dot{x}) = F-\phi(\dot{x})$, where $\phi(\dot{x})$ is the resistance to the motion. Then (1) becomes

$$m\ddot{x} = F-\phi(\dot{x}). \tag{6}$$

Usually $\phi(\dot{x})$ is an increasing function of $\dot{x}$, so there will be a velocity V at which the resistance to motion is equal to the applied force F. That is

$$\phi(V) = F. \tag{7}$$

When $\dot{x}$ has the value V, $\ddot{x}$ is zero by (6), and thus the particle continues to move with constant velocity V. For this reason V is called the *terminal velocity*; clearly it will appear as a natural parameter in many solutions.

Before proceeding to solve problems it is useful to consider the commonly occurring laws of resistance to motion. Resistance proportional to velocity arises from shearing of ideal viscous fluid and has been studied in Chapter IV. It also occurs for the slow fall of small spheres through viscous fluid. If a is the radius of the sphere and ρ and ν the density and kinematic viscosity of the fluid, Stokes's law states that the resistance to motion of the sphere at velocity v is

$$6\pi a\rho\nu v. \tag{8}$$

If ρ_0 is the density of the sphere, its terminal velocity when falling in the fluid under gravity is by (7) and (8)

$$\frac{2(\rho_0-\rho)ga^2}{9\rho\nu}. \tag{9}$$

The law (8) is of considerable importance since it determines the motion of small raindrops and small particles settling in liquid (sedimentation).

(8) is valid for motion in which the fluid is not turbulent. The degree of turbulence of the fluid is determined by the Reynolds number R, a dimensionless quantity which in the present case is

$$R = \frac{2va}{\nu}. \tag{10}$$

The resistance to motion of the sphere is found experimentally to be given by

$$\tfrac{1}{2}\pi\rho a^2 C(R)v^2, \tag{11}$$

where $C(R)$ is a function of R only. If $R < 10^{-1}$, $C(R) = 24/R$, and (11) reduces to (8). For $10^{-1} < R < 10^3$ there is a transition region; while for $10^3 < R < 10^5$, C is very nearly constant so that the resistance to motion is proportional to the square of the velocity.

The variation of $C(R)$ with R is shown in Fig. 48.

Enough has been said to show that even for the simple case of a falling sphere the variation of resistance with velocity is extremely complicated. For other bodies, projectiles, etc., the resistance is usually specified graphically or as a power series in the velocity.

Since resistance proportional to the square of the velocity has been seen above to have some physical significance and leads to fairly simple results, we consider it in the examples below.

Ex. 1. *A particle of mass m falls from rest under gravity and resistance to motion km times the square of the velocity.*

Taking the origin so that $x = 0$ when $t = 0$, and the x-axis vertically downwards, the equation of motion is

$$m\ddot{x} = mg - mk\dot{x}^2. \tag{12}$$

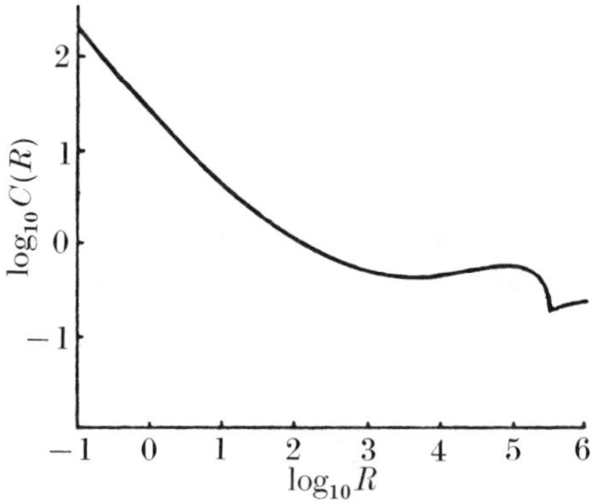

FIG. 48.

The terminal velocity V is given by

$$V^2 = g/k. \tag{13}$$

To find velocity in terms of distance, as in (4) we write (12) in the form

$$v\frac{dv}{dx} = k(V^2 - v^2). \tag{14}$$

Integrating, and using $v = 0$ when $x = 0$, we get

$$\int\limits_0^v \frac{v\,dv}{V^2 - v^2} = kx.$$

That is,
$$-\tfrac{1}{2}\ln\left(\frac{V^2 - v^2}{V^2}\right) = kx.$$

Therefore
$$v^2 = V^2(1 - e^{-2kx}). \tag{15}$$

This shows the way in which $v \to V$ as $x \to \infty$.

Next, to find velocity in terms of time, as in (5) we write (12) in the form

$$\frac{dv}{dt} = k(V^2 - v^2). \tag{16}$$

Since $v = 0$ when $t = 0$ this gives

$$\int\limits_0^v \frac{dv}{V^2 - v^2} = kt.$$

Therefore $$v = V \tanh kVt. \tag{17}$$

To find the position at any time, we eliminate v between (15) and (17). This gives

$$\tanh^2 kVt = 1 - e^{-2kx},$$

or $$x = \frac{1}{k} \ln \cosh kVt. \tag{18}$$

Ex. 2. *A particle is projected vertically upwards with velocity U under gravity and resistance to motion mk times the square of the velocity.*

Choosing the x-axis vertically upwards with the origin at the point of projection, the equation of motion is

$$m\ddot{x} = -mg - mk\dot{x}^2, \tag{19}$$

to be solved with $\dot{x} = U$, $x = 0$, when $t = 0$.

To find velocity as a function of distance we write (19) as

$$v\frac{dv}{dx} = -k(V^2+v^2), \tag{20}$$

where V^2 is defined in (13). The solution of (20) is

$$\int \frac{v\,dv}{V^2+v^2} = -kx+C, \qquad \tfrac{1}{2}\ln(V^2+v^2) = -kx+C.$$

Since $v = U$ when $x = 0$, this gives

$$2kx = \ln\frac{V^2+U^2}{V^2+v^2}. \tag{21}$$

The greatest height attained, which is the value of x when $v = 0$, is

$$\frac{1}{2k}\ln\left(1+\frac{U^2}{V^2}\right). \tag{22}$$

To find velocity as a function of the time we write (19) as

$$\frac{dv}{dt} = -k(V^2+v^2).$$

The solution of this with $v = U$ when $t = 0$ is

$$kVt = \tan^{-1}\frac{U}{V} - \tan^{-1}\frac{v}{V}. \tag{23}$$

After the particle has come to rest at height given by (22), and at time

$$\frac{1}{kV}\tan^{-1}\frac{U}{V}, \tag{24}$$

it commences to fall, and the equation (19) ceases to hold since the resistance to motion now acts upwards.

Ex. 3. *A particle of mass m is blown along the x-axis by a wind of velocity U, the force on the particle being $mk(U-v)^2$, where v is the velocity of the particle. If it starts from rest at $x = 0$ at $t = 0$, find the motion.*

The equation of motion is

$$\frac{dv}{dt} = k(U-v)^2, \tag{25}$$

and its solution with $v = 0$ when $t = 0$ is

$$\frac{1}{U-v}-\frac{1}{U} = kt,$$

or

$$v = \frac{ktU^2}{1+ktU}. \tag{26}$$

To find the position at time t we have from (26)

$$x = \int\limits_0^t \frac{ktU^2\,dt}{1+ktU} = Ut-\frac{1}{k}\ln(1+ktU). \tag{27}$$

57. Non-linear problems in electric circuit theory

In Chapter V the equations of electric circuit theory were discussed on the linear assumptions § 41 (1)–(3). For most purposes these are adequate, and also the general non-linear equations are quite intractable. There is a number of cases in which non-linearity is of great importance, and in which exact solutions of simple special problems can be obtained by methods similar to those of §§ 55, 56. Here we discuss briefly non-linear resistances and iron-cored inductances.

(i) *Non-linear resistance*

Many semi-conductors and electron tubes may be treated as resistances in which the voltage drop V is connected with the current I by the relation

$$V = \phi(I) \tag{1}$$

in place of § 41 (1).

Suppose, for example, that constant voltage E is applied at $t = 0$ to such a resistance in series with an inductance L obeying § 41 (2). Then, as in § 41, the circuit equation is

$$L\frac{dI}{dt}+\phi(I) = E. \tag{2}$$

If the current in the inductance is zero when $t = 0$, the solution of (2) is

$$t = L\int_0^I \frac{dI}{E-\phi(I)}. \tag{3}$$

This can be evaluated for simple forms of $\phi(I)$, the one most used in practice being the power law $I = KV^n$.

(ii) *Iron-cored inductance*

In this case the linear relation § 41 (2) does not hold. It is now most convenient to work with the flux Φ linked with the inductance. The voltage drop V across the inductance is then

$$\frac{d\Phi}{dt} = V. \tag{4}$$

The flux Φ, instead of being LI as in the linear case, is now connected with I by a non-linear relation

$$I = f(\Phi). \tag{5}$$

For example, suppose that constant voltage E is applied at $t = 0$ to an inductance satisfying (4) and (5), in series with a linear resistance R satisfying § 41 (1). Then the circuit relations are

$$\frac{d\Phi}{dt}+Rf(\Phi) = E, \tag{6}$$

and if $I = 0$ when $t = 0$, the solution is

$$t = \int_0^\Phi \frac{d\Phi}{E-Rf(\Phi)}. \tag{7}$$

Clearly other simple problems of types (i) and (ii) can be solved explicitly in the same way. When we come to slightly more general problems, such as the L, R, C circuit, we usually reach second-order non-linear equations which cannot be solved explicitly. For example, the circuit equation for oscillations in a closed L, R, C circuit with linear resistance and iron-cored inductance is

$$\frac{d\Phi}{dt}+RI+\frac{Q}{C} = 0,$$

or

$$\frac{d^2\Phi}{dt^2}+R\frac{dI}{dt}+\frac{1}{C}I = 0, \tag{8}$$

where I and Φ are connected by (5). Taking the simple form

$$I = A\Phi + B\Phi^3 \tag{9}$$

as the approximation next in order of simplicity to the linear one, (8) becomes

$$\frac{d^2\Phi}{dt^2} + R(A+3B\Phi^2)\frac{d\Phi}{dt} + \frac{1}{C}\Phi(A+B\Phi^2) = 0. \tag{10}$$

58. Oscillations of non-linear systems

It was remarked in § 54 that the types of problem discussed in §§ 55–7 are the only important non-linear ones for which exact solutions are obtainable by elementary methods. Because of the difficulty of studying more general non-linear problems, attention has largely been concentrated on what is perhaps the most important problem of this type, namely the effect of small non-linear terms on the oscillations of a system.

Thus we study equations of motion of the type

$$\ddot{x} + n^2x + \epsilon f(x, \dot{x}) = 0, \tag{1}$$

where ϵ is small. If we neglect ϵ altogether, the equation becomes

$$\ddot{x} + n^2x = 0, \tag{2}$$

whose solution,

$$a\sin(nt+\phi), \tag{3}$$

is a harmonic oscillation whose period $2\pi/n$ is independent of the amplitude of the oscillation. We wish to see what effect the small non-linear terms in (1) have on the simple solution (3). It may be remarked that, even if an exact solution could be found, it probably would not be very useful for this purpose (e.g. § 55 (24) is not) so that in any case new methods have to be devised.

First we list some problems of the type envisaged, most of which will be used as examples later:

(i) The motion of a mass m with linear restoring force and resistance to motion proportional to the square of the velocity†

$$\ddot{x} + k\dot{x}|\dot{x}| + n^2x = 0. \tag{4}$$

(ii) The anharmonic oscillator

$$\ddot{x} + n^2x + bx^3 = 0. \tag{5}$$

† Notice that $\dot{x}\,|\dot{x}|$ changes sign with the velocity, so (4) is applicable to both directions of motion whereas § 54 (3) is not.

(iii) The equation § 57 (10) for electrical oscillations in a circuit containing an iron-cored inductance, viz.

$$\ddot{\Phi}+R(A+3B\Phi^2)\dot{\Phi}+\frac{1}{C}\Phi(A+B\Phi^2) = 0. \tag{6}$$

(iv) The non-linear equation § 47 (15) for the tunnel-diode oscillator circuit

$$A\ddot{V}+(P+QV^2)\dot{V}+V+FV^3 = 0. \tag{7}$$

(v) The van der Pol equation

$$\ddot{x}-\epsilon(1-x^2)\dot{x}+n^2x = 0 \tag{8}$$

has been extensively studied in non-linear dynamics. An equivalent equation is Rayleigh's equation

$$\ddot{x}+a\dot{x}+b\dot{x}^3+n^2x = 0. \tag{9}$$

There are various methods of treating (1); we shall first give a sketch of that used by Kryloff and Bogoliuboff† since it leads to a single formula applicable to a wide variety of cases. The other methods of approach will be treated more briefly later.

If we neglect ϵ in (1) we get the solution (3), that is,

$$x = a\sin(nt+\phi), \tag{10}$$

$$\dot{x} = na\cos(nt+\phi), \tag{11}$$

in which a and ϕ are constants. We now attempt to find a solution of (1) in which x and $\dot{x}$ are still given by (10) and (11), but a and ϕ in these are now functions of the time. If x is given by (10) with a and ϕ functions of the time, we have

$$\dot{x} = an\cos(nt+\phi)+\dot{a}\sin(nt+\phi)+a\dot{\phi}\cos(nt+\phi), \tag{12}$$

but since we have assumed that $\dot{x}$ is given by (11), the sum of the last two terms of (12) must be zero, that is

$$\dot{a}\sin(nt+\phi)+a\dot{\phi}\cos(nt+\phi) = 0. \tag{13}$$

Differentiating (11) gives

$$\ddot{x} = -n^2a\sin(nt+\phi)+n\dot{a}\cos(nt+\phi)-na\dot{\phi}\sin(nt+\phi). \tag{14}$$

† *Introduction to Non-Linear Mechanics* (Princeton University Press).

Substituting (10), (11), and (14) in (1), and writing for shortness
$$\psi = nt+\phi, \tag{15}$$
we get
$$n\dot{a}\cos\psi - na\dot{\phi}\sin\psi = -\epsilon f(a\sin\psi, na\cos\psi). \tag{16}$$

Solving (13) and (16) we get

$$\dot{a} = -\frac{\epsilon}{n} f(a\sin\psi, an\cos\psi)\cos\psi, \tag{17}$$

$$\dot{\phi} = \frac{\epsilon}{na} f(a\sin\psi, an\cos\psi)\sin\psi. \tag{18}$$

These equations give the way in which the amplitude a and the phase ϕ of the solution (10) of (1) vary with time. Since both $\dot{a}$ and $\dot{\phi}$ are proportional to the small quantity ϵ, it follows that they are small, that is, that a and ϕ are slowly varying functions of the time. Thus in time $2\pi/n$, $\psi = nt+\phi$ will increase by nearly 2π while a and ϕ will have changed very little. Thus in calculating $\dot{a}$ and $\dot{\phi}$ from (17) and (18) we may, as a first approximation, replace the right-hand sides by their average values over a range 2π in ψ, regarding a as constant when taking the average; that is we take

$$\frac{da}{dt} = -\frac{\epsilon}{2\pi n}\int\limits_0^{2\pi} f(a\sin\psi, an\cos\psi)\cos\psi\, d\psi, \tag{19}$$

$$\frac{d\phi}{dt} = \frac{\epsilon}{2\pi na}\int\limits_0^{2\pi} f(a\sin\psi, an\cos\psi)\sin\psi\, d\psi. \tag{20}$$

It should be emphasized that (17) and (18) are exact; (19) and (20) are simply obtained from them here as approximations which are physically reasonable. For a complete justification of (19) and (20) and higher approximations (the method is to expand the right-hand sides of (17) and (18) in Fourier series (cf. Chap. XI) of which (19) and (20) are the first terms) the reader is referred to Kryloff and Bogoliuboff (loc. cit.).

Ex. 1. *The anharmonic oscillator* (5).

In this case (19) and (20) give

$$\frac{da}{dt} = -\frac{b}{2\pi n}\int\limits_0^{2\pi} a^3\sin^3\psi\cos\psi\,d\psi = 0, \tag{21}$$

$$\frac{d\phi}{dt} = \frac{b}{2\pi na}\int\limits_0^{2\pi} a^3\sin^4\psi\,d\psi = \frac{3ba^2}{8n}. \tag{22}$$

By (21) the amplitude is independent of time, and by (22) the phase increases linearly with time. Thus the solution is

$$x = a\sin nt\left(1+\frac{3ba^2}{8n^2}\right). \tag{23}$$

Thus the period, which is

$$\frac{2\pi}{n}\left(1+\frac{3ba^2}{8n^2}\right)^{-1}, \tag{24}$$

decreases with increasing amplitude.

If we take $b = -\frac{1}{6}n^2$ in (5), we get the approximation to the equation § 55 (13) for the simple pendulum obtained by replacing $\sin\theta$ by $(\theta-\frac{1}{6}\theta^3)$. With this value of b, (24) gives for the period in this case

$$\frac{2\pi}{n}\left(1-\frac{a^2}{16}\right)^{-1} = \frac{2\pi}{n}\left(1+\frac{a^2}{16}\right), \tag{25}$$

neglecting terms in a^4. This agrees with § 55 (26).

Ex. 2. *Equation* (4).

For this equation (19) and (20) give

$$\frac{da}{dt} = -\frac{k}{2\pi n}\int\limits_0^{2\pi} a^2n^2\cos^2\psi|\cos\psi|\,d\psi = -\frac{4ka^2n}{3\pi}, \tag{26}$$

$$\frac{d\phi}{dt} = \frac{k}{2\pi na}\int\limits_0^{2\pi} a^2n^2\cos\psi|\cos\psi|\sin\psi\,d\psi = 0. \tag{27}$$

By (27), ϕ is constant, that is, the period is unaffected by the damping to this approximation. (26) is a differential equation for a, and its solution is

$$\frac{1}{a}-\frac{1}{a_0} = \frac{4knt}{3\pi}, \tag{28}$$

where a_0 is the value of a when $t = 0$. In a half-swing, t increases by π/n, $1/a$ increases by $4k/3$, that is, the amplitude a decreases by $4ka^2/3$, approximately.

Ex. 3. *The van der Pol equation* (8).

We consider the slightly more general equation

$$\ddot{x}-\epsilon(1-\alpha x-\beta x^2)\dot{x}+n^2x = 0, \tag{29}$$

for which (19) and (20) give

$$\frac{da}{dt} = \frac{\epsilon}{2\pi n}\int\limits_0^{2\pi}(1-\alpha a\sin\psi-\beta a^2\sin^2\psi)an\cos^2\psi\,d\psi$$

$$= \tfrac{1}{2}\epsilon a(1-\tfrac{1}{4}\beta a^2), \tag{30}$$

$$\frac{d\phi}{dt} = -\frac{\epsilon}{2\pi na}\int\limits_0^{2\pi}(1-\alpha a\sin\psi-\beta a^2\sin^2\psi)an\cos\psi\sin\psi\,d\psi$$

$$= 0. \tag{31}$$

From (31), ϕ is constant, that is the period is not affected to this approximation. Also, since α does not appear in (30) it does not affect the amplitude to this approximation.

The differential equation (30) for a has solution

$$\int\frac{2da}{a(1-\frac{1}{4}\beta a^2)} = \epsilon t+C,$$

$$\ln\frac{a^2}{1-\frac{1}{4}\beta a^2} = \epsilon t+C,$$

$$\frac{a^2}{1-\frac{1}{4}\beta a^2} = \frac{a_0^2}{1-\frac{1}{4}\beta a_0^2}e^{\epsilon t},$$

where a_0 is the value of a when $t = 0$. Thus, finally,

$$a^2 = \frac{a_0^2 e^{\epsilon t}}{1+\frac{1}{4}\beta a_0^2(e^{\epsilon t}-1)}. \tag{32}$$

(32) shows how the amplitude varies with time. As $t \to \infty$, $a \to 2\beta^{-\frac{1}{2}}$ whatever the initial value of the amplitude, and thus the final oscillation of the system is

$$x = \frac{2}{\beta^{\frac{1}{2}}}\sin(nt+\phi), \tag{33}$$

where ϕ is a constant. It should be observed that the solution tends to (33) *whether the initial amplitude is larger or smaller than* $2\beta^{-\frac{1}{2}}$. The way in which the oscillations build up if a_0 is small is shown in Fig. 49 (*b*).

In the next two examples we illustrate other methods of attack which are often used. They are a little *ad hoc* and have to be used with care as it frequently happens that while they work well in some cases they need modifications in others.

Ex. 4. *Solution in a trigonometric series.*

As an example we consider the anharmonic oscillator with applied force $F\sin\omega t$. The differential equation is

$$\ddot{x}+n^2x+bx^3 = F\sin\omega t. \tag{34}$$

We seek a solution $$x = A\sin\omega t, \tag{35}$$

where A is a constant; substituting this in (34) requires

$$(-\omega^2+n^2)A\sin\omega t+bA^3\sin^3\omega t = F\sin\omega t.$$

Using the result

$$\sin^3\omega t = \tfrac{3}{4}\sin\omega t-\tfrac{1}{4}\sin 3\omega t,$$

this becomes

$$\{(n^2-\omega^2)A+\tfrac{3}{4}bA^3-F\}\sin\omega t-\tfrac{1}{4}bA^3\sin 3\omega t = 0. \tag{36}$$

Clearly this cannot be satisfied exactly for all values of t. But if

$$(n^2-\omega^2)A+\tfrac{3}{4}bA^3 = F, \tag{37}$$

the coefficient of $\sin\omega t$ is zero, so that if we ignore the term in $\sin 3\omega t$ in (36) the amplitude of the solution (35) is given by the solution of the cubic (37). In particular if $\omega = n$, corresponding to resonance if the non-linear term is neglected,

$$A = (4F/3b)^{\frac{1}{3}}, \tag{38}$$

so that the amplitude depends on the cube root of the applied force.

The crude treatment above in which the term $\sin 3\omega t$ in (36) has been neglected may be improved as follows. Instead of (35) we assume a series

$$x = A\sin\omega t+A_2\sin 2\omega t+A_3\sin 3\omega t+..., \tag{39}$$

substitute in (34), and equate the coefficients of the successive trigonometric functions to zero. The first equation so obtained will be (37) which determines A, subsequent equations determine A_2, etc.

The free oscillation corresponding to $F = 0$ in (34) may be studied in the same way, but ω in the substitution (35) must be regarded as an unknown to be determined.

Ex. 5. *The method of iteration.*

We again consider the anharmonic oscillator

$$\ddot{x}+n^2x+bx^3 = 0, \tag{40}$$

where b is small. The first approximation, neglecting the small term, is

$$x = A\sin nt. \tag{41}$$

The general method of iteration consists of substituting the first approximation in the small terms of the original equation and solving again. This process, theoretically, has to be repeated indefinitely, and it has to be proved that the set of results so obtained converges to a definite solution. The difficulty which appears at the outset is that if the effect of the non-linear terms in (40) is to change the period, the assumed first approximation (41) will after a short time cease to be a valid approximation. Thus instead of (41) we can only assume as first approximation

$$x = A\sin mt, \tag{42}$$

where m is unknown but does not differ greatly from n.

Substituting (42) in the small term of (40) gives

$$\begin{aligned}\ddot{x}+n^2x &= -bA^3\sin^3 mt\\ &= -\tfrac{1}{4}bA^3(3\sin mt-\sin 3mt).\end{aligned} \tag{43}$$

We now seek a solution of (43) of the form

$$x = A\sin mt+B\sin 3mt. \tag{44}$$

Substituting (44) in (43) requires

$$\{A(n^2-m^2)+\tfrac{3}{4}bA^3\}\sin mt+\{B(n^2-9m^2)-\tfrac{1}{4}bA^3\}\sin 3mt = 0.$$

That is
$$B = \frac{bA^3}{4(n^2-9m^2)}, \tag{45}$$

$$m^2 = n^2+\tfrac{3}{4}bA^2. \tag{46}$$

(46) shows the way in which the period varies with the amplitude: it agrees with the result (24) found previously.

59. Relaxation oscillations

This is the name given to periodic oscillations of a system in which energy is supplied from outside during part of a period and dissipated within the system in another part of the period, so that the total energy in the system oscillates periodically. Most practical systems in which oscillations are maintained are of this type.

One simple example which has been analysed in detail is the system of § 30, Ex. 4. Here the potential energy stored in the spring increases steadily in the static phase, and this energy is dissipated by friction while slipping occurs. The system has a well-defined period determined largely by the velocity of the moving plane.

An analogous electrical system is shown in Fig. 49 (*a*). This is idealized, but represents in principle the working of many practical circuits. A battery of voltage E is connected to resistance R and capacitance C in series: when the voltage drop across C reaches the value $E_1 < E$, a semiconductor breakdown device D across it suddenly discharges it completely.

The charge Q on the condenser satisfies

$$R\frac{dQ}{dt}+\frac{Q}{C} = E, \tag{1}$$

and if we assume $Q = 0$ when $t = 0$, the solution is

$$Q = CE(1-e^{-t/RC}).$$

The voltage drop across the condenser is E_1 when

$$E - Ee^{-t/RC} = E_1.$$

that is, when

$$t = RC \ln \frac{E}{E-E_1}. \tag{2}$$

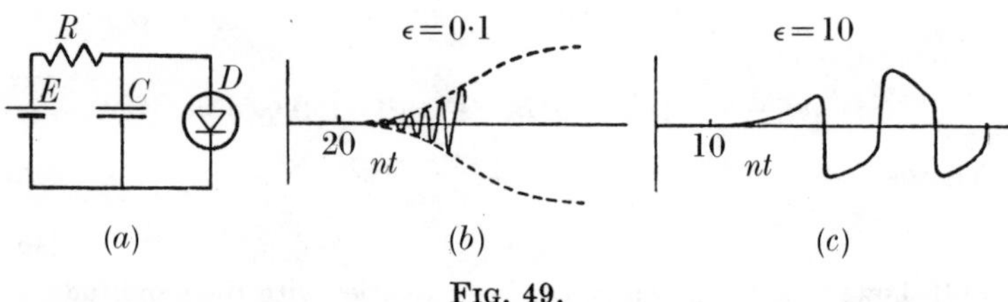

Fig. 49.

The condenser is then discharged, and the process repeats itself with period given by (2).

The most important example of relaxation oscillations is van der Pol's equation

$$\ddot{x} - \epsilon(1-x^2)\dot{x} + n^2x = 0 \tag{3}$$

in which the damping coefficient is negative and the system unstable for small displacements, while for large displacements the damping coefficient becomes positive. Thus small oscillations tend to grow, and large oscillations tend to diminish, and a stable oscillation results. The nature of the solutions of (3) for various values of the parameter ϵ has been exhaustively studied by van der Pol. An approximation to the method of growth and the final steady state for small values of ϵ was found in § 58 and is illustrated in Fig. 49 (*b*). The way in which the oscillations build up for large ϵ is shown in Fig. 49 (*c*). The final wave form is far from sinusoidal—this is a characteristic of such oscillations.

Rayleigh's equation § 58 (9) is equivalent to van der Pol's equation. It represents, for instance, the sound produced by the wind blowing against telephone wires; here the frequency of oscillation is a relaxation effect that is quite different from the natural frequency of the wires.

60. Motion in two or more dimensions

Problems of this type usually are very difficult to handle unless they separate into a number of equations of the types previously discussed. We discuss a number of examples in which this is the case.

Ex. 1. *The simple projectile.*

A particle of mass m is projected at $t = 0$ with velocity u at an angle α to the horizontal. Taking the origin O at the point of projection, and the axes of x and y horizontal and vertical, the equations of motion are

$$m\ddot{x} = 0, \tag{1}$$

$$m\ddot{y} = -mg. \tag{2}$$

Integrating twice and using the initial values $x = y = 0$, $\dot{x} = u\cos\alpha$, $\dot{y} = u\sin\alpha$, we get

$$x = ut\cos\alpha, \tag{3}$$

$$y = ut\sin\alpha - \tfrac{1}{2}gt^2. \tag{4}$$

Eliminating t between (3) and (4) gives the equation of the path

$$y = x\tan\alpha - \frac{gx^2}{2u^2}\sec^2\alpha. \tag{5}$$

Ex. 2. *The projectile of Ex. 1 but with resistance to motion a function $f(v)$ of the velocity.*

Let ψ be the slope of the path at the point (x, y). The resistance $f(v)$ is directed backwards along the tangent to the path, so its components in the x- and y-directions are $-f(v)\cos\psi$ and $-f(v)\sin\psi$. Thus the equations of motion are now

$$m\ddot{x} = -f(v)\cos\psi \tag{6}$$

$$m\ddot{y} = -f(v)\sin\psi - mg, \tag{7}$$

where, if s is the arc measured along the path,

$$v = \dot{s}; \qquad \cos\psi = dx/ds; \qquad \sin\psi = dy/ds. \tag{8}$$

The equations now do not separate into two for x and y except in the case of resistance proportional to velocity in which

$$f(v) = mkv = mk\frac{ds}{dt}$$

which we now consider. In this case (6) and (7) become

$$m\ddot{x} = -mk\frac{ds}{dt}\frac{dx}{ds} = -mk\dot{x} \tag{9}$$

$$m\ddot{y} = -mk\frac{ds}{dt}\frac{dy}{ds}-mg = -mk\dot{y}-mg. \tag{10}$$

Equations (9) and (10) are linear second-order equations. The general solution of (9) is

$$x = A+Be^{-kt};$$

choosing A and B to make $x = 0$, $\dot{x} = u\cos\alpha$, when $t = 0$, we get

$$x = \frac{u\cos\alpha}{k}(1-e^{-kt}). \tag{11}$$

Similarly, from (10) with $y = 0$, $\dot{y} = u\sin\alpha$, when $t = 0$,

$$y = -\frac{gt}{k}+\frac{1}{k}\left(u\sin\alpha+\frac{g}{k}\right)(1-e^{-kt}). \tag{12}$$

Eliminating the time between (11) and (12) gives the equation of the path. From (11)

$$t = -\frac{1}{k}\ln\left(1-\frac{kx}{u\cos\alpha}\right), \tag{13}$$

and substituting (13) and (11) in (12), we get

$$y = \frac{g}{k^2}\ln\left(1-\frac{kx}{u\cos\alpha}\right)+\left(u\sin\alpha+\frac{g}{k}\right)\frac{x}{u\cos\alpha}. \tag{14}$$

Although the linear law of resistance is rather artificial the simple results (11), (12), (14) may be used to illustrate the general effects of resistance to motion. From (11) and (12) it follows that

$$x \to \frac{u\cos\alpha}{k}, \quad y \to -\infty, \quad \text{as } t \to \infty, \tag{15}$$

that is, the curve has a vertical asymptote at this value of x. The range on the horizontal plane, of course, is less than this; it is obtained by putting $y = 0$ in (14) which gives a transcendental equation for x.

Finally, if k is small, we may expand the logarithm in (14) by the logarithmic series. This gives

$$y = \frac{g}{k^2}\left\{-\frac{kx}{u\cos\alpha}-\frac{k^2x^2}{2u^2\cos^2\alpha}-\frac{k^3x^3}{3u^3\cos^3\alpha}-\dots\right\}+ \\ +\left(u\sin\alpha+\frac{g}{k}\right)\frac{x}{u\cos\alpha}$$

$$= x\tan\alpha-\frac{gx^2}{2u^2}\sec^2\alpha-\frac{gkx^3}{3u^3}\sec^3\alpha-\dots. \tag{16}$$

The first two terms of (16) are the path (5) in the absence of resistance. The terms of the series give the change in the path caused by resistance.

Ex. 3. *The motion of a particle of mass m in the xy-plane with restoring force $m\nu^2x$ parallel to the x-axis and mn^2y parallel to the y-axis.*

The equations of motion are

$$\ddot{x}+\nu^2x = 0, \tag{17}$$

$$\ddot{y}+n^2y = 0, \tag{18}$$

corresponding to two simple harmonic motions. With an appropriate choice of the origin of time their general solutions may be written

$$x = a\sin(\nu t+\theta), \tag{19}$$

$$y = b\sin nt. \tag{20}$$

The path of the particle is obtained by eliminating t between (19) and (20). If ν is a simple multiple of n the paths are known as Lissajous figures.

If $\nu = n$ the path is

$$\left(\frac{x}{a}-\frac{y}{b}\cos\theta\right)^2 = \left(1-\frac{y^2}{b^2}\right)\sin^2\theta,$$

or

$$\frac{x^2}{a^2}+\frac{y^2}{b^2}-\frac{2xy}{ab}\cos\theta = \sin^2\theta, \tag{21}$$

an ellipse whose shape and orientation depend on the relative phase θ of the two vibrations.

If $\nu = 2n$ the path is

$$\frac{x}{a} = 2\frac{y}{b}\left(1-\frac{y^2}{b^2}\right)^{\frac{1}{2}}\cos\theta+\left(1-\frac{2y^2}{b^2}\right)\sin\theta. \tag{22}$$

The motion of an electron in electric and magnetic fields.

The force $\mathbf{F}$ on an electron of charge $-e$ in a static electric field $\mathbf{E}$ and a static magnetic field $\mathbf{H}$ is

$$\mathbf{F} = -e\mathbf{E}-\frac{e}{c}\mathbf{v}\wedge\mathbf{H}, \tag{23}$$

where $\mathbf{v}$ is the vector velocity of the electron and c is the velocity of light. Here e and $\mathbf{E}$ are measured in e.s.u. and $\mathbf{H}$ in e.m.u. The force of gravity is usually negligible but may be included if desired.

If the magnetic field is uniform and H is its magnitude, and we choose the z-axis in the direction of this field, the components of $\mathbf{F}$ in the x, y, and z directions become

$$F_x = -eE_x - \frac{eH}{c}\dot{y}, \tag{24}$$

$$F_y = -eE_y + \frac{eH}{c}\dot{x}, \tag{25}$$

$$F_z = -eE_z. \tag{26}$$

Ex. 4. *No electric field. The particle projected from the origin at $t = 0$ with velocity V in a direction which makes an angle θ with the magnetic field.*

Taking the x-axis to be in the plane containing the z-axis and the direction of projection, the initial conditions are

$$x = y = z = 0; \quad \dot{x} = V\sin\theta, \quad \dot{y} = 0, \quad \dot{z} = V\cos\theta, \tag{27}$$

when $t = 0$.

The equations of motion are

$$m\ddot{x} = -\frac{eH}{c}\dot{y}, \tag{28}$$

$$m\ddot{y} = \frac{eH}{c}\dot{x}, \tag{29}$$

$$m\ddot{z} = 0. \tag{30}$$

(30) gives immediately

$$\dot{z} = V\cos\theta, \qquad z = Vt\cos\theta. \tag{31}$$

Writing $$\omega = \frac{eH}{mc}, \qquad u = \dot{x}, \qquad v = \dot{y}, \tag{32}$$

(28) and (29) become

$$Du + \omega v = 0, \tag{33}$$

$$Dv - \omega u = 0. \tag{34}$$

These give $$(D^2+\omega^2)u = 0, \tag{35}$$

the general solution of which is

$$u = A\sin\omega t+B\cos\omega t. \tag{36}$$

Then, by (33)

$$v = -\frac{1}{\omega}Du = -A\cos\omega t+B\sin\omega t. \tag{37}$$

The initial conditions, $u = V\sin\theta$, $v = 0$, when $t = 0$, give $A = 0$, $B = V\sin\theta$, and we get

$$\dot{x} = u = V\sin\theta\cos\omega t, \tag{38}$$

$$\dot{y} = v = V\sin\theta\sin\omega t. \tag{39}$$

It follows that the speed of the electron is

$$(\dot{x}^2+\dot{y}^2+\dot{z}^2)^{\frac{1}{2}} = V. \tag{40}$$

Integrating (38) and (39) and using the initial conditions (27) gives

$$x = \frac{V}{\omega}\sin\theta\sin\omega t, \tag{41}$$

$$y = \frac{V}{\omega}\sin\theta(1-\cos\omega t), \tag{42}$$

and therefore $$x^2+\left(y-\frac{V}{\omega}\sin\theta\right)^2 = \frac{V^2\sin^2\theta}{\omega^2}. \tag{43}$$

The projection of the path on the plane $z = 0$ is thus a circle of radius $(V\sin\theta)/\omega$ and with centre at $(0, (V/\omega)\sin\theta)$. The path of the particle is, by (31), a helix with axis along the direction of the magnetic field described with constant speed V. If $\theta = \frac{1}{2}\pi$, so that the motion is wholly in the plane $z = 0$, the path is a circle of radius (V/ω).

Ex. 5. *No electric field. The particle projected in the xy-plane from the point (a, b) with speed V in a direction inclined at ϕ to the x-axis.*

This is a problem similar to that of Ex. 4, but we shall now solve it by a useful alternative method. The equations of motion are

$$\ddot{x}+\omega\dot{y} = 0, \tag{44}$$

$$\ddot{y}-\omega\dot{x} = 0, \tag{45}$$

where ω is defined in (32). Writing $\zeta = x+iy$ and adding i times (45) to (44) we get

$$\dot{\zeta}-i\omega\zeta = 0. \tag{46}$$

This is a single differential equation for the complex quantity ζ. It has to be solved with the initial values

$$\zeta = a+ib, \qquad \dot{\zeta} = Ve^{i\phi}, \qquad \text{when } t = 0. \tag{47}$$

The general solution of (46) is

$$\zeta = Ae^{i\omega t}+B, \tag{48}$$

where now the arbitrary constants A and B are complex. Using (48) in (47) gives A and B, and we get finally

$$\zeta = a+ib+(V/\omega)[e^{i\omega t}-1]e^{i(\phi-\frac{1}{2}\pi)}. \tag{49}$$

The path of the particle may either be found by writing down the real and imaginary parts of (49) and discussing them as in Ex. 4, or, better, by noticing that for all values of t the points (49) lie on a circle of radius V/ω whose centre is at

$$a+ib-(V/\omega)e^{i(\phi-\frac{1}{2}\pi)}. \tag{50}$$

The speed of the particle is

$$(\dot{x}^2+\dot{y}^2)^{\frac{1}{2}} = |\dot{\zeta}| = V. \tag{51}$$

Another example of the method is given in Ex. 8.

Ex. 6. *The problem of Ex. 4 with, in addition, a constant electric field E parallel to the magnetic field.*

The only change is that (30) is replaced by

$$m\ddot{z} = -eE \tag{52}$$

and thus the electron has constant acceleration in the z direction.

Ex. 7. *Constant electric field E along the x-axis and constant magnetic field H along the z-axis. Zero initial velocity and displacement.*

The equations of motion are now

$$m\ddot{x} = -eE-\frac{eH}{c}\dot{y}, \tag{53}$$

$$m\ddot{y} = \frac{eH}{c}\dot{x}, \tag{54}$$

$$m\ddot{z} = 0. \tag{55}$$

Making the substitution (32) as before, (53) and (54) become

$$Du+\omega v = -\frac{e}{m}E, \tag{56}$$

$$Dv-\omega u = 0. \tag{57}$$

As before we find $$u = A\sin\omega t + B\cos\omega t, \tag{58}$$

and from (56) $$v = -\frac{e}{m\omega}E - A\cos\omega t + B\sin\omega t. \tag{59}$$

The initial conditions $u = v = 0$ when $t = 0$ give

$$B = 0, \qquad A = -eE/m\omega,$$

and we get

$$\dot{x} = u = -\frac{eE}{m\omega}\sin\omega t, \tag{60}$$

$$\dot{y} = v = \frac{eE}{m\omega}(\cos\omega t - 1). \tag{61}$$

Integrating (60) and (61) with $x = y = 0$ when $t = 0$ gives

$$x = \frac{eE}{m\omega^2}(\cos\omega t - 1), \tag{62}$$

$$y = \frac{eE}{m\omega^2}(\sin\omega t - \omega t). \tag{63}$$

The path is therefore a cycloid. The maximum value of x is $2eE/m\omega^2$.

Ex. 8. *An electron of mass m is attracted to a centre of force at the origin by a force λ times its displacement. There is a magnetic field H along the z-axis. Motion in the xy-plane only will be considered.*

The equations of motion are now

$$m\ddot{x} = -\lambda x - \frac{eH}{c}\dot{y}, \tag{64}$$

$$m\ddot{y} = -\lambda y + \frac{eH}{c}\dot{x}. \tag{65}$$

Writing

$$n^2 = \frac{\lambda}{m}, \qquad \omega = \frac{eH}{mc}, \tag{66}$$

$$\zeta = x + iy, \tag{67}$$

and adding i times (65) to (64) gives

$$\ddot{\zeta} - i\omega\dot{\zeta} + n^2\zeta = 0. \tag{68}$$

The general solution of this is

$$\zeta = A\exp\{\tfrac{1}{2}i\omega + i(n^2 + \tfrac{1}{4}\omega^2)^{\frac{1}{2}}\}t + B\exp\{\tfrac{1}{2}i\omega - i(n^2 + \tfrac{1}{4}\omega^2)^{\frac{1}{2}}\}t,$$

where A and B are constants which may be complex. Taking the real part, x has the form

$$x = C\cos[\{(n^2 + \tfrac{1}{4}\omega^2)^{\frac{1}{2}} - \tfrac{1}{2}\omega\}t + D] + E\cos[\{(n^2 + \tfrac{1}{4}\omega^2)^{\frac{1}{2}} + \tfrac{1}{2}\omega\}t + F], \tag{69}$$

where C, D, E, F are real constants.

In the absence of a magnetic field the frequency of the oscillations is $n/2\pi$; the effect of a magnetic field is to introduce in place of this a pair of frequencies which are $(n \pm \frac{1}{2}\omega)/2\pi$, provided ω^2 can be neglected. Problems of this type occur in the classical theory of the Zeeman effect.

Ex. 9. *The motion of a charged particle in the field of a magnetic dipole.*† In this case the field is a function of position, and equation (23) gives

$$m\ddot{x} = -(\dot{y}H_z - \dot{z}H_y)e/c, \tag{70}$$

$$m\ddot{y} = -(\dot{z}H_x - \dot{x}H_z)e/c, \tag{71}$$

$$m\ddot{z} = -(\dot{x}H_y - \dot{y}H_x)e/c. \tag{72}$$

It follows that $\dot{x}\ddot{x}+\dot{y}\ddot{y}+\dot{z}\ddot{z} = 0$, and therefore

$$\dot{x}^2+\dot{y}^2+\dot{z}^2 = V^2, \tag{73}$$

where V^2 is a constant. That is, the speed of a charged particle in any pure magnetic field is constant. If we take the dipole to be of moment μ and to be situated at the origin with its axis along the z-axis, its magnetic field‡ at (x, y, z) has components

$$H_x = 3\mu xzr^{-5}, \qquad H_y = 3\mu yzr^{-5}, \qquad H_z = \mu(3z^2-r^2)r^{-5}, \tag{74}$$

where

$$r^2 = x^2+y^2+z^2. \tag{75}$$

Putting (74) in (70) and (71) we get

$$\ddot{x} = k\{\dot{y}(3z^2-r^2)-3\dot{z}yz\}r^{-5}, \tag{76}$$

$$\ddot{y} = k\{3\dot{z}xz-\dot{x}(3z^2-r^2)\}r^{-5}, \tag{77}$$

where $k = -\mu e/mc$.

Multiplying (77) by x and (76) by y and subtracting gives

$$x\ddot{y}-y\ddot{x} = k\{3z\dot{z}(x^2+y^2)-(3z^2-r^2)(x\dot{x}+y\dot{y})\}r^{-5}. \tag{78}$$

Using the value (75) of r^2 this may be written

$$\frac{d}{dt}(x\dot{y}-y\dot{x})+k\frac{d}{dt}\left\{\frac{x^2+y^2}{(x^2+y^2+z^2)^{\frac{3}{2}}}\right\} = 0. \tag{79}$$

Integrating, using the notation (75) and writing $R^2 = x^2+y^2$, we get

$$x\dot{y}-y\dot{x} = C-\frac{kR^2}{r^3}, \tag{80}$$

where C is a constant of integration. The quantity on the left is proportional to the angular momentum of the particle about the z-axis. If V is the constant speed of the particle and θ is the angle its direction at any point makes with the plane through the particle and the z-axis, the left-hand side of (80) is $VR\sin\theta$, and (80) becomes finally

$$VR\sin\theta = C-\frac{kR^2}{r^3}. \tag{81}$$

This first integral has been extensively studied by Störmer in connexion with the theory of aurorae. Using the fact that $|\sin\theta| < 1$, he found that charged particles incident from space can only reach the earth's surface in certain regions.

† This problem is given as an example of rather complicated manipulation in Cartesians. It is an interesting exercise to study it vectorially.

‡ The field due to a magnetic dipole is calculated in exactly the same way as that for an electric dipole; cf. § 72 (23).

61. Motion on a fixed plane curve

When a particle moves on a fixed curve the forces on the particle in the directions of the tangent and normal to the curve are usually known, and to write down equations of motion we need expressions for the accelerations in these directions.

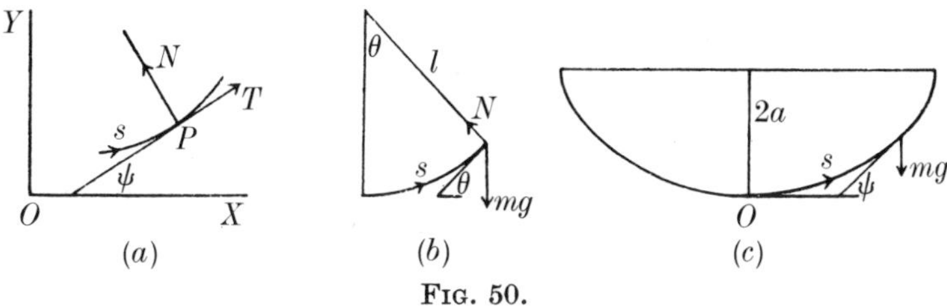

FIG. 50.

Suppose the intrinsic equation of the curve is $s = f(\psi)$, where s is the arc measured along the curve from a fixed point to the point P, and ψ is the angle the tangent at P makes with a fixed direction OX, which we shall take as one of the axes OX, OY of a rectangular coordinate system. The radius of curvature ρ of the curve at P is given by

$$\rho = \frac{ds}{d\psi}. \tag{1}$$

The particle moves along the curve with speed $\dot{s} = ds/dt$. Its components of velocity parallel to OX and OY are

$$\dot{x} = \dot{s}\cos\psi, \tag{2}$$

$$\dot{y} = \dot{s}\sin\psi. \tag{3}$$

Differentiating these we get for its components of acceleration in the directions of OX and OY

$$\ddot{x} = \ddot{s}\cos\psi - \dot{s}\dot{\psi}\sin\psi = \ddot{s}\cos\psi - (\dot{s}^2/\rho)\sin\psi, \tag{4}$$

$$\ddot{y} = \ddot{s}\sin\psi + \dot{s}\dot{\psi}\cos\psi = \ddot{s}\sin\psi + (\dot{s}^2/\rho)\cos\psi, \tag{5}$$

where in (4) and (5) we have used the result

$$\dot{\psi} = \frac{d\psi}{ds}\frac{ds}{dt} = \frac{\dot{s}}{\rho}. \tag{6}$$

The *tangential acceleration* of the particle is by (4) and (5)

$$\ddot{x}\cos\psi + \ddot{y}\sin\psi = \ddot{s}. \tag{7}$$

The *normal acceleration* (in the direction of the inward normal) is

$$-\ddot{x}\sin\psi+\ddot{y}\cos\psi = \dot{s}^2/\rho. \tag{8}$$

If the forces on the particle in the directions of the tangent and inward normal to the curve are T and N, respectively, its equations of motion are

$$m\ddot{s} = T, \tag{9}$$

$$\frac{m\dot{s}^2}{\rho} = N. \tag{10}$$

Ex. 1. *Motion on a smooth vertical circle of radius l: the simple pendulum.*

Measuring s from the lowest point of the circle (Fig. 50 (*b*)) and θ from the downward vertical we have $s = l\theta$, $\psi = \theta$.

The forces on the particle are N, the normal reaction of the circle, and mg, the force of gravity. Thus (9) and (10) give

$$ml\ddot{\theta} = -mg\sin\theta, \tag{11}$$

$$ml\dot{\theta}^2 = N-mg\cos\theta. \tag{12}$$

The integration of (11) has been discussed in § 55, and when $\dot{\theta}^2$ is known, (12) gives N. If the particle is constrained to move in the circle by being attached to O by a light rigid rod freely hinged at O, or if the particle is a bead sliding on a smooth circular wire, N may have either sign. On the other hand, if the particle is connected to O by a flexible string or slides on the inside of a smooth circular cylinder, N must be positive: if at any stage N becomes zero, the particle will leave the circle and the equations (11) and (12) no longer hold.

Ex. 2. *Motion on a rough vertical circle of radius a.*

Suppose the particle to be moving in the direction of increasing θ. Then if the coefficient of friction is μ, frictional force μN acts tangentially in the direction of decreasing θ. The equations of motion (9) and (10) now become

$$ma\ddot{\theta} = -\mu N-mg\sin\theta, \tag{13}$$

$$ma\dot{\theta}^2 = N-mg\cos\theta. \tag{14}$$

Eliminating N these give

$$a\ddot{\theta}+\mu a\dot{\theta}^2 = -g\sin\theta-\mu g\cos\theta. \tag{15}$$

Putting
$$\dot{\theta}^2 = u, \qquad \ddot{\theta} = \frac{1}{2}\frac{du}{d\theta},$$

(15) becomes a linear first-order equation for u, namely

$$\frac{du}{d\theta}+2\mu u = -\frac{2g}{a}(\sin\theta+\mu\cos\theta). \tag{16}$$

Ex. 3. *Motion on a smooth vertical cycloid with its vertex downwards.*

Measuring s from the vertex O (Fig. 50(c)), the intrinsic equation of the cycloid is

$$s = 4a\sin\psi, \tag{17}$$

where $2a$ is the distance from the vertex to the line of cusps.

The equation of motion (9) gives

$$m\ddot{s} = -mg\sin\psi, \tag{18}$$

or, using (17),

$$\ddot{s}+\frac{g}{4a}s = 0. \tag{19}$$

The solution of this is

$$s = A\sin\{t(g/4a)^{\frac{1}{2}}+B\},$$

an oscillation whose period $2\pi(4a/g)^{\frac{1}{2}}$ is independent of its amplitude.

62. Central forces

If the force on a particle consists of attraction or repulsion from a fixed point O, it is convenient to work in plane polar coordinates with this point as origin.

Let (x,y) be the rectangular Cartesian coordinates of the particle P, and let (r,θ) be its polar coordinates with O as origin and θ measured from OX. Then

$$x = r\cos\theta, \tag{1}$$

$$y = r\sin\theta, \tag{2}$$

$$\dot{x} = \dot{r}\cos\theta-r\dot{\theta}\sin\theta, \tag{3}$$

$$\dot{y} = \dot{r}\sin\theta+r\dot{\theta}\cos\theta, \tag{4}$$

$$\ddot{x} = \ddot{r}\cos\theta-2\dot{r}\dot{\theta}\sin\theta-r\dot{\theta}^2\cos\theta-r\ddot{\theta}\sin\theta, \tag{5}$$

$$\ddot{y} = \ddot{r}\sin\theta+2\dot{r}\dot{\theta}\cos\theta-r\dot{\theta}^2\sin\theta+r\ddot{\theta}\cos\theta. \tag{6}$$

The *radial velocity* of P, that is, its component of velocity along OP, is

$$\dot{x}\cos\theta+\dot{y}\sin\theta = \dot{r}, \tag{7}$$

by (3) and (4).

The *transverse velocity* of P, that is, its component of velocity perpendicular to OP in the direction of θ increasing, is

$$\dot{y}\cos\theta-\dot{x}\sin\theta = r\dot{\theta}. \tag{8}$$

The *radial acceleration* of P is

$$\ddot{x}\cos\theta+\ddot{y}\sin\theta = \ddot{r}-r\dot{\theta}^2. \tag{9}$$

The *transverse acceleration* of P is

$$\ddot{y}\cos\theta-\ddot{x}\sin\theta = 2\dot{r}\dot{\theta}+r\ddot{\theta} = \frac{1}{r}\frac{d}{dt}(r^2\dot{\theta}). \tag{10}$$

The problem to be considered is that of the motion of a particle of mass m attracted to a centre of force O by force $mf(r)$ and under no other forces. Let (r,θ) be polar coordinates of the particle in the plane containing O and the direction of projection of the particle: since there are no forces perpendicular to this plane the particle will always remain in it.

The equations of motion are, by (9) and (10),

$$\ddot{r}-r\dot{\theta}^2 = -f(r), \tag{11}$$

$$\frac{1}{r}\frac{d}{dt}(r^2\dot{\theta}) = 0. \tag{12}$$

(12) gives immediately $$r^2\dot{\theta} = h, \tag{13}$$

where h is a constant. Since the transverse velocity of the particle is $r\dot{\theta}$, $mr^2\dot{\theta} = mh$ is the constant angular momentum of the particle about the centre of force.

By using (13) we can eliminate the time from (11) and get the differential equation of the orbit. Instead of working in terms of the radius r it is more convenient to use its reciprocal

$$u = \frac{1}{r}. \tag{14}$$

Then

$$\dot{r} = \frac{d}{dt}\left(\frac{1}{u}\right) = -\frac{1}{u^2}\frac{du}{dt} = -\frac{1}{u^2}\frac{du}{d\theta}\frac{d\theta}{dt} = -h\frac{du}{d\theta}, \tag{15}$$

using (13). Also

$$\ddot{r} = -h\frac{d}{dt}\left(\frac{du}{d\theta}\right) = -h\frac{d^2u}{d\theta^2}\frac{d\theta}{dt} = -h^2u^2\frac{d^2u}{d\theta^2}, \tag{16}$$

again using (13). Substituting (16) and (13) in (11) gives

$$-h^2u^2\frac{d^2u}{d\theta^2}-h^2u^3 = -f(1/u),$$

or $$\frac{d^2u}{d\theta^2}+u = \frac{1}{h^2u^2}f\left(\frac{1}{u}\right). \tag{17}$$

This is the differential equation of the orbit: it may be solved by the methods of § 55.

The general problem is: given the circumstances of projection, find the orbit. Normally, at the instant of projection, $t = 0$, we are given the distance a, the speed V, and the angle β that the direction of motion makes with the outward radius vector. Thus the constant h of (13) is

$$h = Va\sin\beta. \tag{18}$$

The angle ϕ between the tangent and radius vector of the curve at any point is given by

$$\cot\phi = \frac{1}{r}\frac{dr}{d\theta} = u\frac{d}{d\theta}\left(\frac{1}{u}\right) = -\frac{1}{u}\frac{du}{d\theta}. \tag{19}$$

When $t = 0$ we have

$$u = 1/a,\ \phi = \beta,\ \text{and so}\ \frac{du}{d\theta} = -\frac{1}{a}\cot\beta. \tag{20}$$

Thus the initial conditions required for the solution of (17) are known in terms of V, a, and β.

It follows from (19) that when $\phi = 90^\circ$, that is, the particle is moving perpendicular to the radius vector, $du/d\theta = 0$. Points at which this occurs are called apses.

Finally the speed v at any point of the orbit is, by (7) and (8),

$$v^2 = \dot{r}^2 + r^2\dot{\theta}^2 = h^2\left(\frac{du}{d\theta}\right)^2 + h^2u^2, \tag{21}$$

using (15) and (13).

We now consider the solution of (17). First we remark that if the law of force is the inverse square, or the inverse cube, or a combination of the two, namely

$$f(r) = \frac{\mu}{r^2} + \frac{\lambda}{r^3},$$

so that

$$f\left(\frac{1}{u}\right) = \mu u^2 + \lambda u^3,$$

(17) is linear and its general solution can be written down immediately.

For other laws of force, as in § 55 (7), we multiply (17) by $h^2\,du/d\theta$. This gives

$$h^2\frac{d^2u}{d\theta^2}\frac{du}{d\theta} + h^2u\frac{du}{d\theta} = \frac{1}{u^2}f\left(\frac{1}{u}\right)\frac{du}{d\theta}.$$

Integrating gives

$$\tfrac{1}{2}h^2\left(\frac{du}{d\theta}\right)^2+\tfrac{1}{2}h^2u^2 = \int \frac{du}{u^2}f\left(\frac{1}{u}\right)+C$$

$$= -\int f(r)\,dr+C. \tag{22}$$

By (21) this is $$\tfrac{1}{2}v^2+\int f(r)\,dr = C, \tag{23}$$

which is the energy equation, § 73.

In the above we have studied the motion of a particle attracted to a fixed point. The case of practical importance, which is that of two particles of masses m_1 and m_2 attracted to one another by force $f(r)$, can be reduced to this (cf. § 76, Ex. 2). The centre of mass of the two particles moves with constant velocity, and the motion of m_2 relative to m_1 is the same as if m_1 were fixed and the attractive force were $\{(m_1+m_2)/m_1\}f(r)$.

Ex. 1. *The particle is projected from distance a with velocity V at an angle β to the radius vector. The law of force is the attractive inverse square $f(r) = \mu/r^2$.*

The differential equation (17) is

$$\frac{d^2u}{d\theta^2}+u = \frac{\mu}{h^2}, \tag{24}$$

where, by (18), $$h = Va\sin\beta. \tag{25}$$

If we measure θ from the radius vector to the point of projection, it follows from (20) that (24) has to be solved with

$$u = \frac{1}{a}, \qquad \frac{du}{d\theta} = -\frac{1}{a}\cot\beta, \quad \text{when } \theta = 0. \tag{26}$$

The general solution of (24) is

$$u = \frac{\mu}{h^2}\{1+e\cos(\theta+\alpha)\}, \tag{27}$$

where e and α are unknown constants to be determined from (26).

(27) may be written

$$\frac{l}{r} = 1+e\cos(\theta+\alpha), \tag{28}$$

where $$l = h^2/\mu = (Va\sin\beta)^2/\mu. \tag{29}$$

Now the polar equation of a conic of semi-latus rectum l and eccentricity e, referred to a focus as origin and with θ measured from its axis, is

$$\frac{l}{r} = 1+e\cos\theta. \tag{30}$$

Thus the orbit (28) is a conic of semi-latus rectum (29), eccentricity e, and with its axis inclined at α to the radius vector to the point of projection. To find e and α we have from (26) and (28)

$$\frac{l}{a} = 1+e\cos\alpha, \tag{31}$$

$$-\frac{l}{a}\cot\beta = -e\sin\alpha. \tag{32}$$

From (31) and (32)
$$\tan\alpha = \frac{l\cot\beta}{l-a}, \tag{33}$$

$$\begin{aligned} e^2 &= \left(\frac{l}{a}-1\right)^2+\frac{l^2}{a^2}\cot^2\beta \\ &= \frac{l^2}{a^2}\operatorname{cosec}^2\beta-\frac{2l}{a}+1 \\ &= 1-\frac{2V^2a\sin^2\beta}{\mu}+\frac{V^4a^2\sin^2\beta}{\mu^2}. \end{aligned} \tag{34}$$

Now the conic (30) is an ellipse, parabola, or hyperbola according as $e < = > 1$. Thus from (34) the orbit is an ellipse, parabola, or hyperbola according as

$$V^2 < = > \frac{2\mu}{a}. \tag{35}$$

Ex. 2. *A particle is projected with velocity V at a very great distance from an inverse square centre of repulsive force, $f(r) = -\mu/r^2$, in a direction whose perpendicular distance from the centre of force is p. It is required to find the angle ψ between the initial and final directions of its path.*

The equation (17) is

$$\frac{d^2u}{d\theta^2}+u = -\frac{\mu}{h^2}, \tag{36}$$

and its general solution is

$$u = A\cos\theta+B\sin\theta-\mu/h^2. \tag{37}$$

From the circumstances of projection we know that $h = pV$. Also at the point of projection r is very large, u and θ are very small, and

$$p = r\sin\theta = r\theta, \tag{38}$$

very nearly. Since θ is small at the point of projection, we may replace $\sin\theta$ by θ, and $\cos\theta$ by 1, in (37), so that, using (38), (37) becomes

$$\frac{\theta}{p} = A+B\theta-\frac{\mu}{h^2}. \tag{39}$$

It follows that $A = \mu/h^2$, $B = 1/p$, and the equation of the path is

$$u = \frac{\mu}{h^2}(\cos\theta-1)+\frac{1}{p}\sin\theta. \tag{40}$$

From (40)
$$\frac{du}{d\theta} = -\frac{\mu}{h^2}\sin\theta+\frac{1}{p}\cos\theta. \tag{41}$$

Therefore $du/d\theta = 0$ when

$$\tan\theta = \frac{h^2}{\mu p} = \frac{pV^2}{\mu}.$$

By symmetry the required angle ψ is twice this angle, that is

$$\psi = 2\tan^{-1}\frac{pV^2}{\mu}. \tag{42}$$

This result leads to Rutherford's scattering formula. Suppose that there are N scattering centres per unit area of a plane normal to the initial direction of the particle. Then the chance of the initial direction lying between distances p and $p+\delta p$ from one of these is

$$2\pi Np\,\delta p. \tag{43}$$

If $\phi = \pi-\psi$ is the angle of deflexion of the particle, (43) is the chance of the particle being deflected through an angle between ϕ and $\phi+\delta\phi$, and using (42) in the form

$$p = \frac{\mu}{V^2}\cot\tfrac{1}{2}\phi,$$

it becomes

$$\frac{\pi\mu^2 N}{V^4}\cot\tfrac{1}{2}\phi\,\mathrm{cosec}^2\,\tfrac{1}{2}\phi\,\delta\phi. \tag{44}$$

Ex. 3. *Stability of a circular orbit.*

Writing, for convenience, $f(1/u) = \phi(u)$ in (17), this becomes

$$\frac{d^2u}{d\theta^2}+u = \frac{1}{h^2u^2}\phi(u). \tag{45}$$

A circular orbit with $u = b$, $d^2u/d\theta^2 = 0$, is possible under any law of force provided the velocity in the orbit is chosen so that

$$h^2 = \frac{1}{b^3}\phi(b). \tag{46}$$

We have to consider whether such an orbit is stable, that is, if the motion is disturbed slightly, whether the particle will oscillate about the circular orbit or will diverge from it. Suppose the particle is disturbed by a small radial impulse so that h will remain unchanged. In the equation (45) for the orbit put

$$u = b+x, \tag{47}$$

where x is supposed small, and (45) becomes

$$\frac{d^2x}{d\theta^2}+b+x = \frac{1}{h^2(b+x)^2}\phi(b+x) \tag{48}$$

$$= \frac{1}{b^2h^2}\left(1-\frac{2x}{b}+...\right)\{\phi(b)+x\phi'(b)+...\}$$

$$= \frac{\phi(b)}{b^2h^2}+\frac{x}{b^2h^2}\left(\phi'(b)-\frac{2}{b}\phi(b)\right), \tag{49}$$

where to get (49) we have expanded the right-hand side of (48), using Taylor's theorem and retaining only the terms in x. Using (46) this becomes

$$\frac{d^2x}{d\theta^2}+x\left\{3-\frac{b\phi'(b)}{\phi(b)}\right\} = 0. \tag{50}$$

If the coefficient of x in (50) is positive the solution consists of trigonometric terms and the particle oscillates about the circular orbit, that is, the motion is stable. If this coefficient is negative, x contains an exponentially increasing term and the motion is unstable. Thus the condition for stability is

$$3 > \frac{b\phi'(b)}{\phi(b)}. \tag{51}$$

For example, for the inverse nth power attraction $f(r) = \mu r^{-n}$, $\phi(b) = \mu b^n$, (51) requires $n < 3$. Thus circular motion with the inverse nth power law is unstable if $n > 3$.

63. Motion of a particle whose mass varies

If the mass m of a particle is not constant, Newton's law of motion must be used in the form § 53 (10), namely,

$$\frac{d}{dt}(mv) = X, \tag{1}$$

where v is the component of the velocity of the particle in the direction of the x-axis and X is the component of the force on it in this direction.

Integrating (1) with respect to the time from t to $t+\delta t$ gives

$$[mv]_t^{t+\delta t} = X\,\delta t, \tag{2}$$

that is, $X\,\delta t$ is the increase in time δt of the momentum of the particle in the direction of the x-axis.

In many problems of varying mass, for example the motion of a raindrop which grows by absorbing smaller drops with which it collides, or the motion of a rocket, the mass gained or lost has momentum itself, and (1) must be generalized to account for this. Only the case of motion in one dimension will be considered here.

Suppose that at time t the mass is m and its velocity v, and suppose that in time t to $t+\delta t$ it gains mass δm which moves with velocity V. The gain in momentum in the time t to $t+\delta t$ is thus

$$(m+\delta m)(v+\delta v)-mv-V\,\delta m = (v-V)\,\delta m+m\,\delta v.$$

By (2), if force X acts on the particle the gain of momentum must be $X\,\delta t$, and so

$$(v-V)\,\delta m+m\,\delta v = X\,\delta t,$$

or, in the limit as $\delta t \to 0$

$$m\frac{dv}{dt}+(v-V)\frac{dm}{dt} = X, \tag{3}$$

or

$$\frac{d}{dt}(mv)-V\frac{dm}{dt} = X. \tag{4}$$

(4) reduces to (1) only if $V = 0$.

In the case of a rocket the velocity of efflux U of the burnt gases relative to the rocket is known, and in (4) we have $V = v-U$. Thus the equations of motion of a rocket under no forces are

$$\frac{d}{dt}(mv)-(v-U)\frac{dm}{dt} = 0,$$

or

$$m\frac{dv}{dt} = -U\frac{dm}{dt}. \tag{5}$$

If U is constant the solution of this is

$$\frac{v}{U} = -\ln m+C,$$

so that, if the rocket starts from rest with initial mass M,

$$v = U\ln\frac{M}{m}. \tag{6}$$

If m_0 is the mass of fuel carried, so that $m = M-m_0$ when all the fuel has been used, the maximum velocity attained is

$$U\ln\frac{M}{M-m_0}. \tag{7}$$

If the rocket is projected vertically upwards under gravity, (4) becomes

$$m\frac{dv}{dt}+U\frac{dm}{dt} = -mg. \tag{8}$$

This may be written

$$U\frac{dm}{dt} = -m\frac{d(v+gt)}{dt}.$$

The solution of this with $m = M$, $v = 0$, when $t = 0$, is

$$v = U \ln \frac{M}{m} - gt, \tag{9}$$

assuming, as before, that U is constant.

64. Moving axes

Hitherto the position of the particle being studied has always been referred to axes fixed in space. But it is often desirable to

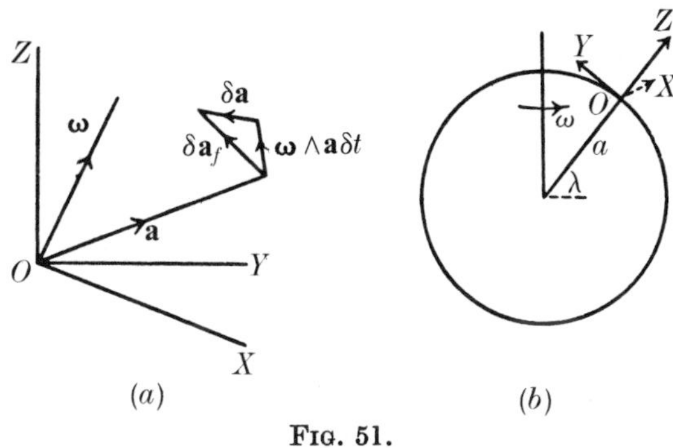

Fig. 51.

use a set of axes which are moving. For example, in problems on motion relative to the earth it is natural to use axes fixed at the point of observation, such as the north and east directions, and these axes, being fixed to the earth, move and rotate with it.

When we come to write down equations of motion, this has to be done relative to fixed axes, and we choose fixed axes along the instantaneous positions of the moving axes at the instant under consideration.

Consider first the case of rotating axes with a fixed origin O. For definiteness we shall take the rotating axes to be right-handed rectangular axes OX, OY, OZ, and we suppose that this system is rotating with angular velocity $\boldsymbol{\omega}$ relative to a set of fixed axes along the instantaneous directions of OX, OY, OZ.

Suppose that $\mathbf{a}$ is a vector specified relative to the moving system OX, OY, OZ, and that we require the value of its time-rate of change relative to the fixed system.

The change $(\delta\mathbf{a})_f$ in the vector $\mathbf{a}$ relative to the fixed axes in an element of time δt will be made up of two parts; (i) the change $\delta\mathbf{a}$ relative to the moving system OX, OY, OZ, and, (ii) the change $\boldsymbol{\omega}\wedge\mathbf{a}\,\delta t$ due to the motion of the system OX, OY, OZ carrying the vector $\mathbf{a}$ with it; cf. Fig. 51 (*a*). That is,

$$(\delta\mathbf{a})_f = \delta\mathbf{a}+\boldsymbol{\omega}\wedge\mathbf{a}\,\delta t. \tag{1}$$

Writing
$$\dot{\mathbf{a}} = \lim_{\delta t\to 0}\frac{\delta\mathbf{a}}{\delta t} \tag{2}$$

for the rate of change of $\mathbf{a}$ relative to the moving system, and taking the limit of (1) as $\delta t \to 0$ we get for the rate of change of $\mathbf{a}$ relative to the fixed system

$$\left(\frac{d\mathbf{a}}{dt}\right)_f = \dot{\mathbf{a}}+\boldsymbol{\omega}\wedge\mathbf{a}. \tag{3}$$

If (a_1, a_2, a_3) and $(\omega_1, \omega_2, \omega_3)$ are the components of $\mathbf{a}$ and $\boldsymbol{\omega}$, the components of (3) are

$$(\dot{a}_1+\omega_2 a_3-\omega_3 a_2,\ \dot{a}_2+\omega_3 a_1-\omega_1 a_3,\ \dot{a}_3+\omega_1 a_2-\omega_2 a_1). \tag{4}$$

We shall use these results to calculate velocities and accelerations: they take into account the effect of rotation of the axes about the origin O. If, in addition, the origin O is in motion, its velocity and acceleration must be added to the values calculated from (3) to get results relative to an origin and axes at rest.

Ex. 1. *Velocities and accelerations in plane polar coordinates.*

Let the polar coordinates of a point be (r,θ): choose OX along the direction to the point from the origin, OY perpendicular to this in the coordinate plane, and OZ perpendicular to the plane. Then the components of the angular velocity of the moving axes are $(0,0,\dot{\theta})$. The point we are interested in is $(r,0,0)$.

By (4) the rate of change of its position is

$$(\dot{r}, r\dot{\theta}, 0), \tag{5}$$

and this is its velocity.

To find its acceleration we need the rate of change of the vector (5), and by (4) this is
$$(\ddot{r}-r\dot{\theta}^2,\ r\ddot{\theta}+2\dot{r}\dot{\theta},\ 0), \tag{6}$$

as in § 62 (9) and (10). Velocities and accelerations in other coordinate systems may be found in the same way.

Ex. 2. *Môtion relative to the earth.*

Consider a point on the earth's surface in latitude λ. Choose axes OZ inclined at $\frac{1}{2}\pi-\lambda$ to the earth's axis; OX easterly, perpendicular to the plane of OZ and the axis; and OY to make a right-handed system (Fig. 51 (*b*)). The earth rotates with angular velocity ω about its axis, and so the components of the angular velocity of the system OX, OY, OZ about their instantaneous positions are

$$(0,\ \omega\cos\lambda,\ \omega\sin\lambda). \tag{7}$$

If the position of a particle is (x,y,z), its components of velocity relative to O along the instantaneous directions of OX, OY, OZ are, by (4),

$$(\dot{x}+\omega z\cos\lambda-\omega y\sin\lambda,\ \dot{y}+\omega x\sin\lambda,\ \dot{z}-\omega x\cos\lambda). \tag{8}$$

To find the acceleration of the particle relative to the origin O we need the components of the rate of change of the vector (8) along the instantaneous directions of the axes. By (4) these are

$$\ddot{x}-2\omega\dot{y}\sin\lambda+2\omega\dot{z}\cos\lambda-\omega^2x, \tag{9}$$

$$\ddot{y}+2\omega\dot{x}\sin\lambda+\omega^2z\sin\lambda\cos\lambda-\omega^2y\sin^2\lambda, \tag{10}$$

$$\ddot{z}-2\omega\dot{x}\cos\lambda-\omega^2z\cos^2\lambda+\omega^2y\sin\lambda\cos\lambda. \tag{11}$$

In (9) to (11) the terms in ω^2x, ω^2y, and ω^2z are negligible.

If the perpendicular from the origin O to the earth's axis is of length p, the origin has acceleration ω^2p towards the axis. Therefore components of acceleration

$$(0,\ \omega^2p\sin\lambda,\ -\omega^2p\cos\lambda) \tag{12}$$

have to be added to (9)–(11) to get accelerations relative to fixed axes.

Since the earth is spheroidal, gravity will act in the plane YOZ: suppose its direction is inclined at a (small) angle α to OZ. Then the equations of motion of a particle of mass m under gravity are (neglecting the small terms in ω^2x, etc.)

$$\ddot{x}-2\omega\dot{y}\sin\lambda+2\omega\dot{z}\cos\lambda = 0, \tag{13}$$

$$\ddot{y}+2\omega\dot{x}\sin\lambda = -g\sin\alpha-\omega^2p\sin\lambda, \tag{14}$$

$$\ddot{z}-2\omega\dot{x}\cos\lambda = -g\cos\alpha+\omega^2p\cos\lambda. \tag{15}$$

The right-hand side of these equations is a force in the direction obtained by compounding the earth's attraction with centrifugal force. If this direction (the direction of apparent gravity) is inclined at $\frac{1}{2}\pi-\theta$ to the axis (θ is the geographical latitude) and we take OZ in this direction and OY correspondingly, the equations (13)–(15) become

$$\ddot{x}-2\omega\dot{y}\sin\theta+2\omega\dot{z}\cos\theta = 0, \tag{16}$$

$$\ddot{y}+2\omega\dot{x}\sin\theta = 0, \tag{17}$$

$$\ddot{z}-2\omega\dot{x}\cos\theta = -g. \tag{18}$$

Suppose the particle falls from rest at the origin when $t = 0$. Integrating (16)–(18) gives

$$\dot{x}-2\omega y\sin\theta+2\omega z\cos\theta = 0, \tag{19}$$

$$\dot{y}+2\omega x\sin\theta = 0, \tag{20}$$

$$\dot{z}-2\omega x\cos\theta = -gt. \tag{21}$$

Using (20) and (21) in (16),

$$\ddot{x}+4\omega^2 x = 2\omega g t\cos\theta.$$

Therefore, using $x = \dot{x} = 0$, when $t = 0$

$$x = -\frac{g\cos\theta}{4\omega^2}\sin 2\omega t+\frac{gt}{2\omega}\cos\theta. \tag{22}$$

Since ωt is small, expanding $\sin 2\omega t$ in (22) gives, very nearly,

$$x = \tfrac{1}{3}g\omega t^3\cos\theta,$$

a small deviation in an easterly direction. A much smaller deviation in a southerly direction can be found from (20).

EXAMPLES ON CHAPTER VII

1. A simple pendulum is set in motion from the downward vertical position with angular velocity $2n$, where $n^2 = g/l$. Show that it has just sufficient energy to reach the upward vertical position, and that the time it takes to reach the angle θ is

$$\frac{1}{n}\ln\tan\left(\frac{\pi}{4}+\frac{\theta}{4}\right).$$

2. The earth's attraction on a particle of mass m at height h above its surface is $mga^2(a+h)^{-2}$, where a is the earth's radius. If a particle is projected vertically upwards with velocity V, show that (neglecting air resistance and the rotation of the earth) its velocity at height h is

$$[aV^2+h(V^2-2ga)]^{\frac{1}{2}}(a+h)^{-\frac{1}{2}}.$$

Show that if $V^2 = 2ga$, the time it takes to reach the height h is

$$2[(a+h)^{\frac{3}{2}}-a^{\frac{3}{2}}]/3Va^{\frac{1}{2}}.$$

3. An elastic sphere of mass m and radius a, moving with velocity V, collides with an equal sphere which is directly in its path. If the force between the spheres is kr^3, where r is the radius of the impression on either sphere, show that the maximum depth of the impression is

$$\left[\frac{5mV^2}{16k(2a)^{\frac{3}{2}}}\right]^{\frac{2}{5}},$$

and discuss the motion. [Cf. § 55, Ex. 4.]

4. The force on a particle of mass m is $max^{-2}-mbx^{-3}$. It is released from rest at $x = X$; show that it oscillates between $x = X$ and $x = Xb/(2Xa-b)$ with period $2\pi aX^3(2Xa-b)^{-\frac{3}{2}}$.

5. The motion of a mass m in the direction of x increasing is resisted by a compressed air spring which provides restoring force $P(1-kx)^{-\gamma}$, where P, k, and γ are constants. If a constant force $P' > P$ is applied to the mass when it is at rest and $x = 0$, show that it comes to rest when

$$kP'x(1-\gamma)+P(1-kx)^{1-\gamma} = P.$$

6. The region $0 < x < a$ contains a space charge of electrons (as in a vacuum tube). The electric potential V in the region satisfies

$$\frac{d^2V}{dx^2} = KJV^{-\frac{1}{2}},$$

where K is a constant and J is the density of electric current (also constant). If $V = 0$ when $x = 0$, $V = V_a$ when $x = a$, and $dV/dx = 0$ when $x = 0$, show that

$$J = 4V_a^{\frac{3}{2}}/(9Ka^2).$$

7. Show that the solution of

$$\frac{d^2v}{dx^2}+\beta e^v = 0,$$

with $dv/dx = 0$ and $v = v_0$ when $x = 0$, is

$$v = v_0-2\ln\cosh x(\tfrac{1}{2}\beta e^{v_0})^{\frac{1}{2}}.$$

This corresponds to motion with a restoring force which increases exponentially with the displacement. In the theory of the thermal breakdown of a dielectric, the temperature v satisfies the above equation with $dv/dx = 0$ when $x = 0$, and $v = 0$ when $x = 1$. Find an equation for $\sqrt{(\frac{1}{2}\beta e^{v_0})}$, where v_0 is the temperature at $x = 0$, and discuss its solution graphically.

8. A particle moves in a straight line under resistance to motion mkv^3, where v is its velocity. If the initial velocity of the particle is V, show that the distance described in time t is

$$\{(2ktV^2+1)^{\frac{1}{2}}-1\}/kV.$$

9. A particle is projected vertically upwards with velocity U in a medium whose resistance varies as the square of the velocity. Show that the particle returns to its starting-point with velocity $UV(U^2+V^2)^{-\frac{1}{2}}$, where V is the terminal velocity.

10. A particle of mass m is attached to a spring of stiffness mn^2, and its motion is resisted by a force mkv^2, where v is its velocity. Derive a linear first-order equation for v^2. Show that if the particle is released from rest at $x = a$, it next comes to rest at $x = b$ given by

$$(1+2ak)e^{-2ak} = (1+2bk)e^{-2bk}.$$

11. A battery of voltage E is applied at $t = 0$ to an inductance L in series with a non-linear resistance for which $I = KV^n$; cf. § 57 (i). If the initial current in the inductance is zero, show that if $n = 3$ the voltage drop across the resistance at time t is given by

$$t = -3KE^2L\left\{\ln\left(1-\frac{V}{E}\right)+\frac{V}{E}\left(1+\frac{V}{2E}\right)\right\},$$

while the corresponding result for the case $n = 3/2$ is

$$t = 3KLE^{\frac{1}{2}}\{\tanh^{-1}(V/E)^{\frac{1}{2}}-(V/E)^{\frac{1}{2}}\}.$$

12. A battery of voltage E is applied at $t = 0$ to a capacitance C in series with a non-linear resistance for which $I = KV^n$. If the condenser is initially uncharged, show that the voltage drop V over the resistance at time t is given by

$$(n-1)KE^{n-1}V^{n-1}t = CE^{n-1}-CV^{n-1}.$$

13. A non-linear inductance in which the flux ϕ is related to the current I by $I = A\phi+B\phi^3$ is in series with a linear resistance R. If the flux is ϕ_0 at time $t = 0$, show that its value at time t is given by

$$\frac{\phi}{(A+B\phi^2)^{\frac{1}{2}}} = \frac{\phi_0}{(A+B\phi_0^2)^{\frac{1}{2}}}e^{-RAt}.$$

14. For the equation of type

$$\ddot{I}-\epsilon(1-\alpha\dot{I}-\beta\dot{I}^2)\dot{I}+n^2I = 0,$$

show that if the amplitude of an oscillation at $t = 0$ is a_0, its value at time t is

$$a_0 e^{\frac{1}{2}\epsilon t}\{1+\tfrac{3}{4}\beta a_0^2 n^2(e^{\epsilon t}-1)\}^{-\frac{1}{2}}.$$

15. Show that the period of oscillation in a circuit containing a non-linear inductance specified by § 57 (9) and (10) is approximately

$$\frac{2\pi}{n}\left(1+\frac{3Ba^2}{8A}\right)^{-1},$$

where $n^2 = A/C$, and a is the amplitude of the oscillation. Discuss the variation of a with time.

16. A particle of mass m is pressed against a plane by a force P, the coefficients of static and dynamic friction between the particle and the plane being μ' and μ respectively, and $\mu' > \mu$. The particle is attached to a spring of stiffness λ, the other end of which is moved with constant velocity V. Discuss the motion of the particle, and show that it will perform stick-slip relaxation oscillations and find their period.

17. A particle is projected vertically upwards from ground level with velocity U into air in which a horizontal wind of velocity u is blowing. Assuming that the wind causes a horizontal force $mk(u-v)^2$ on the particle, where v is the horizontal component of its velocity, and that there is resistance to vertical motion mk times the square of the vertical component of its velocity, show that it reaches the ground again after time T given by

$$T = \frac{1}{kV}\left\{\tan^{-1}\frac{U}{V}+\tanh^{-1}\frac{U}{(U^2+V^2)^{\frac{1}{2}}}\right\},$$

and that its direction of motion makes then an angle

$$\tan^{-1}\{UV(1+kTu)/[kTu^2(U^2+V^2)^{\frac{1}{2}}]\}$$

with the ground, where $V = \sqrt{(g/k)}$.

18. A particle is projected with velocity U at an angle α to the horizontal in a medium in which the resistance is proportional to the

velocity. Show that for the range on a horizontal plane to be a maximum, α must be given by

$$U(V+U\sin\alpha) = V(U+V\sin\alpha)\ln[1+(U/V)\operatorname{cosec}\alpha],$$

where V is the terminal velocity.

19. An electron gun emits a slightly divergent beam of electrons from a point on the z-axis; all of these have the same speed V, but their directions are inclined at angles up to θ with the z-axis. Show that, if a magnetic field H is applied along the z-axis, all the electrons will cross this axis again at points whose distances from the gun lie in the range from $(2\pi Vmc/eH)$ to $(2\pi Vmc/eH)\cos\theta$. Thus if θ is small the electrons can very nearly be focused at a point.

20. An electron is acted on by an electric field $E_0 \sin nt$ along the x-axis, and a magnetic field H along the z-axis. Show that if $\omega \neq n$, where $\omega = eH/mc$, and the electron is emitted at the origin $t = 0$ with zero velocity, its path is

$$x = eE_0\{n\sin\omega t-\omega\sin nt\}/m\omega(\omega^2-n^2),$$

$$y = eE_0\{(\omega^2-n^2)+n^2\cos\omega t-\omega^2\cos nt\}/m\omega n(n^2-\omega^2).$$

Discuss also the case $\omega = n$.

21. An electron moves in the region between two concentric cylinders parallel to the z-axis of radii a and b, $b > a$. It is acted on by a magnetic field H parallel to the z-axis, and by an electrostatic field between the cylinders such that the force on the electron is radial, and its magnitude is k/r, where r is the distance from the z-axis and k is a constant (the components of this force along the x- and y-axes will be kx/r^2 and ky/r^2). Considering only motion in the xy-plane, show that if v is the speed of the electron at any point, two first integrals of the motion are

$$\tfrac{1}{2}mv^2 = k\ln r+C, \quad \text{and} \quad x\dot{y}-y\dot{x} = \tfrac{1}{2}\omega r^2+C',$$

where C and C' are arbitrary constants, and $\omega = eH/mc$.

If $v = 0$ when $r = a$, and when $r = b$ the electron is moving in a direction perpendicular to the radius vector (that is, it reaches the cylinder $r = b$ at grazing incidence), show that

$$H = \frac{2cb}{b^2-a^2}\left(\frac{2mk}{e^2}\ln(b/a)\right)^{\frac{1}{2}}.$$

This is the theory of the cylindrical magnetron, H being the field which is just sufficient to stop electrons from reaching the anode. If the anode potential is V, the constant k above has the value $eV/\ln(b/a)$.

22. Deduce the result of Ex. 21 by working (i) in cylindrical polar coordinates, and (ii) vectorially.

23. A particle is projected with velocity u at an angle α to the horizontal under gravity in a medium in which the resistance to motion is mk times the square of the velocity. If s is the length of the arc of its path

from the point of projection, and ψ is the slope of the tangent at s, show that

$$\dot{x} = u\cos\alpha e^{-ks},$$

$$\sec^3\psi\frac{d\psi}{ds}+(g/u^2)e^{2ks}\sec^2\alpha = 0.$$

Find the intrinsic equation of the path by integrating the last equation.

24. A particle is moving under a central force to the origin. Show that the rate at which area is swept over by the radius vector to the particle (the areal velocity) is $\frac{1}{2}h$, where h is defined in § 62 (13).

If the path of the particle is known to be the ellipse $l = r(1+e\cos\theta)$, show that the attractive force must be mkr^{-2}, where $k = h^2/l$, and that the periodic time is $2\pi a^{\frac{3}{2}}k^{-\frac{1}{2}}$, where a is the semi-major axis of the ellipse.

25. A particle is projected at distance c from a centre of force at the origin with velocity v in a direction inclined at β to the radius vector. If the law of force is an inverse square repulsion $m\mu/r^2$, show that the orbit is a hyperbola of eccentricity

$$\left\{1+\frac{2v^2c\sin^2\beta}{\mu}+\frac{v^4c^2\sin^2\beta}{\mu^2}\right\}^{\frac{1}{2}},$$

and that
$$v^2 = \frac{\mu}{a}-\frac{2\mu}{r},$$

where a is the semi-major axis of the hyperbola.

26. A particle is attracted to the origin O by the force $m\mu r^{-2}+m\lambda r^{-3}$. Show that it is moving perpendicular to the radius vector when

$$\theta = n\pi h(h^2-\lambda)^{-\frac{1}{2}}+C,$$

where C is a constant depending on the circumstances of projection, n is any integer, and h is defined in § 62 (13).

27. The equation of motion of a particle of mass m attracted by a central force $mf(r)$ to an origin O may be written

$$\ddot{\mathbf{r}} = -\mathbf{r}\{f(r)/r\}, \tag{1}$$

where $\mathbf{r}$ is the position vector of the particle relative to the origin O.

(i) Deduce the constancy of the angular momentum $\mathbf{H} = m\mathbf{r}\wedge\dot{\mathbf{r}}$ of the particle about the origin by taking the vector product of (1) with $\mathbf{r}$.

(ii) Deduce the energy equation, § 62 (23), by taking the scalar product of (1) with $\dot{\mathbf{r}}$.

(iii) Show that
$$\mathbf{H}\wedge\ddot{\mathbf{r}} = -mr^2f(r)\frac{d}{dt}\left(\frac{\mathbf{r}}{r}\right).$$

28. Show that the equation of motion of an electron of charge $-e$ and mass m in a magnetic field $\mathbf{H}$ is

$$\ddot{\mathbf{r}} = -(e/mc)\dot{\mathbf{r}}\wedge\mathbf{H}.$$

Deduce that, whatever the nature of $\mathbf{H}$ as a function of position, the speed of the electron is constant.

If $\mathbf{H}$ is independent of position, show that

$$\dot{\mathbf{r}} = -(e/mc)\mathbf{r}\wedge\mathbf{H}+\mathbf{V},$$

and deduce that the particle moves in a helix.

29. Let $\mathbf{r}$ be the position vector of a point on a plane curve whose arc measured from some reference point to the point is s. Show that $\mathbf{t} = d\mathbf{r}/ds$ is a unit vector along the tangent at $\mathbf{r}$ to the curve. If ρ is the radius of curvature of the curve at the point $\mathbf{r}$, show that

$$\frac{d\mathbf{t}}{ds} = \frac{\mathbf{n}}{\rho},$$

where $\mathbf{n}$ is a unit vector along the normal to the curve at the point $\mathbf{r}$.

Show that $$\dot{\mathbf{r}} = \dot{s}\mathbf{t}, \qquad \ddot{\mathbf{r}} = \ddot{s}\mathbf{t}+(\dot{s}^2/\rho)\mathbf{n},$$

and deduce the equations of motion § 61 (9) and (10).

30. A simple pendulum of length l performs small oscillations in latitude θ. If the origin is in its equilibrium position, and the x- and y-axes are chosen as in § 64, Ex. 2, show that $\zeta = x+iy$ satisfies

$$\ddot{\zeta}+2i\omega\dot{\zeta}\sin\theta+n^2\zeta = 0,$$

where $n^2 = g/l$. Show that the path of the pendulum is an ellipse which rotates with constant angular velocity $-\omega\sin\theta$.

31. A particle is projected with velocity u at an angle α to the horizontal in a direction β to the north of east. Show that its deviation to the left of the plane of projection at time t is

$$\{(u\sin\alpha-\tfrac{1}{3}gt)\sin\beta\cos\theta-u\cos\alpha\sin\theta\}\omega t^2,$$

where θ is the latitude and ω the earth's angular velocity.

32. Assuming that, if p is the atmospheric pressure, the components of force in the x and y directions on an element of mass $\rho\,\delta x\delta y$ of air are $-(\partial p/\partial x)\,\delta x\delta y$ and $-(\partial p/\partial y)\,\delta x\delta y$, respectively, where ρ is the density of the air, show that the components of the velocity of the steady wind due to this pressure distribution are

$$-\frac{1}{2\rho\omega\sin\theta}\frac{\partial p}{\partial y} \quad\text{and}\quad \frac{1}{2\rho\omega\sin\theta}\frac{\partial p}{\partial x},$$

where θ is the latitude and ω is the earth's angular velocity. That is (neglecting friction) the wind tends to blow so as to keep the low pressure to its left in the northern hemisphere and to its right in the southern.

33. Show that with a suitable choice of the constants α and β, the substitution $\alpha t = t'$ and $\beta\dot{x} = x'$ reduces Rayleigh's equation

$$\ddot{x}+a\dot{x}+b\dot{x}^3+n^2x = 0$$

to van der Pol's equation § 58 (8).

VIII

RIGID DYNAMICS

65. Moments and products of inertia

SUPPOSE we have a rigid body and a set of fixed rectangular axes OX, OY, OZ. Let OR be a line through the origin whose direction cosines are (l, m, n). We suppose the body to be divided up into small elements of mass, of which a typical one is m at the point P, (x, y, z), of Fig. 52. The moment of inertia I_{OR} of the body about the axis OR is defined as

$$I_{OR} = \sum mPQ^2, \tag{1}$$

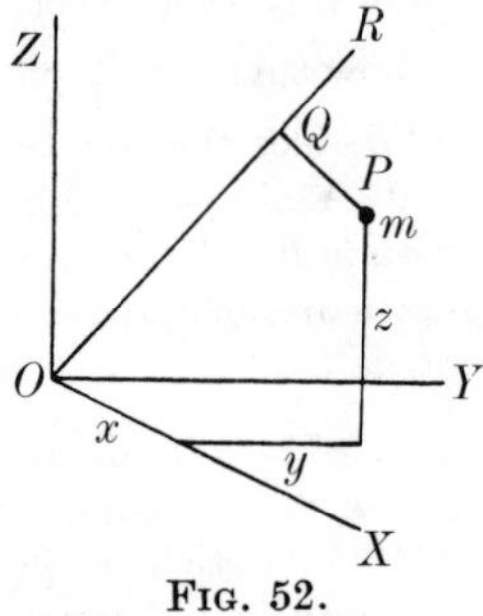

FIG. 52.

where PQ is the perpendicular distance from the particle m to the axis OR, and $\sum$ denotes, here and throughout this chapter, a summation over the elements of mass comprising the body: if the body is a continuous one these sums have to be evaluated by integration by methods described in text-books on the calculus.

If M is the mass of the body, the length k defined by

$$k^2 = I_{OR}/M$$

is called the *radius of gyration* of the body about the axis OR.

For the chosen axes OX, OY, OZ we define six fundamental quantities, namely

$$A = \sum m(y^2+z^2),\quad B = \sum m(z^2+x^2),\quad C = \sum m(x^2+y^2), \tag{2}$$

$$F = \sum myz,\quad G = \sum mzx,\quad H = \sum mxy. \tag{3}$$

A, B, C are by (1) the moments of inertia of the body about the axes OX, OY, OZ, respectively. F, G, H are called the products of inertia of the body for the axes OX, OY, OZ. We proceed to show that, if these six quantities are known, the moment of inertia I_{OR} about any axis can be expressed in terms of them.

From (1)†

$$I_{OR} = \sum m(OP^2 - OQ^2)$$
$$= \sum m\{x^2+y^2+z^2-(lx+my+nz)^2\}$$
$$= \sum m\{x^2(1-l^2)+y^2(1-m^2)+z^2(1-n^2)- \\ -2mnyz-2nlzx-2lmxy\}$$
$$= \sum m\{x^2(m^2+n^2)+y^2(n^2+l^2)+z^2(l^2+m^2)- \\ -2mnyz-2nlzx-2lmxy\}$$
$$= l^2\sum m(y^2+z^2)+m^2\sum m(z^2+x^2)+n^2\sum m(x^2+y^2)- \\ -2mn\sum myz-2nl\sum mzx-2lm\sum mxy$$
$$= Al^2+Bm^2+Cn^2-2Fmn-2Gnl-2Hlm. \tag{4}$$

This is the general result in three dimensions. It is interesting, and important for many purposes, to study the corresponding two-dimensional problem of the variation of the moment of inertia of a lamina in the xy-plane about an axis in this plane inclined at θ to OX. Since the lamina lies in the xy-plane, $z = 0$ for all its particles, and so $F = G = 0$, and $C = A+B$. Also the direction cosines of an axis in the plane will be $l = \cos\theta$, $m = \sin\theta$, $n = 0$. Thus (4) becomes

$$I_\theta = A\cos^2\theta+B\sin^2\theta-2H\sin\theta\cos\theta. \tag{5}$$

To see the way in which the moment of inertia varies with θ we may make a polar plot of I_θ against θ, plotting a distance proportional to I_θ in the direction θ. This is done in Fig. 53 for the lamina of angle section shown dotted, and gives the dumb-bell-shaped figure shown by the full line. To find the maxima and minima of I_θ, we have from (5)

$$\frac{dI_\theta}{d\theta} = (B-A)\sin 2\theta-2H\cos 2\theta, \tag{6}$$

and thus I_θ is stationary when

$$\tan 2\theta = \frac{2H}{B-A}. \tag{7}$$

† Note that m is used here in two senses, namely, as a direction cosine and as the mass of the typical particle. The latter is always written immediately after the sign of summation to avoid confusion.

(7) gives two values of θ at right angles, corresponding to maxima and minima of the moment of inertia. The values of the moments of inertia in these directions are called the *Principal Moments of Inertia* of the lamina, and the directions themselves are called the *Principal Axes of Inertia* of the lamina.

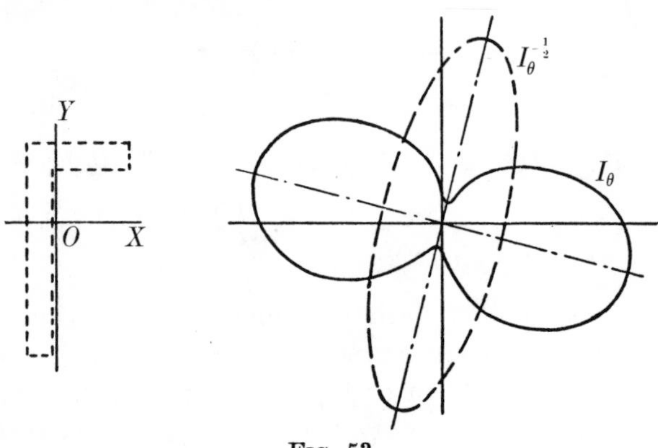

FIG. 53.

These results may be obtained in another way which leads to a very simple geometrical picture. If, instead of plotting I_θ against θ in a polar plot as was done above, we plot $1/\sqrt{(I_\theta)}$, then the rectangular coordinates of the point plotted will be

$$X = \frac{k}{\sqrt{(I_\theta)}}\cos\theta, \qquad Y = \frac{k}{\sqrt{(I_\theta)}}\sin\theta, \tag{8}$$

where k is the constant of proportionality. Substituting these values in (5) it appears that (X, Y) lies on the ellipse

$$AX^2+BY^2-2HXY = k^2. \tag{9}$$

This ellipse is shown dotted in Fig. 53. It is called the *Momental Ellipse* for the lamina. It is known that an ellipse has two principal axes relative to which its equation is

$$A'X'^2+B'Y'^2 = k^2; \tag{10}$$

these axes will be the principal axes of inertia defined previously, and their inclination to the original axes will be given by (7). By inspection of Fig. 53 it appears that the momental ellipse is elongated in the direction of the body, that is, that its major axis, which is the direction of least moment of inertia, lies roughly along the greatest diameter of the body. We may also remark that if the moments of inertia of the lamina in any three directions are equal, the momental ellipse is a circle and therefore the moments of inertia in all directions are equal.

The same theory applies to the three-dimensional case (4). If a point (X, Y, Z) is plotted in the direction (l, m, n) at a distance $k/\surd I$ from the origin so that

$$X = \frac{kl}{\surd I}, \qquad Y = \frac{km}{\surd I}, \qquad Z = \frac{kn}{\surd I}, \tag{11}$$

then X, Y, Z lies on the ellipsoid

$$AX^2+BY^2+CZ^2-2FYZ-2GZX-2HXY = k^2. \tag{12}$$

This is the *momental ellipsoid* for the body. It has three principal axes at right angles, which are the principal axes of inertia of the body. Relative to these, (12) becomes

$$A'X'^2+B'Y'^2+C'Z'^2 = k^2. \tag{13}$$

The moments of inertia of the body in the directions of the principal axes are called its principal moments of inertia.

So far we have regarded A, B, C, F, G, H as given. For complicated bodies they have to be determined by integration. For simple shapes the results of these integrations are summed up in *Routh's rule*, which states that *the moment of inertia about an axis of symmetry of a body of mass M which has three perpendicular axes of symmetry* is

$$\frac{M}{n}\{\text{the sum of the squares of the semi-axes perpendicular to the one considered}\}, \tag{14}$$

where $n = 3$, for a rectangular parallelepiped,

$n = 4$, for a circular lamina,

$n = 5$, for a sphere or ellipsoid.

For example, for a rectangular lamina of sides $2a$ and $2b$ the moments of inertia about the axes of symmetry are

$$\tfrac{1}{3}Mb^2, \quad \tfrac{1}{3}Ma^2, \quad \tfrac{1}{3}M(a^2+b^2), \tag{15}$$

the latter being about the line through the centre perpendicular to the plane of the lamina.

The results obtained by Routh's rule (in which the axes, being axes of symmetry, necessarily pass through the centre of mass of the body) may be extended by the use of simple results connecting the moment of inertia of a body about any axis with the moment of inertia about a parallel axis through the centre of mass.

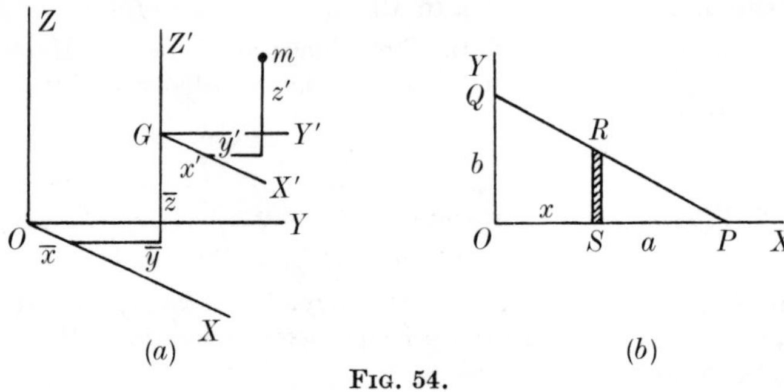

FIG. 54.

Let OX, OY, OZ be any rectangular axes, let $(\bar{x}, \bar{y}, \bar{z})$ and (x, y, z) be the coordinates of the centre of mass G of the body, and the particle of mass m, relative to them. Let GX', GY', GZ' be a set of parallel axes through G, and let (x', y', z') be the coordinates of m relative to them, Fig. 54 (*a*). Then

$$x = \bar{x}+x', \qquad y = \bar{y}+y', \qquad z = \bar{z}+z'. \tag{16}$$

Also, since the centre of mass of the body is at the origin of the (x', y', z') coordinate system, we have from the definition of the centre of mass, cf. § 66 (18),

$$\sum mx' = \sum my' = \sum mz' = 0. \tag{17}$$

Relative to the axes OX, OY, OZ we have

$$\begin{aligned} A = \sum m(y^2+z^2) &= \sum m\{(\bar{y}+y')^2+(\bar{z}+z')^2\} \\ &= (\bar{y}^2+\bar{z}^2)\sum m+2\bar{y}\sum my'+2\bar{z}\sum mz'+ \\ &\qquad\qquad + \sum m(y'^2+z'^2) \\ &= M(\bar{y}^2+\bar{z}^2)+A_{CG}, \end{aligned} \tag{18}$$

by (17), where we have used the fact that $\bar{x}$, $\bar{y}$, $\bar{z}$ are the same for all particles m and so may be taken outside the summations. In (18) M is the total mass of the body, and A_{CG} is the quantity A for the parallel system of axes through the centre of mass.

In the same way

$$F = \sum myz = \sum m(\bar{y}+y')(\bar{z}+z')$$
$$= \bar{y}\bar{z}\sum m+\bar{y}\sum mz'+\bar{z}\sum my'+\sum my'z'$$
$$= M\bar{y}\bar{z}+F_{CG}. \tag{19}$$

Similar results hold for B, C, G, H. These results may be expressed by the statement that a moment or product of inertia relative to any system of rectangular axes is obtained by adding to the corresponding quantity for parallel axes through the centre of mass a *transfer term* which is just the moment or product of inertia relative to the original axes of a particle of mass equal to the total mass of the body and placed at its centre of mass.

As an example we find the moments and product of inertia relative to the axes OX, OY (Fig. 54 (*b*)) of a right-angled triangular lamina of sides a and b.

To find A, the moment of inertia about OX, we consider the strip RS of width δx at x. If ρ is the surface density of the lamina, the moment of inertia of this strip about OX is, by Routh's rule,

$$\tfrac{1}{3}\rho y^3\,\delta x.$$

Here $y = b(a-x)/a$, and, integrating from $x = 0$ to $x = a$ we get

$$A = \frac{\rho b^3}{3a^3}\int\limits_0^a (a-x)^3\,dx = \tfrac{1}{12}\rho ab^3. \tag{20}$$

Interchanging a and b we get

$$B = \tfrac{1}{12}\rho a^3 b. \tag{21}$$

To find the product of inertia we use the fact that, because of its symmetry, the product of inertia of the strip RS about axes parallel to OX and OY through its centre of mass must be zero.

Thus the product of inertia for the strip RS relative to the axes OX, OY consists only of the transfer term

$$\rho y\,\delta x\,x\,\tfrac{1}{2}y.$$

Integrating we get

$$H = \frac{1}{2}\frac{\rho b^2}{a^2}\int\limits_0^a x(a-x)^2\,dx = \tfrac{1}{24}\rho a^2b^2. \tag{22}$$

Thus the momental ellipse (9) for the lamina relative to these axes is

$$b^2x^2+a^2y^2-abxy = \text{constant}. \tag{23}$$

And, by (7), the inclination θ of the principal axes to OX and OY is given by

$$\tan 2\theta = \frac{ab}{a^2-b^2}.$$

66. Fundamental equations

We regard a rigid body as composed of a large number of particles held together by cohesive or *internal* forces acting between them. In addition to these there are *external* forces acting on the particles which may be of two types, either *body* forces such as gravity which act on all particles, or applied forces or reactions which we regard as being applied to certain specified particles. The masses of the particles will always be assumed to be constant.

Let m be the mass of a typical particle of the body and let $\mathbf{r}$ be its position vector relative to a fixed frame of reference with origin O. Let $\mathbf{P}$ and $\mathbf{P}'$, respectively, be the resultant external and internal forces on the particle. Then its equation of motion is

$$\frac{d}{dt}(m\dot{\mathbf{r}}) = \mathbf{P}+\mathbf{P}'. \tag{1}$$

As in § 53 we use $\sum$ to denote a summation over all particles of the body, and write

$$\mathbf{L} = \sum m\dot{\mathbf{r}} \tag{2}$$

for its linear momentum, and

$$\mathbf{H} = \sum \mathbf{r}\wedge m\dot{\mathbf{r}} \tag{3}$$

for its angular momentum about O. Summing (1) over all particles of the body gives

$$\frac{d\mathbf{L}}{dt} = \sum \mathbf{P} + \sum \mathbf{P}'. \tag{4}$$

Also, differentiating (3) and using $\dot{\mathbf{r}}\wedge\dot{\mathbf{r}} = 0$,

$$\frac{d\mathbf{H}}{dt} = \sum \mathbf{r}\wedge\frac{d}{dt}(m\dot{\mathbf{r}}) = \sum \mathbf{r}\wedge\mathbf{P} + \sum \mathbf{r}\wedge\mathbf{P}', \tag{5}$$

using (1).

Some assumption has to be made as to the nature of the internal forces, and here we shall make the simplest possible one, namely, that they consist of forces of attraction or repulsion between the individual particles. Since for each such force on a particle A due to a particle B there is an equal, oppositely directed, reaction force on B due to A, the sum of these, and also the sum of their moments about O, vanishes. And therefore, summing over all particles,

$$\sum \mathbf{P}' = 0, \tag{6}$$

$$\sum \mathbf{r} \wedge \mathbf{P}' = 0. \tag{7}$$

The results (6) and (7) are also true under less restrictive conditions than those assumed above. Using (6) in (4) gives

$$\frac{d\mathbf{L}}{dt} = \sum \mathbf{P}. \tag{8}$$

This is the first fundamental result, and may be stated as:

I. *The rate of change of linear momentum of the body is equal to the vector sum of the external forces acting on it.*

Again, using (7) in (5) gives the second fundamental result:

II. *The rate of change of angular momentum of the body about a fixed origin O is equal to the sum of the moments of the external forces about O; that is,*

$$\frac{d\mathbf{H}}{dt} = \sum \mathbf{r} \wedge \mathbf{P}. \tag{9}$$

If the forces on the body are in static equilibrium and it is initially at rest, $\mathbf{L}$ and $\mathbf{H}$ must remain zero, and (8) and (9) give

$$\sum \mathbf{P} = 0, \qquad \sum \mathbf{r} \wedge \mathbf{P} = 0, \tag{10}$$

which are the conditions of static equilibrium stated in § 53.

If (x, y, z) and (X, Y, Z) are the components of $\mathbf{r}$ and $\mathbf{P}$ relative to right-handed rectangular axes through O, (8) becomes

$$\frac{d}{dt}\sum m\dot{x} = \sum X, \qquad \frac{d}{dt}\sum m\dot{y} = \sum Y, \qquad \frac{d}{dt}\sum m\dot{z} = \sum Z. \tag{11}$$

Also, using § 52 (18), (9) gives

$$\frac{d}{dt}\sum m(y\dot{z}-z\dot{y}) = \sum (yZ-zY), \tag{12}$$

together with two similar equations.

An alternative method of deriving the fundamental equations should also be noted. (1) may be written in the form

$$\mathbf{P}+\mathbf{P}'-m\ddot{\mathbf{r}} = 0. \tag{13}$$

Now in (13), $-m\ddot{\mathbf{r}}$ has the dimensions of a force, and in this sense is referred to as the *reversed effective force* on the particle. The whole of the forces acting on the system of particles which compose the body are now the external, internal, and reversed effective forces, and the equations of motion are obtained by writing down the conditions that these be in static equilibrium. When this is done the internal forces will disappear by (6) and (7), and the result may be stated in the form that the external and reversed effective forces on all particles of the body are in static equilibrium: this is known as *d'Alembert's principle*. In this method the conditions for a set of forces to be in static equilibrium are used, and are assumed to have been derived from statical considerations (cf. Chap. VI, Exs. 11 to 13).

If $M = \sum m$ is the mass of the body, the position vector $\bar{\mathbf{r}}$ of the centre of mass relative to O is defined by

$$M\bar{\mathbf{r}} = \sum m\mathbf{r}. \tag{14}$$

Using this result in (2) gives

$$\mathbf{L} = M\dot{\bar{\mathbf{r}}}, \tag{15}$$

that is, the linear momentum of the body is equal to that of a particle of mass M moving with the velocity of the centre of mass of the body. Using (15) in (8) we get

$$\frac{d}{dt}(M\dot{\bar{\mathbf{r}}}) = \sum \mathbf{P}, \tag{16}$$

which is the third fundamental result and may be stated as:

III. *The centre of mass moves like a particle of mass M placed there and acted on by the vector sum of the external forces on the body.*

Now let $\mathbf{r}'$ be the position vector of the typical particle m relative to the centre of mass, so that

$$\mathbf{r} = \bar{\mathbf{r}}+\mathbf{r}'. \tag{17}$$

From (17) $$\sum m\mathbf{r}' = \sum m\mathbf{r} - \sum m\bar{\mathbf{r}} = 0, \tag{18}$$

using (14). Substituting (17) in (3) we get

$$\begin{aligned}\mathbf{H} &= \sum m(\bar{\mathbf{r}}+\mathbf{r}')\wedge(\dot{\bar{\mathbf{r}}}+\dot{\mathbf{r}}') \\ &= \sum m\bar{\mathbf{r}}\wedge\dot{\bar{\mathbf{r}}} + \sum m\mathbf{r}'\wedge\dot{\bar{\mathbf{r}}} + \sum m\bar{\mathbf{r}}\wedge\dot{\mathbf{r}}' + \sum m\mathbf{r}'\wedge\dot{\mathbf{r}}' \\ &= M\bar{\mathbf{r}}\wedge\dot{\bar{\mathbf{r}}} + \sum m\mathbf{r}'\wedge\dot{\mathbf{r}}',\end{aligned} \tag{19}$$

on taking $\bar{\mathbf{r}}$ and $\dot{\bar{\mathbf{r}}}$ outside the summations over the particles and using (18).

The first term on the right-hand side of (19) is the angular momentum about O of a particle of mass equal to the total mass

M of the body placed at the centre of mass and moving with it: the second term is the angular momentum of the body about its centre of mass. Using (19) and (17) in (9) we get

$$M\frac{d}{dt}(\bar{\mathbf{r}}\wedge\dot{\bar{\mathbf{r}}})+\frac{d}{dt}\sum m\mathbf{r}'\wedge\dot{\mathbf{r}}' = \bar{\mathbf{r}}\wedge\sum\mathbf{P}+\sum\mathbf{r}'\wedge\mathbf{P}.$$

By (16) this reduces to

$$\frac{d}{dt}\sum m\mathbf{r}'\wedge\dot{\mathbf{r}}' = \sum\mathbf{r}'\wedge\mathbf{P}, \tag{20}$$

that is:

IV. *The rate of change of angular momentum about the centre of mass is equal to the sum of the moments of the external forces about the centre of mass.*

The results II, III, IV are those usually needed in solving problems and we shall refer to them shortly as 'motion about a fixed axis', 'motion of the centre of mass', and 'motion about the centre of mass', respectively.

To determine the angular momentum of the body about its centre of mass in (19), suppose that its angular velocity about its centre of mass is $\boldsymbol{\omega}$, so that $\dot{\mathbf{r}}' = \boldsymbol{\omega}\wedge\mathbf{r}'$ and

$$\sum m\mathbf{r}'\wedge\dot{\mathbf{r}}' = \sum m\mathbf{r}'\wedge(\boldsymbol{\omega}\wedge\mathbf{r}') = \boldsymbol{\omega}\sum m(\mathbf{r}'.\mathbf{r}')-\sum m(\boldsymbol{\omega}.\mathbf{r}')\mathbf{r}'. \tag{21}$$

If $(\omega_1, \omega_2, \omega_3)$ and (x', y', z') are the components of $\boldsymbol{\omega}$ and $\mathbf{r}'$ referred to right-handed rectangular axes, those of the angular momentum (21) are

$$(A\omega_1-H\omega_2-G\omega_3,\quad -H\omega_1+B\omega_2-F\omega_3,\quad -G\omega_1-F\omega_2+C\omega_3), \tag{22}$$

where A, B, C, F, G, H are defined in § 65 (2) and (3).

The kinetic energy T of the body may be studied in the same way. It is

$$T = \tfrac{1}{2}\sum m|\dot{\mathbf{r}}|^2 = \tfrac{1}{2}\sum m|\dot{\bar{\mathbf{r}}}+\dot{\mathbf{r}}'|^2 = \tfrac{1}{2}M|\dot{\bar{\mathbf{r}}}|^2+\tfrac{1}{2}\sum m|\dot{\mathbf{r}}'|^2. \tag{23}$$

The first term on the right hand side of (23) is the kinetic energy of a particle of mass M at the centre of mass of the body and moving with it. The second is the kinetic energy of the motion about the centre of mass. With the notation of (22) it is

$$\tfrac{1}{2}\sum m|\dot{\mathbf{r}}'|^2 = \tfrac{1}{2}\sum m\{(\omega_2 z'-\omega_3 y')^2+(\omega_3 x'-\omega_1 z')^2+$$
$$+(\omega_1 y'-\omega_2 x')^2\}$$
$$= \tfrac{1}{2}\{A\omega_1^2+B\omega_2^2+C\omega_3^2-2F\omega_2\omega_3-2G\omega_3\omega_1-$$
$$-2H\omega_1\omega_2\}. \quad (24)$$

The result (24) also gives the kinetic energy of the motion of the body about a fixed origin if $A,..., H$ are the moments and products of inertia relative to fixed axes through this origin, and in the same way (22) gives the angular momentum.

67. Motion about a fixed axis

In this case only the principle II of § 66 is needed. The angular momentum about the axis may be written down from § 66 (22) but we derive it here *ab initio*.

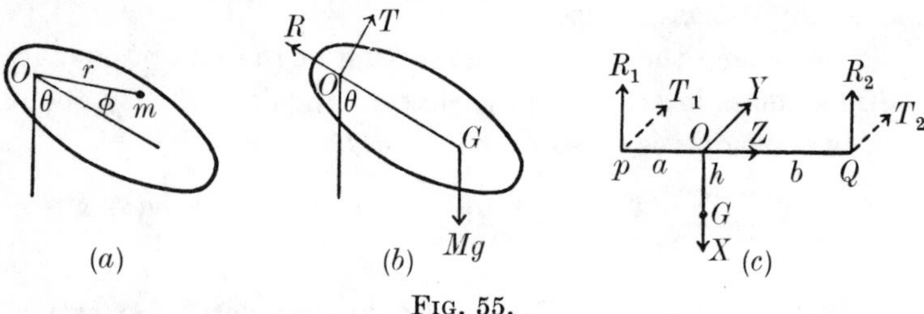

FIG. 55.

Let θ be the angle between a marked plane in the body passing through the axis of rotation and a fixed plane through the axis. Let m be a typical particle of the body, and r its perpendicular distance from the axis. Let the plane through m and the axis make an angle ϕ with the marked plane in the body; cf. Fig. 55 (*a*).

Then the angular momentum of the body about the axis is

$$\sum mr^2\frac{d}{dt}(\theta+\phi) = \sum mr^2\dot{\theta} = \dot{\theta}\sum mr^2 = I\dot{\theta}, \quad (1)$$

since ϕ is independent of the time, and θ is the same for all particles of the body.

The equation of motion § 66 II thus becomes

$$I\ddot{\theta} = \text{couple}. \tag{2}$$

This is the equation often referred to in Chapter IV.

The compound or rigid body pendulum

It is required to find the motion of a rigid body which can oscillate freely about a horizontal axis. Let θ be the angle which the plane through the axis and the centre of mass G makes with the downward vertical. Then if h is the distance of the centre of mass from the axis, the sum of the moments of the external forces (gravity) about the axis in the direction of θ increasing is $-Mgh\sin\theta$. If I is the moment of inertia of the body about the axis, (2) gives

$$I\ddot{\theta} = -Mgh\sin\theta, \tag{3}$$

or, writing $I = Mk^2$,

$$\ddot{\theta}+(gh/k^2)\sin\theta = 0. \tag{4}$$

This is precisely the same as the equation of motion, § 55 (13), of a simple pendulum of length k^2/h. Thus a simple pendulum of this length and the rigid body pendulum, if started with the same values of θ and $\dot{\theta}$, will keep step exactly. The simple pendulum of length k^2/h is called the *simple equivalent pendulum*.

The solution of (4) has been discussed in § 55.

Next we determine the reactions at the axis of rotation. Consider first the case of a lamina which is freely hinged at a point and which oscillates in the vertical plane through the hinge. We wish to find the reaction at the hinge, and we may either seek its radial and transverse components, R and T, Fig. 55 (*b*), or its horizontal and vertical components. Here we shall find the former, and in § 68 the latter, as both methods of procedure are important.

We now have to use the principle III of § 66, namely, that the centre of mass moves like a particle of mass M placed there and acted on by the resultant external force on the body, the external forces being in the present case R, T, and Mg. The equations of motion § 61 (9) and (10) give

$$Mh\ddot{\theta} = T-Mg\sin\theta, \tag{5}$$

$$Mh\dot{\theta}^2 = R-Mg\cos\theta. \tag{6}$$

From (5) and (4) it follows that

$$T = Mg\left(1-\frac{h^2}{k^2}\right)\sin\theta. \tag{7}$$

To determine R we need $\dot{\theta}^2$, and thus have to integrate (4). If ω is the value of $\dot{\theta}$ when $\theta = 0$, (4) gives as in § 55

$$\dot{\theta}^2 = \omega^2+\frac{2gh}{k^2}(\cos\theta-1), \tag{8}$$

and from (6)

$$R = Mg\left(1+\frac{2h^2}{k^2}\right)\cos\theta+Mh\left(\omega^2-\frac{2gh}{k^2}\right). \tag{9}$$

We now consider the more general case of any rigid body which can turn freely about a horizontal axis. It may be remarked that the present theory includes that of a fly-wheel whose centre of mass does not lie on its axis, and that the reactions on the bearings may cause important vibrations.

Suppose that the axis of rotation is supported by two bearings P and Q, and that O, the foot of the perpendicular from the centre of mass G on the axis, is distant a from P and b from Q. Let R_1 and T_1 be the radial and transverse components of the reaction at P, and R_2, T_2 those of the reaction at Q; cf. Fig. 55 (c).

The preceding calculation gives the total radial and transverse components of the reactions, that is,

$$R_1+R_2 = R, \tag{10}$$

$$T_1+T_2 = T, \tag{11}$$

where R and T are given by (9) and (7). To determine the reactions completely we need two more equations, and for these we must use § 66 II or IV. A fundamental difficulty arises here which appears with all irregularly shaped bodies: if we take axes fixed in direction, the moments and products of inertia of the body relative to them change as the body moves. Thus we must take a set of axes fixed in the body and allow for the motion of these axes as in § 64. Take axes OX along OG, OZ along the axis of rotation, and OY to make a right-handed system. Let A, B, I, F, G, H be the moments and products of inertia of the body referred to these axes. The components of the angular velocity of the body along these axes are

$$(0, 0, \dot{\theta}). \tag{12}$$

The components of the angular momentum of the body along these axes are by § 66 (22)

$$(-G\dot{\theta}, -F\dot{\theta}, I\dot{\theta}). \tag{13}$$

Using (12) and (13) in § 64 (4), the components of the rate of change of angular momentum of the body about the instantaneous directions of the axes are

$$(-G\ddot{\theta}+F\dot{\theta}^2, -F\ddot{\theta}-G\dot{\theta}^2, I\ddot{\theta}). \tag{14}$$

The sum of the moments of the external forces about the instantaneous directions of the axes are

$$(T_1 a - T_2 b,\ R_1 a - R_2 b,\ -Mgh\sin\theta). \tag{15}$$

Equating (14) and (15) by § 66 II we get

$$G\ddot{\theta} - F\dot{\theta}^2 = T_2 b - T_1 a, \tag{16}$$

$$-F\ddot{\theta} - G\dot{\theta}^2 = R_1 a - R_2 b. \tag{17}$$

$\ddot{\theta}$ and $\dot{\theta}^2$ are given by (3) and (8), and (10), (11), (16), (17) are four equations for R_1, R_2, T_1, T_2.

68. Motion in two dimensions

In this section we solve a number of two-dimensional problems.

Ex. 1. *A thin rod of mass M and length $2a$ is initially at rest in the vertical position of unstable equilibrium and rotates freely about its lower end O, **Fig. 56** (a). It is required to find the motion and the horizontal and vertical components, F and R, of the reaction at the point of support.*

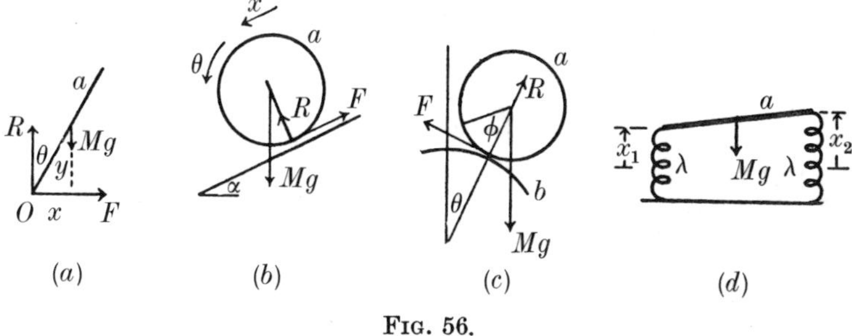

FIG. 56.

By Routh's rule, § 65 (14), the square of the radius of gyration, k^2, of the rod about O is $4a^2/3$. Thus the equation of motion of the rod about O, found as in § 67 (3), is

$$\frac{4a^2}{3}\ddot{\theta} = ga\sin\theta. \tag{1}$$

Integrating as in § 55 we have

$$\frac{2a}{3}\dot{\theta}^2 = g(1-\cos\theta), \tag{2}$$

since we are given $\dot{\theta} = 0$ when $\theta = 0$.

We now consider the motion of the centre of mass in rectangular coordinates x, y. We have

$$x = a\sin\theta, \qquad y = a\cos\theta, \tag{3}$$

$$\dot{x} = a\dot{\theta}\cos\theta, \qquad \dot{y} = -a\dot{\theta}\sin\theta, \tag{4}$$

$$\ddot{x} = a\ddot{\theta}\cos\theta - a\dot{\theta}^2\sin\theta, \tag{5}$$

$$\ddot{y} = -a\ddot{\theta}\sin\theta - a\dot{\theta}^2\cos\theta. \tag{6}$$

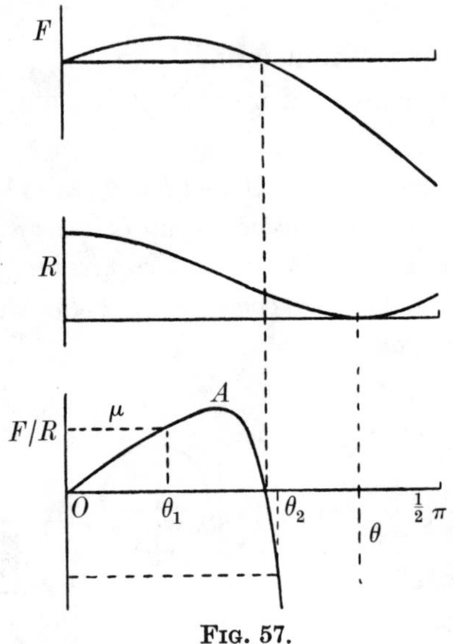

FIG. 57.

Thus motion of the centre of mass, § 66 III, gives

$$M(a\ddot{\theta}\cos\theta - a\dot{\theta}^2\sin\theta) = F, \tag{7}$$

$$M(-a\ddot{\theta}\sin\theta - a\dot{\theta}^2\cos\theta) = R - Mg. \tag{8}$$

Using the values (1) and (2) in these, we get finally

$$F = \frac{3Mg}{4}\sin\theta(3\cos\theta - 2), \tag{9}$$

$$R = \frac{Mg}{4}(3\cos\theta - 1)^2. \tag{10}$$

The way in which F, R, and F/R vary with θ is shown in Fig. 57. F changes sign as θ passes through $\cos^{-1}(2/3)$, so that, for values of θ larger than this, the horizontal component of the reaction is in the opposite direction to the arrow in Fig. 56 (a). The curves will be discussed further in § 69.

Ex. 2. *A cylinder of radius a and moment of inertia Mk^2 rolls down a perfectly rough plane inclined at α to the horizontal* [*Fig. 56* (b)].

Let x be the distance the centre of the cylinder has moved from its initial position, and let θ be the angle through which a marked line on the cylinder has turned from its initial position. Then

$$x = a\theta, \tag{11}$$

and, differentiating,

$$\dot{x} = a\dot{\theta}. \tag{12}$$

(12) might have been written down directly since $\dot{x}-a\dot{\theta}$ is the velocity down the plane of the point of the cylinder instantaneously in contact with the plane, and since the cylinder rolls without slipping this velocity must be zero.

Let F and R be the components of the force on the cylinder at the point of contact; then motion of the centre of mass (§ 66 III) gives

$$M\ddot{x} = Mg\sin\alpha - F, \tag{13}$$

$$0 = R - Mg\cos\alpha. \tag{14}$$

Also motion about the centre of mass (§ 66 IV) gives

$$Mk^2\ddot{\theta} = Fa. \tag{15}$$

Since, by (12), $\ddot{x} = a\ddot{\theta}$, (13) and (15) give

$$\ddot{x} = \frac{ga^2}{k^2+a^2}\sin\alpha, \tag{16}$$

$$F = \frac{Mgk^2}{k^2+a^2}\sin\alpha. \tag{17}$$

The cylinder thus rolls down the plane with constant acceleration.

Ex. 3. *A cylinder of radius a and moment of inertia Mk^2 rolls down the outside of a perfectly rough cylinder of radius b* [*Fig. 56* (c)].

Let θ be the angle the plane containing the axes of the cylinders makes with the vertical, and let ϕ be the angle between a marked plane through the axis of the cylinder of radius a and the plane containing the axes of the cylinders. Then, since the lengths of the arcs of the two circles which have rolled over each other must be equal, we must have

$$b\theta = a\phi + \text{constant},$$

and

$$b\dot{\theta} = a\dot{\phi}. \tag{18}$$

Let F and R be the components of the force on the cylinder of radius a at the point of contact, then, using § 61 (9) and (10), motion of the centre of mass (§ 66 III) gives

$$M(a+b)\ddot{\theta} = Mg\sin\theta - F, \tag{19}$$

$$M(a+b)\dot{\theta}^2 = Mg\cos\theta - R. \tag{20}$$

For motion about the centre of mass (§ 66 IV) we have to specify the position of the marked plane on the cylinder, not relative to the moving plane joining the axes of the cylinders, but relative to a fixed plane such as the vertical with which it makes an angle $(\theta+\phi)$. Thus this equation is

$$Mk^2(\ddot{\theta}+\ddot{\phi}) = Fa. \tag{21}$$

(18), (19), and (21) give

$$(a+b)(a^2+k^2)\ddot{\theta} = ga^2\sin\theta. \tag{22}$$

Ex. 4. *A uniform rod of mass M and length 2a is supported in a horizontal position by two equal springs of stiffness λ at its ends. Discuss the small vertical oscillations of the system; cf. Fig. 56 (d).*

Let x_1 and x_2 be the extensions of the springs from their equilibrium positions. The displacement of the centre of mass is $\frac{1}{2}(x_1+x_2)$ so that motion of the centre of mass, § 66 III, gives

$$\tfrac{1}{2}M(\ddot{x}_1+\ddot{x}_2) = -\lambda x_1 - \lambda x_2. \tag{23}$$

The force of gravity is not included in (23) since x_1 and x_2 are measured from their equilibrium positions.

The small angle through which the rod has turned is $(x_2-x_1)/2a$, so that motion about the centre of mass, § 66 IV, gives

$$\tfrac{1}{6}Ma(\ddot{x}_2-\ddot{x}_1) = -\lambda a x_2 + \lambda a x_1. \tag{24}$$

It follows from (23) that x_1+x_2 oscillates with frequency $(2\lambda/M)^{\frac{1}{2}}/2\pi$, and from (24) that x_1-x_2 oscillates with frequency $(6\lambda/M)^{\frac{1}{2}}/2\pi$.

69. Problems of rolling or sliding

When there is an imperfectly rough contact to be considered, we do not know *a priori* whether there is slipping at this contact. We must make one or other of two assumptions, calculate the motion on this basis, and verify that it satisfies the required conditions: if it does not, the other assumption must be used.

(i) *Assume the point of contact to be at rest.* In this case the components of the reaction at the point of contact along and normal to the surface, F and R, can be calculated and we must have $F < \mu R$, where μ is the coefficient of friction. If F becomes equal to μR slipping commences.

(ii) *Assume the point of contact to be moving.* In this case the tangential and normal components of the reaction, F and R, are connected by $F = \mu R$. The velocity of sliding of the point of contact is then calculated: if at any time this becomes zero, calculations must be continued with the assumption (i).

It should be remarked that the coefficients of friction μ in (i) and (ii) are respectively the static and dynamic coefficients and that these are not necessarily the same [cf. § 30 (iii)], but here we shall for simplicity assume that they are equal.

Ex. 1. *A thin rod of mass M and length $2a$ is initially at rest in the position of unstable equilibrium with its lower end resting on a horizontal plane, the coefficient of friction between the rod and the plane being μ.*

If we assume that the point of contact is at rest, the rod turns about it, and the equations of motion are those of § 68, Ex. 1. F and R have been calculated in § 68 (9) and (10), and they and F/R are graphed in Fig. 57. For the point of contact to remain at rest we must have $|F/R| < \mu$. If μ is less than the maximum at A of the curve F/R, Fig. 57, the point of contact will slip backwards at the angle θ_1. But if μ is greater than this maximum, the point of contact will remain at rest all the time that friction acts forwards, and slipping will take place (forwards) at the angle θ_2 at which $-F/R = \mu$. Slipping must occur before θ reaches the value $\cos^{-1}(1/3)$.

Ex. 2. *A cylinder of radius a and moment of inertia Mk^2 is placed at rest on a plane inclined at θ to the horizontal, the coefficient of friction between the cylinder and the plane being μ.*

This has been discussed on the assumption of *rolling*, that is that the point of contact is at rest, in § 68, Ex. 2. F and R are calculated in § 68 (17) and (14), and if the assumption of rolling is correct we must have

$$\frac{F}{R} = \frac{k^2}{k^2+a^2}\tan\alpha < \mu. \tag{1}$$

If this is not the case we must assume that the point of contact slips: since the cylinder is initially at rest, the point of contact must slip down the plane and the frictional force $F = \mu R$ must act up the plane, Fig. 58 (*a*).

x and θ are now independent, and the equations of motion are

$$M\ddot{x} = Mg\sin\alpha - \mu R, \tag{2}$$

$$0 = R - Mg\cos\alpha, \tag{3}$$

$$Mk^2\ddot{\theta} = \mu Ra. \tag{4}$$

(2) and (3) give
$$\ddot{x} = g(\sin\alpha - \mu\cos\alpha),$$
$$\dot{x} = gt(\sin\alpha - \mu\cos\alpha), \tag{5}$$
since $\dot{x} = 0$ when $t = 0$.

(3) and (4) give
$$\ddot{\theta} = \frac{\mu g a}{k^2}\cos\alpha,$$
$$\dot{\theta} = \frac{\mu g a t}{k^2}\cos\alpha, \tag{6}$$
since $\dot{\theta} = 0$ when $t = 0$.

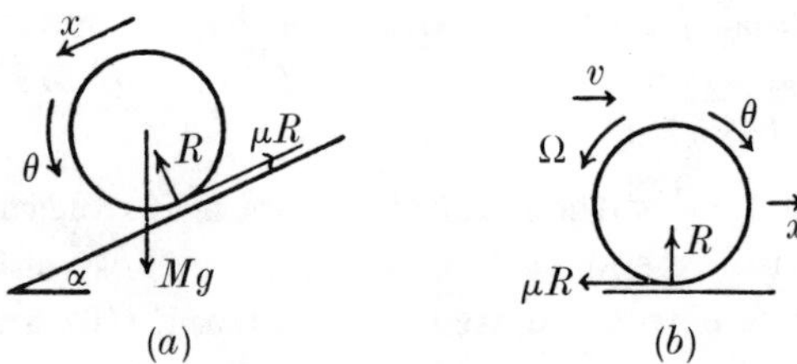

FIG. 58.

(5) and (6) give the velocity and angular velocity of the cylinder, both of which increase steadily. The velocity of the point of contact is
$$\dot{x} - a\dot{\theta} = gt\left\{\sin\alpha - \mu\left(1 + \frac{a^2}{k^2}\right)\cos\alpha\right\}, \tag{7}$$
and thus is zero if
$$\tan\alpha = \mu\left(1 + \frac{a^2}{k^2}\right). \tag{8}$$

This is the transition value between rolling and sliding found in (1).

Ex. 3. *A cylinder of radius a and moment of inertia Mk^2 is placed gently at $t = 0$ on a horizontal table of coefficient of friction μ. It has initial velocity v and back spin Ω [Fig. 58 (b)].*

We measure x in the direction of v from an origin at the initial position of the cylinder, and θ in the direction corresponding to rolling in this direction.

Then when $t = 0$, $\quad \dot{x} = v, \quad \dot{\theta} = -\Omega, \quad$ (9)

and the velocity of the point of contact $\dot{x} - a\dot{\theta}$ is $v + a\Omega$. Since the point of contact initially slips forwards, the frictional force μR must act backwards. The equations of motion are
$$M\ddot{x} = -\mu R, \tag{10}$$
$$Mk^2\ddot{\theta} = \mu Ra, \tag{11}$$
$$0 = R - Mg. \tag{12}$$

From (10) with $\dot{x} = v$ when $t = 0$,
$$\dot{x} = v - \mu g t, \tag{13}$$
that is, the velocity of the cylinder decreases linearly.

From (11) with $\dot{\theta} = -\Omega$ when $t = 0$,

$$k^2\dot{\theta} = -k^2\Omega + \mu gat, \tag{14}$$

that is, the backspin decreases linearly.

The velocity of the point of contact is

$$\dot{x} - a\dot{\theta} = (v + a\Omega) - \mu gt(1 + a^2/k^2). \tag{15}$$

This becomes zero, and so rolling commences, when

$$t = \frac{k^2(v + a\Omega)}{\mu g(a^2 + k^2)}. \tag{16}$$

At this instant the velocity of the centre is by (13)

$$\frac{va^2 - a\Omega k^2}{k^2 + a^2}. \tag{17}$$

If $v > \Omega k^2/a$ this is positive and the cylinder rolls forwards; if $v = \Omega k^2/a$ it comes to rest; and if $v < \Omega k^2/a$ it rolls backwards.

70. Impulsive motion

When impulsive forces act on a system of rigid bodies we assume as in § 39 that the time τ during which they act is so small that the changes in position of the bodies during it are negligible. By integrating the equations of motion over the small time τ the changes in velocity due to the blows can be found.

If $\mathbf{P}$ is a typical impulsive force applied at $t = 0$, we write

$$\mathbf{J} = \int_0^\tau \mathbf{P}\, dt \tag{1}$$

for the impulse of the blow.

We now obtain four results corresponding to § 66 I to IV.

Integrating § 66 (8) over the small time τ gives

$$\mathbf{L}_f - \mathbf{L}_i = \sum \mathbf{J}, \tag{2}$$

where $\mathbf{L}_i$ and $\mathbf{L}_f$ are the linear momenta of the body before and after the blow. Thus:

I. *The change in linear momentum of the body is equal to the vector sum of the impulses of the blows applied to it.*

Again, integrating § 66 (9) over the small time τ and remembering that the changes in the $\mathbf{r}$ in this time are negligible, gives

$$\mathbf{H}_f - \mathbf{H}_i = \sum \mathbf{r} \wedge \mathbf{J}, \tag{3}$$

where $\mathbf{H}_i$ and $\mathbf{H}_f$ are the angular momenta of the body about the origin before and after the blow. That is:

II. *The change in angular momentum about a fixed origin is equal to the vector sum of the moments of the impulses of the blows about the origin.*

In the same way from § 66 (16):

III. *M times the change in linear velocity of the centre of mass of the body is equal to the vector sum of the impulses of the blows.*

And finally from § 66 (20):

IV. *The change in angular momentum about the centre of mass is equal to the vector sum of the moments of the impulses of the blows about the centre of mass.*

Usually III and either II or IV are used to determine the change in the motion. If a system is set in motion by blows, we have to determine the initial values of its velocities and angular velocities by the methods of this section and subsequently to study the motion with these initial conditions as in §§ 68, 69. If the Laplace transformation is used, treating a blow of impulse $\mathbf{J}$ as a force $\mathbf{J}\,\delta(t)$, the preliminary calculation of the initial conditions is avoided, but this method is only available when the equations of motion are linear.

Ex. 1. *A lamina can rotate freely in its own plane about a hinge O, its moment of inertia about an axis through O perpendicular to its plane being Mk^2. It is set in motion by a blow of impulse P in a direction perpendicular to the line joining O and the centre of mass G.*

If ω is the angular velocity of the lamina after the blow, and a is the distance of the line of action of P from the hinge, Fig. 59 (a), we have by II

$$Mk^2\omega = Pa. \tag{4}$$

There will be an impulsive reaction at the hinge O: suppose that X and Y are the components of this impulse in the direction of P and perpendicular to it. Then if h is the distance OG, we have by III

$$Mh\omega = P+X, \tag{5}$$

$$0 = Y. \tag{6}$$

Therefore, using (4),

$$X = P\left(\frac{ah}{k^2}-1\right). \tag{7}$$

If $a = k^2/h$, the length of the simple equivalent pendulum, $X = 0$, so that if the blow is struck at this point (called the

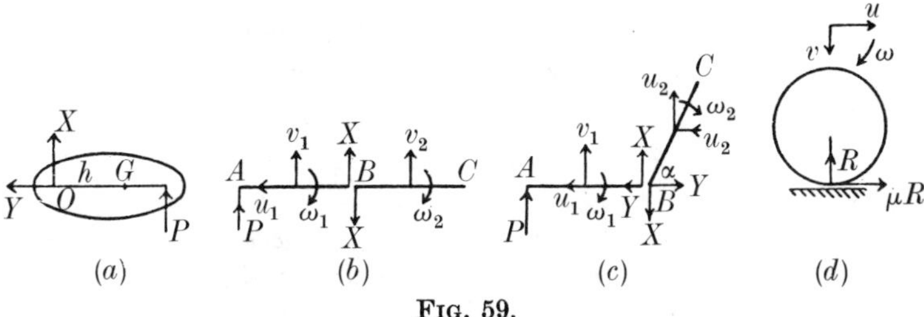

FIG. 59.

centre of percussion for this reason) there is no reaction at the hinge. If $a > k^2/h$ the reaction X is positive, that is, in the direction of the blow; if $a < k^2/h$ it is in the opposite direction.

The case of any body free to rotate about an axis which is carried in bearings may be treated as in § 67.

Ex. 2. *Two equal uniform rods AB, BC, each of length $2a$ and mass M, are freely hinged at B and are at rest in a straight line when a blow of impulse P is struck at A in a direction perpendicular to the rods.*

Let v_1 and v_2 be the velocities of the centres of the rods after the blow, and let ω_1 and ω_2 be the angular velocities of the rods, Fig. 59 (*b*). Since the velocity of B, calculated from the assumed velocity and angular velocity of the rod AB, must be the same as that calculated from BC, we must have

$$v_1 - a\omega_1 = v_2 + a\omega_2. \tag{8}$$

If desired, v_2 in Fig. 59 (*b*) could have been replaced by $v_1 - a\omega_1 - a\omega_2$ and the use of (8) avoided; it is desirable to proceed in this way in more complicated problems.

There will be a blow of impulse X on the rod AB at the hinge, and an equal and opposite blow on BC.

From IV we get

$$\tfrac{1}{3}Ma^2\omega_1 = (P-X)a, \tag{9}$$

$$\tfrac{1}{3}Ma^2\omega_2 = -Xa. \tag{10}$$

Also from III

$$Mv_1 = P+X, \tag{11}$$

$$Mv_2 = -X. \tag{12}$$

(8) to (12) are five algebraic equations for v_1, v_2, ω_1, ω_2, X.

Ex. 3. *The rods in Ex. 2 are inclined at an angle* α.

In this case we must assume components of velocity of the rods, and components of the impulse at the hinge, both along and perpendicular to the rod AB [Fig. 59 (c)].

III and IV now give

$$Mu_1 = Y,$$
$$Mv_1 = P+X,$$
$$X = -Mv_2 = -M(v_1-a\omega_1-a\omega_2\cos\alpha),$$
$$Y = -Mu_2 = -M(u_1-a\omega_2\sin\alpha),$$
$$\tfrac{1}{3}Ma^2\omega_1 = (P-X)a,$$
$$\tfrac{1}{3}Ma^2\omega_2 = -aX\cos\alpha-aY\sin\alpha.$$

Ex. 4. *A sphere of mass M and radius a, spinning about a horizontal axis with angular velocity* $\omega > u/a$ *impinges on a horizontal plane with velocity v towards the plane and u parallel to it, Fig. 59 (d). The coefficient of resistution is e, and the coefficient of friction is* μ.

If R is the impulsive reaction normal to the plane we have from III

$$R = Mv(1+e). \tag{13}$$

At the instant of contact, the horizontal component of the velocity of the point of contact is

$$u-a\omega,$$

which by hypothesis is negative. We thus assume an impulsive frictional force $\mu R = \mu Mv(1+e)$ acting forwards. Then if U and Ω are the horizontal component of the velocity, and the angular velocity, after the impact

$$M(U-u) = \mu R = \mu Mv(1+e), \tag{14}$$
$$\frac{2Ma^2}{5}(\Omega-\omega) = -\mu Ra = -\mu Mva(1+e). \tag{15}$$

Therefore

$$U = u+\mu v(1+e),$$
$$\Omega = \omega-\frac{5\mu v}{2a}(1+e).$$

Thus the horizontal component of the velocity is increased and the angular velocity decreased.

71. The gyrostat†

A gyrostat consists essentially of a solid of revolution which spins about its axis of symmetry. This axis is freely hinged at a point O of itself. The centre of mass G of the solid is on the

† The name gyrostat is usually used for systems of this type in which gravity has to be considered: the system is also that of a top spinning about a fixed point. In the gyroscope the centre of mass is at the origin, so that gravity exerts no moment about O, and the motion of the system caused by given external couples has to be considered. In the usual mounting in concentric rings or gimbals, θ and ψ specify the positions of the rings and so have a fundamental significance.

axis and distant h from the hinge O, so that the force of gravity has a moment about O. We shall consider the effect of gravity alone; if there are, in addition, externally applied forces they are treated in the same way.

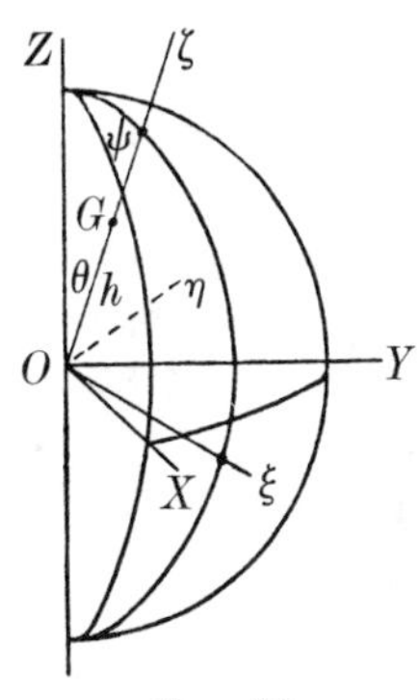

Fig. 60.

First we have to specify the position of the solid in space. Let OX, OY, OZ be fixed rectangular axes through the hinge O with OZ vertical. We specify the position of $O\zeta$, the axis of symmetry of the body, by its spherical polar coordinates relative to the axes OX, OY, OZ: that is, θ is the angle between $O\zeta$ and OZ, and ψ is the angle between the planes $ZO\zeta$ and ZOX. We then take a system of axes $O\xi$, $O\eta$, $O\zeta$ such that $O\xi$ lies in the plane $ZO\zeta$ and makes an angle $\frac{1}{2}\pi+\theta$ with OZ, while $O\eta$ is chosen so that $O\xi$, $O\eta$, $O\zeta$ form a right-handed system of rectangular axes. Finally, let ϕ be the angle between a marked plane in the body through its axis $O\zeta$ and the plane $\xi O\zeta$. The position of any point of the body is then known if θ, ψ, ϕ are known; these are called the Eulerian angles and are used in most problems of this type.

There are many ways of deriving the equations of motion of the gyrostat: in this section we shall use moving axes; alternative deductions using Lagrange's equations, and using the energy and momentum equations, are given in §§ 76, 74.

We take the system $O\xi$, $O\eta$, $O\zeta$ as moving axes; cf. § 64. Their components of angular velocity about their instantaneous directions are

$$(-\dot{\psi}\sin\theta, \dot{\theta}, \dot{\psi}\cos\theta). \tag{1}$$

The components of angular velocity of the body itself in the instantaneous directions of the axes are

$$(-\dot{\psi}\sin\theta, \dot{\theta}, \dot{\phi}+\dot{\psi}\cos\theta). \tag{2}$$

Since $O\zeta$ is the axis of symmetry of the body, its moments of inertia about all axes in the plane $\xi O\eta$ will be the same, say A, and all the products of inertia will vanish. If C is the moment of inertia about $O\zeta$, the moments of inertia about the axes $O\xi$, $O\eta$, $O\zeta$ will be A, A, C, respectively, and the components of the angular momentum of the body along the instantaneous directions of these axes will be

$$\{-A\dot{\psi}\sin\theta, A\dot{\theta}, C(\dot{\phi}+\dot{\psi}\cos\theta)\}. \tag{3}$$

The components of the moment of the external force (gravity) about O are

$$(0, Mgh\sin\theta, 0). \tag{4}$$

The equations of motion are now obtained by equating the components of the rate of change of angular momentum along the instantaneous directions of the axes $O\xi$, $O\eta$, $O\zeta$ to the components of the moment of the external force in these directions. That is, using (3) and (1) in § 64 (4),

$$-A\frac{d}{dt}(\dot{\psi}\sin\theta)-A\dot{\theta}\dot{\psi}\cos\theta+C\dot{\theta}(\dot{\phi}+\dot{\psi}\cos\theta) = 0, \tag{5}$$

$$A\frac{d}{dt}(\dot{\theta})-A\dot{\psi}^2\sin\theta\cos\theta+C\dot{\psi}\sin\theta(\dot{\phi}+\dot{\psi}\cos\theta) = Mgh\sin\theta, \tag{6}$$

$$C\frac{d}{dt}(\dot{\phi}+\dot{\psi}\cos\theta) = 0. \tag{7}$$

(7) gives immediately

$$\dot{\phi}+\dot{\psi}\cos\theta = n, \tag{8}$$

where n is a constant (Cn is the angular momentum about the axis of symmetry, and, in the absence of friction, this is constant).

Using (8), (5) and (6) become

$$A\frac{d}{dt}(\dot{\psi}\sin\theta)+A\dot{\theta}\dot{\psi}\cos\theta-Cn\dot{\theta} = 0, \tag{9}$$

$$A\ddot{\theta}-A\dot{\psi}^2\sin\theta\cos\theta+Cn\,\dot{\psi}\sin\theta = Mgh\sin\theta. \tag{10}$$

Multiplying (9) by $\sin\theta$, it may be written

$$\frac{d}{dt}(A\dot{\psi}\sin^2\theta+Cn\cos\theta) = 0. \tag{11}$$

Therefore, integrating,

$$A\dot{\psi}\sin^2\theta+Cn\cos\theta = H, \tag{12}$$

where H is a constant (which is the constant angular momentum about the vertical). Finally, multiplying (9) by $\dot{\psi}\sin\theta$, (10) by $\dot{\theta}$, and adding gives

$$A\dot{\theta}\ddot{\theta}+A\dot{\psi}\sin\theta\,\frac{d}{dt}(\dot{\psi}\sin\theta)-Mgh\dot{\theta}\sin\theta = 0. \tag{13}$$

Integrating (13) gives

$$\tfrac{1}{2}A(\dot{\theta}^2+\dot{\psi}^2\sin^2\theta)+Mgh\cos\theta = E, \tag{14}$$

where E is a constant. This will be seen in § 76 to be the energy equation.

(8), (12), and (14) are three first integrals of the motion. To study the motion further we eliminate $\dot{\psi}$ from (14) by using (12), and get an equation for θ only. It is a little simpler to work in terms of the new variable

$$x = \cos\theta. \tag{15}$$

Also, for shortness, we write

$$\frac{Cn}{A} = a, \qquad \frac{H}{A} = b, \qquad \frac{Mgh}{A} = c, \qquad \frac{2E}{A} = d, \tag{16}$$

so that (12) and (14) become

$$\dot{\psi} = \frac{b-ax}{1-x^2}, \tag{17}$$

$$\dot{x}^2+\dot{\psi}^2(1-x^2)^2 = (d-2cx)(1-x^2). \tag{18}$$

Using (17) in (18) we get

$$\dot{x}^2 = (d-2cx)(1-x^2)-(b-ax)^2 \tag{19}$$

This is an equation of the form studied in § 55. We remark that, writing $f(x)$ for the right-hand side of (19), $f(x)$ is a cubic in x and so the complete solution of (19) will in general involve elliptic functions [cf. § 55, Ex. 2].

Since $x = \cos\theta$, we are interested in the range $-1 \leqslant x \leqslant 1$ of x, and by (19) both $f(1)$ and $f(-1)$ are negative. But at the point of projection $\dot{x}^2$, and thus $f(x)$, must be positive. Thus $f(x)$ must have two zeros in $-1 \leqslant x \leqslant 1$, and so the motion consists of an oscillation between two fixed values of x or θ.

The most interesting case is that of steady motion in which

these fixed values coincide so that $\theta = \alpha$, constant. Putting $\dot{\theta} = 0$ in (9) we find that $\dot{\psi}$ must be constant, and from (10)

$$A\dot{\psi}^2 \cos\alpha - Cn\dot{\psi} + Mgh = 0. \tag{20}$$

This equation has real roots if

$$C^2n^2 \geqslant 4MghA \cos\alpha. \tag{21}$$

If this condition is satisfied, *steady precession* of this type is possible with $\theta = \alpha$ and with either of the angular velocities

$$\dot{\psi} = \{Cn \pm \sqrt{(C^2n^2 - 4MghA \cos\alpha)}\}/2A \cos\alpha.$$

Since, usually, the angular velocity of spin about the axis is large, $Cn \gg 4MghA \cos\alpha$ and these angular velocities become approximately

$$Cn/A \cos\alpha, \quad \text{'quick precession'}, \tag{22}$$

$$Mgh/Cn, \quad \text{'slow precession'}. \tag{23}$$

These results, and extensions such as the period of small oscillations about steady precession, can also be found from a further study of (19)—the condition for steady precession at $\cos\alpha = x_0$ is that $f(x)$ should have a double zero at x_0.

The gyro-compass

Suppose the gyroscope is pivoted so that its centre of mass coincides with the hinge O of Fig. 60, and suppose further that its axis is constrained to move in a horizontal plane. We consider the effect of the earth's rotation on its motion when it is in latitude λ.

Suppose the axis $O\zeta$ of the gyroscope makes an angle θ with the axis OY of Fig. 51 (*b*), that $O\xi$ lies along OZ, and that $O\eta$ makes a right-handed system, Fig. 61, where the axes OX, OY, OZ are those chosen in § 64, Ex. 2.

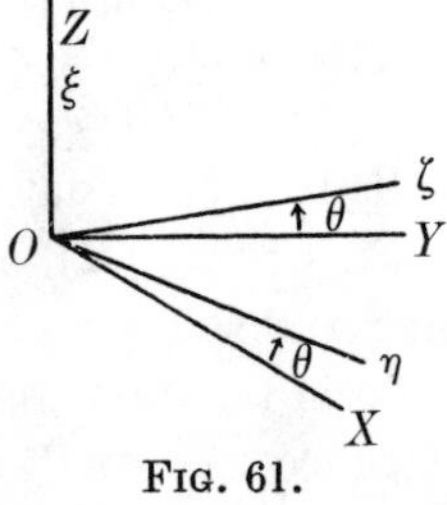

FIG. 61.

The components of the earth's angular velocity in the directions of OX, OY, OZ are by § 64 (7)

$$(0, \omega\cos\lambda, \omega\sin\lambda). \tag{24}$$

Therefore, in the directions $O\xi$, $O\eta$, $O\zeta$ they are

$$(\omega\sin\lambda, \omega\cos\lambda\sin\theta, \omega\cos\lambda\cos\theta). \tag{25}$$

The components of the angular velocity of the system $O\xi$, $O\eta$, $O\zeta$ about their instantaneous directions are

$$(\dot{\theta} + \omega\sin\lambda, \omega\cos\lambda\sin\theta, \omega\cos\lambda\cos\theta). \tag{26}$$

If ϕ and the moments of inertia are defined as before, the components of the angular momentum of the body along the instantaneous directions of $O\xi$, $O\eta$, $O\zeta$ are

$$\{A(\dot\theta+\omega\sin\lambda),\ A\omega\cos\lambda\sin\theta,\ C(\dot\phi+\omega\cos\lambda\cos\theta)\}. \tag{27}$$

There are no couples about the axes $O\xi$ and $O\zeta$. Thus, writing down the rates of change of angular momentum about these axes by using (27) and (26) in § 64 (4), we get

$$A\ddot\theta+C\omega\cos\lambda\sin\theta(\dot\phi+\omega\cos\lambda\cos\theta)-A\omega^2\cos^2\lambda\sin\theta\cos\theta = 0, \tag{28}$$

$$\frac{d}{dt}(\dot\phi+\omega\cos\lambda\cos\theta) = 0. \tag{29}$$

(29) gives $\dot\phi+\omega\cos\lambda\cos\theta = n$, and substituting this in (28) and neglecting the small term in ω^2 gives

$$A\ddot\theta+Cn\omega\cos\lambda\sin\theta = 0. \tag{30}$$

Thus there is equilibrium if $\theta = 0$, that is, if the axis points north, and the period of small oscillations about equilibrium is

$$2\pi\left(\frac{A}{Cn\omega\cos\lambda}\right)^{\frac{1}{2}}. \tag{31}$$

The simple arrangement described above is unsatisfactory owing to the friction introduced by the constraining couple, but the theory of the systems used in practice follows the same lines.

EXAMPLES ON CHAPTER VIII

1. Show that the moments of inertia of a right circular cone of semi-vertical angle α, height h, and mass M are

$$(3/10)Mh^2\tan^2\alpha,\quad (3/20)Mh^2(\tan^2\alpha+4),\quad (3/20)Mh^2(\tan^2\alpha+5\sin^2\alpha)$$

about its axis, a perpendicular to the axis through the vertex, and a slant side, respectively.

2. Show that the moments and products of inertia of a uniform triangle of mass M relative to any axes are the same as those of three particles of mass $M/3$ placed at the mid-points of the sides of the triangle. (Drop a perpendicular from a vertex to the opposite side and take these lines as axes.)

3. A thin wire $ABCD$ whose mass per unit length is ρ is bent in a plane so that $AB = CD = a$, $BC = 2b$, the angles ABC and BCD are 90°, and AB and CD are on opposite sides of BC. Show that the principal axes at the centre of mass are inclined at θ and $\frac{1}{2}\pi-\theta$ to the wire, where

$$\theta = \tfrac{1}{2}\tan^{-1}\frac{3a^2b}{a^3-b^3-3ab^2}.$$

4. l is the length of the simple equivalent pendulum of a rigid body when it oscillates about an axis through O. Show that it will oscillate with the same period about a parallel axis which is distant l from the original axis measured in the direction towards the centre of mass, and

is in the plane of the original axis and the centre of mass. Show also that there are two other axes in this plane about which the period has this value.

5. Show that if the period of small oscillations of a rigid body under gravity about an axis fixed to it is T when the axis is horizontal, it is $T(\sec\theta)^{\frac{1}{2}}$ when the axis is inclined at θ to the horizontal.

6. A fly-wheel of mass M and moment of inertia Mk^2 about its axis has its centre of mass distant h from the axis. Show that, if it is rotating freely with maximum angular velocity ω about a horizontal axis, the horizontal component of the reaction on its bearings is

$$-\frac{3Mgh^2}{k^2}\sin\theta\cos\theta - Mh\left(\omega^2 - \frac{2gh}{k^2}\right)\sin\theta,$$

where θ is the angle the plane containing the axis and the centre of mass makes with the downward vertical.

7. A uniform thin circular disk of mass M and radius a is mounted on a shaft through its centre which makes an angle ϕ with its plane. The shaft is carried in two bearings, each distant b from the centre of the disk, and rotates with angular velocity ω. Show that the reaction of the bearings is a couple of moment

$$\tfrac{1}{4}Ma^2\omega^2\sin\phi\cos\phi$$

in the plane of the shaft and the perpendicular to the disk.

8. In a piston engine AO is the crank of length a, AB is the connecting rod of length b, and the crank rotates uniformly so that the angle AOB is ωt. If the angle ABO is ϕ, show that, neglecting terms in $(a/b)^4$,

$$\cos\phi = 1-(a^2/2b^2)\sin^2\omega t.$$

If the reciprocating parts are of mass M and the connecting rod is uniform and of mass m, show that the reaction on the crankpin in the direction OB is

$$(M+m)\omega^2 a\cos\omega t + (a^2\omega^2/b)(M+\tfrac{1}{2}m)\cos 2\omega t,$$

neglecting friction, and find the reaction in the perpendicular direction.

9. A mass M is hung by two single pulley blocks of mass M'; the wheel in each is of radius a and moment of inertia mk^2. Show that if the rope is allowed to run out freely, the mass M will descend with acceleration

$$\frac{(M+M')g}{M+M'+5mk^2/a^2},$$

neglecting friction and the mass of the rope. Find the corresponding result if friction is included.

10. A solid cylinder of radius b rolls without slipping on the inside of a fixed, hollow, horizontal cylinder of radius a. Discuss the motion, and show that the period of small oscillations about the lowest point is

$$2\pi\{3(a-b)/2g\}^{\frac{1}{2}}.$$

11. A rod OP of length $a+b$, mass M_1, and moment of inertia about O of $M_1 k_1^2$, is freely hinged at O. At P it carries in a frictionless bearing a gear of radius a, mass M, and moment of inertia Mk^2, which meshes with a fixed gear of radius b and centre O. If torque T about O is applied to the rod OP, show that its angular acceleration is

$$T/\{M_1 k_1^2+M(a+b)^2(k^2+a^2)/a^2\}.$$

12. A straight uniform rod slides down in a vertical plane perpendicular to two smooth planes, one of which is vertical and the other horizontal. Find its motion if it is initially at rest at an angle α to the horizontal and with its ends in contact with the planes. Show that contact with the vertical plane ceases when the inclination of the rod to the horizontal is

$$\sin^{-1}\{(2/3)\sin\alpha\}.$$

13. A uniform sphere is projected up an imperfectly rough plane inclined at an angle α to the horizontal, the coefficient of friction being μ. If, initially, its velocity is V and it is not rotating, show that the distance traversed up the plane while there is slipping at the point of contact is

$$\frac{2V^2(6\mu\cos\alpha+\sin\alpha)}{g(7\mu\cos\alpha+2\sin\alpha)^2}.$$

14. Two equal uniform rods AB, BC, each of mass m, are freely hinged at B and rest on a smooth horizontal table folded together so that A and C are touching. The end A is pulled away from C by a blow of impulse P. Show that the initial velocity of C is $P/2m$.

15. A circular cylinder of radius a whose centre of mass is at a distance h from its geometrical centre rolls on a rough horizontal plane. Write down the equations of motion and show that the period of small oscillations about the position of stable equilibrium is

$$2\pi(k^2/gh)^{\frac{1}{2}},$$

where k is the radius of gyration of the cylinder about the generator nearest to the centre of mass.

16. A rectangular plate of sides $2a$ and $2b$ and mass M is supported on four equal springs of stiffness λ and performs small oscillations. Show that the normal modes consist of a vertical motion of frequency n/π, and oscillations about the axes through the centre of mass parallel to the sides with frequency $(n\sqrt{3})/\pi$, where $n^2 = \lambda/M$.

17. A mass $M/6$ is connected by a spring of stiffness λ to one end of a uniform rod of mass M and length $2a$. The rod is freely hinged at its centre, and its other end is connected to a fixed point by a spring of stiffness λ. In the equilibrium position the springs are perpendicular to the rod, and the springs and rod are in the same plane. Show that the natural frequencies of small oscillations are, writing $n^2 = \lambda/M$,

$$(6\pm 3\sqrt{2})^{\frac{1}{2}}n/2\pi,$$

and find the normal modes of oscillation.

18. A uniform log AB of length $2l$ is pushed with velocity V on to a roller rotating with constant angular velocity ω. The other end of the log rests on a horizontal plane at the same height as the top of the roller, and the direction of the log is perpendicular to that of the roller. If the coefficients of friction between the log and the plane, and the log and the roller, both have the same value μ, show that when the mid-point of the log is on the roller its velocity is

$$\{V^2+2\mu lg(2\ln 2-1)\}^{\frac{1}{2}},$$

provided friction at the roller always acts forwards. Discuss also the cases in which this assumption is not true.

19. A uniform rectangular plate $ABCD$ is freely hinged at A and B, and is struck a blow P normal to its plane at D. Show that the reactions at A and B are $-P/4$ and $3P/4$, respectively.

20. A uniform rod AB of mass M and length a is freely hinged at B. An equal rod is freely hinged at a point on the perpendicular to AB at B and rests against the rod AB, making an angle of 45° with it. The rod AB is struck at a distance b from B by a blow P normal to it and in the plane of the rods; show that, if the contact between them is smooth and the rods remain in contact, the initial angular velocity of AB is $3Pb/2Ma^2$. Discuss the validity of the assumptions made.

21. A gyroscope is pivoted so that it rotates about its centre of mass. Find its equations of motion if couples G_ξ and G_η are applied to it about the axes $O\xi$ and $O\eta$. Show that, if ψ is kept constant, G_ξ is proportional to $\dot{\theta}$, and thus the gyroscope can be used to generate the differential coefficient of a function mechanically.

22. Discuss the motion of a nearly vertical gyrostat as follows. Replace $\sin\theta$ by θ and $\cos\theta$ by 1 in the equations of motion, § 71 (9) and (10). By adding i times the first of these to the second, show that $\zeta = \theta e^{i\psi}$ satisfies

$$A\ddot{\zeta}-Cni\dot{\zeta}-Mgh\zeta = 0.$$

Show that the path of a point on the axis of the gyrostat is an ellipse which rotates with angular velocity $Cn/2A$.

23. If a ship is sailing with a velocity whose northerly component is v, show that a gyro-compass will point to the west of north by an amount

$$\frac{v}{\omega a\cos\theta},$$

where a is the earth's radius and θ is the latitude of the ship.

24. If $\mathbf{H}$ is the angular momentum of a body relative to an origin O' which moves with velocity $\mathbf{v}$ relative to a fixed origin O, and $\mathbf{L}$ is the linear momentum of the body, show that

$$\dot{\mathbf{H}} = \mathbf{G}-\mathbf{v}\wedge\mathbf{L},$$

where $\mathbf{G}$ is the sum of the moments of the external forces about O'. Discuss the motion of a cylinder rolling down an inclined plane by taking the point of contact as the origin O'.

25. A rigid body moves about a fixed point O. Taking the principal axes of inertia at O as a system of moving axes, show that if ω_1, ω_2, ω_3 are the components of the angular velocity of this system about their instantaneous directions, and G_1, G_2, G_3 are the components of the moment of the external forces about O in these directions, the equations of motion (Euler's equations) are

$$A\dot{\omega}_1-(B-C)\omega_2\omega_3 = G_1, \text{ etc.}$$

IX

THE ENERGY EQUATION AND LAGRANGE'S EQUATIONS

72. Potential energy

WE consider first a single particle in a field of force. The field is supposed to be a vector field, that is, at each point whose position vector is $\mathbf{r}$ there is a force $\mathbf{P}$ on the particle, where $\mathbf{P}$ is a given function of $\mathbf{r}$. Let (x, y, z) and (X, Y, Z) be the components of $\mathbf{r}$ and $\mathbf{P}$ relative to fixed rectangular axes.

If the particle is at $\mathbf{r}$ and is given a small displacement $\delta\mathbf{r}$, the forces of the field do work

$$\mathbf{P}.\delta\mathbf{r} = X\,\delta x + Y\,\delta y + Z\,\delta z \tag{1}$$

on the particle. This conception needs stating a little more precisely. We think of the particle as being placed in the field and held there by some external agency which applies to it a force which just balances the forces exerted by the field on it. When the particle is 'given a small displacement', it is implied that the forces due to the external agency are relaxed slightly so that the particle can move a small distance, but they are always maintained almost balancing the field forces, and the process is carried out infinitely slowly so that the particle gains no momentum. The amount of work (1) is then done by the field forces, and it is absorbed by the agency which holds the particle in position.

Now suppose that p_1 is some path joining two points A and B, let δs be the element of arc of this path, and let $\mathbf{t}$ be a unit vector along its tangent at any point. Then the displacement $\delta\mathbf{r}$ in (1) is $\mathbf{t}\,\delta s$, and the work done by the forces of the field on the particle when it is made to move in the manner described above from A to B along the path p_1 may be written in any of the forms

$$\int\limits_{p_1}\left(X\frac{dx}{ds}+Y\frac{dy}{ds}+Z\frac{dz}{ds}\right)ds = \int\limits_{p_1}\mathbf{P}.\mathbf{t}\,ds = \int\limits_{p_1}\mathbf{P}.d\mathbf{r}, \tag{2}$$

where the integrals are taken from A to B along the path p_1. If this path is retraced from B to A, the same amount of work would have to be done by the external agency which holds the particle in position, since it has to exert force $-\mathbf{P}$ at the point $\mathbf{r}$. Thus in going from A to B and returning by the same path, neither the field forces nor those of the external agency do any net amount of work.

Clearly this need not be the case if the particle returns from B to A by a different path p_2; in this case the net amount of work done by the field forces in the round trip is

$$\int\limits_{p_1} \mathbf{P}.d\mathbf{r} - \int\limits_{p_2} \mathbf{P}.d\mathbf{r}, \tag{3}$$

and this amount of work is available to the external agency. By repeating the cycle an indefinitely large amount of work can be obtained.

The field of forces is called conservative if this cannot be done. Clearly from (3) the condition for this is that the integral (2) be independent of the path from A to B for all points A and B. That is, the work done by the field forces on the particle in going from A to B must depend on the points A and B only. Let S be some point chosen as a standard position, then this work is

$$\int\limits_A^B \mathbf{P}.d\mathbf{r} = \int\limits_A^B \left(X\frac{dx}{ds}+Y\frac{dy}{ds}+Z\frac{dz}{ds}\right)ds = V(A)-V(B), \tag{4}$$

where $$V(A) = \int\limits_A^S \mathbf{P}.d\mathbf{r} = \int\limits_A^S \left(X\frac{dx}{ds}+Y\frac{dy}{ds}+Z\frac{dz}{ds}\right)ds. \tag{5}$$

This quantity $V(A)$ is called the potential energy at the point A in the field. It is the work done by the forces of the field in taking the particle by any path from the point A to the standard position S; it is also equal to the work which has to be done by an external agency in taking the particle from S to A.

When the potential energy is known, the forces of the field can be calculated. For by (4) the work done by the forces of the field in going from (x,y,z) to $(x+\delta x, y+\delta y, z+\delta z)$ is

$$V(x,y,z)-V(x+\delta x, y+\delta y, z+\delta z)$$
$$= -\frac{\partial V}{\partial x}\,\delta x-\frac{\partial V}{\partial y}\,\delta y-\frac{\partial V}{\partial z}\,\delta z-\ldots, \tag{6}$$

and by (1) it is $\qquad X\,\delta x+Y\,\delta y+Z\,\delta z.$

Thus we have

$$X = -\frac{\partial V}{\partial x}, \qquad Y = -\frac{\partial V}{\partial y}, \qquad Z = -\frac{\partial V}{\partial z}. \tag{7}$$

It will appear below that it is often simpler to calculate the potential energy of a system and then find the forces by differentiating in this way than to calculate the forces directly. Also the same method holds in other coordinate systems; for example, if the position of the particle is specified by plane polar coordinates (r,θ) the potential energy is $V(r,\theta)$, and if R, Θ are the radial and transverse components of force on the particle the work done by these in a small displacement $(\delta r, r\,\delta\theta)$ is

$$R\,\delta r+r\Theta\,\delta\theta = V(r,\theta)-V(r+\delta r,\theta+\delta\theta)$$
$$= -\frac{\partial V}{\partial r}\,\delta r-\frac{\partial V}{\partial \theta}\,\delta\theta\ldots,$$

and thus $$R = -\frac{\partial V}{\partial r}, \qquad \Theta = -\frac{\partial V}{r\,\partial\theta}. \tag{8}$$

For an assemblage of particles, or a rigid body, the potential energy is the sum of the potential energies of the several particles. In this connexion it should be remarked that a number of common types of force do no work in a displacement of the system. These are:

(i) the reaction at a smooth surface;

(ii) the reaction at a frictionless hinge;

(iii) tension or compression in an inextensible rod connecting two bodies;

(iv) the reaction at a rolling contact.

In (i) the reaction is perpendicular to the displacement, while in (iv) the displacement is zero; thus no work is done in either case. In (ii) and (iii) there are equal, oppositely directed, actions and reactions on the two bodies and the displacements are the same so the net work is zero.

We now calculate the potential energy in some systems of practical interest.

(i) *Gravity*. Taking the z-axis vertically upwards and the reference position as $z = 0$ we have

$$Z = -mg,$$

$$V = \int_z^0 (-mg)\,dz = mgz. \tag{9}$$

(ii) *A spring of stiffness* λ. Considering displacements along the x-axis, and taking the reference position at the point $x = 0$ where the spring is unstrained

$$X = -\lambda x,$$

$$V = \int_x^0 (-\lambda x)\,dx = \tfrac{1}{2}\lambda x^2. \tag{10}$$

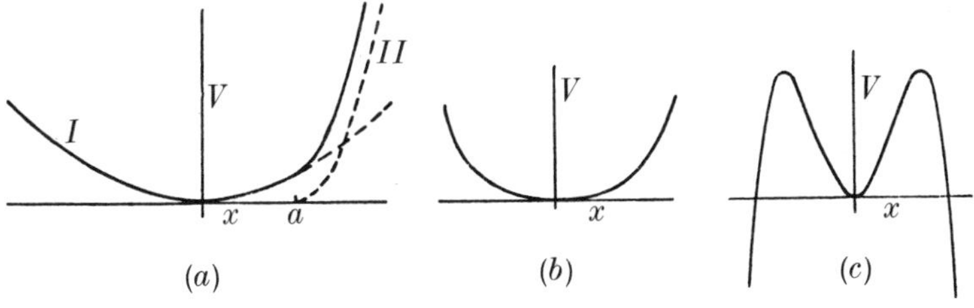

FIG. 62.

(iii) *A spring of stiffness* λ *which comes into contact with another spring of stiffness* λ_1 *when* $x = a$. If $x \leqslant a$ the potential energy is given by (10). If $x > a$ an amount $\frac{1}{2}\lambda_1(x-a)^2$, corresponding to the energy stored in the second spring, has to be added. These quantities are shown in curves I and II of Fig. 62 (a).

(iv) *The anharmonic oscillator* § 55 (28) *in which* $X = -n^2x-bx^3$. Taking the reference position as $x = 0$ we have

$$V = \int_x^0 (-n^2x-bx^3)\,dx = \tfrac{1}{2}n^2x^2+\tfrac{1}{4}bx^4. \tag{11}$$

If $b > 0$ the potential energy increases steadily as $|x|$ increases, Fig. 62 (b). If $b < 0$ it vanishes when $x = (-2n^2/b)^{\frac{1}{2}}$ and has a maximum $(-n^4/4b)$ when $x = (-n^2/b)^{\frac{1}{2}}$; cf. Fig. 62 ($c$).

(v) *The anharmonic oscillator* § 55 (29) *for which* $X = -n^2x-ax^2$. Here

$$V = \tfrac{1}{2}n^2x^2+\tfrac{1}{3}ax^3. \tag{12}$$

If, for example, $a > 0$, the curve of V increases steadily for positive x. For negative x there is a zero at $x = -3n^2/2a$, and a maximum of $n^6/6a^2$ at $x = -n^2/a$ (Fig. 63 (a)).

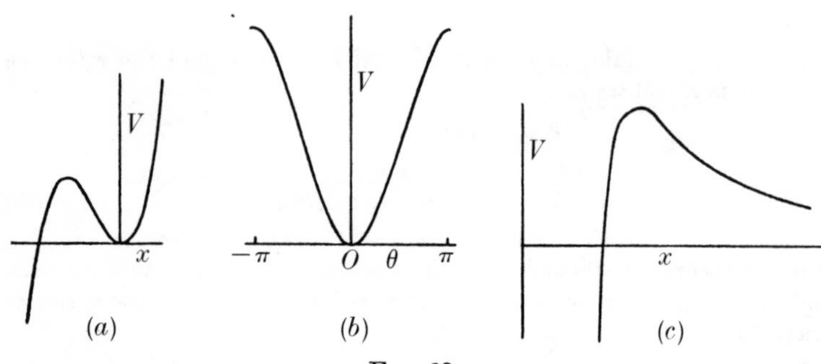

FIG. 63.

(vi) *The simple pendulum.* Taking the reference level as the lowest point of the swing, we have $z = l(1-\cos\theta)$ in (9) and so, cf. Fig. 63 (b),

$$V = mgl(1-\cos\theta). \tag{13}$$

(vii) *Inverse square attraction or repulsion.* Suppose the particle is attracted to the point $x = 0$ by force μ/x^2. The reference position is taken as $x = \infty$. Then

$$V = \int\limits_x^\infty \left(-\frac{\mu}{x^2}\right)dx = -\frac{\mu}{x}. \tag{14}$$

For repulsion the sign is changed.

(viii) *Inverse square repulsion and inverse fifth-power attraction.* The force

$$X = \frac{\mu}{x^2}-\frac{a}{x^5} \tag{15}$$

is attractive at small distances and repulsive at large ones.

Taking the reference position as $x = \infty$,

$$V = \int\limits_x^\infty \left(\frac{\mu}{x^2}-\frac{a}{x^5}\right)dx = \frac{\mu}{x}-\frac{a}{4x^4}. \tag{16}$$

The potential energy has a maximum when $x = (a/\mu)^{\frac{1}{3}}$. It is sketched in Fig. 63 (c). Curves of this sort have been used to represent atomic nuclei.

(ix) *The potential energy of a particle of mass m in the field of a spherical shell of surface density σ and radius a, each element of area dS of which attracts the particle with force $\gamma m\sigma\, dS/R^2$, where R is the distance between the element of area and the particle, and γ is a constant.*

Suppose m is distant $r > a$ from the centre of the shell. The potential energy of m due to the strip of the shell between the angles θ and $\theta+\delta\theta$, all of which is at distance

$$R = \{a^2+r^2-2ar\cos\theta\}^{\frac{1}{2}}$$

from m, is by (14)
$$-\frac{2\pi a^2 m\sigma\gamma\sin\theta\,\delta\theta}{\{a^2+r^2-2ar\cos\theta\}^{\frac{1}{2}}}.$$

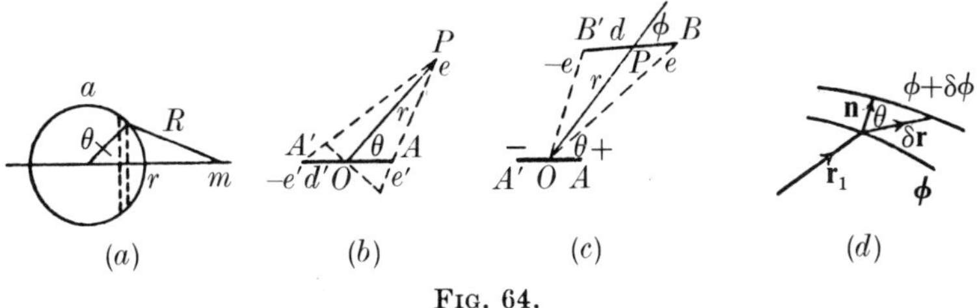

FIG. 64.

Thus the potential energy of m in the field of the whole shell is

$$V = -2\pi a^2 m\sigma\gamma\int\limits_0^\pi \frac{\sin\theta\,d\theta}{\{a^2+r^2-2ar\cos\theta\}^{\frac{1}{2}}} = -\frac{2\pi am\sigma\gamma}{r}\left[\{a^2+r^2-2ar\cos\theta\}^{\frac{1}{2}}\right]_0^\pi$$

$$= -\frac{2\pi am\sigma\gamma}{r}\{(r+a)-(r-a)\} \quad (17)$$

$$= -\frac{4\pi a^2\sigma\gamma m}{r} = -\frac{\gamma Mm}{r}, \quad (18)$$

where M is the mass of the shell. Thus the potential energy of the mass m is the same as if the mass of the shell were concentrated at its centre. The same result holds for a solid sphere.

If the mass m is inside the shell, so that $r < a$, the calculation remains the same to (17) in which $r-a$ is to be replaced by $a-r$, and we get finally in place of (18)

$$-\frac{\gamma Mm}{a}. \quad (19)$$

(x) *Potential energy in the field of a dipole.* An electric dipole of moment μ' consists of a charge e' at A and a charge $-e'$ at A', the distance $AA' = 2d'$ being very small and the product $2e'd' = \mu'$ being finite, Fig. 64 (*b*). We calculate the potential energy of a charge e in the field of this dipole: it is attracted to A' with force $ee'/A'P^2$, and repelled from A with force ee'/AP^2, so by (14) its potential energy is

$$V = \frac{ee'}{AP} - \frac{ee'}{A'P}. \quad (20)$$

We may specify the position of e by its distance $r = OP$ from the dipole, and its angular position θ from the direction $A'A$ of the dipole.

Since d' is very small we may neglect d'^2 when it occurs, and so, to this approximation,

$$AP = r - d'\cos\theta; \qquad A'P = r + d'\cos\theta.$$

Then from (20)

$$V = \frac{ee'}{r-d'\cos\theta} - \frac{ee'}{r+d'\cos\theta} = \frac{2ee'd'\cos\theta}{r^2} = \frac{e\mu'\cos\theta}{r^2}. \tag{21}$$

By (8) the radial and transverse components of the force on e are

$$R = \frac{2e\mu'\cos\theta}{r^3}, \qquad \Theta = \frac{e\mu'\sin\theta}{r^3}. \tag{22}$$

In rectangular Cartesians with the z-axis in the direction of the dipole we have $\cos\theta = z/r$ in (21), and the components of the force on e are

$$X = -\frac{\partial V}{\partial x} = \frac{3e\mu' zx}{r^5}, \qquad Y = \frac{3e\mu' zy}{r^5}, \qquad Z = \frac{e\mu'(3z^2-r^2)}{r^5}. \tag{23}$$

(xi) *The potential energy of a dipole of moment μ in the field of a dipole of moment μ'.* Suppose the second dipole is at P, (r,θ), relative to the first, and that its direction makes an angle ϕ with OP. We may suppose it to be composed of charges e at B and $-e$ at B', such that

$$PB = PB' = d \quad \text{and} \quad 2de = \mu,$$

d being very small; cf. Fig. 64 (c).

Then by (21) the potential energy is

$$\begin{aligned} V &= \frac{e\mu'\cos BOA}{BO^2} - \frac{e\mu'\cos B'OA}{B'O^2} \\ &= \frac{e\mu'\{\cos\theta + (d/r)\sin\phi\sin\theta\}}{(r+d\cos\phi)^2} - \frac{e\mu'\{\cos\theta - (d/r)\sin\phi\sin\theta\}}{(r-d\cos\phi)^2} \\ &= \frac{\mu\mu'}{r^3}(\sin\theta\sin\phi - 2\cos\theta\cos\phi), \end{aligned} \tag{24}$$

where, throughout, we have neglected terms in d^2.

Finally we define the gradient of a scalar function of position and discuss its relation to the theory given above. Suppose ϕ is a single-valued continuous scalar function of position $\mathbf{r}$, that is, ϕ has a given numerical value at each point $\mathbf{r}$, and if $\phi+\delta\phi$ is its value at $\mathbf{r}+\delta\mathbf{r}$, $\delta\phi \to 0$ as $\delta\mathbf{r} \to 0$. If this is the case, a set of *equipotential surfaces* can be drawn on each of which the value of ϕ is constant; there is one such surface through each point.

Suppose that ϕ and $\phi+\delta\phi$ are the values of the function at $\mathbf{r}_1$ and $\mathbf{r}_1+\delta\mathbf{r}$, and that $\mathbf{n}$ is a unit vector normal to the equipotential surface through $\mathbf{r}_1$, Fig. 64 (d). If $\partial\phi/\partial n$ is the rate of change of ϕ at $\mathbf{r}_1$ in the direction of $\mathbf{n}$, the gradient of ϕ is defined as the vector

$$\operatorname{grad}\phi = \frac{\partial\phi}{\partial n}\mathbf{n}. \tag{25}$$

The rate of change of ϕ in the direction of $\delta\mathbf{r}$ is

$$\frac{\partial\phi}{\partial r} = \lim_{\delta r\to 0}\frac{\delta\phi}{\delta r} = \frac{\partial\phi}{\partial n}\cos\theta,$$

where θ is the angle between $\mathbf{n}$ and $\delta\mathbf{r}$. It follows that the rate of change of ϕ is greatest in the direction of $\mathbf{n}$, and also that the magnitude of the component of the vector (25) in the direction of $\delta\mathbf{r}$ is $\partial\phi/\partial r$. In particular, $\partial\phi/\partial x$, $\partial\phi/\partial y$, and $\partial\phi/\partial z$ are the magnitudes of the components of $\operatorname{grad}\phi$ in the directions of the x-, y-, and z-axes of a rectangular system, and if $\mathbf{i}$, $\mathbf{j}$, $\mathbf{k}$ are unit vectors in the directions of these axes,

$$\operatorname{grad}\phi = \mathbf{i}\frac{\partial\phi}{\partial x}+\mathbf{j}\frac{\partial\phi}{\partial y}+\mathbf{k}\frac{\partial\phi}{\partial z}. \tag{26}$$

The potential energy V in a conservative field of force is a scalar function of position, and by (7) the force $\mathbf{P}$ at any point is connected with it by

$$\mathbf{P} = -\operatorname{grad} V. \tag{27}$$

73. The energy equation: applications

The equation of motion for a particle of mass m moving in the direction of the x-axis under force $f(x)$ in that direction is

$$m\ddot{x} = f(x). \tag{1}$$

In § 55 (5) a first integral of this equation was found to be

$$\tfrac{1}{2}mv^2 = \int^x f(x)\,dx + C, \tag{2}$$

where v is the velocity of the particle, and C is a constant.

Now $\frac{1}{2}mv^2$ is the kinetic energy T of the particle, and by § 72 (5) its potential energy V referred to a standard position S is

$$V = \int_x^S f(x)\,dx. \tag{3}$$

Thus (2) may be written

$$T+V = \text{constant}, \tag{4}$$

or if T_0 and V_0 are the initial values of T and V,

$$T+V = T_0+V_0. \tag{5}$$

This is the energy equation for the particle. It could have been written down immediately in many of the problems previously discussed and the equations of motion found by differentiating it. But perhaps its most important use is to give a simple picture of the nature of the motion.

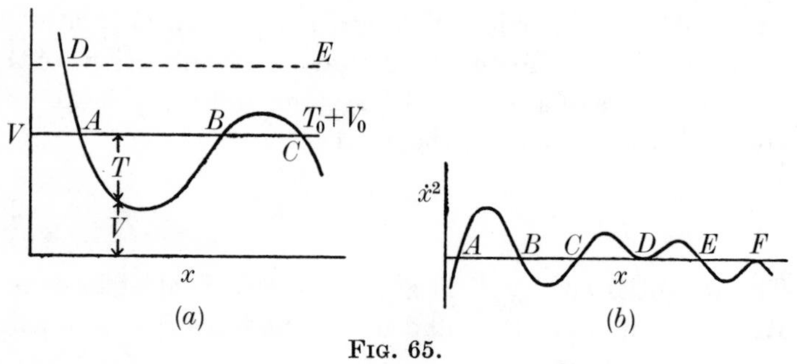

Fig. 65.

Suppose the curve of potential energy V as a function of x is drawn, and we plot on it a horizontal line of ordinate T_0+V_0, e.g. the line ABC in Fig. 65 (a). Then by (5) the amount by which this line is above the curve of V at any value of x is the kinetic energy of the particle at that point. Thus between A and B the kinetic energy is positive; at A and B the kinetic energy is zero, and the particle comes to rest. The particle cannot penetrate into the region to the left of A, or into the region BC, since its total energy is less than the potential energy in these regions. Thus if it is set in motion at a point between A and B with this total energy it will move between A and B (the nature of the motion will be studied further below) and it cannot get into the region to the right of C where motion is also possible with this total energy. On the other hand, if the total energy is given by DE, Fig. 65 (a), there is positive kinetic energy, and motion is possible, for all values of x to the right of D. The types of motion possible with the potential energy curves of Figs. 62 and 63 may be seen easily from these considerations.

To study the motion more closely it is a little more convenient to use (5) in the form $T = T_0+V_0-V$, which gives $\dot{x}^2$ as a function of x. Suppose this curve is as shown in Fig. 65 (b). Motion is only possible in the regions in which $\dot{x}^2$ is positive, such as AB, CDE. We consider now what happens at points such as A, B, at which the velocity is zero.

Suppose that at B, $x = b$, $\dot{x}^2$ has a simple zero as in Fig. 65 (b). Then

$$\dot{x}^2 = (b-x)\phi(x), \tag{6}$$

where $\phi(b) > 0$. From (6), by differentiating,

$$2\dot{x}\ddot{x} = -\dot{x}\,\phi(x)+(b-x)\phi'(x)\dot{x},$$

$$\ddot{x} = -\tfrac{1}{2}\phi(x)+\tfrac{1}{2}(b-x)\phi'(x). \tag{7}$$

It follows from (7) that when $x = b$, $\ddot{x} = -\frac{1}{2}\phi(b)$ and so is finite and negative. Thus at the point b, where the velocity is zero, there is a negative acceleration and the particle commences to move backwards. Similarly when it comes to rest at A there is a forwards acceleration. Thus the particle oscillates between A and B.

Next we consider the nature of the motion near a point D, $x = d$, Fig. 65 (b), where $\dot{x}^2$ has a double zero. In this case

$$\dot{x}^2 = (d-x)^2\psi(x), \tag{8}$$

where $\psi(d) > 0$. Differentiating (8)

$$2\dot{x}\ddot{x} = -2(d-x)\psi(x)\dot{x}+(d-x)^2\psi'(x)\dot{x},$$

$$\ddot{x} = -(d-x)\psi(x)+\tfrac{1}{2}(d-x)^2\psi'(x). \tag{9}$$

Thus $\ddot{x} = 0$ when $x = d$. Also from (8)

$$\frac{dx}{dt} = (d-x)[\psi(x)]^{\frac{1}{2}},$$

and thus the time taken to reach the point $x = d$ is

$$t = \int\limits_x^d \frac{dx}{(d-x)[\psi(x)]^{\frac{1}{2}}}. \tag{10}$$

The integral (10) is divergent, so the particle would take an infinite time to reach $x = d$. This case occurs, for example, when a pendulum has just sufficient energy to reach the upward vertical [cf. Fig. 63 (*b*)]. In the critical case of a double zero the two regions CD and DE of Fig. 65 (*b*) are thus effectively separated; if the total energy of the particle is increased slightly, $\dot{x}^2$ will become positive at D and motion takes place in the whole region CE.

Finally, a point such as F, Fig. 65 (*b*), corresponds to the particle at rest in a position of equilibrium.

The above discussion, based on the behaviour of $\dot{x}^2$ as a function of x, is quite general and not confined to the one-dimensional motion of a particle for which it was made. In more complicated systems specified by several parameters an equation of type $\dot{x}^2 = f(x)$ is usually obtained after some integrations and the elimination of some of the parameters in favour of a chosen one, x, and this equation is discussed as above. For example, an equation of this type was found in § 71 (19) for the motion of a gyroscope, and many properties of the gyroscope can be found from it, e.g. the condition for steady precession which is the condition that § 71 (19) should have a point such as F, Fig. 65 (*b*).

The energy equation has been derived above for a particle moving in one dimension. We now derive it for the general motion of a system of particles or rigid bodies subject to the assumption that any constraints imposed on the system are independent of the time.† Let m at $\mathbf{r}$ be the typical particle of § 66, then as in § 66 (1) the equations of motion of this particle are

$$m\ddot{\mathbf{r}} = \mathbf{P}+\mathbf{P}'. \tag{11}$$

Taking the scalar product of both sides with $\dot{\mathbf{r}}$, and summing over all particles of the system gives

$$\sum m\ddot{\mathbf{r}}.\dot{\mathbf{r}} = \sum \mathbf{P}.\dot{\mathbf{r}} + \sum \mathbf{P}'.\dot{\mathbf{r}}. \tag{12}$$

† The discussion below is not valid if the system is subject to moving constraints, for example, for a particle in a rotating tube or a body rolling on a rotating plane. The reason for this is that in calculating potential energy the displacements have to be consistent with the constraints; for example, for a particle in a tube the displacement must be along and not perpendicular to the tube. If the tube is rotating, the velocity of the particle will have a component perpendicular to the tube because of its rotation, and the displacements in (13) are taken to be proportional to the velocity.

Integrating (12) with respect to the time gives

$$\tfrac{1}{2}\sum m\dot{\mathbf{r}}.\dot{\mathbf{r}} = \sum\int \mathbf{P}.\dot{\mathbf{r}}\,dt + \sum\int \mathbf{P}'.\dot{\mathbf{r}}\,dt$$
$$= \sum\int \mathbf{P}.d\mathbf{r} + \sum\int \mathbf{P}'.d\mathbf{r}. \tag{13}$$

The first term on the right-hand side of (13) is, except for an arbitrary constant, $-V$, where V is the potential energy of the system in the field of force. The second term vanishes, since, as remarked in § 72, the internal forces between two particles, etc., taken together, do no work in a small displacement. Thus (13) gives the energy equation

$$T+V = \text{constant}. \tag{14}$$

The result that the total energy of a system moving under conservative forces is constant is referred to as the principle of conservation of energy. There are many important cases in which it does not hold:

(i) If the forces are not conservative.

(ii) If the system is subject to moving constraints.

(iii) If there is resistance to motion depending on the velocity. In this case it is shown in § 77 that the total energy diminishes steadily.

(iv) If energy is supplied to the system from outside. For example, it was seen in § 59 that in relaxation oscillations the total energy in the system oscillates periodically, instead of remaining constant as in oscillations under conservative forces.

74. The use of conservation of energy and conservation of momentum

As remarked in § 73, a first integral of the equations of motion can be written down by the use of the energy equation if it is applicable.

Also, if there is no component of external force on a system in a fixed direction, the rate of change of momentum of the system in this direction is zero, and thus the momentum of the system in this direction is constant.

Again, if there is no component of external couple on a system about a fixed axis, the rate of change of angular momentum about this axis is zero, and so the angular momentum of the system about this axis is constant.

These results are known as the principles of conservation of momentum and conservation of angular momentum, and are immediate consequences of the equations of motion. By using them it is often possible to avoid writing down and integrating the equations of motion of the system: thus the integrals § 71 (8) and § 71 (12) of the equations of motion of the gyroscope could have been written down by conservation of angular momentum, and the integral § 71 (14) by conservation of energy. But the process of writing down the fundamental equations of motion and integrating them is very little longer and gives more complete and logically connected information about the problem.

The principles of conservation of energy and momentum are particularly useful when a complete solution is not desired, and also when the motion of a system is suddenly changed by blows.

Ex. *A uniform cylinder of mass M and radius a, rolling with velocity v along a horizontal plane, comes to a kerb of height b* ($< a$) *parallel to its axis. Find the velocity necessary for it to surmount the kerb.*

We assume that the generator of the cylinder which meets the edge of the kerb becomes fixed there, and that the cylinder rotates about this line as axis. The generator is fixed by blows applied to it, and, since these have no moment about it, the angular momentum about it is unchanged.

The moment of inertia of the cylinder about its axis is $\frac{1}{2}Ma^2$, so its angular momentum about the kerb before the blows is by § 66 (19)

$$Mv(a-b)+\tfrac{1}{2}Mav. \tag{1}$$

If the angular velocity of the cylinder about the edge of the kerb after the blows is Ω, its angular momentum about this axis is

$$\tfrac{3}{2}Ma^2\Omega. \tag{2}$$

Equating (1) and (2),

$$\Omega = \frac{v(3a-2b)}{3a^2}. \tag{3}$$

The cylinder will surmount the kerb if its new kinetic energy,

$$\tfrac{3}{4}Ma^2\Omega^2 = \frac{Mv^2(3a-2b)^2}{12a^2},$$

is greater than the required potential energy Mgb, that is, if

$$v^2 > \frac{12ga^2b}{(3a-2b)^2}. \tag{4}$$

To complete the solution it is necessary to verify that the cylinder remains in contact with the kerb during the motion, and for this the reaction at the point of contact must be determined as in § 67.

75. Generalized coordinates

Suppose $q_1,\ldots, q_n$ are n independent quantities in terms of which it is possible to specify the position of every particle of a dynamical system. $q_1,\ldots, q_n$ are called *generalized coordinates*, and n is called the number of *degrees of freedom* of the system.

For example a single particle needs three coordinates to specify its position so has three degrees of freedom. Cartesian coordinates (x, y, z), spherical polar coordinates (r, θ, ϕ), or any other system may be chosen as the generalized coordinates.

A rigid body needs six coordinates to specify its position, say three to fix the position of a point of it, two to specify the direction of an axis through that point, and one to specify a rotation about that axis. Alternatively its position may be specified by the positions of three points; for this nine coordinates are required, but there are three relations between these since the distances between the points are fixed so that three of the nine coordinates can be eliminated leaving six as before. We shall always suppose that such an elimination has been carried out† so that $q_1,\ldots, q_n$ are independent.

Clearly a system of equations of motion expressed in terms of the generalized coordinates is needed. These equations, Lagrange's equations, will be established in § 76; we first derive some properties of generalized coordinates which are needed for the proof.

Suppose that (x, y, z) are the coordinates of a particular particle of the system referred to fixed rectangular axes, then x is a function of $q_1,\ldots, q_n$ which we shall write in the functional form

$$x = x(q_1,\ldots, q_n), \tag{1}$$

with similar expressions for y and z.

† There are important cases in which this cannot be done because the extra coordinates are connected by differential and not algebraic equations. Such systems are called non-holonomic; they arise in problems involving rolling contacts. The present theory does not apply to them although it is easily extended to do so.

Differentiation of (1) with respect to the time gives

$$\dot{x} = \frac{\partial x}{\partial q_1}\dot{q}_1 + \dots + \frac{\partial x}{\partial q_n}\dot{q}_n. \tag{2}$$

$\dot{q}_1, \dots, \dot{q}_n$ are called *generalized velocities*. From (2)

$$\frac{\partial \dot{x}}{\partial \dot{q}_r} = \frac{\partial x}{\partial q_r} \quad (r = 1, \dots, n). \tag{3}$$

Also

$$\begin{aligned}\frac{d}{dt}\left(\frac{\partial x}{\partial q_r}\right) &= \frac{\partial^2 x}{\partial q_r\,\partial q_1}\dot{q}_1 + \dots + \frac{\partial^2 x}{\partial q_r\,\partial q_n}\dot{q}_n \\ &= \frac{\partial}{\partial q_r}\left\{\frac{\partial x}{\partial q_1}\dot{q}_1 + \dots + \frac{\partial x}{\partial q_n}\dot{q}_n\right\} = \frac{\partial \dot{x}}{\partial q_r}.\end{aligned} \tag{4}$$

As usual if we are dealing with a rigid body, we suppose (x, y, z) to be the rectangular coordinates of an element of mass m and use $\sum$ to denote a summation over all such elements. The kinetic energy T is

$$\begin{aligned}T &= \tfrac{1}{2}\sum m(\dot{x}^2 + \dot{y}^2 + \dot{z}^2) \\ &= \frac{1}{2}\sum m\left\{\left(\frac{\partial x}{\partial q_1}\dot{q}_1 + \dots + \frac{\partial x}{\partial q_n}\dot{q}_n\right)^2 + \right. \\ &\qquad\qquad \left. + \left(\frac{\partial y}{\partial q_1}\dot{q}_1 + \dots\right)^2 + \left(\frac{\partial z}{\partial q_1}\dot{q}_1 + \dots\right)^2\right\}.\end{aligned} \tag{5}$$

Multiplying out, this may be written in the form

$$\begin{aligned}T &= a_{11}\dot{q}_1^2 + 2a_{12}\dot{q}_1\dot{q}_2 + a_{22}\dot{q}_2^2 + \dots \\ &= \sum_{r=1}^{n}\sum_{s=1}^{n} a_{rs}\dot{q}_r\dot{q}_s,\end{aligned} \tag{6}$$

where the coefficients a_{rs} are functions of $q_1, \dots, q_n$ which can be written down from (5). If they are written out in full it appears that $a_{rs} = a_{sr}$.

The quantity p_r defined by

$$p_r = \frac{\partial T}{\partial \dot{q}_r} = \sum_{s=1}^{n} a_{rs}\dot{q}_s \tag{7}$$

is called the generalized momentum corresponding to the coordinate q_r. This definition may be regarded as being suggested by the result that, for linear motion of a particle of mass m with velocity v, the linear momentum

$$mv = \frac{d}{dv}(\tfrac{1}{2}mv^2). \tag{8}$$

Since the equations of motion of the particle in (8) involve $d(mv)/dt$, we may expect to have to study d/dt of the quantities in (7), that is,

$$\frac{d}{dt}\left(\frac{\partial T}{\partial \dot{q}_r}\right). \tag{9}$$

To do this consider

$$\frac{d}{dt}\left\{\frac{\partial}{\partial \dot{q}_r}(\tfrac{1}{2}\dot{x}^2)\right\} = \frac{d}{dt}\left(\dot{x}\frac{\partial \dot{x}}{\partial \dot{q}_r}\right)$$

$$= \frac{d}{dt}\left(\dot{x}\frac{\partial x}{\partial q_r}\right) \tag{10}$$

$$= \ddot{x}\frac{\partial x}{\partial q_r} + \dot{x}\frac{d}{dt}\left(\frac{\partial x}{\partial q_r}\right)$$

$$= \ddot{x}\frac{\partial x}{\partial q_r} + \dot{x}\frac{\partial \dot{x}}{\partial q_r} \tag{11}$$

$$= \ddot{x}\frac{\partial x}{\partial q_r} + \frac{\partial}{\partial q_r}(\tfrac{1}{2}\dot{x}^2), \tag{12}$$

where we have used (3) in (10), and (4) in (11).

Adding the corresponding equations for y and z, multiplying by the mass m of the typical particle, and summing over all particles, we get from (12)

$$\frac{d}{dt}\left(\frac{\partial T}{\partial \dot{q}_r}\right) - \frac{\partial T}{\partial q_r} = \sum m\left(\ddot{x}\frac{\partial x}{\partial q_r} + \ddot{y}\frac{\partial y}{\partial q_r} + \ddot{z}\frac{\partial z}{\partial q_r}\right). \tag{13}$$

76. Lagrange's equations

Suppose a conservative dynamical system is specified as in § 75 by generalized coordinates $q_1,\ldots, q_n$. As before, let (x, y, z) be the position of the element of mass m, and let (X, Y, Z) be the total force on this particle (including both external and internal forces, cf. § 66). Then its equations of motion are

$$m\ddot{x} = X, \qquad m\ddot{y} = Y, \qquad m\ddot{z} = Z. \tag{1}$$

Multiplying these by $\partial x/\partial q_r$, $\partial y/\partial q_r$, $\partial z/\partial q_r$, respectively, adding, and summing over all elementary particles, gives

$$\sum m\left(\ddot{x}\,\frac{\partial x}{\partial q_r}+\ddot{y}\,\frac{\partial y}{\partial q_r}+\ddot{z}\,\frac{\partial z}{\partial q_r}\right)=\sum\left(X\,\frac{\partial x}{\partial q_r}+Y\,\frac{\partial y}{\partial q_r}+Z\,\frac{\partial z}{\partial q_r}\right). \tag{2}$$

The left-hand side of this has been transformed in § 75 (13). We now have to consider the right-hand side. Suppose a small change δq_r is made in q_r, all the other q being left unchanged (this is possible since by hypothesis the q are independent); the resulting changes in x, y, z will be

$$\frac{\partial x}{\partial q_r}\,\delta q_r,\quad \frac{\partial y}{\partial q_r}\,\delta q_r,\quad \frac{\partial z}{\partial q_r}\,\delta q_r,$$

and the total work done will be

$$\sum\left(X\,\frac{\partial x}{\partial q_r}+Y\,\frac{\partial y}{\partial q_r}+Z\,\frac{\partial z}{\partial q_r}\right)\delta q_r, \tag{3}$$

on summing over all particles of the system.† Now this work is also the decrease in potential energy of the system, that is

$$-\frac{\partial V}{\partial q_r}\,\delta q_r. \tag{4}$$

Therefore, equating (3) and (4),

$$\sum\left(X\,\frac{\partial x}{\partial q_r}+Y\,\frac{\partial y}{\partial q_r}+Z\,\frac{\partial z}{\partial q_r}\right)=-\frac{\partial V}{\partial q_r}, \tag{5}$$

and, using (5) and § 75 (13) in (2), we get finally

$$\frac{d}{dt}\left(\frac{\partial T}{\partial \dot{q}_r}\right)-\frac{\partial T}{\partial q_r}=-\frac{\partial V}{\partial q_r}. \tag{6}$$

r is unspecified in this, and the set of n equations (6) for $r = 1,\ldots, n$ are *Lagrange's equations.*

The form (6) will be sufficient for the present applications, but there are simple generalizations‡ to: (i) cases in which the forces are not conservative; (ii) cases in which the time appears

† The internal forces, which have been specifically included in (2), disappear in the summation since equal and opposite contributions come from the two particles between which any internal force acts. Also forces such as reactions at smooth surfaces or hinges, or at rolling contacts, do no work.

‡ Cf. Whittaker, *Analytical Dynamics* (Cambridge).

explicitly, i.e. $x = x(q_1,..., q_n, t)$; (iii) systems with redundant coordinates connected by algebraic equations; (iv) systems with redundant coordinates connected by differential equations (non-holonomic systems).

If the system is not conservative we define the generalized force corresponding to the coordinate q_r by

$$\sum \left(X\frac{\partial x}{\partial q_r}+Y\frac{\partial y}{\partial q_r}+Z\frac{\partial z}{\partial q_r}\right)\delta q_r = Q_r\,\delta q_r,$$

and (6) becomes

$$\frac{d}{dt}\left(\frac{\partial T}{\partial \dot{q}_r}\right)-\frac{\partial T}{\partial q_r} = Q_r. \tag{7}$$

A great advantage of Lagrange's equations for the solution of dynamical problems is that forces which do no work, such as reactions at smooth hinges, do not appear. In writing down equations of motion in the ordinary way these usually have to be considered.

The equations may be extended to cover the case in which there is resistance to motion k times the velocity. In this case (1) are replaced by

$$m\ddot{x}+k\dot{x} = X, \qquad m\ddot{y}+k\dot{y} = Y, \qquad m\ddot{z}+k\dot{z} = Z, \tag{8}$$

and (2) is replaced by

$$\sum m\left(\ddot{x}\frac{\partial x}{\partial q_r}+...\right)+\sum k\left(\dot{x}\frac{\partial x}{\partial q_r}+\dot{y}\frac{\partial y}{\partial q_r}+\dot{z}\frac{\partial z}{\partial q_r}\right) = \sum\left(X\frac{\partial x}{\partial q_r}+...\right). \tag{9}$$

Now by § 75 (3)

$$\sum k\left(\dot{x}\frac{\partial x}{\partial q_r}+\dot{y}\frac{\partial y}{\partial q_r}+\dot{z}\frac{\partial z}{\partial q_r}\right) = \sum k\left(\dot{x}\frac{\partial \dot{x}}{\partial \dot{q}_r}+\dot{y}\frac{\partial \dot{y}}{\partial \dot{q}_r}+\dot{z}\frac{\partial \dot{z}}{\partial \dot{q}_r}\right)$$

$$= \frac{\partial F}{\partial \dot{q}_r}, \tag{10}$$

where

$$F = \tfrac{1}{2}\sum k(\dot{x}^2+\dot{y}^2+\dot{z}^2). \tag{11}$$

Proceeding as before, Lagrange's equations become

$$\frac{d}{dt}\left(\frac{\partial T}{\partial \dot{q}_r}\right)-\frac{\partial T}{\partial q_r}+\frac{\partial F}{\partial \dot{q}_r} = -\frac{\partial V}{\partial q_r} \quad (r = 1,..., n). \tag{12}$$

The function F defined in (11) is called the *dissipation function*. Using § 75 (2) it appears that it has the form

$$F = \sum_{r=1}^{n}\sum_{s=1}^{n} b_{rs}\,\dot{q}_r\,\dot{q}_s, \tag{13}$$

where the coefficients b_{rs} are functions of $q_1,..., q_n$, and $b_{rs} = b_{sr}$. It follows

T

from (13), either by writing out the expressions explicitly or by using Euler's theorem on homogeneous functions, that

$$\sum_{r=1}^{n} \frac{\partial F}{\partial \dot{q}_r} \dot{q}_r = 2F. \tag{14}$$

Also, it follows in the same way from § 75 (6) that

$$\sum_{r=1}^{n} \frac{\partial T}{\partial \dot{q}_r} \dot{q}_r = 2T. \tag{15}$$

We now proceed to study the way in which the total energy of the system diminishes because of the resistance to motion. Multiplying the equations (12) by $\dot{q}_1,..., \dot{q}_n$, respectively, and adding gives

$$\sum_{r=1}^{n} \dot{q}_r \frac{d}{dt}\left(\frac{\partial T}{\partial \dot{q}_r}\right) - \sum_{r=1}^{n} \frac{\partial T}{\partial q_r} \dot{q}_r + \sum_{r=1}^{n} \frac{\partial F}{\partial \dot{q}_r} \dot{q}_r + \sum_{r=1}^{n} \frac{\partial V}{\partial q_r} \dot{q}_r = 0. \tag{16}$$

Now $$\frac{dT}{dt} = \sum_{r=1}^{n} \frac{\partial T}{\partial q_r} \dot{q}_r + \sum_{r=1}^{n} \frac{\partial T}{\partial \dot{q}_r} \ddot{q}_r. \tag{17}$$

Using (17) and (14) in (16) gives

$$\sum_{r=1}^{n} \dot{q}_r \frac{d}{dt}\left(\frac{\partial T}{\partial \dot{q}_r}\right) + \sum_{r=1}^{n} \frac{\partial T}{\partial \dot{q}_r} \ddot{q}_r - \frac{dT}{dt} + 2F + \frac{dV}{dt} = 0.$$

Therefore $$\frac{d}{dt} \sum_{r=1}^{n} \left(\frac{\partial T}{\partial \dot{q}_r} \dot{q}_r\right) - \frac{dT}{dt} + \frac{dV}{dt} + 2F = 0.$$

Therefore, using (15), $$\frac{d}{dt}(T+V) = -2F. \tag{18}$$

Thus $2F$ measures the rate at which the total energy of the system is being dissipated by friction. If $F = 0$, (18) is the equation of conservation of energy.

Ex. 1. *The gyroscope.*

The system is specified in § 71. The components of the angular velocity of the body along the instantaneous directions of the axes are by § 71 (2)

$$(-\dot{\psi} \sin\theta,\ \dot{\theta},\ \dot{\phi} + \dot{\psi}\cos\theta). \tag{19}$$

The moments of inertia of the body about these axes are A, A, C. Thus, by § 66 (24)

$$T = \tfrac{1}{2}A(\dot{\psi}^2 \sin^2\theta + \dot{\theta}^2) + \tfrac{1}{2}C(\dot{\phi} + \dot{\psi}\cos\theta)^2. \tag{20}$$

Also the potential energy is $V = Mgh\cos\theta$.

Thus Lagrange's equations for θ, ψ, and ϕ respectively are

$$\frac{d}{dt}(A\dot{\theta})-A\dot{\psi}^2\sin\theta\cos\theta+C\dot{\psi}(\dot{\phi}+\dot{\psi}\cos\theta)\sin\theta = Mgh\sin\theta, \tag{21}$$

$$\frac{d}{dt}\{A\dot{\psi}\sin^2\theta+C\cos\theta(\dot{\phi}+\dot{\psi}\cos\theta)\} = 0, \tag{22}$$

$$\frac{d}{dt}(\dot{\phi}+\dot{\psi}\cos\theta) = 0. \tag{23}$$

(23) gives
$$\dot{\phi}+\dot{\psi}\cos\theta = n. \tag{24}$$

(21) to (23) are the same as § 71 (6), (11), (7). Also, using (24) in (20), the energy equation becomes

$$\tfrac{1}{2}A(\dot{\psi}^2\sin^2\theta+\dot{\theta}^2)+Mgh\cos\theta = \text{constant}, \tag{25}$$

which is § 71 (14).

Ex. 2. *The motion of two free, attracting particles.*

Let m_1 and m_2 be the masses of the particles, and let (x_1, y_1, z_1), (x_2, y_2, z_2), and (X, Y, Z) be respectively the coordinates of the two particles and their centre of mass, relative to fixed rectangular axes. Write

$$x = x_2-x_1, \qquad y = y_2-y_1, \qquad z = z_2-z_1. \tag{26}$$

Then, writing

$$\mu_1 = m_1/(m_1+m_2), \qquad \mu_2 = m_2/(m_1+m_2),$$

$$x_1 = X-\mu_2 x, \qquad x_2 = X+\mu_1 x,$$

with similar expressions for y_1, z_1, y_2, z_2. The kinetic energy T is

$$T = \tfrac{1}{2}m_1\{(\dot{X}-\mu_2\dot{x})^2+(\dot{Y}-\mu_2\dot{y})^2+(\dot{Z}-\mu_2\dot{z})^2\}+$$
$$+\tfrac{1}{2}m_2\{(\dot{X}+\mu_1\dot{x})^2+(\dot{Y}+\mu_1\dot{y})^2+(\dot{Z}+\mu_1\dot{z})^2\}. \tag{27}$$

The potential energy V is a function of x, y, and z. Taking X, Y, Z, x, y, z as coordinates, Lagrange's equations give

$$\ddot{X} = \ddot{Y} = \ddot{Z} = 0,$$

that is, the centre of mass moves with constant velocity. The other three equations are

$$m_1\mu_2\ddot{x} = -\frac{\partial V}{\partial x}; \qquad m_1\mu_2\ddot{y} = -\frac{\partial V}{\partial y}; \qquad m_1\mu_2\ddot{z} = -\frac{\partial V}{\partial z}, \tag{28}$$

that is, the motion of m_2 relative to m_1 is the same as if m_1 were fixed and the potential energy were

$$\frac{m_1+m_2}{m_1}V. \tag{29}$$

Ex. 3. *The problem of § 36.*

Using the result § 72 (10) we have

$$T = \tfrac{1}{2}M_1\dot{x}_1^2+\tfrac{1}{2}M_2\dot{x}_2^2,$$

$$V = \tfrac{1}{2}\lambda_1 x_1^2+\tfrac{1}{2}\lambda_2(x_2-x_1)^2.$$

Therefore Lagrange's equations give

$$M_1 \ddot{x}_1 = -\lambda_1 x_1 + \lambda_2(x_2 - x_1),$$
$$M_2 \ddot{x}_2 = -\lambda_2(x_2 - x_1),$$

in agreement with § 36 (1) and (2).

Ex. 4. *Two equal uniform rods OA, AB of length $2a$ and mass M are freely hinged together at A, and OA is freely hinged at a fixed point O. They oscillate in a plane under gravity.*

Let θ and ϕ be the angles which the rods make with the vertical. Then the horizontal and vertical displacements of G, the centre of mass of AB, are

$$x = 2a\sin\theta + a\sin\phi,$$
$$y = 2a\cos\theta + a\cos\phi.$$

Therefore

$$\dot{x}^2 + \dot{y}^2 = (2a\dot{\theta}\cos\theta + a\dot{\phi}\cos\phi)^2 + (2a\dot{\theta}\sin\theta + a\dot{\phi}\sin\phi)^2$$
$$= 4a^2\dot{\theta}^2 + a^2\dot{\phi}^2 + 4a^2\dot{\theta}\dot{\phi}\cos(\theta - \phi).$$

The kinetic energy of OA is $(2Ma^2/3)\dot{\theta}^2$, and that of the motion of AB about its centre of mass is $(Ma^2/6)\dot{\phi}^2$. Using § 66 (23) for the kinetic energy of AB, we have for the kinetic energy T of the system

$$T = \frac{2Ma^2}{3}\dot{\theta}^2 + \frac{Ma^2}{6}\dot{\phi}^2 + \tfrac{1}{2}M\{4a^2\dot{\theta}^2 + a^2\dot{\phi}^2 + 4a^2\dot{\theta}\dot{\phi}\cos(\theta - \phi)\}$$
$$= \frac{8Ma^2}{3}\dot{\theta}^2 + \frac{2Ma^2}{3}\dot{\phi}^2 + 2Ma^2\dot{\theta}\dot{\phi}\cos(\theta - \phi). \tag{30}$$

Also the potential energy, measured from O, is

$$V = -Mga\{3\cos\theta + \cos\phi\}. \tag{31}$$

Lagrange's equations then give

$$\frac{d}{dt}\left\{\frac{16Ma^2}{3}\dot{\theta} + 2Ma^2\dot{\phi}\cos(\theta - \phi)\right\} + 2Ma^2\dot{\theta}\dot{\phi}\sin(\theta - \phi) = -3Mga\sin\theta, \tag{32}$$

$$\frac{d}{dt}\left\{\frac{4Ma^2}{3}\dot{\phi} + 2Ma^2\dot{\theta}\cos(\theta - \phi)\right\} - 2Ma^2\dot{\theta}\dot{\phi}\sin(\theta - \phi) = -Mga\sin\phi. \tag{33}$$

Ex. 5. *Lagrange's equations and electric circuit theory.*

The equations § 41 (5) and (4) for an L, R, C circuit with no applied voltage are

$$L\dot{I} + RI + \frac{1}{C}Q = 0, \tag{34}$$

$$\dot{Q} = I. \tag{35}$$

Multiplying (34) by I and using (35) gives

$$\frac{d}{dt}\left(\tfrac{1}{2}LI^2 + \frac{1}{2C}Q^2\right) + RI^2 = 0. \tag{36}$$

Now RI^2 is the rate of dissipation of energy in the circuit, so (36) may be regarded as the energy equation if we identify

$$\tfrac{1}{2}LI^2+\frac{1}{2C}Q^2$$

with the total energy in the circuit. $\tfrac{1}{2}LI^2$ is regarded as kinetic energy associated with current I in an inductance, and $(1/2C)Q^2$ as the potential energy associated with charge Q on a condenser.

Further, we may regard Q as a generalized coordinate specifying the electrical state of the circuit and $\dot{Q} = I$ as the corresponding generalized velocity. Taking

$$T = \tfrac{1}{2}LI^2, \qquad V = \frac{1}{2C}Q^2, \qquad F = \tfrac{1}{2}RI^2, \tag{37}$$

Lagrange's equations (12) give

$$\frac{d}{dt}(LI)+RI+\frac{1}{C}Q = 0,$$

in agreement with (34).

The general equations of electric circuit theory may be obtained in this way. It has the advantage that in systems such as electric motors which contain moving masses the kinetic and potential energies of these may be included and the electrical and mechanical parts of the system considered together.

77. Small oscillations about statical equilibrium

Lagrange's equations of motion, § 76 (6), are

$$\frac{d}{dt}\left(\frac{\partial T}{\partial \dot{q}_r}\right)-\frac{\partial T}{\partial q_r} = -\frac{\partial V}{\partial q_r} \quad (r = 1, 2,..., n). \tag{1}$$

At the position of equilibrium these must be satisfied with $\dot{q}_1 = ... = \dot{q}_n = 0$, so that all the terms on the left-hand sides vanish and (1) require

$$\frac{\partial V}{\partial q_1} = ... = \frac{\partial V}{\partial q_n} = 0. \tag{2}$$

These are the conditions to be satisfied at a position of equilibrium. They also express the fact that the potential energy V is to be stationary at such a point. If it has a minimum there the equilibrium will be stable. For if the potential energy has a minimum V_0 at the position of equilibrium, and the system is given a small amount of kinetic energy T_0, the motion must be confined to the small region about the position of equilibrium in which $V < T_0+V_0$. That is, the system executes a small oscillation about the position of equilibrium and so is stable. This argument is essentially the extension to n dimensions of the ideas of § 73.

Ex. 1. *The dipole μ of § 72 (xi) is freely hinged at a point on the axis of the dipole μ' and distant r from it.*

In § 72 (24) we have $\theta = 0$ and so

$$V = -\frac{2\mu\mu'}{r^3}\cos\phi. \tag{3}$$

There is equilibrium when $dV/d\phi = 0$, that is, when $\phi = 0$ and when $\phi = \pi$. Now

$$\frac{d^2V}{d\phi^2} = \frac{2\mu\mu'}{r^3}\cos\phi. \tag{4}$$

When $\phi = 0$, (4) is positive, V has a minimum, and so the equilibrium is stable.

When $\phi = \pi$, (4) is negative, V has a maximum, and the equilibrium is unstable.

Suppose, now, that a position of equilibrium has been found by solving the set of equations (2) and that we wish to study small oscillations of the system about it. It is convenient to change the origin of coordinates to the position of equilibrium so that $q_1,\ldots, q_n$ vanish in the position of equilibrium and are small throughout the motion. The potential energy V can then be written in the form

$$V = \sum_{r=1}^{n}\sum_{r=1}^{n} c_{rs}\, q_r\, q_s \tag{5}$$

which contains only terms of the second degree in the coordinates. There will be no terms of the first degree since $\partial V/\partial q_r = 0$ when $q_1 = \ldots = q_n = 0$, and a constant term can be neglected since it will not affect the equations (1). Terms of the third and higher degrees in the q_r are neglected since we are assuming that these are small. In the same way the kinetic energy T, § 75 (6), becomes

$$T = \sum_{r=1}^{n}\sum_{s=1}^{n} a_{rs}\,\dot{q}_r\,\dot{q}_s, \tag{6}$$

where now the a_{rs} may be taken to be constants with the values for the equilibrium position, since allowing for their variation with position would introduce terms of the third degree in the q and $\dot{q}$ and these are negligible.

Thus the equations (1) give

$$\sum_{s=1}^{n}(a_{rs}\ddot{q}_s+c_{rs}q_s) = 0 \quad (r = 1,..., n), \tag{7}$$

which are a system of n linear differential equations in n unknowns.

If there is resistance to motion proportional to velocity there will be a dissipation function F as in § 76 (13) given by

$$F = \sum_{r=1}^{n}\sum_{s=1}^{n} b_{rs}\dot{q}_r\dot{q}_s, \tag{8}$$

where the b_{rs} are constants with the values at the equilibrium position. Lagrange's equations in the form § 76 (12) then give

$$\sum_{r=1}^{n}(a_{rs}\ddot{q}_s+b_{rs}\dot{q}_s+c_{rs}q_s) = 0 \quad (r = 1,..., n). \tag{9}$$

Ex. 2. *Small oscillations of the system of § 76, Ex. 4, about the vertical.* Retaining only terms of the second degree in the small quantities $\theta, \phi, \dot{\theta}, \dot{\phi}$ in § 76 (30) and (31) these become

$$T = \frac{8Ma^2}{3}\dot{\theta}^2+\frac{2Ma^2}{3}\dot{\phi}^2+2Ma^2\dot{\theta}\dot{\phi}, \tag{10}$$

$$V = Mga(-4+\tfrac{3}{2}\theta^2+\tfrac{1}{2}\phi^2). \tag{11}$$

Therefore, by Lagrange's equations (1),

$$\frac{16Ma^2}{3}\ddot{\theta}+2Ma^2\ddot{\phi}+3Mga\theta = 0,$$

$$\frac{4Ma^2}{3}\ddot{\phi}+2Ma^2\ddot{\theta}+Mga\phi = 0.$$

Or, writing $n^2 = g/a$,

$$(16D^2+9n^2)\theta+6D^2\phi = 0,$$

$$6D^2\theta+(4D^2+3n^2)\phi = 0,$$

a pair of simultaneous linear differential equations.

EXAMPLES ON CHAPTER IX

1. A mass m is attached to a fixed point by a spring of stiffness λ. When its displacement is a it comes in contact with another spring of stiffness λ'. If it is set in motion from its equilibrium position by a blow of impulse P, show that it will come to rest when

$$\lambda x^2+\lambda'(x-a)^2 = P^2/m,$$

if $P > a(\lambda m)^{\frac{1}{2}}$.

Which root of the equation is to be taken?

2. A rod of mass m is freely hinged to a fixed point on a horizontal plane. Its centre of mass is distant l from the hinge, and its moment

of inertia about the hinge is I. The rod is held in a nearly vertical position by four equal springs, each of stiffness k and unstrained length $\surd(a^2+b^2)$, attached to it at a distance a from the hinge, and attached symmetrically to the plane at distances b from the hinge. Show that the period of small oscillations about the vertical is

$$2\pi I^{\frac{1}{2}}\left\{\frac{2ka^2b^2}{a^2+b^2}-mgl\right\}^{-\frac{1}{2}}.$$

3. A uniform rod is hung in a horizontal position by two parallel strings of length l (bifilar suspension). Show that the period of small oscillations of the rod in which its centre moves in a vertical straight line is

$$2\pi\surd(l/3g).$$

4. A body rests in equilibrium on a rough surface, its centre of mass being a distance h vertically above the point of contact of the surfaces, and their common tangent plane being horizontal and the bodies on opposite sides of it. If ρ_1 and ρ_2 are the radii of curvature of the surfaces at the point of contact show that the equilibrium is stable if

$$\frac{1}{h} > \frac{1}{\rho_1}+\frac{1}{\rho_2}.$$

5. A particle of mass m is at the point (x, y) in the field of a uniform thin rod $-a < x < a$, every element δx of which attracts the mass m with force $\gamma m\rho\,\delta x/r^2$, where γ is a constant, ρ is the density of the rod, and r is the distance between m and the element δx. Show that the potential energy of m is

$$\gamma m\rho \ln\frac{(x-a)+[(x-a)^2+y^2]^{\frac{1}{2}}}{(x+a)+[(x+a)^2+y^2]^{\frac{1}{2}}}.$$

6. If heat is supplied at the constant rate of Q units per unit time at a point in an infinite medium of thermal conductivity K, the steady temperature at a point distant r from the point of supply is known to be $Q/2\pi Kr$.

Show that if heat is supplied at the constant rate q units per unit time per unit area over a square of side $2a$, the steady temperature at the centre of the square is

$$\frac{4qa}{K\pi}\ln(1+\surd 2),$$

and the average temperature over the square is

$$\frac{4qa}{K\pi}\left\{\frac{1-\surd 2}{3}+\ln(1+\surd 2)\right\}.$$

7. A, B, C are the principal moments of inertia of a body of mass M referred to its centre of mass O as origin. r is the distance of the typical particle of mass m from O, and x is the projection of r along any axis OX. Show that

$$\sum mr^2 = \tfrac{1}{2}(A+B+C),$$
$$\sum mx^2 = \tfrac{1}{2}(A+B+C)-I,$$

where I is the moment of inertia of the body about OX.

Deduce that the potential energy V of a particle of mass M' at a point on OX distant R from O, where R is large compared with the linear dimensions of the body, is

$$V = -\gamma M' \sum m(R^2+r^2-2Rx)^{-\frac{1}{2}}$$
$$= -\gamma M' \left\{\frac{M}{R}+\frac{A+B+C-3I}{2R^3}+\dots\right\}.$$

8. Four small magnets, each of moment μ, are placed at the corners of a square of side $2a$ and all point in the direction of a side of it. Show that a small magnet of moment μ' placed at the centre of the square will point in the same direction, and if its moment of inertia is I it will oscillate about this position with frequency

$$(\mu\mu'/Ia^3\sqrt{2})^{\frac{1}{2}}/2\pi.$$

9. Four small magnets are pivoted at the corners of a square of side $2a$ and oscillate under the influence they exert on each other. If their magnetic moments are μ and their moments of inertia are I, show that the natural frequencies of the system are

$$\frac{1}{2\pi}\left\{\frac{3\mu^2(2+2^{-\frac{3}{2}})}{8Ia^3}\right\}^{\frac{1}{2}}, \qquad \frac{1}{2\pi}\left\{\frac{\mu^2(3-2^{-\frac{3}{2}})}{8Ia^3}\right\}^{\frac{1}{2}}, \qquad \frac{1}{2\pi}\left\{\frac{3\mu^2}{16Ia^3 2^{\frac{1}{2}}}\right\}^{\frac{1}{2}}.$$

10. A particle of mass m is connected to the four corners of a square by four equal springs of stiffness λ. If x and y are its displacements from the centre of the square in the directions of the diagonals, and z is its displacement perpendicular to the plane of the square, show that its potential energy for small displacements is

$$2\lambda(x^2+y^2)(1-k)+2\lambda z^2(1-2k)+\text{constant},$$

where k is the ratio of the unstretched length of a spring to a diagonal of the square. Find the natural frequencies of small oscillations of the mass.

11. A uniform rod of length $2l$ has one end attached to a fixed point O by a light string of length $5l/12$. Show that the natural frequencies of small oscillations in a vertical plane through O are

$$(3g/5l)^{\frac{1}{2}}/2\pi \quad \text{and} \quad (3g/l)^{\frac{1}{2}}/\pi.$$

12. A truck of total mass M has two axles carrying wheels of radius a, the moment of inertia of each axle and its wheels being I. Show that the acceleration of the truck when rolling down a plane inclined at α to the horizontal is

$$\frac{Mg\sin\alpha}{M+2I/a^2}.$$

13. A particle moves from A to B along any path in either free or constrained motion in a field of force of potential V, there being no

resistance to the motion. Show that if V_A and V_B are the values of its potential energy at A and B, and v_A and v_B are its speeds there,

$$\tfrac{1}{2}mv_B^2 - \tfrac{1}{2}mv_A^2 = V_A - V_B.$$

Show that if the particle is charged the same result still holds if there is a magnetic field present.

14. Discuss the nature of the motion under a central force by treating § 62 (22) as in § 73. Show that for the attractive force μr^{-n}, where $n > 3$, a circular orbit is unstable.

15. Show that the kinetic energy of a gyroscope mounted in gimbals may be put in the form

$$2T = (A+A_1)\dot{\theta}^2 + \{(A+C_1)\sin^2\theta + A_1\cos^2\theta + I\}\dot{\psi}^2 + C(\dot{\phi}+\dot{\psi}\cos\theta)^2,$$

where I and A_1 are the moments of inertia of the outer and inner rings about their axes, and C_1 is the moment of inertia of the inner ring about a perpendicular to its plane, and the other symbols have their usual meanings.

Deduce the equations of motion of the system.

16. Discuss the motion of a gyrostat by treating § 71 (19) as in § 73. Writing $f(x)$ for the right-hand side of § 71 (19), show that the condition for steady precession with $x = x_0$ is

$$f(x_0) = f'(x_0) = 0,$$

and deduce § 71 (20). Show that the period of small oscillations about steady precession at x_0 with angular velocity ω is

$$2\pi\{-\tfrac{1}{2}f''(x_0)\}^{-\frac{1}{2}} = 2\pi\{(a-2\omega x_0)^2 + \omega^2(1-x_0^2)\}^{-\frac{1}{2}}.$$

17. A uniform rod OA of length $2a$ is freely hinged at the point O and oscillates about it under gravity, θ being its inclination to the vertical and ψ the angle between the vertical plane through the rod and a fixed plane. Find the equations of motion of the rod, and show that a steady motion with $\theta = \alpha$, $\dot{\psi} = \omega$, where $\omega^2 = (3g/4a)\sec\alpha$ is possible.

Discuss small oscillations about steady motion by putting $\theta = \alpha + x$, $\dot{\psi} = \omega + y$, and neglecting squares and products of x, y, $\dot{x}$, $\dot{y}$. Show that the period of these oscillations is

$$2\pi\left\{\frac{4a\cos\alpha}{3g(1+3\cos^2\alpha)}\right\}^{\frac{1}{2}}.$$

18. An engine governor consists of the rod OA of Ex. 17 which is hung from a system of moment of inertia I which rotates about a vertical axis. In order to keep the angular velocity of the system constant at the value ω of Ex. 17, a couple $k(\alpha-\theta)$ is applied about the vertical axis when the inclination of OA to the vertical is θ. Show that the equations of motion of the system are

$$\ddot{\theta} - \dot{\psi}^2\sin\theta\cos\theta = -(3g/4a)\sin\theta,$$

$$(4ma^2/3)\{\ddot{\psi}\sin^2\theta + 2\dot{\psi}\dot{\theta}\sin\theta\cos\theta\} + I\ddot{\psi} = k(\alpha-\theta).$$

Study the effect of oscillations about steady motion, and show that the frequency equation for these is a cubic with one real positive root, so that the system is unstable (as discussed here neglecting friction).

19. One end A of a uniform rod of length $2a$ and mass M is rotated with constant angular velocity ω in a horizontal circle of centre O and radius b ($< a$). The rod is hinged at A so that it can move freely in the plane of OA and the vertical. If θ is the angle between the rod and the downward vertical at A, measured in the direction towards the downward vertical at O, show that the kinetic energy of the rod is

$$M\{4a^2\dot{\theta}^2+3\omega^2(b-a\sin\theta)^2+a^2\omega^2\sin^2\theta\}/6.$$

Deduce that there is always a position of equilibrium of the rod in the range $\frac{1}{2}\pi < \theta < \pi$ and another in the range $3\pi/2 < \theta < 2\pi$, and that if ω is sufficiently large there are two in the range $0 < \theta < \frac{1}{2}\pi$. Show that the position of equilibrium in the range $\frac{1}{2}\pi < \theta < \pi$ is unstable (Lagrange's equations may be used).

20. Show that the components of grad V in the directions r, θ, z of cylindrical coordinates (the 'direction θ' is the direction in which the point specified by (r,θ,z) moves when θ is increased, r and z being kept constant, etc., see Fig. 91 (a)) are

$$\left(\frac{\partial V}{\partial r},\ \frac{1}{r}\frac{\partial V}{\partial \theta},\ \frac{\partial V}{\partial z}\right).$$

Show that the components of grad V in the directions r, θ, ϕ of spherical polar coordinates (cf. Fig. 91 (b)) are

$$\left(\frac{\partial V}{\partial r},\ \frac{1}{r}\frac{\partial V}{\partial \theta},\ \frac{1}{r\sin\theta}\frac{\partial V}{\partial \phi}\right).$$

X

BOUNDARY VALUE PROBLEMS

78. Introductory

THE problems of dynamics and electric circuit theory are initial value problems in which we have to find the solution of a set of differential equations which is valid for all times $t > 0$ and which satisfies certain initial conditions at $t = 0$.

In this chapter we study some boundary value problems for ordinary linear differential equations in which we have to find a solution of a differential equation which is valid in a region, say $0 < x < l$, and which has to satisfy conditions at both ends of the region. The region may be the infinite one $x > 0$, but in this case there will be conditions to be satisfied as $x \to \infty$ (e.g. that the solution remain finite), whereas in initial value problems there are no conditions on the behaviour of the solution as $t \to \infty$: this is determined solely by the differential equation and the conditions at $t = 0$.

In problems involving partial differential equations, Chapter XV, it will appear that in many cases these have to be solved with both initial and boundary conditions.

In this chapter we shall discuss the boundary value problems arising in the theory of the deflexion of beams. These illustrate very well the new types of phenomenon which arise. In §§ 79, 80 a brief discussion of the differential equation and boundary conditions is given, and in the subsequent sections various types of boundary value problem arising from them are studied.

79. Bending moment and shear force

We shall usually consider beams which when undeflected are straight and horizontal. The x-axis will be taken along the beam, and the y-axis vertically downwards† so that deflexions and forces are positive when in a downward direction. Usually

† This corresponds to using left-handed axes which is not very desirable, but it is obviously convenient to have deflexion and load positive when measured in their commonest direction. Right-handed axes with OY upwards are also used.

the beam will be of length l and will be supported in some way at its ends $x = 0$ and $x = l$.

We assume that all forces on the beam act in a vertical direction: the extension to horizontal forces is made in § 86. The loads on the beam may be of two types: (i) concentrated loads W applied at definite points (the reactions at the supports come into this category), and (ii) distributed loads w per unit length, where w is a prescribed function of x.

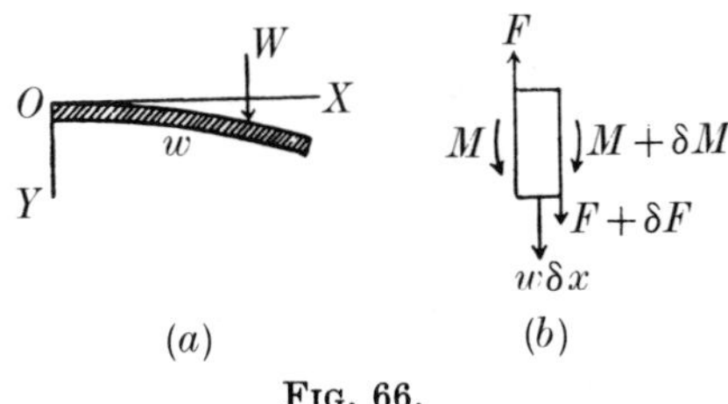

FIG. 66.

We now define two fundamental quantities, shear force and bending moment.

The *shear force* F at a point x of the beam is the resultant of all the forces on the beam to the right of the point x, measured positively in the direction OY. The shear force is thus discontinuous at a concentrated load: this fact will often be used later to calculate the reactions at the supports of a beam.

The *bending moment* M at a point x of the beam is the sum of the moments about x (in the direction from OX to OY) of all the forces to the right of x.

All the forces to the right of the point x are then statically equivalent to a force F and a couple M applied at x.

If the beam has a distributed load w in a region, there are important relations connecting w, F, and M. To derive these, consider the equilibrium of the element of length δx of the beam between x, where the shear force and bending moment are F and M, and $x+\delta x$, where they are $F+\delta F$ and $M+\delta M$. The load on the element is $w\,\delta x$. The forces on the element are shown in Fig. 66 (*b*); the forces on the beam to the right of $x+\delta x$ exert force $F+\delta F$ and couple $M+\delta M$ at $x+\delta x$ on the element, while, since the forces to the right of x exert force F and couple M

at x, the portion of the beam to the left of x exerts equal and opposite reactions on the element. The conditions of equilibrium give, neglecting squares of small quantities,

$$\delta F + w\,\delta x = 0,$$
$$\delta M + F\,\delta x = 0.$$

In the limit as $\delta x \to 0$ these become

$$\frac{dF}{dx} = -w, \tag{1}$$

$$\frac{dM}{dx} = -F, \tag{2}$$

$$\frac{d^2M}{dx^2} = w. \tag{3}$$

These hold in any portion of a beam free from concentrated loads.

The graphs of F and M against x are known as shear force and bending moment diagrams. In simple cases these may be calculated by the methods of pure statics (for example, when a beam is freely hinged at its ends at the same level), but in more complicated cases, such as a beam with its ends clamped or a beam which runs continuously over several supports, the reactions at the supports and thus the shear and bending moment diagrams cannot be determined by statics alone since the elastic properties of the beam enter into the problem. It will be seen in § 80 that a knowledge of M is sufficient for the simpler design problems of engineering, and it appears that in many cases the theory of deflexion of beams has to be used to calculate M although the deflexion itself is not of great interest to engineers.

Finally it should be remarked that, since the equations of this section and the next are linear, the principle of superposition holds, that is, the shear, bending moment, deflexion, etc., of a beam due to a number of superposed loads are the sum of the values for the separate loads. In particular, if a beam carries a distributed and several concentrated loads, we may make calculations for the distributed and concentrated loads separately and add the results.

In Fig. 67 bending moment and shear force diagrams for four cases are shown. In Figs. 67 (*a*) and (*b*) the beam is freely hinged at its ends, $x = 0$ and $x = l$; it carries a concentrated load W at $x = a$ in Fig. 67 (*a*), and a uniform load w per unit length in Fig. 67 (*b*). Calculations for these cases are made in Exs. 1 and 2 below. In Fig. 67 (*d*) and (*c*) the corresponding

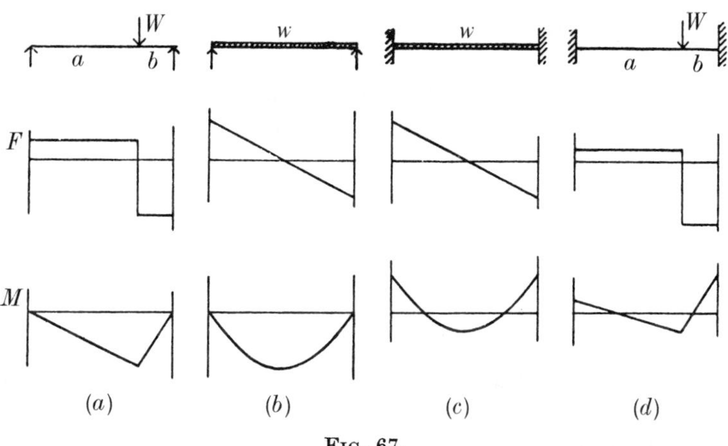

FIG. 67.

cases for a beam which is clamped horizontally at its ends at the same level are given—these cannot be calculated by the methods of the present section and are discussed in §§ 82 and 81, respectively.

Ex. 1. *The beam of length l is freely hinged at its ends and carries a concentrated load W at $x = a$.*

Taking moments about the points $x = l$ and $x = 0$ gives for the reactions R_1 and R_2 at the supports

$$R_1 = W(l-a)/l, \qquad R_2 = Wa/l. \tag{4}$$

The shear is

$$F = -R_2 = -Wa/l \qquad (a < x < l),$$
$$F = W-R_2 = W(l-a)/l \quad (0 < x < a).$$

The bending moment is

$$M = -R_2(l-x) = -Wa(l-x)/l \qquad (a < x < l),$$
$$M = W(a-x)-R_2(l-x) = -W(l-a)x/l \quad (0 < x < a).$$

F and M are graphed in Fig. 67 (*a*).

Ex. 2. *The beam of length l is freely hinged at its ends and carries a uniform load w per unit length.*

Clearly we could determine the reactions, and proceed as in Ex. 1. As an alternative which is useful for distributed loads, we use (3) and the fact that $M = 0$ at the ends $x = 0$ and $x = l$ where the beam is freely hinged. We have

$$\frac{d^2M}{dx^2} = w. \tag{5}$$

Integrating,
$$\frac{dM}{dx} = wx+A, \tag{6}$$

where A is an unknown constant. Integrating (6) gives

$$M = \tfrac{1}{2}wx^2+Ax+B. \tag{7}$$

The conditions $M = 0$ when $x = 0$ and $x = l$ give

$$B = 0, \qquad A = -\tfrac{1}{2}wl,$$

and finally
$$M = -\tfrac{1}{2}wx(l-x). \tag{8}$$

Then by (2)
$$F = \tfrac{1}{2}w(l-2x). \tag{9}$$

F and M are shown in Fig. 67 (b).

80. The differential equation for the deflexion of a beam

To calculate the deflexion we need further information which comes from the theory of elasticity. We consider a portion of the beam and suppose it to be bent to a large radius of curvature by couples M applied to its ends, Fig. 68 (a).

The beam is regarded as being composed of fibres which exert no influence on their neighbours, and it is assumed that a plane section of the beam remains plane after bending. Let AB and $A'B'$ be two sections of the beam which before bending were perpendicular to its direction, and which after bending intersect at a small angle θ in a line through C, parallel to the direction of the couples M and perpendicular to the plane of the paper in Fig. 68 (a). The fibres at AA' will have been extended and those at BB' will have been compressed, so there must be some intermediate fibres OO' whose length is unchanged: the surface in the beam containing these fibres is called the *neutral axis*.

Suppose Fig. 68 (b) is a section through the beam in the plane through AB parallel to the direction of the couples M, and suppose that we take OX and OY as axes in this plane, OX being through O, Fig. 68 (a), and parallel to the direction of the couples M. Then all fibres in the line OX will have their lengths unchanged.

Let R be the radius of curvature of the fibres OO' of the neutral axis, so that $OO' = R\theta$, and this is the unstretched length of all fibres between the planes AB and $A'B'$. Now

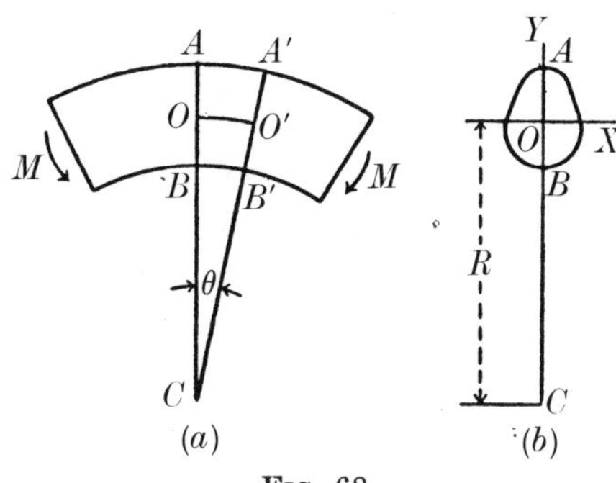

FIG. 68.

consider the fibre at (x, y) in the plane XOY. The stretched length is $(R+y)\theta$ so that its extension is $y\theta$. Therefore, if E is Young's modulus, the tension in it is

$$E\frac{y\theta}{R\theta} = \frac{Ey}{R}. \tag{1}$$

The forces across the cross-section may be determined by combining the effects (1) of the individual fibres.

The total force across the cross-section in the direction of the beam is

$$\frac{E}{R}\iint y\,dydx, \tag{2}$$

and since by hypothesis this is zero we must have

$$\iint y\,dxdy = 0, \tag{3}$$

and thus the 'centre of gravity' of the cross-section must lie in the neutral axis OX.

The sum of the moments about OX of the forces across the cross-section is

$$\frac{E}{R}\iint y^2\,dydx = \frac{EI}{R}, \tag{4}$$

where I is the 'moment of inertia' of the cross-section about OX. Since the sum of the moments must be equal to M, this gives the fundamental relation

$$\frac{EI}{R} = M. \tag{5}$$

Finally, the sum of the moments about OY of the forces across the cross-section is

$$\frac{E}{R}\iint xy\,dxdy. \tag{6}$$

This vanishes if OX is an axis of symmetry. Now the position of O in OX has never been defined, so that if the cross-section has a vertical axis of symmetry we may take this to be OY and again (6) vanishes. If the cross-section has neither a horizontal nor a vertical axis of symmetry, (6) does not vanish. Now there is no couple on the beam about OY, so for the theory to hold (6) should be zero: in the unsymmetrical cases the couple (6) causes the beam to twist and the simple theory is inadequate.

If y_0 is the greatest distance of any fibre from the neutral axis, and f_0 is the stress in this extreme fibre, we have from (1) and (5)

$$f_0 = \frac{Ey_0}{R} = \frac{My_0}{I}, \tag{7}$$

and so f_0 is determined in terms of M. A knowledge of f_0 is required for the design of beams.

The theory above is known as the Bernoulli–Euler theory of flexure and, while it is approximate, its results are sufficiently near to those deduced from the accurate theory of elasticity for most practical purposes.

(5) gives the radius of curvature of a beam bent by couples M applied to its ends, so that M is constant along the beam. In an actual beam in which M is a varying function of x, (5) will give the radius of curvature at a point of the beam in terms of the bending moment M at that point. The radius of curvature R is given in terms of the deflexion by

$$\frac{1}{R} = \frac{d^2y/dx^2}{\{1+(dy/dx)^2\}^{\frac{3}{2}}}, \tag{8}$$

and in the problems we shall consider the slope of the beam is small and $(dy/dx)^2$ negligible so that (8) may be replaced by

$$\frac{1}{R} = \frac{d^2y}{dx^2}. \tag{9}$$

For problems involving large deflexions the complete expression (8) must be used and an awkward non-linear equation results, which, however, can be solved exactly in some simple cases (cf. Ex. 21 at the end of this chapter).

From (9) and (5) we get finally

$$EI\frac{d^2y}{dx^2} = M, \tag{10}$$

which is the differential equation for the deflexion. E and I are known, and M is supposed to have been determined by the methods of § 79.

For a distributed load w, (10) may be combined with § 79 (3) to give

$$\frac{d^2}{dx^2}\left(EI\frac{d^2y}{dx^2}\right) = w, \tag{11}$$

or, if E and I are independent of x,

$$EI\frac{d^4y}{dx^4} = w. \tag{12}$$

The differential equation to be solved for y is thus one of (10) to (12). The connexion between the bending moment and deflexion is given by (10), and, at points free from concentrated loads, the shear F is by § 79 (2)

$$F = -EI\frac{d^3y}{dx^3}, \tag{13}$$

provided E and I are independent of x.

The differential equations have to be solved with boundary conditions depending on the nature of the supports at the ends. The most common such conditions are:

(i) *A freely hinged end.* Here y is prescribed (usually zero) and $M = 0$. That is, by (10),

$$\frac{d^2y}{dx^2} = 0, \qquad y \text{ given}. \tag{14}$$

(ii) *A clamped end.* Here y and dy/dx are prescribed (both usually zero). That is

$$\frac{dy}{dx} \text{ given,} \qquad y \text{ given.} \tag{15}$$

(iii) *A free end.* Here $M = 0$, and $F = 0$ unless there is a concentrated load at the end. That is, by (10) and (13),

$$\frac{d^2y}{dx^2} = \frac{d^3y}{dx^3} = 0. \tag{16}$$

If there is a concentrated load at the end, F is prescribed.

(iv) *An elastic support.* Suppose the beam is freely hinged and supported by a reaction k times the deflexion y. Since the shear at the end is $-R$, where R is the reaction, the boundary conditions are

$$M = 0, \qquad R = ky, \tag{17}$$

or

$$\frac{d^2y}{dx^2} = 0, \qquad EI\frac{d^3y}{dx^3} - ky = 0. \tag{18}$$

81. Distributed loads

Problems of distributed loads on uniform beams are usually best solved by integrating § 80 (12) with the appropriate boundary conditions.

As an example we consider first *a uniform beam of length l, carrying a uniform load w per unit length, which is freely hinged at $x = 0$ and $x = l$.*

Writing D for d/dx we have to solve

$$EID^4y = w, \tag{1}$$

with, by § 80 (14)

$$y = D^2y = 0, \quad \text{when } x = 0 \text{ and } x = l.$$

Integrating (1) gives

$$EID^3y = wx + A,$$

where A is an unknown constant. Integrating again,

$$EID^2y = \tfrac{1}{2}wx^2 + Ax + B, \tag{2}$$

and since $D^2y = 0$ when $x = 0$, $B = 0$, so that (2) becomes

$$EID^2y = \tfrac{1}{2}wx^2 + Ax. \tag{3}$$

Integrating this gives

$$EIDy = \tfrac{1}{6}wx^3+\tfrac{1}{2}Ax^2+C, \tag{4}$$

where C is an unknown constant, and integrating again we get

$$EIy = \tfrac{1}{24}wx^4+\tfrac{1}{6}Ax^3+Cx+H. \tag{5}$$

The condition $y = 0$ when $x = 0$ requires $H = 0$. The remaining constants A and C are found from the conditions $y = D^2y = 0$ when $x = l$. These require

$$\tfrac{1}{24}wl^4+\tfrac{1}{6}Al^3+Cl = 0,$$
$$\tfrac{1}{2}wl+A = 0.$$

Solving for A and C and substituting in (5) we get finally

$$y = \frac{wx}{24EI}(x^3-2lx^2+l^3). \tag{6}$$

Ex. 1. *A uniform beam of length l, carrying a uniform load w per unit length, is clamped horizontally at the same level at its ends $x = 0$ and $x = l$.*

Here we have to solve (1) with the conditions

$$y = Dy = 0, \quad \text{when } x = 0, \tag{7}$$
$$y = Dy = 0, \quad \text{when } x = l. \tag{8}$$

Integrating (1) four times as before and using (7), by virtue of which the additive constants of the last two integrations vanish, we get

$$EIy = \tfrac{1}{24}wx^4+\tfrac{1}{6}Ax^3+\tfrac{1}{2}Bx^2, \tag{9}$$

where A and B are arbitrary constants. The conditions (8) then give

$$\tfrac{1}{24}wl^4+\tfrac{1}{6}Al^3+\tfrac{1}{2}Bl^2 = 0,$$
$$\tfrac{1}{6}wl^3+\tfrac{1}{2}Al^2+Bl = 0.$$

Therefore
$$A = -\tfrac{1}{2}wl, \qquad B = \tfrac{1}{12}wl^2, \tag{10}$$

and so
$$y = \frac{wx^2(l-x)^2}{24EI}. \tag{11}$$

We can now find the bending moment and shear force at any point of the beam; as remarked in § 79, this cannot be done from purely statical considerations.

The bending moment is

$$\begin{aligned} M &= EID^2y \\ &= \tfrac{1}{2}wx^2+Ax+B \\ &= \tfrac{1}{12}wl^2-\tfrac{1}{2}wx(l-x). \end{aligned} \tag{12}$$

The shear force at any point is by § 80 (13)

$$\begin{aligned} F &= -EID^3y \\ &= -wx-A \\ &= w(\tfrac{1}{2}l-x). \end{aligned} \tag{13}$$

When $x = l$, $F = -\frac{1}{2}wl$, and therefore the reaction at l is $\frac{1}{2}wl$: this could have been inferred by symmetry and (13) deduced. Bending moment and shear force diagrams for this problem are shown in Fig. 67 (c).

Ex. 2. *A uniform beam of length l, carrying a uniform load w per unit length, is clamped horizontally at $x = 0$, and at $x = l$ it is freely hinged to a yielding support which provides reaction k times the deflexion.*

We have to solve (1) with $y = Dy = 0$ when $x = 0$, and, by § 80 (17) and (18),

$$D^2y = 0, \quad EID^3y - ky = 0, \qquad \text{when } x = l. \tag{14}$$

As in Ex. 1, (9) we find

$$EIy = \tfrac{1}{24}wx^4 + \tfrac{1}{6}Ax^3 + \tfrac{1}{2}Bx^2, \tag{15}$$

where A and B are to be determined from (14). This requires

$$\tfrac{1}{2}wl^2 + Al + B = 0,$$

$$EI(wl + A) - k(\tfrac{1}{24}wl^4 + \tfrac{1}{6}Al^3 + \tfrac{1}{2}Bl^2) = 0.$$

Solving these for A and B and substituting in (15) gives y.

Ex. 3. *A cantilever of length l has its end $x = 0$ clamped horizontally and its end $x = l$ free. It carries a uniform load w per unit length. Its cross-section is constant in $0 < x < a$, and has a different constant value in $a < x < l$.*

Suppose that

$$\frac{1}{EI} = \alpha \qquad (0 < x < a),$$

$$\frac{1}{EI} = \alpha + \beta \qquad (a < x < l).$$

These may be written as in § 17

$$\frac{1}{EI} = \alpha + \beta H(x - a). \tag{16}$$

Since I is variable, the differential equation must be taken in the form § 80 (10), that is

$$D^2y = \{\alpha + \beta H(x - a)\}M, \tag{17}$$

where the bending moment M, calculated as in § 79, is

$$M = \tfrac{1}{2}w(l - x)^2. \tag{18}$$

Using (18) in (17) we have to solve

$$\begin{aligned} D^2y &= \tfrac{1}{2}\alpha w(l-x)^2 + \tfrac{1}{2}\beta w(l-x)^2 H(x-a) \\ &= \tfrac{1}{2}\alpha w(l-x)^2 + \tfrac{1}{2}\beta w\{(x-a)^2 + 2(x-a)(a-l) + (l-a)^2\}H(x-a), \end{aligned} \tag{19}$$

with $y = Dy = 0$ when $x = 0$.

Integrating (19), using § 17 (5) for the integral of the Heaviside function, gives

$$\begin{aligned} Dy &= \tfrac{1}{2}\alpha w(l^2x - lx^2 + \tfrac{1}{3}x^3) + \\ &\quad + \tfrac{1}{2}\beta w\{\tfrac{1}{3}(x-a)^3 + (x-a)^2(a-l) + (x-a)(l-a)^2\}H(x-a), \end{aligned}$$

where the additive constant is zero since $Dy = 0$ when $x = 0$.

Integrating again, and using $y = 0$ when $x = 0$, we get finally

$$y = \tfrac{1}{2}\alpha w(\tfrac{1}{2}l^2x^2 - \tfrac{1}{3}lx^3 + \tfrac{1}{12}x^4) +$$
$$+\tfrac{1}{2}\beta w\{\tfrac{1}{12}(x-a)^4 + \tfrac{1}{3}(x-a)^3(a-l) + \tfrac{1}{2}(x-a)^2(l-a)^2\}H(x-a)$$
$$= \frac{\alpha w x^2}{24}(6l^2 - 4lx + x^2) +$$
$$+\frac{\beta w(x-a)^2}{24}\{6(l-a)^2 - 4(l-a)(x-a) + (x-a)^2\}H(x-a). \quad (20)$$

The use of the Heaviside function in this way avoids the necessity of treating the two parts $0 < x < a$ and $a < x < l$ of the beam separately. It may be used in the same way when the load changes discontinuously at a point.

82. Concentrated loads. The Green's function

Concentrated loads are a little more complicated to study than distributed loads. To illustrate the difficulties and the new ideas involved we consider in detail the case of a light uniform beam $0 < x < l$, clamped horizontally at the same level at its ends, and carrying a concentrated load W at $x = a$.

As in § 81, Ex. 1, we cannot determine the bending moment and shear by statical considerations, so we have to use the differential equation § 80 (12), but now the two regions

$$0 < x < a \quad \text{and} \quad a < x < l$$

must be treated separately.

Since we are neglecting the weight of the beam, $w = 0$ in § 80 (12) and this becomes

$$D^4 y = 0 \quad (0 < x < a), \quad (1)$$

with

$$y = Dy = 0, \quad \text{when } x = 0. \quad (2)$$

The solution of (1) satisfying (2), found as in § 81, is

$$y = \tfrac{1}{6}Ax^3 + \tfrac{1}{2}Bx^2, \quad (3)$$

where A and B are unknown constants.

Also in $a < x < l$ the differential equation is

$$D^4 y = 0 \quad (a < x < l), \quad (4)$$

with

$$y = Dy = 0, \quad \text{when } x = l. \quad (5)$$

The solution of (4) which satisfies (5) is

$$y = \tfrac{1}{6}C(l-x)^3 + \tfrac{1}{2}H(l-x)^2, \quad (6)$$

where C and H are unknown constants. The four unknowns

A, B, C, H are to be found from the conditions at $x = a$. Firstly, at $x = a$ the values of the deflexion y, slope Dy, and bending moment $M = EID^2y$ of the beam, calculated from (3), must be equal to their values calculated from (6). These conditions give

$$\tfrac{1}{6}Aa^3+\tfrac{1}{2}Ba^2 = \tfrac{1}{6}C(l-a)^3+\tfrac{1}{2}H(l-a)^2, \tag{7}$$

$$\tfrac{1}{2}Aa^2+Ba = -\tfrac{1}{2}C(l-a)^2-H(l-a), \tag{8}$$

$$Aa+B = C(l-a)+H. \tag{9}$$

Also, as we pass through the point $x = a$ from right to left, the shear force F increases by W. And, by § 80 (13),

$$F = -EID^3y. \tag{10}$$

Thus the value of (10) calculated from (3) when $x = a$ must be greater by W than the value calculated from (6), that is

$$-EIA = EIC+W. \tag{11}$$

(7), (8), (9), and (11) are four equations for A, B, C, H. Solving we get

$$A = -\frac{W(l-a)^2(l+2a)}{EIl^3}, \qquad B = \frac{Wa(l-a)^2}{EIl^2},$$

and
$$y = \frac{Wx^2(l-a)^2}{6EIl^3}\{3al-x(l+2a)\} \quad (0 < x < a). \tag{12}$$

In the same way, solving for C and H we find

$$y = \frac{Wa^2(l-x)^2}{6EIl^3}\{(3l-2a)x-la\} \quad (a < x < l). \tag{13}$$

(13) may be obtained from (12) by interchanging a and x.

The bending moment $M = EID^2y$ is given by

$$\left.\begin{aligned} M &= \frac{W(l-a)^2}{l^3}\{al-x(l+2a)\} \quad (0 < x < a) \\ M &= \frac{Wa^2}{l^3}\{al-2l^2+x(3l-2a)\} \quad (a < x < l) \end{aligned}\right\}. \tag{14}$$

The shear force $F = -EID^3y$ is given by

$$\left.\begin{aligned} F &= \frac{W(l-a)^2(l+2a)}{l^3} \quad (0 < x < a) \\ F &= -\frac{Wa^2(3l-2a)}{l^3} \quad (a < x < l) \end{aligned}\right\}. \tag{15}$$

Bending moment and shear force diagrams for this problem are shown in Fig. 67 (d).

The solution for a concentrated load with given boundary conditions is of fundamental theoretical importance, for by using it the solution for any distributed load with the same boundary conditions can be written down immediately. Write $G(a,x)$ for the deflexion at x due to a unit concentrated load at $x = a$ given by (12) and (13): the deflexion of the same beam with the same boundary conditions and a distributed load $w(x)$ may be obtained by superposing the effects of concentrated loads $w(a)\,\delta a$ in the region $a < x < a+\delta a$, so that it is

$$\int\limits_0^l G(a,x)w(a)\,da. \tag{16}$$

The same remark applies to any other boundary conditions. $G(a,x)$ is called the Green's function for the given equation and boundary conditions, and these results are special cases of a very general theory discussed further in § 102. It may be remarked that it follows from this theory that the result $G(x,a) = G(a,x)$ which appeared in (12) and (13) is true in general: from the present point of view this states that the deflexion at a due to a load W at x is equal to the deflexion at x due to a load W at a.

83. Concentrated loads. Use of the δ function

In § 17 it was remarked that a concentrated load W at $x = a$ could be treated as a distributed load

$$w = W\,\delta(x-a). \tag{1}$$

We now discuss the problem of § 82 in this way. We have to solve

$$EID^4y = W\,\delta(x-a) \quad (0 < x < l), \tag{2}$$

with

$$y = Dy = 0, \quad \text{when } x = 0 \text{ and } x = l. \tag{3}$$

Integrating (2) and using § 17 (10) gives

$$EID^3y = W\,H(x-a)+A, \tag{4}$$

where A is an arbitrary constant. Integrating (4), using § 17 (4), gives

$$EID^2y = W(x-a)H(x-a)+Ax+B. \tag{5}$$

Integrating (5), using § 17 (5), we get

$$EIDy = \tfrac{1}{2}W(x-a)^2H(x-a)+\tfrac{1}{2}Ax^2+Bx+C. \tag{6}$$

The constant C must be zero since (3) requires $Dy = 0$ when $x = 0$, and $H(x-a) = 0$ when $x = 0$.

Finally, integrating again gives

$$EIy = \tfrac{1}{6}W(x-a)^3H(x-a)+\tfrac{1}{6}Ax^3+\tfrac{1}{2}Bx^2, \tag{7}$$

the arbitrary constant being zero by (3).

The conditions $y = Dy = 0$ when $x = l$ require by (6) and (7)

$$\tfrac{1}{2}W(l-a)^2+\tfrac{1}{2}Al^2+Bl = 0,$$

$$\tfrac{1}{6}W(l-a)^3+\tfrac{1}{6}Al^3+\tfrac{1}{2}Bl^2 = 0.$$

Therefore

$$A = -\frac{W(l-a)^2(l+2a)}{l^3}, \qquad B = \frac{Wa(l-a)^2}{l^2}.$$

And, finally,

$$EIy = \tfrac{1}{6}W(x-a)^3H(x-a)+\frac{Wx^2(l-a)^2}{6l^3}\{3al-(l+2a)x\}, \tag{8}$$

or, writing out the function $H(x-a)$ explicitly,

$$EIy = \frac{Wx^2(l-a)^2}{6l^3}\{3al-(l+2a)x\} \quad (0 < x < a), \tag{9}$$

$$EIy = \frac{Wa^2(l-x)^2}{6l^3}\{(3l-2a)x-la\} \quad (a < x < l). \tag{10}$$

These are the results § 82 (12) and (13).

84. The beam on an elastic foundation

Suppose that deflexion of the beam in the direction OY of Fig. 66 (a) is resisted by a force ky per unit length of the beam,† and that the beam carries a load w per unit length. The effect of the elastic foundation is to add a term $-ky$ to the load, and, for a uniform beam in a region free from concentrated loads, the differential equation § 80 (12) for the deflexion becomes

$$EI\frac{d^4y}{dx^4}+ky = w. \tag{1}$$

† This implies that the support exerts a tension if y becomes negative. Frequently this is not the case.

Writing $$4\omega^4 = k/EI, \tag{2}$$
this may be written $$(D^4+4\omega^4)y = \frac{w}{EI}. \tag{3}$$

As an example we consider a uniform semi-infinite beam $x > 0$ resting on such a foundation with the end $x = 0$ freely hinged at zero deflexion.

The boundary conditions are
$$y = D^2y = 0, \quad \text{when } x = 0, \tag{4}$$
$$y \text{ to remain finite as } x \to \infty. \tag{5}$$

As in § 13, Ex. 5, the general solution of (3) with constant w is
$$y = e^{\omega x}(A\cos\omega x+B\sin\omega x)+ \\ +e^{-\omega x}(C\cos\omega x+H\sin\omega x)+w/k. \tag{6}$$

In order that y may remain finite as $x \to \infty$ we must have $A = B = 0$.

The conditions at $x = 0$ give
$$C+w/k = 0,$$
$$H = 0,$$
and thus the solution is
$$y = \frac{w}{k}\{1-e^{-\omega x}\cos\omega x\}. \tag{7}$$

85. A continuous beam resting on several supports at the same level

In the other sections of this chapter the beams have been assumed to be supported at two points only. The problem of a uniform continuous beam resting on a number of frictionless supports at the same level and carrying a uniformly distributed load w per unit length can also be treated fairly simply.

Let A, B, C,... be consecutive supports and let $AB = a$, $BC = b$. Take the origin at B with the x-axis along BC. Then in BC we have to solve
$$EID^4y = w, \tag{1}$$
with $$y = 0 \quad \text{when } x = 0 \text{ and when } x = b. \tag{2}$$

Integrating (1) four times in the usual way and using $y = 0$ when $x = 0$ we get

$$EIy = \tfrac{1}{24}wx^4+\tfrac{1}{6}Px^3+\tfrac{1}{2}Qx^2+Rx, \tag{3}$$

where P, Q, and R are arbitrary constants. The condition $y = 0$ when $x = b$ gives

$$\tfrac{1}{24}wb^4+\tfrac{1}{6}Pb^3+\tfrac{1}{2}Qb^2+Rb = 0. \tag{4}$$

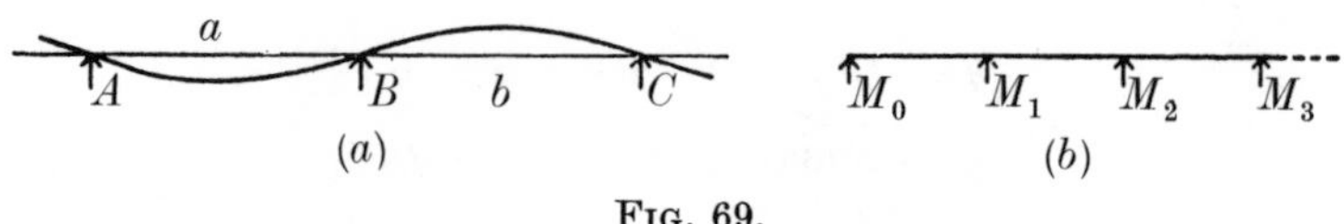

FIG. 69.

Also, if M is the bending moment at x,

$$M = EID^2y = \tfrac{1}{2}wx^2+Px+Q. \tag{5}$$

Now let M_A, M_B, M_C,... be the unknown bending moments at the supports. Putting $x = 0$ and $x = b$ in (5) gives

$$M_B = Q, \tag{6}$$

$$M_C = \tfrac{1}{2}wb^2+Pb+Q. \tag{7}$$

Also, if i_B is the slope of the beam at $x = 0$,

$$\begin{aligned} EIi_B &= R, \\ &= -\tfrac{1}{24}wb^3-\tfrac{1}{6}Pb^2-\tfrac{1}{2}Qb, \\ &= \tfrac{1}{24}wb^3-\tfrac{1}{6}bM_C-\tfrac{1}{3}bM_B, \end{aligned} \tag{8}$$

where we have used (4), and then (6) and (7).

Now suppose, taking B as origin again, that we take the x-axis in the direction BA. The bending moments calculated in this way will be equal to those calculated with the axis in the other direction, so that (8) still holds with the appropriate changes of notation, *except* that, as can be seen from Fig. 69 (*a*), the sign of i_B is changed. Therefore we get from (8)

$$-EIi_B = \tfrac{1}{24}wa^3-\tfrac{1}{6}aM_A-\tfrac{1}{3}aM_B. \tag{9}$$

Adding (8) and (9) gives finally

$$aM_A+2(a+b)M_B+bM_C = \tfrac{1}{4}w(a^3+b^3). \tag{10}$$

This is Clapeyron's *theorem of three moments*. It is an algebraic equation connecting the bending moments at three consecutive supports. Using it and the conditions at the first and last of the supports, sufficient equations can be written down to determine the bending moments at all the supports. Since P, Q, and R in (3) can be expressed in terms of the bending moments by (4), (6), and (7), the deflexion at any point of the beam then follows. The reaction at the support B, which is the difference between the shears to the right and left of B, is

$$\tfrac{1}{2}w(a+b)+\left(\frac{1}{a}+\frac{1}{b}\right)M_B-\frac{1}{a}M_A-\frac{1}{b}M_C. \tag{11}$$

(10) and (11) and their many generalizations to more complicated systems are of the greatest importance in practice.

Ex. *A uniform semi-infinite beam rests on supports at*

$$x = na \quad (n = 0, 1, ...).$$

Let M_n be the bending moment at the nth support, Fig. 69 (*b*), then (10) gives

$$M_{n+2}+4M_{n+1}+M_n = \tfrac{1}{2}wa^2, \tag{12}$$

to be solved with

$$M_0 = 0, \tag{13}$$

and the condition that M_n remains finite as $n \to \infty$.

We seek a solution of (12) of the form

$$M_n = A+Bk^n. \tag{14}$$

Substituting (14) in (12) we must have

$$A = \tfrac{1}{12}wa^2, \tag{15}$$

$$k^2+4k+1 = 0. \tag{16}$$

The solution of (16) is

$$k = -2\pm\sqrt{3}. \tag{17}$$

For M_n to remain finite as $n \to \infty$ we must have $|k| < 1$, and thus must choose the value $-2+\sqrt{3}$ of (17). Therefore a solution of (12) is

$$M_n = \frac{wa^2}{12}+B(-2+\sqrt{3})^n. \tag{18}$$

The condition $M_0 = 0$ gives B, and we get finally

$$M_n = \frac{wa^2}{12}\{1-(-2+\sqrt{3})^n\}. \tag{19}$$

86. A beam with transverse loads and axial tension or compression

Suppose that a uniform beam lies along the x-axis, as in Fig. 70, and is freely hinged at its ends $x = 0$ and $x = l$. Suppose that it carries a uniformly distributed load w per unit length in the direction of OY, and that in addition there is tension T along the x-axis.

We take the differential equation in the form § 80 (10),

$$EID^2y = M, \tag{1}$$

to be solved with $y = 0$ when $x = 0$ and $x = l$. The bending moment M at the point x contains a term Ty, due to the axial tension, as well as the term $-\frac{1}{2}wx(l-x)$, calculated as in § 79 (8) transverse load.

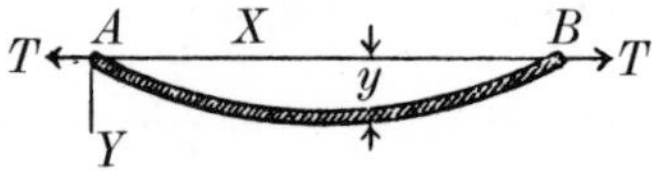

FIG. 70.

Therefore $$M = Ty-\tfrac{1}{2}wx(l-x),$$

and the differential equation (1) becomes

$$D^2y-\frac{T}{EI}y = -\frac{w}{2EI}x(l-x). \tag{2}$$

The general solution of this, found as in § 14 (ii), is

$$y = A\sinh\alpha x+B\sinh\alpha(l-x)-\frac{w}{\alpha^2T}+\frac{wlx}{2T}-\frac{wx^2}{2T}, \tag{3}$$

where $$\alpha^2 = T/EI. \tag{4}$$

The conditions $y = 0$ when $x = 0$ and $x = l$ give

$$B\sinh\alpha l = w/\alpha^2T,$$

$$A\sinh\alpha l = w/\alpha^2T.$$

Therefore, finally,

$$y = \frac{w\{\sinh\alpha x+\sinh\alpha(l-x)\}}{\alpha^2T\sinh\alpha l}-\frac{w}{\alpha^2T}+\frac{wx(l-x)}{2T}$$

$$= \frac{w}{\alpha^2T}\left\{\frac{\cosh\alpha(\frac{1}{2}l-x)}{\cosh\frac{1}{2}\alpha l}-1\right\}+\frac{wx(l-x)}{2T}. \tag{5}$$

Other cases are treated similarly. If there is axial compression instead of tension, trigonometrical functions appear in place of the hyperbolic functions.

87. Column formulae. Eigenvalue problems

In the theory of columns we consider a beam with axial compression and no transverse loading.

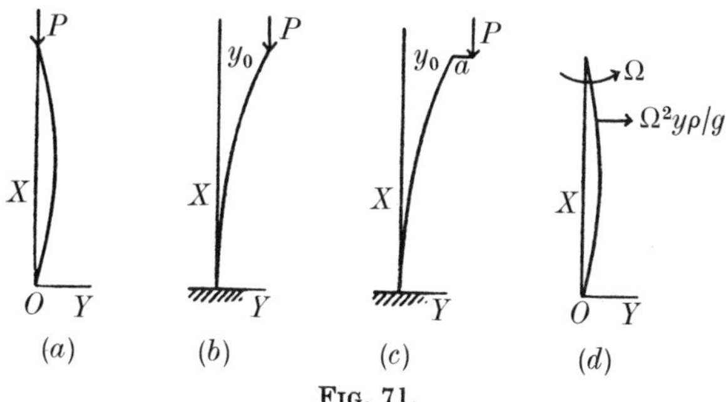

Fig. 71.

The most important case is that of a column of length l, freely hinged at its ends, with axial compression P, Fig. 71 (a). The bending moment M at the point x of the column is $-Py$, as in § 86, and the differential equation for the deflexion becomes

$$EID^2y = -Py,$$

or
$$(D^2+\omega^2)y = 0, \tag{1}$$

where
$$\omega^2 = P/EI. \tag{2}$$

This has to be solved with $y = D^2y = 0$, when $x = 0$ and $x = l$.

The general solution of (1) is

$$y = A\sin\omega x+B\cos\omega x. \tag{3}$$

The condition $y = D^2y = 0$ when $x = 0$ requires $B = 0$, so that (3) becomes

$$y = A\sin\omega x, \tag{4}$$

and the condition $y = D^2y = 0$ when $x = l$ requires

$$A\sin\omega l = 0. \tag{5}$$

Thus we must have *either* $A = 0$, in which case the solution is $y = 0$ for all x and the beam is undeflected, *or*

$$\sin\omega l = 0, \tag{6}$$

that is,
$$\omega l = n\pi \quad (n = 1, 2,...). \tag{7}$$

Using (2), (7) gives

$$P = \frac{n^2\pi^2 EI}{l^2} \quad (n = 1, 2,...). \tag{8}$$

Thus, unless P has one of the values (8), the only solution of the differential equation and boundary conditions is $y = 0$ for all x. If P has the value $n^2\pi^2 EI/l^2$, another solution is

$$y = A\sin\frac{n\pi x}{l}, \tag{9}$$

where A is arbitrary.

From the practical point of view this result may be interpreted in the following way: if $P < \pi^2 EI/l^2$, the first of the values (8), the only solution is $y = 0$ and the column is undeflected. Further discussion shows that this solution is stable, that is, if the column is slightly bent it will return to the straight form. When P reaches the value $\pi^2 EI/l^2$, another solution also becomes possible in which the form of the column is $A\sin \pi x/l$, and now this solution is stable and the undeflected position unstable: thus if the column is slightly disturbed it will assume a bent position and for practical purposes will collapse. The value $\pi^2 EI/l^2$ is known as the Euler load or crippling load, and the above theory is the Euler theory.

From the theoretical point of view the problem is that of a homogeneous linear differential equation which contains a parameter ω^2 and which has to be solved subject to homogeneous boundary conditions. In such problems there is always a trivial solution which is zero for all values of x, and non-zero solutions can be found only for a discrete set of values of the parameter. These values are called the eigenvalues, the corresponding solutions are the eigenfunctions, and such problems are referred to as eigenvalue problems.

Ex. 1. *A column is clamped vertically at its base $x = 0$, and is free at $x = l$.*

Let y_0 be the deflexion at the top of the column (Fig. 71 (*b*)).

The bending moment M at the point x is $P(y_0-y)$, so the differential equation for the deflexion is

$$EID^2y = P(y_0-y),$$

or

$$(D^2+\omega^2)y = \omega^2 y_0, \tag{10}$$

where ω^2 is defined in (2). (10) has to be solved with $y = Dy = 0$ when $x = 0$, and $y = y_0$ when $x = l$. Its solution which satisfies the conditions at $x = 0$ is

$$y = y_0(1-\cos\omega x). \tag{11}$$

If, in addition, we are to have $y = y_0$ when $x = l$, we must have

$$\cos\omega l = 0,$$

that is
$$\omega l = \frac{(2n-1)\pi}{2} \quad (n = 1, 2,...),$$

or
$$P = \frac{(2n-1)^2\pi^2 EI}{4l^2} \quad (n = 1, 2,...). \tag{12}$$

The crippling load, corresponding to $n = 1$ in (12), is $\pi^2 EI/4l^2$.

It is interesting to consider the problem from another point of view. One objection to the Euler theory is that in practice it is impossible to ensure that the load lies precisely along the axis of the column, and it is not clear what effect a small eccentricity of loading may have. We show that if it is allowed for, Euler's result still applies.

Suppose, then, that the load P is applied a small distance a from the axis of the column, Fig. 71 (c). The only change is that (10) has to be replaced by

$$(D^2+\omega^2)y = \omega^2(y_0+a), \tag{13}$$

which has to be solved as before with $y = Dy = 0$, when $x = 0$, and with $y = y_0$ when $x = l$. The solution of (13) for which $y = Dy = 0$ when $x = 0$ is

$$y = (y_0+a)(1-\cos\omega x), \tag{14}$$

and the condition $y = y_0$ when $x = l$ requires

$$y_0 = (y_0+a)(1-\cos\omega l). \tag{15}$$

Solving (15) for y_0 and substituting in (14), gives

$$y = \frac{a(1-\cos\omega x)}{\cos\omega l}. \tag{16}$$

This problem, of course, is not an eigenvalue problem: there is a finite deflexion for any load. But it follows from (16), because of the denominator $\cos\omega l$, that the deflexion becomes very large as P tends to any of the values (12), and this is effectively the same result as before.

Ex. 2. *A column is freely hinged at its ends. Lateral displacement is resisted by a force per unit length equal to k times the displacement.*

In this case we have as in § 86 (1)

$$EID^2y = -Py+M', \tag{17}$$

where M' is the bending moment at x due to the transverse loading $-ky$ at x. Differentiating (17) twice, and using § 79 (3), gives

$$EID^4y+PD^2y+ky = 0. \tag{18}$$

This has to be solved with

$$y = D^2y = 0, \quad \text{when } x = 0, \tag{19}$$

$$y = D^2y = 0, \quad \text{when } x = l. \tag{20}$$

The problem may be treated as above, but the eigenvalues may be found very simply by noting that

$$\sin\frac{n\pi x}{l} \quad (n = 1, 2,...)$$

satisfies (19) and (20), and it also satisfies (18) if

$$\frac{EIn^4\pi^4}{l^4}-\frac{Pn^2\pi^2}{l^2}+k = 0,$$

that is, if
$$P = \frac{EIn^2\pi^2}{l^2}+\frac{kl^2}{n^2\pi^2}. \tag{21}$$

Ex. 3. *Shaft whirling.*

Another simple and important type of eigenvalue problem arises in the following way. Suppose a uniform, straight shaft rotates with angular velocity Ω about its axis and is carried in bearings at $x = 0$ and $x = l$ which may be regarded as free hinges.

Suppose the shaft is in a bent position as in Fig. 71 (*d*). There is a transverse loading of $\Omega^2 y\rho/g$ per unit length on the shaft caused by the centrifugal or reversed effective forces, where ρ is the mass per unit length of the shaft.†

The differential equation for the deflexion y is

$$EID^4y-\Omega^2\rho y/g = 0,$$

or
$$(D^4-k^4)y = 0, \tag{22}$$

where
$$k^4 = \Omega^2\rho/EIg. \tag{23}$$

(22) has to be solved with $y = D^2y = 0$, when $x = 0$ and when $x = l$. Its general solution is

$$y = A\sin kx+B\cos kx+C\sinh kx+D\cosh kx. \tag{24}$$

The boundary condition at $x = 0$ requires

$$B+D = 0,$$
$$B-D = 0,$$

so that $B = D = 0$. The conditions at $x = l$ now give

$$A\sin kl+C\sinh kl = 0,$$
$$A\sin kl-C\sinh kl = 0.$$

Thus we must have either $A = C = 0$, corresponding to the undeflected position, or
$$\sin kl\sinh kl = 0. \tag{25}$$

This gives
$$ki = n\pi \quad (n = 1, 2,...)$$

or
$$\Omega^2 = \frac{n^4\pi^4EIg}{\rho l^4} \quad (n = 1, 2,...). \tag{26}$$

At these angular velocities deflected positions are possible (and the undeflected positions are unstable) and the shaft 'whirls'.

† The units in which the various quantities are measured have not been specified above. In the usual engineering practice gravitational units are used for the loads, and therefore the reversed effective force is divided by g.

Ex. 4. *Critical size of a nuclear reactor.*

A somewhat different example of an eigenvalue problem arises in connexion with the critical size of a nuclear reactor.

The neutron flux ϕ in the reactor is a measure of the intensity of radiation. For a reactor operating at steady power, ϕ satisfies the equation

$$\tfrac{1}{3}\lambda\,\nabla^2\phi - A\phi + S = 0, \tag{27}$$

where the constants λ and A are respectively the effective distance a neutron travels between collisions and the 'absorption cross-section' of the reactor. S is the rate of production of neutrons. During fission each capture of a neutron results in the production of k (> 1) new neutrons, so that $S = Ak\phi$.

For a spherical reactor the operator $\nabla^2 = \dfrac{d^2}{dr^2}+\dfrac{2}{r}\dfrac{d}{dr}$, and so (27) may be written

$$\frac{d^2\phi}{dr^2}+\frac{2}{r}\frac{d\phi}{dr}+B^2\phi = 0, \tag{28}$$

where $B^2 = 3(k-1)A/\lambda$.

The substitution $u = r\phi$ reduces (28) to

$$\frac{d^2u}{dr^2}+B^2u = 0$$

which has the general solution

$$u = P\sin(Br)+Q\cos(Br),$$

whence

$$\phi = \frac{P}{r}\sin(Br)+\frac{Q}{r}\cos(Br). \tag{29}$$

Since the neutron flux at the centre $r = 0$ of the sphere must be finite, we take $Q = 0$, so

$$\phi = \frac{P}{r}\sin(Br). \tag{30}$$

The second boundary condition is that the flux ϕ is zero at an effective boundary $r = a$ which is slightly larger than the real radius of the sphere. Putting $\phi = 0$ at $r = a$ in (30) leads either to $P = 0$, that is, there is no nuclear reaction, or else

$$\sin(Ba) = 0,$$

that is,

$$B = \pi/a,\ 2\pi/a,\dots.$$

Only the first eigenvalue $B = \pi/a$ is of practical significance. It leads to the solution

$$\phi = \frac{P}{r}\sin\!\left(\frac{\pi r}{a}\right)$$

where P can take on *any* value. We therefore talk of the reactor going 'critical' when the radius $a = \pi/B$; a is called the critical radius of the sphere.

If the sphere is made of Uranium 235, the above analysis is a simplified description of an atom bomb.

EXAMPLES ON CHAPTER X

In all these examples I refers to the moment of inertia of the cross-section of the beam and E is Young's modulus.

1. A beam of length l, freely hinged at its ends, carries a distributed load which increases linearly from zero at the ends to w per unit length at the mid-point of the beam. Sketch the bending moment and shear diagrams, and show that the maximum bending moment in the beam is $wl^2/12$.

2. A uniform beam of length $2l$ and weight w per unit length is rested symmetrically on two supports distant $2a$ apart. Sketch the bending moment and shear diagrams, and discuss the variation of the bending moments at the middle of the beam and at the supports with a. Find the maximum bending moment in the beam, and show that this is least if $a = (2-\sqrt{2})l$.

3. Show that when two unequal concentrated loads a fixed distance apart are traversed along a beam supported at the ends, the maximum bending moment occurs when the heavier load and the centre of gravity of the two loads are equidistant from the centre of the span, provided that both loads are then on the span.

4. Show that the maximum bending moments which can be withstood by beams of square and circular cross-section of the same area are in the ratio $2\sqrt{\pi}/3$.

5. Find the deflexion of a light uniform cantilever clamped horizontally at $x = 0$ and free at $x = l$ for

(i) a uniformly distributed load w per unit length,

(ii) a concentrated load W at $x = l$.

If the cantilever carries a uniformly distributed load w per unit length and is propped at $x = l$ to the same level as that at $x = 0$, show, by combining the two above results, or directly, that the reaction at the prop is $3wl/8$.

6. A beam of length l is freely hinged at the same level at its ends and carries a uniformly distributed load w per unit length. At its mid-point it rests on a yielding support which provides a reaction of k times the deflexion. Show that the reaction at this support is

$$\frac{5kwl^4}{384EI+8kl^3}.$$

7. A tapered cantilever of rectangular section has constant width but its depth decreases linearly to zero at $x = l$. If w_0 is its weight per unit length, and I_0 the moment of inertia of its cross-section at the origin, show that its deflexion under its own weight is

$$w_0 l^2x^2/12EI_0.$$

8. Show that the potential energy of the portion of the beam between the planes AB, $A'B'$, of Fig. 68 (a) is $EI\theta/2R$, and deduce that the potential energy of the whole beam is

$$\frac{1}{2}\int EI\left(\frac{d^2y}{dx^2}\right)^2 dx.$$

Show by integrating by parts that for a beam $0 < x < l$ this may be put in the form

$$\frac{1}{2}\left[Fy+M\frac{dy}{dx}\right]_0^l+\tfrac{1}{2}\int\limits_0^l wy\,dx.$$

9. A light beam of length $a+b$ is freely hinged at its ends and carries a mass M at the point a. Find the deflexion of the beam at a, and show that the natural frequency of oscillation of M is

$$\left\{\frac{3EI(a+b)g}{Ma^2b^2}\right\}^{\frac{1}{2}}\Big/2\pi.$$

10. A light uniform beam of length l is clamped horizontally at the same level at both ends. It carries a constant load w per unit length for $0 < x < \frac{1}{2}l$ and no load in the region $\frac{1}{2}l < x < l$. Show that the deflexion is given by

$$EIy = -\tfrac{13}{192}wlx^3+\tfrac{11}{384}wl^2x^2+\tfrac{1}{24}wx^4$$

for $0 < x < \frac{1}{2}l$, and find its value in $\frac{1}{2}l < x < l$.

11. A light uniform beam of length l carries a load which is zero at its ends and increases linearly up to w per unit length at its mid-point. The beam is clamped horizontally at $x = 0$ and freely hinged at the same level at $x = l$. Show that the deflexion at its mid-point is

$$\frac{53wl^4}{15360EI}.$$

12. A uniform beam of length l and weight w per unit length is freely hinged at its ends and subject to axial compression P. Show that its deflexion at the point x is, writing $\omega^2 = P/EI$,

$$\frac{EIw}{P^2}\left\{\frac{\cos\omega(\frac{1}{2}l-x)}{\cos\frac{1}{2}\omega l}-1\right\}-\frac{wx(l-x)}{2P}.$$

13. A light uniform beam of length l is freely hinged at its ends and subject to axial compression P. It also carries a concentrated transverse load W at $x = a$. Show that the deflexion is

$$\frac{W\sin\omega(l-a)\sin\omega x}{\omega P\sin\omega l}-\frac{W(l-a)x}{Pl},$$

if $0 < x < a$, where $\omega^2 = P/EI$.

14. A light semi-infinite beam $x > 0$ is attached to an elastic foundation which provides restoring force k times the deflexion. It is free at $x = 0$ and carries a concentrated load W at $x = a$. Show that its deflexion at $x = 0$ is

$$\frac{We^{-\omega a}}{2EI\omega^3}\cos\omega a,$$

where $\omega^4 = k/4EI$.

15. Find the form taken by the equation of three moments when two of the supports approach one another (corresponding to a beam clamped horizontally at a point). Deduce the results of § 81, Ex. 1, in this way.

Show that if a uniform beam of length $2l$ is clamped horizontally at $x = 0$ and is supported at this level at $x = l$ and $x = 2l$, the reaction at the support at $x = l$ is

$$8wl/7.$$

16. A uniform beam of length nl rests on $n+1$ equally spaced supports. Show that the bending moment at the rth support is

$$\tfrac{1}{12}wl^2\left\{1+\frac{k^{1-r}(1-k^n)-k^{r-1}(1-k^{-n})}{k^n-k^{-n}}\right\},$$

where $k = -2+\sqrt{3}$ and w is the weight per unit length of the beam.

17. A light beam of length l is freely hinged at its ends. Find the deflexion produced by unit force applied at its mid-point perpendicular to the beam.

If a mass M is attached to the centre of the beam show that its natural frequency is

$$\left(\frac{48gEI}{Ml^3}\right)^{\frac{1}{2}}\Big/2\pi.$$

If the beam carrying the mass M rotates about its axis, show that the angular velocity of whirling is 2π times the above result.

18. A column is clamped vertically at $x = 0$ and freely hinged at $x = l$. Show that the crippling load is $EI\alpha^2/l^2$, where α is the smallest (non-zero) root of

$$\tan\alpha = \alpha.$$

19. Deduce the result § 87 (21) by studying the solution of the differential equation and boundary conditions.

20. A uniform straight shaft of length l and mass ρ per unit length rotates with angular velocity ω and is subject to axial compression P. Show that, if the ends are freely hinged, the whirling speed is given by

$$\left(\frac{P^2}{4E^2I^2}+\frac{\rho\omega^2}{gEI}\right)^{\frac{1}{2}}+\frac{P}{2EI}=\frac{\pi^2}{l^2}.$$

21. Show that for *large* deflexions of a thin wire caused by tension T, the deflexion y at the point x satisfies

$$\rho y = EI/T,$$

where ρ is the radius of curvature at x. Show that an integral of this equation with $y = 0$ when $\psi = 0$ is

$$y = 2(EI/T)^{\frac{1}{2}}\sin\tfrac{1}{2}\psi,$$

where ψ is the slope of the tangent to the wire at any point.

Deduce that, if the origin of x is chosen at $\psi = \pi$,

$$x = (EI/T)^{\frac{1}{2}}\{\ln\tan\tfrac{1}{4}\psi+2\cos\tfrac{1}{2}\psi\},$$

and sketch the curve.

XI

FOURIER SERIES AND INTEGRALS

88. Introductory. Periodic functions

PERIODIC functions arise in many branches of applied mathematics. In electric circuit theory functions which are periodic in the time occur: for example, the voltage applied to a circuit by an alternator is periodic but usually not sinusoidal, while in modern practice periodic voltages with forms such as 'square wave', 'saw-tooth', etc., cf. Fig. 76, are common.

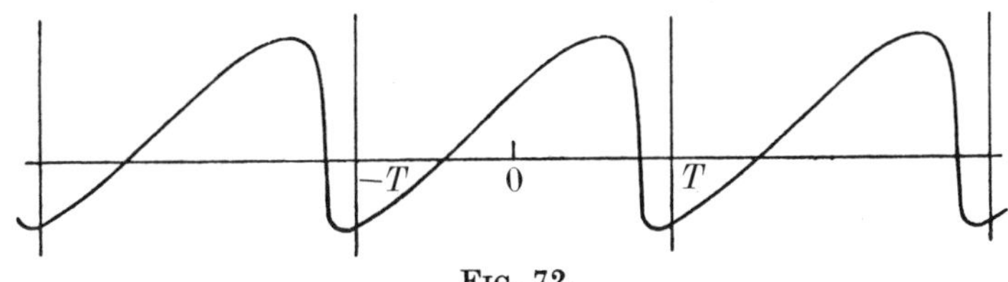

FIG. 72.

Because of the importance of functions which are periodic in the time we shall take t as the independent variable in §§ 88–90 and say $f(t)$ is periodic with period $2T$ if

$$f(t+2rT) = f(t) \quad (r = \pm 1, \pm 2,...). \tag{1}$$

Thus if the value of $f(t)$ is known in any interval of width $2T$ it is known for all t. We shall take $-T \leqslant t \leqslant T$ for the region† in which values of $f(t)$ are given; cf. Fig. 72.

A number of trigonometric functions are known which have period $2T$; these are the set of even‡ functions

$$\cos\frac{n\pi t}{T} \quad (n = 0, 1, 2,...), \tag{2}$$

of which the case $n = 0$ is the constant unity, and the set of odd functions

$$\sin\frac{n\pi t}{T} \quad (n = 1, 2,...). \tag{3}$$

† The period is taken to be $2T$ and the origin in the middle of the region (instead, for example, of the origin at one end of the region which might seem more natural) for convenience in the discussion of odd and even functions in § 89. The final results can easily be put in the appropriate form for functions of period T; cf. Ex. 6.

‡ A function is even if $f(t) = f(-t)$ for all t. It is odd if $f(-t) = -f(t)$.

The complete set of functions (2) and (3) possesses the property that the integral from $-T$ to T of the product of any two different members of the set is zero. This property is known as *orthogonality* over the region $(-T, T)$. To prove it we have

$$\int_{-T}^{T} \cos\frac{n\pi t}{T}\,dt = \int_{-T}^{T} \sin\frac{n\pi t}{T}\,dt = 0 \quad (n = 1, 2, 3,...), \tag{4}$$

$$\int_{-T}^{T} \cos\frac{n\pi t}{T}\sin\frac{m\pi t}{T}\,dt = \frac{1}{2}\int_{-T}^{T}\left\{\sin\frac{(m+n)\pi t}{T}+\sin\frac{(m-n)\pi t}{T}\right\}dt = 0, \tag{5}$$

for all m and n;

$$\int_{-T}^{T} \cos\frac{n\pi t}{T}\cos\frac{m\pi t}{T}\,dt = \frac{1}{2}\int_{-T}^{T}\left\{\cos\frac{(m+n)\pi t}{T}+\cos\frac{(m-n)\pi t}{T}\right\}dt$$

$$= 0, \quad \text{if } m \neq n; \tag{6}$$

and similarly

$$\int_{-T}^{T} \sin\frac{n\pi t}{T}\sin\frac{m\pi t}{T}\,dt = 0, \quad \text{if } m \neq n. \tag{7}$$

The integrals from $-T$ to T of the squares of the functions (2) and (3), however, do not vanish. They are

$$\int_{-T}^{T} \sin^2\frac{n\pi t}{T}\,dt = \frac{1}{2}\int_{-T}^{T}\left\{1-\cos\frac{2n\pi t}{T}\right\}dt = T \quad (n = 1, 2,...), \tag{8}$$

$$\int_{-T}^{T} \cos^2\frac{n\pi t}{T}\,dt = T \quad (n = 1, 2, 3,...), \tag{9}$$

$$\int_{-T}^{T} dt = 2T. \tag{10}$$

It should be noticed that in (10), which corresponds to (9) with $n = 0$, $2T$ occurs in place of T.

It is natural to inquire whether *any* periodic function $f(t)$ with period $2T$, such as that of Fig. 72, can be represented in terms of the set of trigonometric functions (2) and (3) of period $2T$, that is, whether there exists an expansion

$$f(t) = a_0 + \sum_{n=1}^{\infty} a_n \cos\frac{n\pi t}{T} + \sum_{n=1}^{\infty} b_n \sin\frac{n\pi t}{T}, \tag{11}$$

where the a_n and b_n are constants.

We first show that, if such an expansion exists and it is permissible to integrate the infinite series in it term by term, the coefficients a_0, a_n, and b_n can be found very simply. First, if we integrate (11) with respect to t from $-T$ to T, all the integrals on the right-hand side except the first vanish by (4), and we get

$$\int_{-T}^{T} f(t)\,dt = a_0 \int_{-T}^{T} dt = 2Ta_0. \tag{12}$$

Therefore
$$a_0 = \frac{1}{2T} \int_{-T}^{T} f(t')\,dt'. \tag{13}$$

In (13), t' has been written for the variable of integration in the definite integral in place of t as in (12). This will also be done in (14) and (15) below in order to avoid confusion with t in subsequent work.

To find a_m we multiply both sides of (11) by $\cos m\pi t/T$ and integrate from $-T$ to T. All terms on the right-hand side will vanish except one, by (4), (5), and (6), and we get

$$\int_{-T}^{T} f(t) \cos\frac{m\pi t}{T}\,dt = a_m \int_{-T}^{T} \cos^2\frac{m\pi t}{T}\,dt = Ta_m,$$

using (9). Therefore, again replacing t by t' in the definite integral

$$a_m = \frac{1}{T} \int_{-T}^{T} f(t') \cos\frac{m\pi t'}{T}\,dt' \quad (m = 1, 2, 3,...). \tag{14}$$

Similarly

$$b_m = \frac{1}{T} \int_{-T}^{T} f(t') \sin\frac{m\pi t'}{T}\,dt' \quad (m = 1, 2, 3,...). \tag{15}$$

In the same way, if we square both sides of (11) and integrate with respect to t from $-T$ to T, the integrals of all products of different functions on the right-hand side are zero, and, using (8), (9), and (10) for the integrals of squares of trigonometric functions, we get

$$\frac{1}{2T}\int\limits_{-T}^{T}[f(t)]^2\,dt = a_0^2+\tfrac{1}{2}\sum_{r=1}^{\infty}a_r^2+\tfrac{1}{2}\sum_{r=1}^{\infty}b_r^2. \tag{16}$$

These quantities a_0, a_m, b_m can be found for any integrable function $f(t)$ and are called its *Fourier constants*. It will appear from later results that they provide a complete alternative specification of the function, that is, if the Fourier constants of the function $f(t)$ are known it is just as completely specified as by the usual statement of the value of $f(t)$ for each value of t in $(-T, T)$. In some problems it is more useful to specify a function by its Fourier constants than in the ordinary way.

The process of finding the Fourier constants for a function given numerically or graphically is called Fourier analysis or harmonic analysis: there are many ways of doing this arithmetically, also mechanical devices have been invented for finding the Fourier constants of a function whose graph is given, and for finding the graph of a function whose Fourier constants are given.

The series on the right-hand side of (11) with the values (13), (14), and (15) of a_0, a_n, and b_n inserted is known as the *Fourier series* for the function $f(t)$. It may be written as

$$\frac{1}{2T}\int\limits_{-T}^{T}f(t')\,dt'+\frac{1}{T}\sum_{n=1}^{\infty}\cos\frac{n\pi t}{T}\int\limits_{-T}^{T}f(t')\cos\frac{n\pi t'}{T}\,dt'+$$
$$+\frac{1}{T}\sum_{n=1}^{\infty}\sin\frac{n\pi t}{T}\int\limits_{-T}^{T}f(t')\sin\frac{n\pi t'}{T}\,dt', \tag{17}$$

or, combining the last two series, as

$$\frac{1}{2T}\int\limits_{-T}^{T}f(t')\,dt'+\frac{1}{T}\sum_{n=1}^{\infty}\int\limits_{-T}^{T}f(t')\cos\frac{n\pi(t-t')}{T}\,dt'. \tag{18}$$

The above argument has not *proved* that (17) or (18) is equal to $f(t)$. It has merely been a 'plausibility' argument† to intro-

† Such arguments are often not true and must always be supplemented by careful pure-mathematical discussion. For example, the result would have been equally plausible if the constant term a_0 in (11) had been omitted.

duce the Fourier constants and the series (17). To prove the result suggested in (11) it is necessary to show that the series (17) is convergent and that its sum is $f(t)$. Whether this will be the case or not depends on the nature of the function $f(t)$, and it is important to allow $f(t)$ to be fairly general: in particular, since discontinuous functions appear so often in applications, these must be considered. The restrictions usually placed on $f(t)$ are

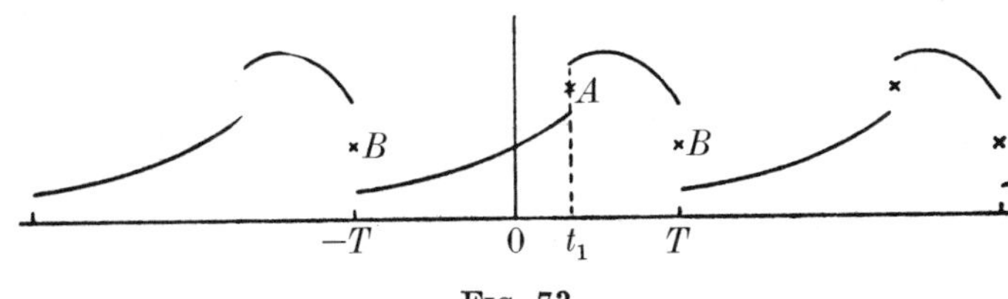

FIG. 73.

that it satisfy 'Dirichlet's conditions', a simple form of which, adequate for our purpose, is that $f(t)$ should be continuous in $(-T, T)$, except at a finite number of ordinary discontinuities, and that it should have only a finite number of maxima and minima in this region.

Fourier's theorem then states that, if $f(t)$ satisfies Dirichlet's conditions, the series (17) *is convergent and its sum is*

(i) *$f(t)$ at all points where this function is continuous,* (19)

(ii) *$\frac{1}{2}\{f(t+0)+f(t-0)\}$ at points where $f(t)$ is discontinuous,*† (20)

(iii) *$\frac{1}{2}\{f(T-0)+f(-T+0)\}$ at $t = \pm T$.* (21)

Thus if $f(t)$ is the function shown in Fig. 73, the sum of the series has the value A when $t = t_1$, and the value B when $t = \pm T$, and for all other values of t is given by the graph. It is in this sense that the expansion (11) is true.

The proof of Fourier's theorem is rather long‡ and will not be given here.

We also state without proof the 'uniqueness theorem' that if

† We write $f(t+0)$ for $\lim_{\tau \to t+0} f(\tau)$ and $f(t-0)$ for $\lim_{\tau \to t-0} f(\tau)$.

‡ Cf. Carslaw, *Fourier's Series and Integrals* (Macmillan); Churchill, *Fourier Series and Boundary Value Problems* (McGraw-Hill).

two Fourier series can be found for the same function the coefficients in these must be equal.

Finally, it should be remarked that series of the general type

$$\sum_{n=1}^{\infty}\left\{a_n \cos\frac{n\pi t}{T}+b_n \sin\frac{n\pi t}{T}\right\}, \tag{22}$$

are called trigonometric series and are perhaps, next to power series $\sum a_n t^n$, the most important type of series in mathematics.

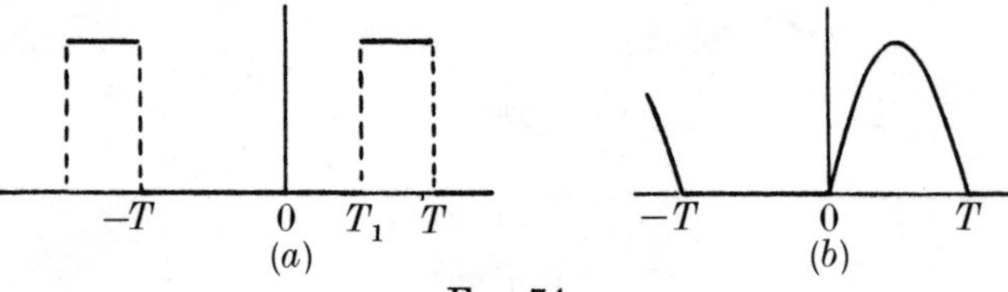

FIG. 74.

If a_n and b_n in (22) are the Fourier constants (14) and (15) of some function $f(t)$, the series becomes a Fourier series, but there are many trigonometric series of type (22) which are not the Fourier series of any function.

Ex. 1. *The repeated pulse of width* $(T-T_1)$ *of Fig. 74(a).*
In this case

$$f(t) = 0 \quad (-T < t < T_1),$$
$$f(t) = 1 \quad (T_1 < t < T).$$

Here (13), (14), and (15) give

$$a_0 = \frac{1}{2T}\int_{T_1}^{T} dt = \frac{T-T_1}{2T},$$

$$a_m = \frac{1}{T}\int_{T_1}^{T} \cos\frac{m\pi t'}{T}\,dt' = -\frac{1}{m\pi}\sin\frac{m\pi T_1}{T},$$

$$b_m = \frac{1}{T}\int_{T_1}^{T} \sin\frac{m\pi t'}{T}\,dt' = \frac{1}{m\pi}\left\{(-1)^{m+1}+\cos\frac{m\pi T_1}{T}\right\}.$$

Thus the Fourier series for $f(t)$ is

$$\frac{T-T_1}{2T}-\frac{1}{\pi}\sum_{m=1}^{\infty}\frac{1}{m}\sin\frac{m\pi T_1}{T}\cos\frac{m\pi t}{T}+$$
$$+\frac{1}{\pi}\sum_{m=1}^{\infty}\frac{1}{m}\left\{(-1)^{m+1}+\cos\frac{m\pi T_1}{T}\right\}\sin\frac{m\pi t}{T}. \tag{23}$$

By Fourier's theorem, (19)–(21), the sum of this series is 0 if $-T < t < T_1$; 1 if $T_1 < t < T$; and $\frac{1}{2}$ when $t = T_1$ or $t = \pm T$.

Ex. 2. *The half-wave rectified sine wave, Fig. 74(b).*

Here
$$f(t) = 0 \quad (-T \leqslant t \leqslant 0),$$
$$f(t) = \sin\frac{\pi t}{T} \quad (0 \leqslant t \leqslant T).$$

By (13) and (14), respectively, we have

$$a_0 = \frac{1}{2T}\int\limits_0^T \sin\frac{\pi t'}{T}\,dt' = \frac{1}{\pi},$$

$$a_m = \frac{1}{T}\int\limits_0^T \sin\frac{\pi t'}{T}\cos\frac{m\pi t'}{T}\,dt'$$

$$= \frac{1}{2T}\int\limits_0^T \left\{\sin\frac{(m+1)\pi t'}{T} - \sin\frac{(m-1)\pi t'}{T}\right\}dt'$$

$$\left.\begin{aligned} &= \frac{1}{2\pi}\left\{\frac{1-\cos(m+1)\pi}{m+1} - \frac{1-\cos(m-1)\pi}{m-1}\right\}, &\text{if } m > 1\\ &= 0, &\text{if } m = 1\end{aligned}\right\}. \tag{24}$$

It follows from (24) that $a_m = 0$ if m is odd, while if m is even, say $m = 2r$,

$$a_{2r} = \frac{1}{2\pi}\left\{\frac{2}{2r+1} - \frac{2}{2r-1}\right\} = -\frac{2}{\pi(4r^2-1)}. \tag{25}$$

Similarly by (15)

$$b_m = \frac{1}{T}\int\limits_0^T \sin\frac{\pi t'}{T}\sin\frac{m\pi t'}{T}\,dt'$$

$$= \frac{1}{2T}\int\limits_0^T \left\{\cos\frac{(m-1)\pi t'}{T} - \cos\frac{(m+1)\pi t'}{T}\right\}dt',$$

so that $b_m = \frac{1}{2}$ if $m = 1$, and is zero if $m > 1$.

Therefore, finally,

$$f(t) = \frac{1}{\pi} + \tfrac{1}{2}\sin\frac{\pi t}{T} - \frac{2}{\pi}\sum_{r=1}^{\infty}\frac{1}{(4r^2-1)}\cos\frac{2r\pi t}{T}. \tag{26}$$

89. Odd and even functions: Fourier sine and cosine series

If the function $f(t)$ of § 88 is either odd or even, important simplifications occur.

Suppose, first, that $f(t)$ is even so that $f(-t) = f(t)$ for all t,

Fig. 75 (a), and, in addition, of course, $f(t)$ is periodic with period $2T$. It follows that $f(T-0) = f(T+0)$, so that the function must† be continuous at $\pm T$.

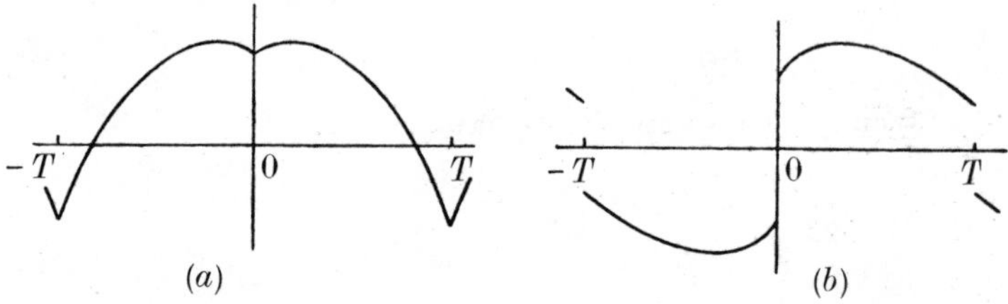

FIG. 75.

In this case from § 88 (13), (14), and (15), using the fact that $f(t)$ is even,

$$a_0 = \frac{1}{2T}\int_{-T}^{T} f(t')\,dt' = \frac{1}{T}\int_{0}^{T} f(t')\,dt', \tag{1}$$

$$a_m = \frac{1}{T}\int_{-T}^{T} f(t')\cos\frac{m\pi t'}{T}\,dt' = \frac{2}{T}\int_{0}^{T} f(t')\cos\frac{m\pi t'}{T}\,dt', \tag{2}$$

$$b_m = \frac{1}{T}\int_{-T}^{T} f(t')\sin\frac{m\pi t'}{T}\,dt' = 0, \tag{3}$$

and the series

$$\frac{1}{T}\int_{0}^{T} f(t')\,dt' + \frac{2}{T}\sum_{n=1}^{\infty}\cos\frac{n\pi t}{T}\int_{0}^{T} f(t')\cos\frac{n\pi t'}{T}\,dt', \tag{4}$$

is convergent, its sum being $f(t)$ at points where the function is continuous and $\frac{1}{2}\{f(t+0)+f(t-0)\}$ at points of discontinuity.

The series (4) is called a *Fourier cosine series*; its coefficients a_0 and a_n could have been determined directly by the method of § 88 instead of quoting the results of that section.

† Except, of course, for the trivial possibility $f(T+0) = f(T-0) \neq f(T)$, which we exclude.

If the function $f(t)$ is odd, so that $f(-t) = -f(t)$, Fig. 75 (*b*), we have from § 88 (13), (14), and (15)

$$a_m = 0 \quad (m = 0, 1, 2,...), \tag{5}$$

$$b_m = \frac{2}{T}\int\limits_0^T f(t')\sin\frac{m\pi t'}{T}\,dt', \tag{6}$$

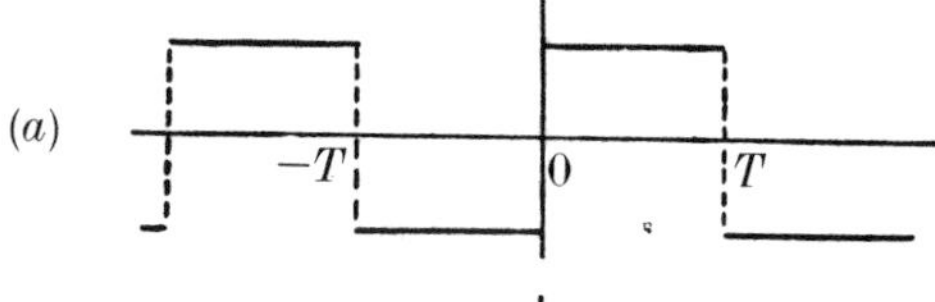

(b) −T 0 T

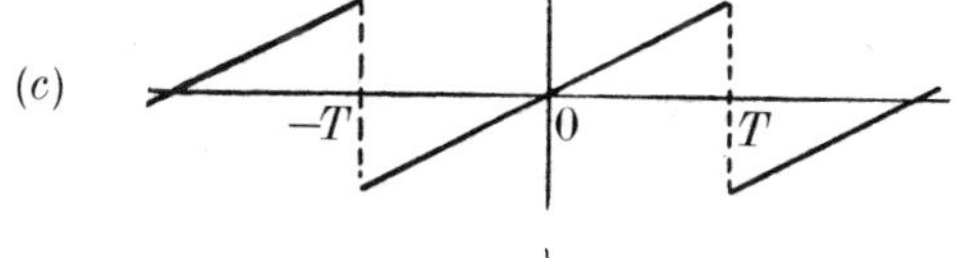

(d) −T 0 T

FIG. 76.

and the series

$$\frac{2}{T}\sum_{n=1}^{\infty}\sin\frac{n\pi t}{T}\int\limits_0^T f(t')\sin\frac{n\pi t'}{T}\,dt', \tag{7}$$

is convergent, its sum being $f(t)$ at points where this function is continuous and $\frac{1}{2}\{f(t+0)+f(t-0)\}$ at points where it is discontinuous. In particular, if $f(t)$ is discontinuous at $t = 0$, the sum of the series must be zero there since $f(+0) = -f(-0)$; also, if $f(t)$ is discontinuous when $t = T$, the sum of the series must be zero when $t = T$ since

$$f(T-0) = -f(-T+0) = -f(T+0),$$

cf. Fig. 75 (*b*). The series (7) is called a *Fourier sine series*, and the remark above establishes that the sum of such a series is zero when $t = 0$ and $t = T$.

Ex. 1. *The 'square wave' of Fig. 76 (a) defined by*

$$f(t) = 1 \qquad (0 < t < T),$$
$$= -1 \quad (-T < t < 0).$$

This is an odd function, and so by (6),

$$b_m = \frac{2}{T}\int\limits_0^T \sin\frac{m\pi t'}{T}\,dt' = \frac{2}{m\pi}\{1-(-1)^m\}. \tag{8}$$

Thus b_m is zero if m is even, and if m is odd, say $m = 2r+1$,

$$b_{2r+1} = \frac{4}{\pi(2r+1)}.$$

Thus, by (7) the Fourier series for $f(t)$ is

$$\frac{4}{\pi}\sum_{r=0}^{\infty}\frac{1}{2r+1}\sin\frac{(2r+1)\pi t}{T}. \tag{9}$$

The sum of the series (9) is zero when $t = 0$ or T, and unity if $0 < t < T$.

It is of some importance to study the way in which a Fourier series such as (9) converges to its sum. When such a series is used in practice it is usually hoped that the results derived from the first few terms will give an adequate approximation to the result. In Fig. 77 the sums of 1, 3, and 6 terms of the series (9) are graphed: it appears that 6 terms give a reasonable approximation to the function *except*† near the ends of the interval where it has a maximum which is rather large. The example chosen is rather an unfavourable one, but it serves to illustrate the fact that the convergence of the series is often rather slow for practical purposes, particularly when discontinuous functions are involved.

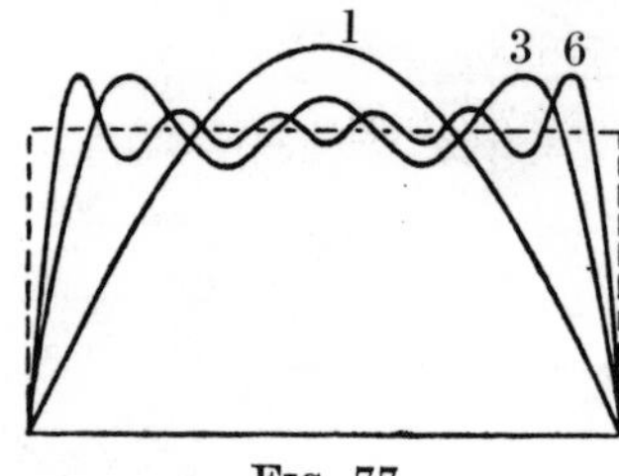

FIG. 77.

Ex. 2. *The function of Fig. 76 (b) defined by*

$$f(t) = 1, \quad -\tfrac{1}{2}T < t < \tfrac{1}{2}T,$$
$$f(t) = -1, \quad \textit{when } -T < t < -\tfrac{1}{2}T \textit{ and } \tfrac{1}{2}T < t < T.$$

In this case $f(t)$ is an even function of t, and by (1) and (2)

$$a_0 = 0,$$

$$a_m = \frac{2}{T}\int\limits_0^{\frac{1}{2}T}\cos\frac{m\pi t'}{T}\,dt' - \frac{2}{T}\int\limits_{\frac{1}{2}T}^{T}\cos\frac{m\pi t'}{T}\,dt' = \frac{4}{m\pi}\sin\tfrac{1}{2}m\pi.$$

† This maximum moves towards the ends of the interval as the number of terms is increased but it does not disappear. This is an illustration of the Gibbs phenomenon, caused by non-uniform convergence of the series.

Therefore $a_{2r} = 0$, and

$$a_{2r+1} = \frac{4(-1)^r}{(2r+1)\pi} \quad (r = 0, 1,...).$$

Thus the Fourier series for $f(t)$ is

$$\frac{4}{\pi}\sum_{r=0}^{\infty}\frac{(-1)^r}{(2r+1)}\cos\frac{(2r+1)\pi t}{T}. \tag{10}$$

The function $f(t)$ of this example and the last are the same except for a shift of origin. This illustrates the fact that, since any point may be taken as origin, many apparently different expressions for the same periodic function may be obtained. (9) reduces to (10) on putting $t = \frac{1}{2}T+t'$.

Ex. 3. *The saw-tooth wave, Fig. 76 (c),*

$$f(t) = t \quad (-T < t < T).$$

Here $f(t)$ is an odd function of t, and from (6)

$$\begin{aligned} b_m &= \frac{2}{T}\int_0^T t'\sin\frac{m\pi t'}{T}\,dt' \\ &= \frac{2}{T}\left[-\frac{Tt'}{m\pi}\cos\frac{m\pi t'}{T}\right]_0^T+\frac{2}{m\pi}\int_0^T\cos\frac{m\pi t'}{T}\,dt' \\ &= \frac{2T}{m\pi}(-1)^{m-1}. \end{aligned}$$

Therefore the Fourier series for $f(t)$ is

$$\frac{2T}{\pi}\sum_{n=1}^{\infty}\frac{(-1)^{n-1}}{n}\sin\frac{n\pi t}{T}. \tag{11}$$

Ex. 4. *The function of Fig. 76 (d)*

$$f(t) = t \qquad (0 < t < T),$$
$$f(t) = -t \quad (-T < t < 0).$$

This is an even function, and proceeding in the usual way we find

$$f(t) = \frac{T}{2}-\frac{4T}{\pi^2}\sum_{r=0}^{\infty}\frac{1}{(2r+1)^2}\cos\frac{(2r+1)\pi t}{T}. \tag{12}$$

Ex. 5. *Summation of important infinite series.*

The sums of a number of important series may be found by inserting particular values in Fourier series.

For example, the sum of the series (9) when $t = \frac{1}{2}T$ is 1, so putting $t = \frac{1}{2}T$ in (9) gives

$$\sum_{r=0}^{\infty}\frac{(-1)^r}{2r+1} = \frac{\pi}{4}. \tag{13}$$

The sum of the series (12) when $t = 0$ is zero, so

$$\sum_{r=0}^{\infty} \frac{1}{(2r+1)^2} = \frac{\pi^2}{8}. \tag{14}$$

Putting $t = 0$ and $t = \frac{1}{2}T$, respectively, in § 88 (26) gives

$$\sum_{r=1}^{\infty} \frac{1}{4r^2-1} = \frac{1}{2}, \tag{15}$$

and

$$\sum_{r=1}^{\infty} \frac{(-1)^r}{4r^2-1} = \frac{1}{2}-\frac{\pi}{4}. \tag{16}$$

90. The Fourier series of a function defined in $(-T, T)$ or $(0, T)$

In §§ 88 and 89 we have discussed the representation of periodic functions by Fourier series. But in the theory the fact

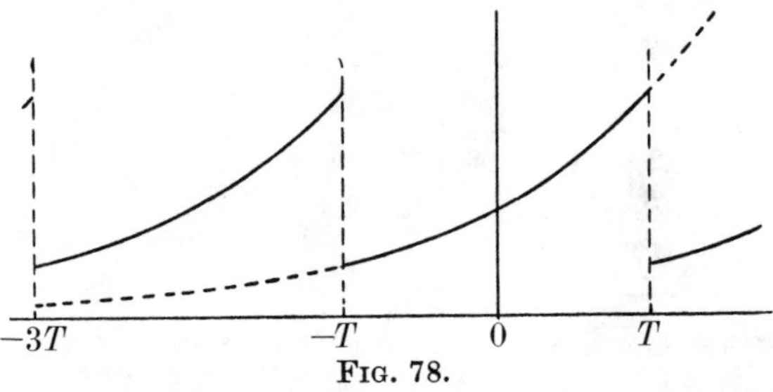

FIG. 78.

that the function was periodic was never used; in particular, in the integrals § 88 (13)–(15) for the Fourier constants only the values of $f(t)$ in the region $(-T, T)$ appear. Thus Fourier's theorem § 88 (19)–(21) may be regarded as a statement about a function $f(t)$ defined in $(-T, T)$: namely that the series § 88 (17) is convergent for all values of t between $-T$ and T, and that its sum is $f(t)$ at points where the function is continuous and $\frac{1}{2}\{f(t+0)+f(t-0)\}$ at points where it is discontinuous. But since the terms of the series § 88 (17) are all periodic with period $2T$, the sum of the series is periodic also, and thus repeats the set of values of $f(t)$ in the range $(-T, T)$, irrespective of whether $f(t)$ has any values, or different values, outside this range..

For example, if $f(t) = e^t$, $-T < t < T$, its Fourier series repeats periodically the portion of the graph of e^t between $-T$ and T as in Fig. 78.

In the Fourier cosine and sine series of § 89 (4) and (7) only the values of $f(t)$ in the region $(0, T)$ are needed. The sums of these series are respectively even and odd functions of t as well as being periodic with period $2T$.

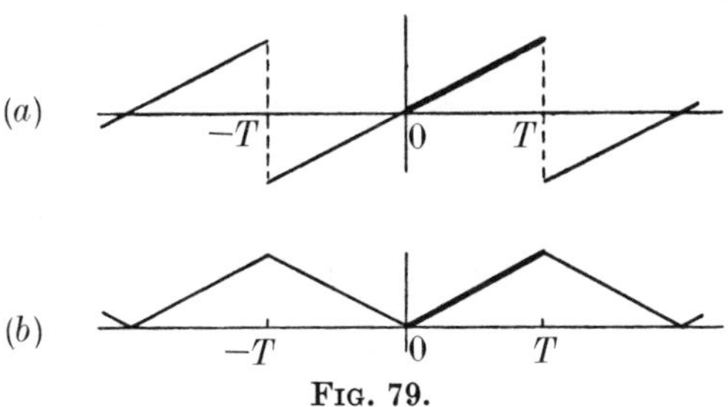

FIG. 79.

Thus suppose we form the cosine series of the function defined by $f(t) = t$ in $0 < t < T$, the sum of the series, being an even function, must be symmetrical about $t = 0$ giving Fig. 79 (*b*). If we form the sine series of the same function, it gives Fig. 79 (*a*).

91. Fourier series in electric circuit theory

If a periodic voltage $V(t)$ of period $2\pi/\omega$ is applied to a circuit, we express it by its Fourier series

$$V(t) = \sum_{n=1}^{\infty} a_n \cos n\omega t + \sum_{n=1}^{\infty} b_n \sin n\omega t \tag{1}$$

$$= \sum_{n=1}^{\infty} A_n \sin(n\omega t + \phi_n), \tag{2}$$

as in §§ 88, 89, and treat each term separately. Thus, for example, using the result of § 44 (11), the steady current in an L, R, C circuit due to the voltage (2) is

$$\sum_{n=1}^{\infty} \frac{A_n}{Z_n} \sin(n\omega t + \phi_n - \theta_n), \tag{3}$$

where
$$Z_n = \left\{\left(nL\omega - \frac{1}{nC\omega}\right)^2 + R^2\right\}^{\frac{1}{2}}, \tag{4}$$

$$\theta_n = \tan^{-1}\left\{\left(nL\omega - \frac{1}{nC\omega}\right)\Big/R\right\}. \tag{5}$$

It is frequently necessary to know the mean value of the square of a periodic quantity, or the mean value of the product of two such quantities. These can be written down as in § 88 (16).

Suppose V is the voltage applied to a circuit, and I the current at its point of application, and that

$$V = a_0 + \sum_{n=1}^{\infty} a_n \cos\frac{n\pi t}{T} + \sum_{n=1}^{\infty} b_n \sin\frac{n\pi t}{T}, \tag{6}$$

$$I = a_0' + \sum_{n=1}^{\infty} a_n' \cos\frac{n\pi t}{T} + \sum_{n=1}^{\infty} b_n' \sin\frac{n\pi t}{T}. \tag{7}$$

Then by § 88 (16)

$$\frac{1}{2T}\int_{-T}^{T} V^2\,dt = a_0^2 + \tfrac{1}{2}\sum_{r=1}^{\infty}(a_r^2+b_r^2). \tag{8}$$

In the same way

$$\frac{1}{2T}\int_{-T}^{T} VI\,dt = a_0 a_0' + \tfrac{1}{2}\sum_{r=1}^{\infty}(a_r a_r' + b_r b_r'), \tag{9}$$

which is the mean rate at which energy is being supplied to the circuit.

Similar analysis occurs in the theory of rectifiers. Suppose that voltage given by (6) is applied to a rectifier with a non-linear characteristic

$$I = f(V). \tag{10}$$

Using (6) in (10) gives by Taylor's theorem

$$I = f\left\{a_0 + \sum_{n=1}^{\infty}\left(a_n\cos\frac{n\pi t}{T} + b_n\sin\frac{n\pi t}{T}\right)\right\}$$

$$= f(a_0) + f'(a_0)\sum_{n=1}^{\infty}\left(a_n\cos\frac{n\pi t}{T} + b_n\sin\frac{n\pi t}{T}\right) +$$

$$+ \tfrac{1}{2}f''(a_0)\left\{\sum_{n=1}^{\infty}\left(a_n\cos\frac{n\pi t}{T} + b_n\sin\frac{n\pi t}{T}\right)\right\}^2 + \dots.$$

$f(a_0)$ is the current due to the steady component a_0 of the voltage V, so writing I' for the change of current, $I - f(a_0)$, due to the periodic part of (6), we get for the mean change of current, using § 88 (4)–(9),

$$\frac{1}{2T}\int_{-T}^{T} I'\,dt = \tfrac{1}{4}f''(a_0)\sum_{n=1}^{\infty}(a_n^2+b_n^2)+\dots,$$
$$= \tfrac{1}{2}(\overline{v^2})f''(a_0)+\dots,$$

where $(\overline{v^2})$ is by (6) and (8) the mean of the square of $V-a_0$, the departure of the voltage V from its steady value.

92. Fourier series in mechanical problems

As an example we treat the slider-crank mechanism whose theory is fundamental for the study of reciprocating engines.

Suppose that a crank OP of length R rotates with constant angular velocity ω, and that the point A, which is connected to P by a connecting rod of length r, is constrained to move in a straight line through O. If the connecting rod is very long the motion of A is very nearly simple harmonic, but for shorter connecting rods the departure from this is of importance in problems of engine balancing.

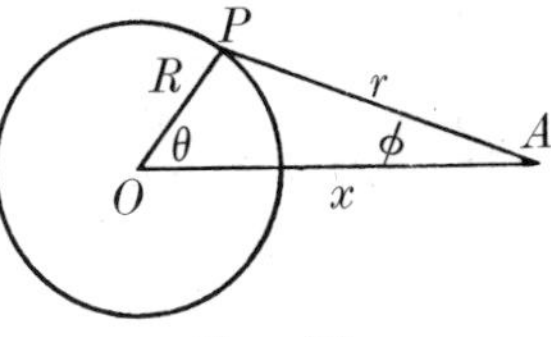

Fig. 80.

Writing $\omega t = \theta$, measuring θ from the line OA, we see that $OA = x$ is an even function of θ with period 2π and thus may be expanded in the Fourier cosine series

$$x = a_0+\sum_{n=1}^{\infty} a_n\cos n\theta. \tag{1}$$

From the triangle OAP it follows that

$$\sin\phi = k\sin\theta, \tag{2}$$

where

$$k = R/r. \tag{3}$$

Also
$$x = R\cos\theta+r\cos\phi = R\cos\theta+r\sqrt{(1-k^2\sin^2\theta)}. \tag{4}$$

The coefficients a_0, a_1,... in (1) may now be found from § 89 (1) and (2). Thus

$$a_0 = \frac{1}{\pi}\int_0^{\pi}\{R\cos\theta+r\sqrt{(1-k^2\sin^2\theta)}\}\,d\theta = \frac{2r}{\pi}\int_0^{\frac{1}{2}\pi}\sqrt{(1-k^2\sin^2\theta)}\,d\theta, \tag{5}$$

$$a_1 = \frac{2}{\pi}\int_0^{\pi} R\cos^2\theta\,d\theta+\frac{2}{\pi}\int_0^{\pi} r\cos\theta\sqrt{(1-k^2\sin^2\theta)}\,d\theta = R. \tag{6}$$

Also, for $m > 1$,

$$a_m = \frac{2R}{\pi}\int_0^{\pi}\cos\theta\cos m\theta\,d\theta+\frac{2r}{\pi}\int_0^{\pi}\cos m\theta\sqrt{(1-k^2\sin^2\theta)}\,d\theta,$$

so that
$$a_{2n+1} = 0 \quad (n = 1, 2,...), \tag{7}$$
$$a_{2n} = \frac{4r}{\pi}\int\limits_0^{\frac{1}{2}\pi} \cos 2n\theta\sqrt{(1-k^2\sin^2\theta)}\, d\theta \quad (n = 1, 2,...). \tag{8}$$

Thus, finally, replacing θ by ωt we have
$$x = a_0 + R\cos\omega t + \sum_{n=1}^{\infty} a_{2n}\cos 2n\omega t, \tag{9}$$
where a_0 and a_{2n} are given by (5) and (8). It appears that only even harmonics are present.

There are no simple expressions for a_0 and a_{2n} in terms of elementary functions, for example a_0 involves the elliptic integral $E(k, \frac{1}{2}\pi)$; cf. § 55 (20). But since $k < 1$ we may expand the square root in (8) by the binomial theorem and interchange the orders of integration and summation: since in most practical systems $k < \frac{1}{4}$ the series obtained in this way is rapidly convergent. Thus we have from (8)
$$a_{2n} = \frac{4r}{\pi}\int\limits_0^{\frac{1}{2}\pi} \cos 2n\theta\{1-\tfrac{1}{2}k^2\sin^2\theta-\tfrac{1}{8}k^4\sin^4\theta...\}\, d\theta$$
$$= -\frac{2rk^2}{\pi}\int\limits_0^{\frac{1}{2}\pi} \cos 2n\theta\{\sin^2\theta+\tfrac{1}{4}k^2\sin^4\theta+...\}\, d\theta. \tag{10}$$

Therefore
$$a_2 = \tfrac{1}{4}rk^2+\tfrac{1}{16}rk^4+.... \tag{11}$$

In general, since
$$\left.\begin{aligned}\int\limits_0^{\frac{1}{2}\pi} \cos 2n\theta\sin^{2m}\theta\, d\theta &= 0 && (m < n)\\ &= (-1)^n\frac{\pi}{2^{2n+1}} && (m = n)\end{aligned}\right\}, \tag{12}$$
$$a_{2n} = (-1)^{n+1}\frac{1.3...(2n-3)}{2^{3n-1}n!}rk^{2n}+..., \tag{13}$$
and thus the amplitude of the term in $\cos 2n\omega t$ is proportional to k^{2n}.

For example, if k is so small that k^6 is negligible, the departure of A from its mean position is
$$R\left\{\cos\omega t+(\tfrac{1}{4}k+\tfrac{1}{16}k^3)\cos 2\omega t-\frac{k^3}{64}\cos 4\omega t\right\}.$$

93. Fourier series in boundary value problems

In this section we give some examples of the use of Fourier series in deflexion of beams. Further examples in connexion with partial differential equations will be given in § 124.

Ex. 1. *A uniform beam is freely hinged at $x = 0$ and $x = l$, and carries a load $f(x)$.*

We have to solve

$$D^4 y = \frac{1}{EI} f(x) \tag{1}$$

with
$$y = D^2 y = 0 \quad \text{when } x = 0 \text{ and } x = l. \tag{2}$$

If we assume a sine series for y,

$$y = \sum_{n=1}^{\infty} a_n \sin\frac{n\pi x}{l}, \tag{3}$$

the boundary conditions (2) at $x = 0$ and $x = l$ will all be satisfied, since it was shown in § 89 that the sum of a sine series is zero when $x = 0$ and $x = l$. Now suppose that the sine series for $f(x)$ is

$$f(x) = \sum_{n=1}^{\infty} b_n \sin\frac{n\pi x}{l}, \tag{4}$$

where, by § 89 (6),
$$b_n = \frac{2}{l}\int_0^l f(x')\sin\frac{n\pi x'}{l}\,dx'. \tag{5}$$

Substituting (3) and (4) in (1) requires

$$\sum_{n=1}^{\infty} \frac{n^4\pi^4}{l^4} a_n \sin\frac{n\pi x}{l} = \frac{1}{EI}\sum_{n=1}^{\infty} b_n \sin\frac{n\pi x}{l}. \tag{6}$$

Comparing coefficients† of $\sin n\pi x/l$ on the two sides gives

$$a_n = \frac{l^4}{EIn^4\pi^4} b_n. \tag{7}$$

Thus the deflexion y is given by

$$y = \frac{2l^3}{EI\pi^4}\sum_{n=1}^{\infty}\frac{1}{n^4}\sin\frac{n\pi x}{l}\int_0^l f(x')\sin\frac{n\pi x'}{l}\,dx'. \tag{8}$$

If $f(x) = w$, constant, (8) becomes

$$y = \frac{4wl^4}{EI\pi^5}\sum_{r=0}^{\infty}\frac{1}{(2r+1)^5}\sin\frac{(2r+1)\pi x}{l}. \tag{9}$$

If $f(x) = W\delta(x-a)$, a concentrated load W at $x = a$, (8) becomes

$$y = \frac{2Wl^3}{EI\pi^4}\sum_{n=1}^{\infty}\frac{1}{n^4}\sin\frac{n\pi x}{l}\sin\frac{n\pi a}{l}. \tag{10}$$

† This is justified by the uniqueness theorem stated towards the end of § 88. It should be remarked also that arguments involving the differentiation of Fourier series need pure-mathematical justification: this can be supplied for the cases considered here.

The solutions (9) and (10) are extremely useful in practice because they converge very rapidly.

Ex. 2. *The problem of Ex. 1 except that in addition the beam rests on an elastic foundation giving restoring force per unit length of k times the displacement.*

The differential equation (1) is now replaced by

$$EID^4y = -ky+f(x),$$

or

$$(D^4+\omega^4)y = \frac{1}{EI}f(x), \tag{11}$$

where $\omega^4 = k/EI$.

Assuming the sine series (3) and (4) for y and $f(x)$, respectively, and substituting in (11) gives

$$\sum_{n=1}^{\infty}\left(\omega^4+\frac{n^4\pi^4}{l^4}\right)a_n\sin\frac{n\pi x}{l} = \frac{1}{EI}\sum_{n=1}^{\infty}b_n\sin\frac{n\pi x}{l}.$$

Hence

$$a_n = \frac{l^4}{EI(l^4\omega^4+n^4\pi^4)}b_n,$$

and, finally,

$$y = \frac{2l^3}{EI}\sum_{n=1}^{\infty}\frac{1}{(l^4\omega^4+n^4\pi^4)}\sin\frac{n\pi x}{l}\int_0^l f(x')\sin\frac{n\pi x'}{l}\,dx'. \tag{12}$$

94. Double and multiple Fourier series

Such series are often required in the solution of partial differential equations: for example, a double Fourier series is used in the theory of deflexion of rectangular plates in much the same way that an ordinary Fourier series was used in the theory of the deflexion of beams in § 93; cf. Ex. 18.

Here we give only a brief sketch of the theory of the double sine series: other types of series involving both sines and cosines may be obtained in the same way, or the whole theory may be developed *ab initio* along the lines of §§ 88, 89.

Suppose $f(x,y)$ is defined in the region $0 < x < a$, $0 < y < b$, then, as in § 89, we can expand $f(x,y)$ in a sine series in x

$$f(x,y) = \sum_{n=1}^{\infty}b_n(y)\sin\frac{n\pi x}{a}, \tag{1}$$

in which the coefficients

$$b_n(y) = \frac{2}{a}\int_0^a f(x',y)\sin\frac{n\pi x'}{a}\,dx', \tag{2}$$

are functions of y.

Since $b_n(y)$ is a function of y in $0 < y < b$ we can expand it in the sine series

$$b_n(y) = \sum_{m=1}^{\infty} c_{m,n} \sin\frac{m\pi y}{b}, \tag{3}$$

where

$$c_{m,n} = \frac{2}{b}\int_0^b b_n(y')\sin\frac{m\pi y'}{b}\,dy'$$

$$= \frac{4}{ab}\int_0^b \sin\frac{m\pi y'}{b}\,dy' \int_0^a f(x',y')\sin\frac{n\pi x'}{a}\,dx' \tag{4}$$

$$= \frac{4}{ab}\int_0^b\int_0^a f(x',y')\sin\frac{m\pi y'}{b}\sin\frac{n\pi x'}{a}\,dx'dy'. \tag{5}$$

Thus, finally, with this value of $c_{m,n}$ we have the double sine series

$$\sum_{m=1}^{\infty}\sum_{n=1}^{\infty} c_{m,n}\sin\frac{n\pi x}{a}\sin\frac{m\pi y}{b} \tag{6}$$

whose sum is $f(x,y)$ at every point in the rectangle $0 < x < a$, $0 < y < b$. As in § 89 it follows that the sum is zero on the boundaries of the rectangle and that, outside it, it is periodic and odd in x and y.

Ex. *To expand $f(x,y) = 1$ in $0 < x < a$, $0 < y < b$ in a double sine series.*

As in § 89, Ex. 1, the sine series for 1 in $0 < x < a$ is

$$\frac{4}{\pi}\sum_{n=0}^{\infty}\frac{1}{2n+1}\sin\frac{(2n+1)\pi x}{a} = 1.$$

Expanding each of the coefficients (again constants) in a sine series in $0 < y < b$ we have

$$\frac{16}{\pi^2}\sum_{n=0}^{\infty}\sum_{m=0}^{\infty}\frac{1}{(2m+1)(2n+1)}\sin\frac{(2n+1)\pi x}{a}\sin\frac{(2m+1)\pi y}{b} = 1. \tag{7}$$

95. Fourier integrals and Fourier transforms

We begin with the Fourier series § 88 (11) for a function $f(t)$ defined in the interval $(-T, T)$,

$$f(t) = a_0 + \sum_{n=1}^{\infty} a_n \cos\left(\frac{n\pi t}{T}\right) + \sum_{n=1}^{\infty} b_n \sin\left(\frac{n\pi t}{T}\right).$$

This can be written in the complex form

$$f(t) = a_0 + \frac{1}{2}\sum_{n=1}^{\infty} a_n(e^{in\pi t/T} + e^{-in\pi t/T}) + \frac{1}{2i}\sum_{n=1}^{\infty} b_n(e^{in\pi t/T} - e^{-in\pi t/T})$$

or
$$f(t) = \sum_{n=-\infty}^{\infty} A_n e^{in\pi t/T}, \tag{1}$$

where the complex coefficients are given by

$$A_0 = a_0, \quad A_n = \tfrac{1}{2}(a_n - ib_n),$$

and
$$A_{-n} = \tfrac{1}{2}(a_n + ib_n) \quad \text{for } n = 1, 2, \ldots.$$

Substituting for a_0, a_n, and b_n by § 88 (13), (14), and (15) leads to the single expression

$$A_n = \frac{1}{2T}\int_{-T}^{T} f(t')e^{-in\pi t'/T}\,dt' \tag{2}$$

for all n.

When the finite interval $(-T, T)$ tends to the infinite interval $(-\infty, \infty)$ then the Fourier series tends to the Fourier integral. We therefore consider what happens to (1) as $T \to \infty$. Writing $h = \pi/T$ and substituting for A_n in (1) by (2),

$$f(t) = \frac{1}{2\pi}\sum_{n=-\infty}^{\infty} he^{inht}\int_{-T}^{T} f(t')e^{-inht'}\,dt'. \tag{3}$$

Now the integral $\int_{-\infty}^{\infty} \phi(\omega)\,d\omega$ may be defined as the limit as $h \to 0$ of the sum $\sum_{n=-\infty}^{\infty} h\phi(nh)$. It follows that in the limit as $T \to \infty$, or $h \to 0$, the series (3) tends to the double integral

$$f(t) = \frac{1}{2\pi}\int_{-\infty}^{\infty} e^{i\omega t}\,d\omega \int_{-\infty}^{\infty} f(t')e^{-i\omega t'}\,dt'. \tag{4}$$

This is Fourier's integral for the function $f(t)$. The above discussion is only illustrative, but the exact theory shows that *if $f(t)$ satisfies Dirichlet's conditions, and the integral*

$$\int_{-\infty}^{\infty} |f(t)|\, dt \tag{5}$$

exists, the double integral (4) *has the value $f(t)$ where this function is continuous, and the value*

$$\tfrac{1}{2}\{f(t+0)+f(t-0)\}$$

at points where $f(t)$ is discontinuous.

The Fourier integral representation of a non-periodic function plays the same part in the theory of aperiodic phenomena that the Fourier series plays in the theory of periodic phenomena. There is, however, one important restriction, namely that $f(t)$ must be such that the integral (5) is convergent. This is not the case for such common functions as $\sin \omega t$ or a constant, though the theory can be extended to include them.

If $f(t)$ is an odd function of t, (4) simplifies into

$$f(t) = \frac{2}{\pi}\int_{0}^{\infty} \sin \omega t\, d\omega \int_{0}^{\infty} f(t')\sin \omega t'\, dt', \tag{6}$$

while if $f(t)$ is an even function of t, it simplifies to

$$f(t) = \frac{2}{\pi}\int_{0}^{\infty} \cos \omega t\, d\omega \int_{0}^{\infty} f(t')\cos \omega t'\, dt', \tag{7}$$

these being known as Fourier's sine and cosine integrals respectively. In (6) and (7), as always, $f(t)$ is to be replaced on the left by $\frac{1}{2}\{f(t+0)+f(t-0)\}$ at points of discontinuity.

If we write $$F(\omega) = \int_{-\infty}^{\infty} e^{-i\omega t'} f(t')\, dt', \tag{8}$$

then (4) states that

$$f(t) = \frac{1}{2\pi}\int_{-\infty}^{\infty} e^{i\omega t} F(\omega)\, d\omega. \tag{9}$$

$F(\omega)$ defined in (8) is called the *Fourier transform*† of $f(t)$; it is obtained from $f(t)$ by multiplying by $e^{-i\omega t}$ and integrating from $-\infty$ to ∞, just as the Fourier constants of a function defined in $(-T, T)$ are obtained by multiplying by $\sin n\pi t/T$ or $\cos n\pi t/T$ and integrating from $-T$ to T. By (9), if the Fourier transform $F(\omega)$ of a function $f(t)$ is known, the function can be found by a similar integration. $F(\omega)$ and $f(t)$ are referred to as a 'Fourier pair'.

(8) and (9) are the usual form in which Fourier transforms are applied to boundary value problems. In initial value problems a further simplification is often possible, since usually $f(t) = 0$ if $t < 0$. If this is the case, (8) becomes

$$F(\omega) = \int\limits_0^\infty e^{-i\omega t'} f(t')\, dt', \tag{10}$$

while (9) gives

$$\left.\begin{aligned} \frac{1}{2\pi}\int\limits_0^\infty \{e^{i\omega t}F(\omega)+e^{-i\omega t}F(-\omega)\}\, d\omega &= f(t), \quad \text{if } t > 0 \\ &= 0, \quad \text{if } t < 0 \end{aligned}\right\}. \tag{11}$$

If, in addition, $f(t)$ is real, it follows from (10) that $F(-\omega)$ is the conjugate of $F(\omega)$, and (11) becomes

$$\left.\begin{aligned} \frac{1}{\pi}\mathbf{R}\int\limits_0^\infty e^{i\omega t}F(\omega)\, d\omega &= f(t), \quad \text{if } t > 0 \\ &= 0, \quad \text{if } t < 0 \end{aligned}\right\}. \tag{12}$$

The similarity of (10) to the Laplace transform § 18 (1) is evident, $i\omega$ simply appearing in place of p. But it must be remembered that the integral (10) is often not convergent for common functions. We may state that, *if the Fourier transform of a function $f(t)$ exists it is just $\bar{f}(i\omega)$* and thus may be written down from the table of Laplace transforms.

For example, if

$$f(t) = e^{-at}, \text{ then } \bar{f}(p) = \frac{1}{p+a} \quad \text{and} \quad F(\omega) = \frac{1}{a+i\omega}. \tag{13}$$

† An extensive table of Fourier transforms is given by Campbell and Foster, *Fourier Integrals for Practical Applications*, Bell System Technical Monograph B–584 (1931). The Fourier transform is often defined as $(2\pi)^{-\frac{1}{2}}F(\omega)$.

Similarly, if $f(t) = e^{-at}\sin bt$, then $\bar{f}(p) = \dfrac{b}{b^2+(p+a)^2}$

and
$$F(\omega) = \frac{b}{b^2+(a+i\omega)^2}. \tag{14}$$

Two properties of the Fourier transform are important in practical applications.

(i) *The shift theorem.* If $F(\omega)$ is the Fourier transform of $f(t)$, then the transform of $f(t-\tau)$ is $e^{-i\omega\tau}F(\omega)$.

The proof follows directly from (8). The Fourier transform of $f(t-\tau)$ is
$$\int\limits_{-\infty}^{\infty} e^{-i\omega t'}f(t'-\tau)\,dt' = \int\limits_{-\infty}^{\infty} e^{-i\omega(t'+\tau)}f(t')\,dt' = e^{-i\omega\tau}F(\omega).$$

(ii) *The convolution theorem.* If $F(\omega)$ and $G(\omega)$ are the Fourier transforms of $f(t)$ and $g(t)$ respectively, then the product $F(\omega)G(\omega)$ is the transform of the convolution integral
$$\int\limits_{-\infty}^{\infty} f(\tau)g(t-\tau)\,d\tau.$$

Again, by (8), the transform of the convolution integral is
$$\int\limits_{-\infty}^{\infty} e^{-i\omega t'}\,dt' \int\limits_{-\infty}^{\infty} f(\tau)g(t'-\tau)\,d\tau = \int\limits_{-\infty}^{\infty} f(\tau)\,d\tau \int\limits_{-\infty}^{\infty} e^{-i\omega t'}g(t'-\tau)\,dt'$$
on interchanging the order of integration. But by the shift theorem
$$\int\limits_{-\infty}^{\infty} e^{-i\omega t'}g(t'-\tau)\,dt' = e^{-i\omega\tau}G(\omega).$$
Therefore the transform of the convolution integral is
$$\int\limits_{-\infty}^{\infty} f(\tau)e^{-i\omega\tau}G(\omega)\,d\tau = G(\omega)\int\limits_{-\infty}^{\infty} e^{-i\omega\tau}f(\tau)\,d\tau = G(\omega)F(\omega).$$

96. Applications of Fourier transforms

Fourier transforms are widely used in electric circuit theory, communications, the study of wave phenomena and the analysis of time series, and the smoothing and filtering of data. Preliminary to illustrating some of these applications, we will derive the Fourier transforms of a rectangular pulse and a Dirac delta function.

The rectangular pulse $p(t)$ of Fig. 81 (a) is the function

$$p(t) = 0, \quad t \leqslant -T \quad \text{and} \quad t \geqslant T$$
$$= 1, \quad -T < t < T.$$

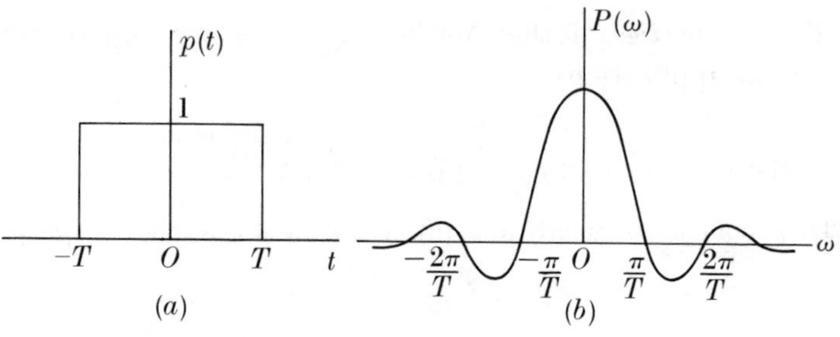

FIG. 81.

Its Fourier transform is

$$P(\omega) = \int_{-\infty}^{\infty} e^{i\omega t'} p(t')\, dt' = \int_{-T}^{T} e^{i\omega t'}\, dt' = (2 \sin \omega T)/\omega. \qquad (1)$$

$P(\omega)$ is plotted against ω in Fig. 81 (b). Each value of ω corresponds to a vibration of frequency $\omega/2\pi$, and $P(\omega)$ is then the amplitude of that particular vibration. Fig. 81 (b) thus shows the frequency spectrum of the rectangular pulse. In particular, we notice that the higher amplitudes of $P(\omega)$ all lie in the range $-\pi/T < \omega < \pi/T$ which corresponds to a frequency spread

$$\Delta f = \frac{2\pi}{T2\pi} = \frac{1}{T}. \qquad (2)$$

Δf is called the 'bandwidth' of the pulse and (2) shows that this is inversely proportional to the duration of the pulse. This result is of importance in communication theory.

If we divide $p(t)$ by $2T$ and then let $T \to 0$, $p(t)$ will tend to the delta function $\delta(t)$. The Fourier transform of the delta function is thus

$$\lim_{T \to 0} \frac{\sin \omega T}{\omega T} = 1. \qquad (3)$$

The delta function is often referred to as the 'unit impulse' in circuit and communication theory.

We now consider how the Fourier transform can be used to find the response of an electric circuit to an aperiodic voltage $f(t)$ applied for $t > 0$. The transform $F(\omega)$ of this voltage represents the amplitudes (and phase shifts) of all possible frequencies of vibration; in particular, the complex voltage of the component of frequency $\omega/2\pi$ will be

$$\frac{1}{\pi} F(\omega)\, d\omega. \tag{4}$$

If the voltage is applied to a circuit of impedance $z(i\omega)$ the steady complex current of period $2\pi/\omega$ will be

$$F(\omega)\, d\omega/\pi z(i\omega), \tag{5}$$

and the whole current, on adding all frequencies, is

$$\frac{1}{\pi} \mathbf{R} \int\limits_0^\infty e^{i\omega t} \frac{F(\omega)\, d\omega}{z(i\omega)}. \tag{6}$$

This procedure is exactly analogous to that of § 91 for a periodic voltage—the effects of all the harmonics are calculated separately and then combined. This has an advantage over the Laplace transformation technique of § 49 because of this analogy, but results obtained by Fourier transforms are often harder to evaluate.

An alternative approach to the circuit problem is the following. Let $H(\omega)$ be the complex current corresponding to a unit voltage impulse $\delta(t)$. This voltage has, as we saw in (3), a perfectly flat frequency spectrum. It follows that the complex current corresponding to a complex voltage $F(\omega)$ is just

$$F(\omega)H(\omega). \tag{7}$$

$H(\omega)$ is thus the transfer function for the circuit and can be obtained from the transfer function, defined in terms of the Laplace transformation in § 50, by replacing p by $i\omega$. The current as a function of time can then be obtained from (7), making use of the convolution theorem.

Ex. *The ideal low-pass filter.*

This is defined as having

$$z(i\omega) = \infty, \qquad \text{if } \omega > \omega_0$$
$$= Ze^{i\pi\omega/\omega_0}, \qquad \text{if } \omega < \omega_0,$$

where Z is a constant. In this case the current I given by (6) is

$$I = \frac{1}{\pi Z} \mathbf{R} \int_0^{\omega_0} e^{i\omega(t-\pi/\omega_0)} F(\omega)\, d\omega. \tag{8}$$

Suppose that we wish to calculate the response of this filter to the rectangular pulse

$$f(t) = 0, \quad \text{for } t < 0 \text{ and } t > T$$
$$= 1, \quad \text{for } 0 < t < T.$$

Comparing this with the pulse $p(t)$ of Fig. 81 (a), by the shift theorem and (1),

$$F(\omega) = (2/\omega)e^{-i\omega T/2} \sin \tfrac{1}{2}\omega T,$$

and so by (8)

$$I = \frac{2}{\pi Z} \mathbf{R} \int_0^{\omega_0} e^{i\omega(t-\pi/\omega_0-T/2)} \sin \tfrac{1}{2}\omega T \,\frac{d\omega}{\omega}$$

$$= \frac{2}{\pi Z} \int_0^{\omega_0} \cos \omega(t-\pi/\omega_0-T/2) \sin \tfrac{1}{2}\omega T \,\frac{d\omega}{\omega}$$

$$= \frac{1}{\pi Z} \int_0^{\omega_0} \{\sin \omega(t-\pi/\omega_0) + \sin \omega(T-t+\pi/\omega_0)\} \frac{d\omega}{\omega}$$

$$= \frac{1}{\pi Z} \{\text{Si}(\omega_0 t-\pi) + \text{Si}(\omega_0 T-\omega_0 t+\pi)\}, \tag{9}$$

where $\text{Si}\, x = \int_0^x \frac{\sin u}{u}\, du$ is a tabulated function.

The effect of a unit voltage applied for $t > 0$ may be found by letting $T \to \infty$ in (9). Using the result $\text{Si}(\infty) = \frac{1}{2}\pi$, (9) becomes in this case

$$I = \frac{1}{Z}\left\{\frac{1}{2} + \frac{1}{\pi} \text{Si}(\omega_0 t-\pi)\right\}. \tag{10}$$

Notice that this result could not have been obtained by taking $f(t) = 1$ for $t > 0$, since $f(t)$ would then not have satisfied § 95 (5).

We next derive an important result in communication theory, known as *the sampling theorem*.

We start with a function $f(t)$ defined in $(-\infty, \infty)$ and suppose that it is sampled at regular time intervals T, that is, we know

$$f_n = f(nT) \tag{11}$$

for all integers n, both positive and negative. Our aim is to find an interval T which enables us to reconstruct the function $f(t)$ from the sampled values f_n. We could then write

$$f(t) = \sum_{n=-\infty}^{\infty} f_n c_n(t) \tag{12}$$

where the coefficients $c_n(t)$ would have to be independent of f.

The only other assumption we will make is that the Fourier transform $F(\omega)$ of $f(t)$ is zero for all values of ω greater than ω_0 or less than $-\omega_0$, that is, the frequency spectrum of $f(t)$ is zero for $|\omega| > \omega_0$. We can then expand $F(\omega)$ in a Fourier *series* in the interval $(-\omega_0, \omega_0)$. Using the complex form of § 95 (1) and (2),

$$F(\omega) = \sum_{n=-\infty}^{\infty} A_n e^{in\pi\omega/\omega_0}, \tag{13}$$

where
$$A_n = \frac{1}{2\omega_0} \int_{-\omega_0}^{\omega_0} F(\omega) e^{-in\pi\omega/\omega_0}\, d\omega. \tag{14}$$

By § 95 (9),

$$f(t) = \frac{1}{2\pi} \int_{-\infty}^{\infty} e^{i\omega t} F(\omega)\, d\omega = \frac{1}{2\pi} \int_{-\omega_0}^{\omega_0} e^{i\omega t} F(\omega)\, d\omega. \tag{15}$$

Substituting for $F(\omega)$ by (13) and interchanging the order of integration and summation,

$$f(t) = \frac{1}{2\pi} \sum_{n=-\infty}^{\infty} A_n \int_{-\omega_0}^{\omega_0} e^{i\omega t} e^{in\pi\omega/\omega_0}\, d\omega = \frac{\omega_0}{\pi} \sum_{n=-\infty}^{\infty} A_n \frac{\sin(\omega_0 t + n\pi)}{\omega_0 t + n\pi}. \tag{16}$$

Now by (11) and (15),

$$f_n = f(nT) = \frac{1}{2\pi} \int_{-\omega_0}^{\omega_0} e^{i\omega nT} F(\omega)\, d\omega. \tag{17}$$

Comparing (14) and (17) we see that

$$f_n = \frac{\omega_0}{\pi} A_{-n} \tag{18}$$

if we choose
$$T = \pi/\omega_0. \tag{19}$$

In that case, (16) can be written

$$f(t) = \sum_{n=-\infty}^{\infty} f_n \frac{\sin(\omega_0 t - n\pi)}{\omega_0 t - n\pi}$$

which is precisely the form of (12).

The sampling theorem thus states that if the sampling interval is $T = \pi/\omega_0$, the function $f(t)$ can be reconstructed completely from the sampled values f_n. If the sampling interval is smaller than π/ω_0, some of the information is redundant, while if the interval is larger than π/ω_0, the reconstructed function will be a distortion of $f(t)$. The sampling theorem is fundamental to communication engineering and the design of automatic control systems.

Other concepts of importance in the transmission and interpretation of data are *power spectra, auto-correlation, and cross-correlation.*

If the function $f(t)$ represents a time signal defined in the infinite interval $(-\infty, \infty)$, the auto-correlation $R(t)$ of the signal is defined by

$$R(t) = \lim_{T\to\infty} \frac{1}{T} \int_{-T/2}^{T/2} f(\tau)f(\tau+t)\, d\tau. \tag{20}$$

For example, it follows from (20) that the auto-correlation of a constant signal $f(t) = c$ is $R(t) = c^2$.

The power spectrum $G(\omega)$ of a time signal $f(t)$ can be defined by

$$G(\omega) = \lim_{T\to\infty} \frac{1}{T} \left| \int_{-T/2}^{T/2} e^{-i\omega t} f(t)\, dt \right|^2. \tag{21}$$

The Wiener–Khintchine theorem states that $R(t)$ and $G(\omega)$ form a Fourier pair (see Ex. 24). If follows that the power spectrum of the constant signal $f(t) = c$ is the delta function $G(\omega) = 2\pi c^2\delta(\omega)$. The case $G(\omega) =$ constant for all ω is important. The function $f(t)$ then represents a statistical or stochastic signal† of a type known as 'white noise'.

The cross-correlation $R_{12}(t)$ of two functions f_1 and f_2 is defined by

$$R_{12}(t) = \lim_{T\to\infty} \frac{1}{T} \int_{-T/2}^{T/2} f_1(\tau)f_2(\tau+t)\, d\tau, \tag{22}$$

and can be used, for instance, to test whether a signal f_2 received from a transmitter on the moon is just noise, or a definite signal f_1 plus noise. The Fourier transform of $R_{12}(t)$ is called the cross-power spectrum. When $t = 0$, equations (20) and (22) are the equivalent, in the infinite interval $(-\infty, \infty)$, of § 91 (8) and (9).

† The definitions (20), (21), and (22) are valid (subject to pure-mathematical considerations) if $f(t)$ is a deterministic function, but need to be qualified if $f(t)$ is stochastic. (See, for example, Papoulis, *Probability, Random Variables and Stochastic Processes*, McGraw-Hill, 1965, chapters 9 and 10.)

EXAMPLES ON CHAPTER XI

1. If $f(t) = |\sin\omega t|$, full wave rectified alternating current, show that

$$f(t) = \frac{2}{\pi} - \frac{4}{\pi}\sum_{r=1}^{\infty} \frac{1}{(4r^2-1)}\cos 2r\omega t.$$

2. If

$$f(t) = 1 \qquad (0 < t < T_1),$$
$$f(t) = 0 \qquad (T_1 < t < T-T_1),$$
$$f(t) = -1 \qquad (T_1-T < t < T),$$

show that the cosine series for $f(t)$ is

$$\frac{4}{\pi}\sum_{r=0}^{\infty} \frac{1}{2r+1}\sin\frac{(2r+1)\pi T_1}{T}\cos\frac{(2r+1)\pi t}{T}.$$

3. Show that the sine series for $\sin^2\omega t$ in $0 < t < \pi/\omega$ is

$$\frac{8}{\pi}\sum_{r=0}^{\infty} \frac{1}{(2r+1)\{4-(2r+1)^2\}}\sin(2r+1)\omega t.$$

4. By expanding $\sin mx$ and $\cos mx$ in sine and cosine series, respectively, in $0 < x < \pi$, show that

$$\sin mx = \frac{2}{\pi}\sin m\pi \sum_{r=1}^{\infty} \frac{(-1)^{r-1} r\sin rx}{r^2-m^2},$$

$$\cos mx = \frac{2}{\pi}\sin m\pi\left\{\frac{1}{2m} + \sum_{r=1}^{\infty} \frac{(-1)^{r-1} m\cos rx}{r^2-m^2}\right\}.$$

5. By expanding $x+x^2$ in a series of sines and cosines in $-\pi < x < \pi$, show that

$$x+x^2 = \frac{\pi^2}{3} + \sum_{n=1}^{\infty} (-1)^n\left\{\frac{4}{n^2}\cos nx - \frac{2}{n}\sin nx\right\} \quad (-\pi < x < \pi),$$

and

$$\sum_{n=1}^{\infty} \frac{1}{n^2} = \frac{\pi^2}{6}.$$

6. Show that the series of sines and cosines which represents $f(x)$ in $0 < x < l$ is

$$a_0 + \sum_{n=1}^{\infty} \left(a_n\cos\frac{2n\pi x}{l} + b_n\sin\frac{2n\pi x}{l}\right),$$

where $$a_0 = \frac{1}{l}\int_0^l f(x')\,dx', \qquad a_n = \frac{2}{l}\int_0^l f(x')\cos\frac{2n\pi x'}{l}\,dx',$$

$$b_n = \frac{2}{l}\int\limits_0^l f(x')\sin\frac{2n\pi x'}{l}\,dx'.$$

7. If
$$V = (1-e^{-\alpha t}) \quad (0 < t < T),$$
and V is periodic with period T (this is the voltage drop over the condenser in the circuit of Fig. 49 (a)), show that

$$V = 1+\frac{1}{\alpha T}(e^{-\alpha T}-1)+$$
$$+(e^{-\alpha T}-1)\sum_{n=1}^{\infty}\frac{2\alpha T\cos(2n\pi t/T)+4n\pi\sin(2n\pi t/T)}{\alpha^2T^2+4n^2\pi^2}.$$

8. Show that the sine series for the function $\delta(t-t_0)$ in $0 < t < T$ is

$$\delta(t-t_0) = \frac{2}{T}\sum_{n=1}^{\infty}\sin\frac{n\pi t}{T}\sin\frac{n\pi t_0}{T}.$$

This series, of course, is not convergent, but correct results may often be obtained very simply by using it. Deduce § 93 (10) in this way.

9. Show that if p and q are positive integers less than n, and $p \neq q$,

$$\sum_{r=1}^{n-1}\sin\frac{rp\pi}{n}\sin\frac{rq\pi}{n} = 0,$$

and that if $p = q$ the sum has the value $\frac{1}{2}n$.

The values $y_1, y_2,..., y_{n-1}$ of a function at the points $t = sT/n$ $(s = 1, 2,..., n-1)$, are known; show that

$$y = \sum_{r=1}^{n-1} b_r\sin\frac{r\pi t}{T},$$

where
$$b_r = \frac{2}{n}\sum_{s=1}^{n-1} y_s\sin\frac{rs\pi}{n}$$

passes through all these points. (The condition that y should pass through the $(n-1)$ given points gives $n-1$ equations for $b_1,..., b_{n-1}$ which are solved by using the first result given.) This gives a sine series of period $2T$ which passes through $n-1$ given points.

10. Show that if

$$y = a_0+\sum_{r=1}^{n}\left(a_r\cos\frac{2r\pi t}{T}+b_r\sin\frac{2r\pi t}{T}\right)$$

passes through the points $(0, y_0), (\alpha, y_1),..., (2n\alpha, y_{2n})$, where $\alpha = T/(2n+1)$,

$$a_0 = \frac{1}{(2n+1)}\sum_{s=0}^{2n} y_s, \qquad a_r = \frac{2}{2n+1}\sum_{s=0}^{2n} y_s \cos\frac{2\pi rs}{(2n+1)},$$

$$b_r = \frac{2}{2n+1}\sum_{s=1}^{2n} y_s \sin\frac{2\pi rs}{(2n+1)}.$$

11. A uniform beam of length l is freely hinged at its ends. Show that the bending moment M in it is

$$-\frac{4wl^2}{\pi^3}\sum_{r=0}^{\infty}\frac{1}{(2r+1)^3}\sin\frac{(2r+1)\pi x}{l}$$

if the beam carries a uniformly distributed load w per unit length, and

$$-\frac{2Wl}{\pi^2}\sum_{n=1}^{\infty}\frac{1}{n^2}\sin\frac{n\pi x}{l}\sin\frac{n\pi a}{l}$$

if it carries a concentrated load W at $x = a$.

12. Assuming the result

$$\int_0^\infty \frac{\sin x}{x}\,dx = \frac{\pi}{2},$$

deduce that

$$\begin{aligned}\frac{1}{2}+\frac{1}{\pi}\int_0^\infty \frac{\sin \omega t}{\omega}\,d\omega &= 1, \quad \text{if } t > 0,\\ &= \tfrac{1}{2}, \quad \text{if } t = 0,\\ &= 0, \quad \text{if } t < 0.\end{aligned}$$

This gives a representation of the unit function $H(t)$.

13. Show that if $f(t) = 0$ if $t < 0$,
$f(t) = e^{-\alpha t}$ if $t > 0$,

where $\alpha > 0$, Fourier's integral theorem § 95 (12) gives

$$\begin{aligned}\frac{1}{\pi}\int_0^\infty \frac{\alpha\cos\omega t+\omega\sin\omega t}{\alpha^2+\omega^2}\,d\omega &= e^{-\alpha t} \quad (t > 0),\\ &= 0 \quad (t < 0).\end{aligned}$$

Deduce that

$$\int_0^\infty \frac{\omega\sin\omega t}{\alpha^2+\omega^2}\,d\omega = \int_0^\infty \frac{\alpha\cos\omega t}{\alpha^2+\omega^2}\,d\omega = \frac{\pi}{2}e^{-\alpha t}.$$

14. Show that the Fourier transform of the function $f(t)$ which is zero for $t < 0$ and has the value $te^{-\alpha t}$ for $t > 0$ is

$$\frac{1}{(\alpha+i\omega)^2}.$$

If this voltage is applied to an L, R, C circuit, show that the amplitude of the component of frequency $\omega/2\pi$ of the current is the real part of

$$\frac{1}{\pi(\alpha+i\omega)^2\{(Li\omega-i/C\omega)+R\}}.$$

15. If
$$f(t) = \cos\omega_0 t \quad (|t| < T),$$
$$= 0 \quad (|t| > T),$$

show from § 95 (9) that

$$f(t) = \frac{1}{\pi}\int_0^\infty \left\{\frac{\sin(\omega_0-\omega)T}{\omega_0-\omega}+\frac{\sin(\omega_0+\omega)T}{\omega_0+\omega}\right\}\cos\omega t\, d\omega.$$

This shows that if radiation of frequency $\omega_0/2\pi$ is emitted for time $2T$, the amplitude of the component of frequency $\omega/2\pi$ is

$$\frac{1}{\pi}\left\{\frac{\sin(\omega_0-\omega)T}{\omega_0-\omega}+\frac{\sin(\omega_0+\omega)T}{\omega_0+\omega}\right\}.$$

Discuss the variation of this with T, and show that as T is increased the relative importance of frequencies near ω_0 increases steadily.

16. Voltage $e^{-\alpha t}$ is applied to the ideal low-pass filter of § 96 for $t > 0$. Find the current in it, and deduce § 96 (10) by letting $\alpha \to 0$. [Use the last integral evaluated in Ex. 13.]

17. Show that a function of x which is defined (and satisfies Dirichlet's conditions) in $0 < x < l$ may be expanded in the form

$$f(x) = \sum_{n=0}^{\infty} a_n \cos\frac{(2n+1)\pi x}{2l},$$

where
$$a_n = \frac{2}{l}\int_0^l f(x')\cos\frac{(2n+1)\pi x'}{2l}\, dx'.$$

[Expand the function defined by $f(x)$ in $0 < x < l$ and $-f(2l-x)$ in $l < x < 2l$ in a cosine series in $0 < x < 2l$.]

18. The deflexion z of a uniform plate carrying a load w per unit area satisfies

$$\frac{\partial^4 z}{\partial x^4}+2\frac{\partial^4 z}{\partial x^2\partial y^2}+\frac{\partial^4 z}{\partial y^4} = \frac{w}{D},$$

where D is a constant depending on the material and thickness of the plate. For a rectangular plate $0 < x < a$, $0 < y < b$, simply supported at its edges, z has to satisfy

$$z = \frac{\partial^2 z}{\partial y^2} = 0, \quad \text{when } y = 0 \text{ and } y = b,$$

$$z = \frac{\partial^2 z}{\partial x^2} = 0, \quad \text{when } x = 0 \text{ and } x = a.$$

If w is constant, show by using § 94 (7) that

$$z = \frac{16w}{\pi^6 D} \sum_{m=0}^{\infty} \sum_{n=0}^{\infty} \frac{\sin[(2m+1)\pi x/a]\sin[(2n+1)\pi y/b]}{(2m+1)(2n+1)[(2m+1)^2/a^2+(2n+1)^2/b^2]^2}.$$

19. An odd function $f(t)$, defined in the interval $(-\pi, \pi)$, is approximated by either:

(i) the first N terms of the Fourier sine series

$$\sum_{n=1}^{N} b_n \sin nt;$$

or (ii) the first N terms of a *different* sine series $\sum_{n=1}^{N} \beta_n \sin nt$, where the coefficients β_n are calculated according to some unspecified algorithm.

If $F(t)$ is the approximation to $f(t)$ in $(-\pi, \pi)$, 'goodness of fit' is measured by the integral

$$I = \int_{-\pi}^{\pi} \{F(t)-f(t)\}^2\,dt.$$

Show that for the sine series in (ii),

$$I = \int_{-\pi}^{\pi} \{f(t)\}^2\,dt + \pi \sum_{n=1}^{N} \beta_n^2 - 2\pi \sum_{n=1}^{N} \beta_n b_n.$$

Hence find I for the truncated Fourier series in (i) and prove that the truncated Fourier series is a better fit than any other N term sine series.

20. If the inverse Fourier transform of $F(\omega)$ is $f(t)$, show that the inverse transform of $F(\omega - p)$ is $e^{ipt}f(t)$. [Cf. the shift theorem of § 95.]

21. If $F(\omega) = \delta(\omega)$, show that $f(t) = 1/2\pi$.

Hence, if

$$F(\omega) = \pi\{\delta(\omega-\omega_0)+\delta(\omega+\omega_0)\},$$

show that $f(t) = \cos \omega_0 t$.

$f(t) = \{1+g(t)\}\cos \omega_0 t$ represents an amplitude modulation of the carrier wave $\cos \omega_0 t$. If the Fourier transform of $g(t)$ is $G(\omega)$, show that the transform of $f(t)$ is

$$F(\omega) = \tfrac{1}{2}\{G(\omega-\omega_0)+\delta(\omega-\omega_0)+G(\omega+\omega_0)+\delta(\omega+\omega_0)\}.$$

[Hint: use Ex. 20.]

22. Show that the response of the ideal low-pass filter of § 96 to a unit impulse is

$$I = \frac{\omega_0}{\pi Z} \frac{\sin(\omega_0 t - \pi)}{\omega_0 t - \pi}.$$

23. Show that the Fourier transform of $f(t) = e^{-t^2/4\alpha}$ is

$$F(\omega) = 2(\pi\alpha)^{\frac{1}{2}} e^{-\omega^2\alpha}.$$

(A function of the form e^{-x^2} is called 'Gaussian'. This example shows that the Fourier transform of a Gaussian function is also Gaussian.)

[Use $\int\limits_{-\infty}^{\infty} e^{-x^2}\,dx = \sqrt{\pi}$.]

24. Use the convolution theorem of § 95 to show that the functions $R(t)$ and $G(\omega)$ in § 96 (21) and (20) form a Fourier pair. Assume that

$$G(\omega) = G(-\omega).$$

25. If $f(t)$ is the sequence of pulses $f(t) = \sum\limits_{n=-\infty}^{\infty} \delta(t-nh)$, show that the auto-correlation of $f(t)$ is

$$R(t) = \frac{1}{h}f(t).$$

Hence show that the power spectrum is

$$G(\omega) = \frac{2\pi}{h^2}\sum_{n=-\infty}^{\infty} \delta\Big(\omega-\frac{2\pi n}{h}\Big).$$

26. If the Fourier transform of $f(t)$ is $F(\omega)$, show that the Fourier transform of df/dt is $i\omega F(\omega)$ and that of d^2f/dt^2 is $-\omega^2F(\omega)$.

XII

ORDINARY LINEAR DIFFERENTIAL EQUATIONS WITH VARIABLE COEFFICIENTS

97. Introductory

In this chapter we consider the general linear equation of the nth order with coefficients which are functions of x,

$$\{f_1(x)D^n+f_2(x)D^{n-1}+...+f_n(x)\}y = f(x). \tag{1}$$

In a few special cases the solution of this equation may be made to depend on the solution of types previously studied. For example the 'Euler' equation

$$\{p_0(a+bx)^n D^n+p_1(a+bx)^{n-1}D^{n-1}+...+p_n\}y = f(x), \tag{2}$$

where $p_0, p_1,..., p_n$ are constants, is reduced to a linear equation by the substitution

$$a+bx = e^t, \tag{3}$$

cf. Ex. 2 on Chapter II.

Again, if the operator can be factorized the solution is reduced to the solution of a chain of first-order linear equations. For example the equation

$$\{xD^2+(x^2+2)D+2x\}y = 0, \tag{4}$$

can be written

$$(D+x)(xD+2)y = 0, \tag{5}$$

where it must be remembered that the operators are not commutative. To solve (5) put

$$(xD+2)y = v, \tag{6}$$

so that (5) becomes

$$(D+x)v = 0,$$

and the solution of this is

$$v = Ae^{-\frac{1}{2}x^2}. \tag{7}$$

Substituting (7) in (6) gives the first-order equation for y,

$$(xD+2)y = Ae^{-\frac{1}{2}x^2}.$$

Methods such as these are clearly only applicable to limited and rather unimportant classes of equation to which the equations which arise most commonly in applied mathematics do not belong. The most important of these latter are second-order equations in which the coefficients $f_1(x)$, $f_2(x)$, $f_3(x)$ are polynomials in x. The method adopted for the solution of these is to seek a solution in the form of an infinite series

$$y = x^c(a_0+a_1 x+a_2 x^2+...). \tag{8}$$

By substituting this series in the equation and equating the coefficients of the successive powers of x to zero we obtain an equation, the *indicial equation*, for c, and a set of equations for the coefficients a_0, a_1, a_2,.... Naturally matters such as the existence of solutions of this type need careful discussion. Here only the process of solution will be given; for its justification the reader is referred to the works on differential equations listed in the footnote † below.

Since the equations we are considering are linear, the general solution of (1) consists of a linear combination of n independent solutions with arbitrary coefficients. In the examples of the second order which we shall consider, two independent solutions are required and usually they correspond to two different values of c in (8): occasionally the method gives only one solution, and a second is obtained by devices due to Frobenius.† In §§ 98 and 99 we shall obtain the series solutions of the two most important equations, Bessel's and Legendre's, and give a brief sketch of their properties.

The equations studied in §§ 98–100 are homogeneous: methods for solving inhomogeneous equations are given in §§ 101, 102. Approximate solutions are discussed in § 103. Finally, in § 104 equations with periodic coefficients are discussed briefly.

98. Bessel's equation‡ of order ν

This is
$$\frac{d^2y}{dx^2}+\frac{1}{x}\frac{dy}{dx}+\left(1-\frac{\nu^2}{x^2}\right)y = 0, \tag{1}$$
where ν is any (real) number, fractional or integral: the usual

† Cf. Piaggio, *Differential Equations* (Bell); Ince, *Ordinary Differential Equations* (Longmans).

‡ The standard works on Bessel functions are Watson, *Bessel Functions* (Cambridge); Gray and Mathews, *Treatise on Bessel Functions* (Macmillan); McLachlan, *Bessel Functions for Engineers* (Oxford).

convention is to use ν for a fraction, and to replace it by n if it is integral. ν is called the order of the equation.

To solve (1) we assume that an expression of type

$$y = x^c(a_0+a_1x+a_2x^2+...) \tag{2}$$

satisfies (1), and seek to find c and the successive coefficients $a_0, a_1,...$. Substituting (2) in (1) gives

$$\begin{aligned} c(c-1)a_0x^{c-2}+(c+1)ca_1x^{c-1}+(c+2)(c+1)a_2x^c+... \\ +ca_0x^{c-2}+(c+1)a_1x^{c-1}+ \qquad (c+2)a_2x^c+... \\ + \qquad a_0x^c+... \\ -\nu^2a_0x^{c-2}-\nu^2a_1x^{c-1}- \nu^2a_2x^c-... = 0. \end{aligned}$$

Equating the coefficients of x^{c-2}, x^{c-1},... to zero we find

$$(c^2-\nu^2)a_0 = 0, \tag{3}$$

$$\{(c+1)^2-\nu^2\}a_1 = 0, \tag{4}$$

$$\{(c+2)^2-\nu^2\}a_2+a_0 = 0, \tag{5}$$

$$\{(c+3)^2-\nu^2\}a_3+a_1 = 0, \tag{6}$$

$$\{(c+4)^2-\nu^2\}a_4+a_2 = 0, \tag{7}$$

. . . .

We may assume $a_0 \neq 0$, since taking $a_0 = 0$ is equivalent to changing the value of c. Then (3) gives

$$c = \pm\nu. \tag{8}$$

For either of these values of c, (4) gives $a_1 = 0$, then by (6), $a_3 = 0$, and so on, that is

$$a_1 = a_3 = a_5 = ... = 0, \tag{9}$$

so all the odd coefficients vanish. The even coefficients come successively from (5), (7),...

$$a_2 = \frac{a_0}{\nu^2-(c+2)^2},$$

$$a_4 = \frac{a_2}{\nu^2-(c+4)^2} = \frac{a_0}{\{\nu^2-(c+2)^2\}\{\nu^2-(c+4)^2\}},$$

and so on.

Using these results, the solution for $c = \nu$ becomes

$$a_0 x^\nu\left\{1-\frac{x^2}{(\nu+2)^2-\nu^2}+\frac{x^4}{\{(\nu+2)^2-\nu^2\}\{(\nu+4)^2-\nu^2\}}-...\right\}$$

$$= a_0 x^\nu\left\{1-\frac{(\frac{1}{2}x)^2}{\nu+1}+\frac{(\frac{1}{2}x)^4}{(\nu+1)(\nu+2)2!}-...\right\}$$

$$= a_0 x^\nu\left\{1+\sum_{r=1}^{\infty}\frac{(-1)^r(\frac{1}{2}x)^{2r}}{(\nu+1)(\nu+2)...(\nu+r)r!}\right\}. \tag{10}$$

Taking the negative sign, $c = -\nu$, in (8) gives the same result except that ν is replaced by $-\nu$. Thus if ν is not an integer we have found two independent solutions of (1) which behave like x^ν and $x^{-\nu}$, respectively, as $x \to 0$.

If ν is a positive integer n, or is zero, this procedure only leads to one solution. For if $\nu = 0$, (3) only gives one value of c, namely $c = 0$; while if $\nu = n$, taking $c = n$ gives the series (10) with $\nu = n$, that is,

$$a_0 x^n\left\{1+\sum_{r=1}^{\infty}\frac{(-1)^r(\frac{1}{2}x)^{2r}}{(n+1)...(n+r)r!}\right\}, \tag{11}$$

but if $c = -n$, the coefficient of a_{2n} in the chain of equations (3), (4),... vanishes; this requires that $a_{2n-2} = ... = a_2 = 0$, and if the procedure is carried through in detail the result is found merely to be a constant multiple of the result for $c = n$.

Those cases in which the order is zero or a positive integer are in fact much the most important, and their solutions are tabulated for a very wide range of n and x. The function chosen to study and tabulate† is a constant multiple, namely $1/(2^n a_0 n!)$, of the series (11); the notation $J_n(x)$ is used for it so that

$$J_n(x) = \sum_{r=0}^{\infty}\frac{(-1)^r(\frac{1}{2}x)^{n+2r}}{r!(n+r)!}. \tag{12}$$

This is called the Bessel function of the first kind of order n. When $n = 0$ it becomes

$$J_0(x) = \sum_{r=0}^{\infty}\frac{(-1)^r(\frac{1}{2}x)^{2r}}{(r!)^2}. \tag{13}$$

† Tables are given in the works quoted above, also Jahnke, Emde, and Lösch, *Tables of Higher Functions* (McGraw-Hill).

When $x = 0$ we have from (12) and (13)

$$J_0(0) = 1, \quad \text{but} \quad J_n(0) = 0 \quad (n = 1, 2,\ldots). \tag{14}$$

We now derive some important properties of $J_n(x)$. From (12), assuming that the infinite series may be differentiated term by term, we get for $n \geqslant 1$,

$$\begin{aligned} \frac{d}{dx}\{x^n J_n(x)\} &= \frac{d}{dx}\left\{\sum_{r=0}^{\infty} \frac{(-1)^r x^{2n+2r}}{2^{n+2r} r!(n+r)!}\right\} \\ &= \sum_{r=0}^{\infty} \frac{(-1)^r x^{2n+2r-1}}{2^{n+2r-1} r!(n+r-1)!} \\ &= x^n \sum_{r=0}^{\infty} \frac{(-1)^r (\frac{1}{2}x)^{2r+(n-1)}}{r!(n+r-1)!} \\ &= x^n J_{n-1}(x). \end{aligned} \tag{15}$$

Therefore, writing $J'_n(x)$ for $(d/dx)J_n(x)$,

$$xJ'_n(x)+nJ_n(x) = xJ_{n-1}(x). \tag{16}$$

In the same way

$$\frac{d}{dx}\{x^{-n}J_n(x)\} = -x^{-n}J_{n+1}(x), \tag{17}$$

and so

$$xJ'_n(x)-nJ_n(x) = -xJ_{n+1}(x). \tag{18}$$

Adding and subtracting (16) and (18) gives

$$2J'_n(x) = J_{n-1}(x)-J_{n+1}(x), \tag{19}$$

$$\frac{2n}{x}J_n(x) = J_{n-1}(x)+J_{n+1}(x). \tag{20}$$

(19) and (20) are called the recurrence relations. It appears from (20) that if we know $J_0(x)$ and $J_1(x)$ for any value of x, $J_2(x)$ can be calculated, and then $J_3(x)$ and so on. (19) gives the differential coefficient $J'_n(x)$ if $n \geqslant 1$. If $n = 0$, it follows immediately by differentiating (13) that

$$J'_0(x) = -J_1(x). \tag{21}$$

For large values of x the Bessel functions can be shown to oscillate steadily with decreasing amplitude; in fact

$$J_n(x) = \sqrt{\left(\frac{2}{\pi x}\right)}\cos(x-\tfrac{1}{2}n\pi-\tfrac{1}{4}\pi)+\mathbf{O}(x^{-\frac{3}{2}}), \tag{22}$$

where $\mathbf{O}(x^{-\frac{3}{2}})$ is written for a term which decreases like $x^{-\frac{3}{2}}$ when x becomes large.

The graphs of $J_0(x)$, $J_1(x)$, and $J_2(x)$ are shown in Fig. 82 (*a*).

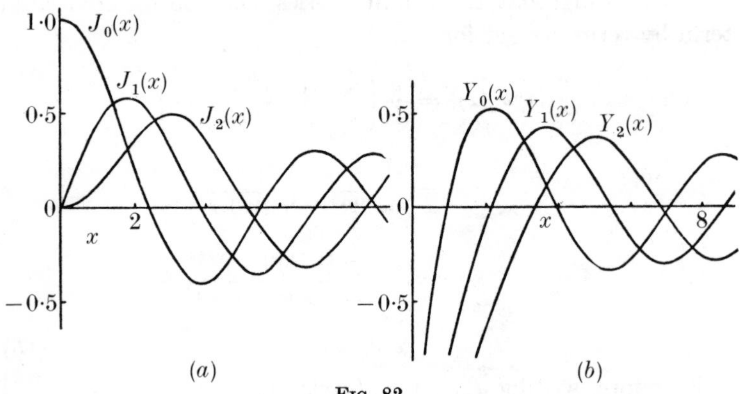

FIG. 82.

The zeros of the Bessel functions are of importance in practice and are tabulated. The first few zeros of $J_0(x)$ are 2·4048..., 5·5200..., 8·6537.... .

Before considering the second solution of Bessel's equation of positive integral order, we return to the solutions (10) of the equation of fractional order. These can be expressed in a form similar to (12) by the use of the gamma function.

The gamma function $\Gamma(\nu)$ is given by

$$\Gamma(\nu) = \int\limits_0^\infty e^{-x}x^{\nu-1}\,dx, \tag{23}$$

if $\nu > 0$. Integrating (23) by parts gives, if $\nu > 1$,

$$\Gamma(\nu) = (\nu-1)\Gamma(\nu-1). \tag{24}$$

$\Gamma(\nu)$ is tabulated in most books of tables for values of ν between 1 and 2: its graph in this region is shown in Fig. 83. By repeated application of (24) the gamma function of any argument may be expressed in terms of the function whose argument lies between 1 and 2.

If $\nu = n$, a positive integer, (24) gives

$$\Gamma(n) = (n-1)\Gamma(n-1) = (n-1)(n-2)...1.\Gamma(1) = (n-1)!. \tag{25}$$

Thus the gamma function provides a generalization of $n!$ to non-integral n.

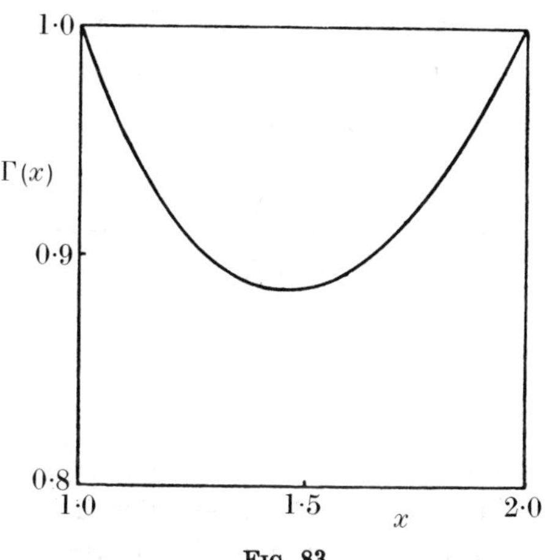

FIG. 83.

Two other properties which are frequently needed may be quoted here; firstly

$$\Gamma(\tfrac{1}{2}) = \pi^{\frac{1}{2}}, \tag{26}$$

and secondly that gamma functions of negative argument may be expressed in terms of those of positive argument by the relation

$$\Gamma(\nu)\Gamma(1-\nu) = \frac{\pi}{\sin\nu\pi}. \tag{27}$$

As $\nu \to 0$, or to a negative integer, $\Gamma(\nu) \to \infty$. The representation (23) does not hold for gamma functions of negative argument, but (24) remains true.

We now define the Bessel function $J_\nu(x)$ of order ν by

$$J_\nu(x) = \sum_{r=0}^{\infty} \frac{(-1)^r(\frac{1}{2}x)^{\nu+2r}}{r!\Gamma(\nu+r+1)}. \tag{28}$$

By (25) this definition agrees with (12) if ν is a positive integer n. Also by (24) it can be seen that it is a constant multiple of (10). Thus if ν is fractional, $J_\nu(x)$ and $J_{-\nu}(x)$ are a pair of independent solutions of (1), and so its general solution is

$$y = AJ_\nu(x)+BJ_{-\nu}(x), \tag{29}$$

where A and B are arbitrary constants. Frequently, however, the linear combination $Y_\nu(x)$ of $J_\nu(x)$ and $J_{-\nu}(x)$ defined by

$$Y_\nu(x) = \frac{J_\nu(x)\cos\nu\pi - J_{-\nu}(x)}{\sin\nu\pi}, \tag{30}$$

is taken as the second solution of (1) so that its general solution is

$$y = CJ_\nu(x)+DY_\nu(x). \tag{31}$$

$Y_\nu(x)$ is called the Bessel function of the second kind of order ν. The main reason for its introduction is that it provides a second solution for the equation of integral order. If ν is zero or a positive integer, both the numerator and denominator of (30) vanish, but it can be shown that

$$\lim_{\nu\to n} Y_\nu(x) \tag{32}$$

exists. We write this $Y_n(x)$, and it is the required second solution for functions of integral order. These Bessel functions of the second kind are of less importance than those of the first kind since

$$Y_\nu(x) \to -\infty \quad \text{and} \quad Y_n(x) \to -\infty \quad \text{as} \quad x \to 0, \tag{33}$$

and for this reason they are excluded from the solutions of problems in applied mathematics for regions including the origin $x = 0$. The graphs of $Y_0(x)$, $Y_1(x)$, and $Y_2(x)$, showing this behaviour as $x \to 0$, are given in Fig. 82 (*b*).

$Y_n(x)$ satisfies the same recurrence relations (19), (20), (21) as $J_n(x)$, and $Y_\nu(x)$ and $J_\nu(x)$ also satisfy these relations with n replaced by ν for all ν. Finally the important formula

$$J_\nu(x)Y'_\nu(x) - Y_\nu(x)J'_\nu(x) = \frac{2}{\pi x} \tag{34}$$

may be quoted.

The most important functions of fractional order are those of orders $\frac{1}{2}$ and $\frac{1}{3}$. By (28) and (24)

$$J_{\frac{1}{2}}(x) = \frac{2^{\frac{1}{2}}}{x^{\frac{1}{2}}\Gamma(\frac{1}{2})}\left\{x - \frac{x^3}{3!} + \frac{x^5}{5!} - \dots\right\}$$

$$= \sqrt{\left(\frac{2}{\pi x}\right)}\sin x, \tag{35}$$

using (26) and the series for $\sin x$. Similarly

$$J_{-\frac{1}{2}}(x) = \sqrt{\left(\frac{2}{\pi x}\right)}\cos x, \tag{36}$$

and other functions of half-integral order may be expressed in terms of these by the recurrence relation (20).

The importance of the Bessel functions of order $\frac{1}{3}$ arises from the fact that the solution of the equation

$$\frac{d^2y}{dx^2}+xy = 0 \tag{37}$$

is
$$y = Ax^{\frac{1}{2}}J_{\frac{1}{3}}(\tfrac{2}{3}x^{\frac{3}{2}})+Bx^{\frac{1}{2}}J_{-\frac{1}{3}}(\tfrac{2}{3}x^{\frac{3}{2}}). \tag{38}$$

A number of second-order equations, of which (37) is the most important, can be transformed into Bessel's equation by an appropriate change of variables and thus their solutions can be expressed in terms of tabulated functions.

Finally the equation

$$\frac{d^2y}{dx^2}+\frac{1}{x}\frac{dy}{dx}-\left(1+\frac{\nu^2}{x^2}\right)y = 0, \tag{39}$$

which corresponds to (1) with x replaced by ix, should be mentioned. It is called the modified Bessel equation and is of almost equal importance to Bessel's equation. Its theory may be developed in the same way.

The solution analogous to $J_\nu(x)$, which is called the modified Bessel function of the first kind of order ν and is denoted by $I(x)$, is

$$I_\nu(x) = \sum_{r=0}^{\infty}\frac{(\frac{1}{2}x)^{\nu+2r}}{r!\Gamma(\nu+r+1)}; \tag{40}$$

it is proportional to $J_\nu(ix)$. The modified Bessel function of the second kind, corresponding to $Y_\nu(x)$, is defined by

$$K_\nu(x) = \tfrac{1}{2}\pi\frac{I_{-\nu}(x)-I_\nu(x)}{\sin\nu\pi}. \tag{41}$$

For integral or zero order, the solutions are

$$I_n(x) = \sum_{r=0}^{\infty}\frac{(\frac{1}{2}x)^{n+2r}}{r!(n+r)!} \tag{42}$$

and
$$K_n(x) = \lim_{v\to n}K_\nu(x). \tag{43}$$

As $x \to 0$, $I_0(x) \to 1$; $I_\nu(x) \to 0$, $\nu \neq 0$; and $K_\nu(x) \to \infty$. But the behaviour of $I_\nu(x)$ and $K_\nu(x)$ as $x \to \infty$ is fundamentally different from that of $J_\nu(x)$ and $Y_\nu(x)$: it is

$$I_\nu(x) = \frac{e^x}{(2\pi x)^{\frac{1}{2}}}\left\{1+\mathbf{O}\left(\frac{1}{x}\right)\right\}, \tag{44}$$

$$K_\nu(x) = \left(\frac{\pi}{2x}\right)^{\frac{1}{2}} e^{-x}\left\{1+\mathbf{O}\left(\frac{1}{x}\right)\right\}. \tag{45}$$

99. Legendre's equation†

The differential equation is

$$(1-x^2)\frac{d^2y}{dx^2}-2x\frac{dy}{dx}+n(n+1)y = 0, \tag{1}$$

where n is zero or any positive integer. The case of fractional n may be treated in the same way.

As before, we seek a solution of type

$$y = x^c(a_0+a_1x+a_2x^2+\ldots), \tag{2}$$

with $a_0 \neq 0$. Substituting in (1) gives

$$\begin{aligned}
&c(c-1)a_0x^{c-2}+\\
&+(c+1)ca_1x^{c-1}+(c+2)(c+1)a_2x^c+(c+3)(c+2)a_3x^{c+1}+\ldots-\\
&\qquad\qquad -c(c-1)a_0x^c \qquad -(c+1)ca_1x^{c+1}-\ldots-\\
&\qquad\qquad -2ca_0x^c \qquad -2(c+1)a_1x^{c+1}-\ldots+\\
&\qquad\qquad +n(n+1)a_0x^c \qquad +n(n+1)a_1x^{c+1}+\ldots = 0.
\end{aligned}$$

Equating coefficients of the powers of x to zero we get

$$c(c-1)a_0 = 0, \tag{3}$$

$$c(c+1)a_1 = 0, \tag{4}$$

$$(c+1)(c+2)a_2-\{c(c+1)-n(n+1)\}a_0 = 0, \tag{5}$$

$$(c+2)(c+3)a_3-\{(c+1)(c+2)-n(n+1)\}a_1 = 0, \tag{6}$$

$$(c+3)(c+4)a_4-\{(c+2)(c+3)-n(n+1)\}a_2 = 0, \tag{7}$$

. . . .

Since $a_0 \neq 0$ it follows from (3) that c must be zero or unity.

† Macrobert, *Spherical Harmonics* (Methuen); Hobson, *Spherical and Ellipsoidal Harmonics* (Cambridge); Byerly, *Fourier's Series and Spherical Harmonics* (Ginn).

If $c = 0$, (4) is satisfied, so a_1 is unspecified. Putting $c = 0$ in (5), (6),... we get

$$a_2 = -\frac{n(n+1)}{2!}a_0, \qquad a_4 = \frac{(n-2)n(n+1)(n+3)}{4!}a_0, \tag{8}$$

$$a_3 = -\frac{(n-1)(n+2)}{3!}a_1, \qquad a_5 = \frac{(n-3)(n-1)(n+2)(n+4)}{5!}a_1. \tag{9}$$

Thus with $c = 0$ we get for the solution of (1)

$$a_0\left\{1-\frac{n(n+1)}{2!}x^2+\frac{(n-2)n(n+1)(n+3)}{4!}x^4-...\right\}+$$
$$+a_1\left\{x-\frac{(n-1)(n+2)}{3!}x^3+\frac{(n-3)(n-1)(n+2)(n+4)}{5!}x^5-...\right\}. \tag{10}$$

The solution (10) consists of arbitrary constant multiples of two independent series, and so is the general solution of (1). The other choice $c = 1$ from (3) merely gives the second series in (10).

If n is even, the coefficients of x^{n+2} and the higher powers of x in the first series in (10) are zero, and so the series reduces to a polynomial of degree n in x. If $n = 0$, only the first term, 1, remains; if $n = 2$, the series reduces to $1-3x^2$, and so on. Similarly if n is odd, the second series in (10) reduces to a polynomial.

Thus in either case the solution of (1) consists of one infinite series and one polynomial. We define $P_n(x)$, the *Legendre Polynomial* or Legendre coefficient of degree n, to be a constant multiple of this polynomial, namely

$$P_n(x) = (-1)^{\frac{1}{2}n}\frac{1.3.5...(n-1)}{2.4...n}\times$$
$$\times\left\{1-\frac{n(n+1)}{2!}x^2+\frac{(n-2)n(n+1)(n+3)}{4!}x^4-...\right\},$$

if n is even, (11)

$$P_n(x) = (-1)^{\frac{1}{2}(n-1)}\frac{1.3.5...n}{2.4...(n-1)}\left\{x-\frac{(n-1)(n+2)}{3!}x^3+...\right\},$$

if n is odd. (12)

If the polynomials (11) and (12) are written in the reverse order, beginning with the terms in x^n, they both take the form

$$P_n(x) = \frac{(2n)!}{2^n(n!)^2}\left\{x^n - \frac{n(n-1)}{2(2n-1)}x^{n-2} + \frac{n(n-1)(n-2)(n-3)}{2.4(2n-1)(2n-3)}x^{n-4} - \ldots\right\}. \tag{13}$$

This leads immediately to *Rodrigues's formula*

$$P_n(x) = \frac{1}{2^n n!}\frac{d^n}{dx^n}(x^2-1)^n, \tag{14}$$

as may be verified by expanding $(x^2-1)^n$ by the binomial theorem and differentiating n times.

A constant multiple of the infinite series solution of (1) is called $Q_n(x)$, the Legendre function of the second kind of order n. It is of less importance than the Legendre polynomials since $Q_n(x) \to \infty$ as $x \to \pm 1$, and because of this has to be excluded from the solution of many physical problems.

The first few Legendre polynomials are

$$P_0(x) = 1, \tag{15}$$

$$P_1(x) = x, \tag{16}$$

$$P_2(x) = \tfrac{1}{2}(3x^2-1), \tag{17}$$

$$P_3(x) = \tfrac{1}{2}(5x^3-3x); \tag{18}$$

they are obtained easily by Rodrigues's formula.

Next we derive the important result

$$(1-2hx+h^2)^{-\frac{1}{2}} = \sum_{n=0}^{\infty} h^n P_n(x), \tag{19}$$

that is, that the Legendre polynomials $P_n(x)$ defined above are the coefficients of h^n in the expansion of the function on the left of (19). To derive (19), assuming that h is sufficiently small to ensure convergence, we expand the left-hand side of (19) by the binomial theorem and rearrange the resulting series. In this way we get

$$(1-2hx+h^2)^{-\frac{1}{2}} = 1+\tfrac{1}{2}(2x-h)h+\frac{1.3}{2.4}(2x-h)^2h^2+...+$$
$$+\frac{1.3...(2n-1)}{2.4...2n}(2x-h)^nh^n+...$$
$$= 1+xh+\tfrac{1}{2}(3x^2-1)h^2+\tfrac{1}{2}(5x^3-3x)h^3+...$$
$$= P_0(x)+hP_1(x)+h^2P_2(x)+...,$$

as required.

The importance of (19) in applied mathematics arises from the fact that it gives an expansion of $1/R$ in ascending or descending powers of r, where R is the distance between the points whose polar coordinates are $(r, 0)$ and (a, θ), so that

$$\frac{1}{R} = \frac{1}{\{r^2+a^2-2ra\cos\theta\}^{\frac{1}{2}}} = \frac{1}{r}\sum_{n=0}^{\infty}\left(\frac{a}{r}\right)^n P_n(\mu) \quad (r > a), \tag{20}$$

$$= \frac{1}{a}\sum_{n=0}^{\infty}\left(\frac{r}{a}\right)^n P_n(\mu) \quad (r < a), \tag{21}$$

where $\mu = \cos\theta$.

Many important properties of the Legendre polynomials follow from (19). For example, differentiating it with respect to h gives

$$(x-h)(1-2hx+h^2)^{-\frac{3}{2}} = \sum_{n=1}^{\infty} nh^{n-1}P_n(x).$$

Therefore

$$(x-h)\sum_{n=0}^{\infty} h^nP_n(x) = (1-2hx+h^2)\sum_{n=1}^{\infty} nh^{n-1}P_n(x). \tag{22}$$

Equating coefficients of h^n in (22) gives

$$(n+1)P_{n+1}(x)-(2n+1)xP_n(x)+nP_{n-1}(x) = 0, \tag{23}$$

which is the recurrence relation connecting the polynomials of degrees $n-1$, n, and $n+1$.

Next we evaluate some important integrals involving Legendre polynomials. Writing D for d/dx, consider

$$2^nn!\int_{-1}^{1} P_n(x)x^m\,dx = \int_{-1}^{1} x^m\{D^n(x^2-1)^n\}\,dx$$

$$= [x^mD^{n-1}(x^2-1)^n]_{-1}^{1}-m\int_{-1}^{1} x^{m-1}\{D^{n-1}(x^2-1)^n\}\,dx, \tag{24}$$

where Rodrigues's formula (14) has been used. Since

$$D^{n-1}(x^2-1)^n$$

has x^2-1 as a factor, the integrated part in (24) vanishes at both limits. If the process of integrating by parts is continued the final result is zero if $m < n$, and if $m = n$ it is

$$\begin{aligned}(-1)^n n! \int\limits_{-1}^{1} (x^2-1)^n\,dx &= n! \int\limits_{-\frac{1}{2}\pi}^{\frac{1}{2}\pi} \cos^{2n+1}\theta\,d\theta \\ &= 2(n!)\frac{2n(2n-2)...2}{(2n+1)(2n-1)...3} \\ &= \frac{2^{2n+1}(n!)^3}{(2n+1)!}. \qquad (25)\end{aligned}$$

Since $P_n(x)$ is a polynomial of degree n in x, it follows from these results and (13) that

$$\int\limits_{-1}^{1} P_n(x)P_m(x)\,dx = 0 \quad (m < n), \qquad (26)$$

$$\begin{aligned}\int\limits_{-1}^{1} [P_n(x)]^2\,dx &= \frac{(2n)!}{2^n(n!)^2}\int\limits_{-1}^{1} P_n(x)x^n\,dx \\ &= \frac{2}{2n+1}. \qquad (27)\end{aligned}$$

Since m and n in (26) are interchangeable, this result holds also for $m > n$. It is analogous to the results of § 88 for trigonometric functions and may be stated in the form that $P_n(x)$ and $P_m(x)$ are orthogonal in $-1 \leqslant x \leqslant 1$. Using (26) and (27), a function $f(x)$, defined in $-1 \leqslant x \leqslant 1$, may be expanded in a series of Legendre polynomials. Assume

$$f(x) = \sum_{n=0}^{\infty} a_n P_n(x). \qquad (28)$$

Multiplying (28) by $P_m(x)$, integrating with respect to x from -1 to 1, and using (26) and (27), we get

$$a_m = \frac{2m+1}{2}\int\limits_{-1}^{1} f(x)P_m(x)\,dx. \qquad (29)$$

This corresponds exactly to the procedure used in § 88 for determining the coefficients in a Fourier series, and, as in that case, the processes used have to be justified carefully from the pure-mathematical point of view.

Finally we consider the equation

$$(1-x^2)\frac{d^2y}{dx^2}-2x\frac{dy}{dx}+\left[n(n+1)-\frac{m^2}{1-x^2}\right]y = 0, \tag{30}$$

where m and n are integers.

Putting $$y = (1-x^2)^{\frac{1}{2}m}z, \tag{31}$$

(30) becomes

$$(1-x^2)\frac{d^2z}{dx^2}-2(m+1)x\frac{dz}{dx}+[n(n+1)-m(m+1)]z = 0, \tag{32}$$

and this equation is satisfied by

$$\frac{d^mv}{dx^m}, \tag{33}$$

where v is any solution of Legendre's equation (1). (30) is known as the associated Legendre equation, and its solutions

$$P_n^m(x) = (1-x^2)^{\frac{1}{2}m}\frac{d^mP_n(x)}{dx^m}, \tag{34}$$

as the associated Legendre functions of the first kind. For the important case in which $x = \cos\theta$ the first few of these are

$$P_0^0 = 1, \qquad P_1^0 = \cos\theta, \qquad P_1^1 = \sin\theta,$$
$$P_2^0 = \tfrac{1}{2}(3\cos^2\theta-1), \qquad P_2^1 = 3\sin\theta\cos\theta, \qquad P_2^2 = 3\sin^2\theta.$$

100. Schrödinger's equation for the hydrogen atom

This is, in effect, a rather more complicated problem on solution in series. The problem is to find the conditions on k under which the differential equation

$$\frac{d^2R}{d\rho^2}+\frac{2}{\rho}\frac{dR}{d\rho}+\left(-\frac{1}{4}+\frac{k}{\rho}-\frac{l(l+1)}{\rho^2}\right)R = 0, \tag{1}$$

where l is zero or a positive integer, will have a solution which remains finite as $\rho \to 0$ and as $\rho \to \infty$. This is an eigenvalue problem: the values of k found will be the eigenvalues, and the corresponding solutions the eigenfunctions. (1) has, of course, solutions for all values of k, but these do not satisfy the required conditions.

First we make the change of variable

$$R = ve^{-\frac{1}{2}\rho}, \tag{2}$$

in (1). This is suggested by the fact that if ρ is so large that $1/\rho$ is negligible (1) becomes

$$\frac{d^2R}{d\rho^2} - \tfrac{1}{4}R = 0,$$

and the solution of this which is finite as $\rho \to \infty$ is $e^{-\frac{1}{2}\rho}$.

Substituting (2) in (1) gives

$$\frac{d^2v}{d\rho^2} + \left(\frac{2}{\rho} - 1\right)\frac{dv}{d\rho} + \left\{\frac{k-1}{\rho} - \frac{l(l+1)}{\rho^2}\right\}v = 0. \tag{3}$$

We seek a solution

$$v = \rho^c \sum_{n=0}^{\infty} a_n \rho^n \tag{4}$$

of this. Substituting (4) in (3) gives

$$\begin{aligned}
c(c-1)a_0\rho^{c-2} &+ (c+1)ca_1\rho^{c-1} + (c+2)(c+1)a_2\rho^c + \ldots + \\
+2ca_0\rho^{c-2} &+ 2(c+1)a_1\rho^{c-1} \qquad +2(c+2)a_2\rho^c + \ldots - \\
&-ca_0\rho^{c-1} \qquad -(c+1)a_1\rho^c - \ldots + \\
&+(k-1)a_0\rho^{c-1} \qquad +(k-1)a_1\rho^c + \ldots - \\
-l(l+1)a_0\rho^{c-2} &- l(l+1)a_1\rho^{c-1} \qquad -l(l+1)a_2\rho^c + \ldots = 0.
\end{aligned}$$

Equating the coefficients of the powers of ρ to zero we get

$$c(c+1) - l(l+1) = 0, \tag{5}$$

$$\{(c+1)(c+2) - l(l+1)\}a_1 - \{c-(k-1)\}a_0 = 0, \tag{6}$$

$$\{(c+2)(c+3) - l(l+1)\}a_2 - \{(c+1)-(k-1)\}a_1 = 0, \tag{7}$$

$$\cdot \quad \cdot \quad \cdot \quad \cdot \quad \cdot \quad \cdot \quad \cdot$$

$$\{(c+n)(c+n+1) - l(l+1)\}a_n - \{(c+n-1)-(k-1)\}a_{n-1} = 0. \tag{8}$$

(5) gives $c = l$ or $c = -l-1$. In order to have v finite as $\rho \to 0$ we must choose the solution with $c = l$. Then the other equations give

$$a_n = \frac{l+n-k}{n(n+2l+1)}a_{n-1} \quad (n = 1, 2, \ldots), \tag{9}$$

from which the solution (4) can be written down.

Now for large n it follows from (9) that

$$\frac{a_n}{a_{n-1}} = \frac{1}{n}$$

approximately. Thus when n is large the terms of the series (4) behave like those of e^{ρ}, and R, given by (2), behaves like $e^{\frac{1}{2}\rho}$, and so tends to infinity as $\rho \to \infty$ and thus does not satisfy the required conditions. The only exception to this is the case in which k is an integer, $l+n$. Then by (9), a_n and all subsequent coefficients in (4) vanish, so that v becomes a polynomial in ρ of degree $l+n-1$, and $R \to 0$ as $\rho \to \infty$ as required. Thus the conditions are only satisfied if

$$k = l+n, \quad \text{where } n = 1, 2, 3, \ldots.$$

Since in the problem of the hydrogen atom† k is

$$\left\{-\frac{2\pi^2 m Z^2 e^4}{Eh^2}\right\}^{\frac{1}{2}}, \tag{10}$$

this leads to the formula for the energy levels of the hydrogen atom

$$E = -\frac{2\pi^2 m e^4 Z^2}{h^2(l+n)^2} \quad (n = 1, 2, \ldots). \tag{11}$$

101. Inhomogeneous equations. Variation of parameters

The solution of an inhomogeneous linear differential equation with variable coefficients, just as in the case of constant coefficients, consists of the sum of a particular integral and the complementary function. The complementary function is found by methods such as those of the preceding sections. Variation of parameters is a method for finding a particular integral when the complementary function is known.

† Schrödinger's equation for an electron of charge $-e$ and mass m in the field of a nucleus of charge Ze is

$$\nabla^2\psi + \frac{8\pi^2 m}{h^2}\left(E + \frac{Ze^2}{r}\right)\psi = 0;$$

where h is Planck's constant and E is the total energy of the electron. If, as in § 128, we seek a solution of type

$$\psi = R(\rho) P_l^m(\cos\theta) \begin{matrix}\cos\\ \sin\end{matrix} m\phi,$$

where $\rho = 4\pi r(-2mE/h^2)^{\frac{1}{2}}$, the equation (1) for R results with k given by (10).

Suppose the differential equation is

$$D^2y+P(x)\,Dy+Q(x)y = R(x), \tag{1}$$

and that u and v are two independent solutions of the corresponding homogeneous equation, so that

$$D^2u+P\,Du+Qu = 0, \tag{2}$$

$$D^2v+P\,Dv+Qv = 0. \tag{3}$$

We seek a solution of (1) of the form

$$y = \phi u+\psi v, \tag{4}$$

where ϕ and ψ are functions of x to be determined. Differentiating (4) gives

$$Dy = \phi\,Du+\psi\,Dv+u\,D\phi+v\,D\psi, \tag{5}$$

and if we require†

$$Dy = \phi\,Du+\psi\,Dv, \tag{6}$$

we must have by (5)

$$u\,D\phi+v\,D\psi = 0. \tag{7}$$

Differentiating (6) gives

$$D^2y = \phi\,D^2u+\psi\,D^2v+D\phi Du+D\psi Dv. \tag{8}$$

Substituting (6), (8), and (4) in (1), and using (2) and (3), gives

$$D\phi Du+D\psi Dv = R(x). \tag{9}$$

(7) and (9) are a pair of equations for $D\phi$ and $D\psi$. Solving them we have

$$(u\,Dv-v\,Du)\,D\phi = -R(x)v, \tag{10}$$

$$(u\,Dv-v\,Du)\,D\psi = R(x)u. \tag{11}$$

Integrating (10) and (11) gives ϕ and ψ, and the complete solution of (1) is finally

$$y = Au+Bv+\phi u+\psi v, \tag{12}$$

where A and B are arbitrary constants. The method can be extended to equations of higher order.

102. Inhomogeneous equations with boundary conditions. The Green's function

The method of variation of parameters given in the preceding section is an analytical device for finding a particular integral

† That is, we make Dy have the form it would have if ϕ and ψ were constants. This is often done in problems of this type; cf. § 58 (10) and (11) for another example.

of any inhomogeneous second-order equation. In this section we give a method for finding the complete solution, satisfying given boundary conditions, of an inhomogeneous second-order equation of the form†

$$\frac{d}{dx}\left\{p(x)\frac{dy}{dx}\right\}-q(x)y = r(x), \tag{1}$$

in terms of a special solution of the corresponding homogeneous equation and the same boundary conditions. This special solution is called the Green's function and it has a simple physical interpretation. Suppose we have to find a solution of (1) with the boundary conditions

$$a\,Dy+by = 0, \quad \text{when } x = 0, \tag{2}$$

$$a'\,Dy+b'y = 0, \quad \text{when } x = l, \tag{3}$$

where a, b, a', b' are constants, and D is written for d/dx.

Let $G(x,\xi)$ be the solution of the homogeneous equation corresponding to (1) which satisfies the boundary conditions (2) and (3), and which is continuous at $x = \xi$ but has a discontinuous first derivative at $x = \xi$ such that

$$\lim_{\epsilon\to 0}\,[p(x)\,DG]_{x=\xi-\epsilon}^{x=\xi+\epsilon} = -1. \tag{4}$$

$G(x,\xi)$ then satisfies

$$D\{p(x)\,DG\}-q(x)G = 0, \tag{5}$$

$$a\,DG+bG = 0 \quad (x = 0), \tag{6}$$

$$a'\,DG+b'G = 0 \quad (x = l). \tag{7}$$

Multiplying (1) by G, (5) by y, and subtracting, gives

$$GD\{p(x)\,Dy\}-yD\{p(x)\,DG\} = r(x)\,G.$$

Integrating with respect to x from $x = 0$ to $x = l$ gives

$$\int\limits_0^l \{GD[p(x)\,Dy]-yD[p(x)\,DG]\}\,dx = \int\limits_0^l r(x)\,G(x,\xi)\,dx. \tag{8}$$

Because of the discontinuity of DG at $x = \xi$, the integral on the left has to be split into integrals from 0 to $\xi-\epsilon$ and from

† The equations of §§ 98, 99 can be put in this form.

$\xi+\epsilon$ to l. Integrating by parts it becomes

$$[G(x,\xi)\,p(x)\,Dy-y(x)\,p(x)\,DG(x,\xi)]_{\xi+\epsilon}^{l}+$$

$$+[G(x,\xi)p(x)\,Dy-y(x)p(x)\,DG(x,\xi)]_0^{\xi-\epsilon} = \int\limits_0^l r(x)G(x,\xi)\,dx. \quad (9)$$

The terms on the left vanish at the limits l and 0 by (2), (3), (6), and (7), and we are left with

$$\lim_{\epsilon\to 0} y(\xi)[p(x)\,DG(x,\xi)]_{\xi-\epsilon}^{\xi+\epsilon} = \int\limits_0^l r(x)G(x,\xi)\,dx,$$

or, using (4),
$$y(\xi) = -\int\limits_0^l r(x)\,G(x,\xi)\,dx. \quad (10)$$

Thus when $G(x,\xi)$ has been found, the required value of the solution y at any point is found by simple integration. It can be shown† that the Green's function $G(x,\xi)$ is a symmetrical function of x and ξ, that is,

$$G(x,\xi) = G(\xi,x). \quad (11)$$

The theory can be extended to equations of order n; in this case G and all its derivatives up to $D^{n-2}G$ are to be continuous at $x = \xi$, and $D^{n-1}G$ is to be discontinuous there.

The fourth-order equation for deflexion of beams has been studied in § 82, and $G(x,\xi)$ in that case was found to be the deflexion at ξ due to a unit concentrated load at x.

103. Reduction to the normal form. Approximate solutions

If the change of variables

$$y = z\exp\{-\tfrac{1}{2}\int P(x)\,dx\} \quad (1)$$

is made in the second-order linear equation

$$\frac{d^2y}{dx^2}+P(x)\frac{dy}{dx}+Q(x)y = R(x), \quad (2)$$

† By the type of argument leading to (9) except that in place of $G(x,\xi)$ and y two Green's functions $G(x,\xi)$ and $G(x,\eta)$ are used.

the resulting equation for z is

$$\frac{d^2z}{dx^2}+\left\{Q(x)-\frac{1}{2}\frac{dP(x)}{dx}-\tfrac{1}{4}[P(x)]^2\right\}z = R(x)\exp\left\{\tfrac{1}{2}\int P(x)\,dx\right\}. \tag{3}$$

This is called the *Normal Form* of (1). By using it, it is often possible to decide whether two different equations can be transformed into one another.

Since any second-order equation may be put into this form we are led to consider the equation

$$\frac{d^2z}{dx^2}-\phi z = 0, \tag{4}$$

where ϕ may be a complicated function of x. It is often useful to have an analytical approximation to the solution of this equation. We consider first the case in which ϕ is a slowly varying function of x which does not vanish in the range of x in which we are interested.

Making the substitution

$$z = \exp\left\{\int^x \eta\,dx\right\}, \tag{5}$$

suggested by the form of the solution of the first-order linear equation, we get from (4) the differential equation for η

$$\frac{d\eta}{dx}+\eta^2 = \phi. \tag{6}$$

If ϕ is slowly varying as we have assumed, $d\eta/dx$ will be small, and as a first approximation $\eta = \pm\phi^{\frac{1}{2}}$. Taking the positive sign, and substituting $\eta = \phi^{\frac{1}{2}}$ in the small term $d\eta/dx$ in (6), gives for the second approximation

$$\eta^2 = \phi-\frac{\phi'}{2\phi^{\frac{1}{2}}},$$

where ϕ' is written for $d\phi/dx$. Taking the square root we get

$$\eta = \phi^{\frac{1}{2}}-\frac{\phi'}{4\phi}, \tag{7}$$

to the same approximation, and (5) gives

$$z = \exp\left\{\int^x \left(\phi^{\frac{1}{2}}-\frac{\phi'}{4\phi}\right)dx\right\} = [\phi(x)]^{-\frac{1}{4}}\exp\left\{\int^x [\phi(x)]^{\frac{1}{2}}\,dx\right\}. \tag{8}$$

Approximate solutions of this type have been much used in quantum mechanics.

An entirely different type of solution arises near a zero of $\phi(x)$. Suppose that $\phi(x)$ has a simple zero at $x = a$, so that

$$\phi(x) = (x-a)\phi'(a), \tag{9}$$

near this point. Putting $\xi = x-a$, $\phi'(a) = k$, (4) becomes

$$\frac{d^2z}{d\xi^2}-k\xi z = 0. \tag{10}$$

If $\xi < 0$ and $k > 0$, putting $\xi = -\zeta$ the solution of (10) is by § 98 (38)

$$A\zeta^{\frac{1}{2}}J_{\frac{1}{3}}(\tfrac{2}{3}k^{\frac{1}{2}}\zeta^{\frac{3}{2}})+B\zeta^{\frac{1}{2}}J_{-\frac{1}{3}}(\tfrac{2}{3}k^{\frac{1}{2}}\zeta^{\frac{3}{2}}), \tag{11}$$

while in the same way if $\xi > 0$ and $k > 0$ it is

$$C\xi^{\frac{1}{2}}I_{\frac{1}{3}}(\tfrac{2}{3}k^{\frac{1}{2}}\xi^{\frac{3}{2}})+E\xi^{\frac{1}{2}}K_{\frac{1}{3}}(\tfrac{2}{3}k^{\frac{1}{2}}\xi^{\frac{3}{2}}). \tag{12}$$

These solutions are often required in a study of the behaviour of waves near their point of reflection.

104. Linear differential equations with periodic coefficients

The theory of these equations is relatively difficult and cannot be given here. At the same time a surprisingly large number of mechanical problems involve differential equations of this type, and it is desirable to indicate the new phenomena which arise. This is done below, briefly, and without proof.

We consider Mathieu's equation†

$$\frac{d^2y}{dt^2}+(\lambda+\gamma\cos\omega t)y = 0, \tag{1}$$

in which λ and γ are constants and so the coefficient of y is periodic with period $2\pi/\omega$. We remark first that, since the equation is linear, the general solution will as usual be a sum of arbitrary constant multiples of two linearly independent solutions. Secondly, while it is true that (1) has solutions with period $2\pi/\omega$, it is not true (as might perhaps be supposed) that all solutions of (1) are periodic: in fact here we shall discuss only the

† There is no standard notation for this equation: many different ones have been used. For an account of Mathieu functions see, for example, Whittaker and Watson, *Modern Analysis* (Cambridge); McLachlan, *Mathieu Functions* (Oxford).

non-periodic solutions. The equation (1) also arises in the solution of Laplace's and Maxwell's equations for an elliptic boundary, and in such problems it is the periodic solutions which are of interest.

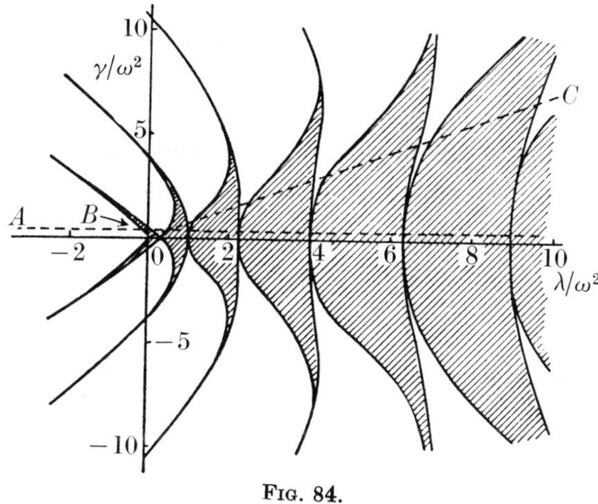

FIG. 84.

There is a theorem (Floquet's theorem) which states that the general solution of (1) has the form

$$y = A\,\phi(t)e^{\mu t}+B\,\phi(-t)e^{-\mu t}, \tag{2}$$

where A and B are arbitrary constants, $\phi(t)$ is a periodic function of t with period $2\pi/\omega$, and μ is a constant depending on λ and γ.

The way in which μ depends on these quantities is shown in Fig. 84 in which the (λ, γ)-plane is divided into shaded and unshaded regions.

If the point whose abscissa is λ/ω^2 and whose ordinate is γ/ω^2 falls in a shaded region, μ is pure imaginary, and the solution (2) is finite for all values of t. These shaded regions are called the *stable* regions for the equation.

On the other hand, if the point $(\lambda/\omega^2, \gamma/\omega^2)$ falls in an unshaded region, μ is complex, and thus one or other of the terms in (2) will increase without limit as t increases; the solution is unstable, and the unshaded regions in Fig. 84 are called the unstable regions for the equation.

A great deal of information can be obtained from a study of this figure. For example, as ω^2 is increased from 0 to ∞, λ/ω^2 will decrease from ∞ to 0 if λ is positive, and thus the point $(\lambda/\omega^2, \gamma/\omega^2)$ will pass through a number of unstable regions: for instance, if $\lambda/\omega^2 = \frac{1}{4}$, so that $\omega = 2\sqrt{\lambda}$ and therefore the frequency $\omega/2\pi$ is twice the natural frequency of the system with $\gamma = 0$, the system is usually unstable. These points are discussed more fully in the examples below.

Equation (1) may be regarded as the equation of motion of a mass attached to a spring whose stiffness varies harmonically; similar results occur if the stiffness varies in any periodic fashion, or if the mass or damping coefficient varies periodically.

Ex. 1. *The inverted pendulum with vertical harmonic motion of the support.*

Suppose the pendulum to consist of a mass m at the end of a light rigid rod of length l. Let θ be the inclination of the rod to the upward vertical, and ξ the upward displacement of the point of support.

The equations for vertical and horizontal motion of the mass m are

$$m\frac{d^2}{dt^2}\{\xi+l\cos\theta\} = -mg+P\cos\theta,$$

$$m\frac{d^2}{dt^2}(l\sin\theta) = P\sin\theta,$$

where P is the stress in the rod. These give

$$\ddot{\xi}-l\ddot{\theta}\sin\theta-l\dot{\theta}^2\cos\theta = -g+(P/m)\cos\theta,$$

$$l\ddot{\theta}\cos\theta-l\dot{\theta}^2\sin\theta = (P/m)\sin\theta.$$

Eliminating P we get

$$\ddot{\theta}-\left(\frac{g}{l}+\frac{\ddot{\xi}}{l}\right)\sin\theta = 0. \tag{3}$$

For small oscillations we replace $\sin\theta$ by θ.

If $\xi = a\cos\omega t$, (3) becomes

$$\ddot{\theta}-\left(n^2-\frac{a\omega^2}{l}\cos\omega t\right)\theta = 0, \tag{4}$$

where $n^2 = g/l$.

This corresponds to (1) with $\lambda = -n^2$ and $\gamma = a\omega^2/l$.

Suppose, for example, that a/l, the ratio of the amplitude of the vibration to the length of the pendulum, is $\frac{1}{2}$. Then $\gamma/\omega^2 = \frac{1}{2}$, and this is the ordinate of the line AB in Fig. 84. λ/ω^2, the abscissa in Fig. 84, is $-n^2/\omega^2$, and as ω is increased the point $(\gamma/\omega^2, \lambda/\omega^2)$ moves to the right along the line AB and finally enters a stable region at the point B. Thus if the frequency of oscillation is increased sufficiently, an inverted pendulum can be stabilized.

Ex. 2. *The hanging pendulum with vertical harmonic motion of the support.*

If the pendulum hangs in the usual way and its point of support is given a downwards vertical motion $a\cos\omega t$, the equation of motion, found as above, is

$$\ddot{\theta}+\left(n^2+\frac{a\omega^2}{l}\cos\omega t\right)\theta = 0. \tag{5}$$

This corresponds to (1) with $\lambda = n^2$, $\gamma = a\omega^2/l$. We have a positive ordinate as before, but $\lambda/\omega^2 = n^2/\omega^2$ is now positive, and is large for small ω. As ω is increased this decreases, and, for the ordinate shown, passes mostly through stable regions, but crosses narrow bands of instability near $n^2/\omega^2 = \frac{1}{4}, 1, \frac{9}{4},\ldots$, corresponding to $\omega = 2n, n, \frac{2}{3}n,\ldots$. In these regions the motion is unstable, and oscillations of large amplitude can be excited.

Ex. 3. *Forced oscillations in an L, C circuit due to a harmonically varying capacitance.*

The differential equation § 41 (6) for the charge on the condenser is

$$L\frac{d^2Q}{dt^2}+\frac{Q}{C} = 0. \tag{6}$$

Suppose the capacitance is varied in such a way that

$$\frac{1}{C} = \frac{1}{C_0}+\frac{1}{C_1}\cos\omega t.$$

Then (6) becomes

$$\frac{d^2Q}{dt^2}+\left(\frac{1}{LC_0}+\frac{1}{LC_1}\cos\omega t\right)Q = 0. \tag{7}$$

Writing $\lambda = 1/LC_0$, $\gamma = 1/LC_1$, we regain (1). Suppose now that the ratio $C_0/C_1 = k$ is fixed so that $\gamma = k\lambda$, and that ω is steadily increased from zero. The point $(\lambda/\omega^2, \gamma/\omega^2)$ then travels in towards the origin along the line CO of Fig. 84, passing through a number of unstable regions. When ω is such that the point lies in one of these, the circuit will oscillate.

EXAMPLES ON CHAPTER XII

1. Show that

$$\frac{d}{dx}\{x^\nu J_\nu(x)\} = x^\nu J_{\nu-1}(x), \qquad \frac{d}{dx}\{x^{-\nu}J_\nu(x)\} = -x^{-\nu}J_{\nu+1}(x),$$

$$\int x^{\nu+1}J_\nu(x)\,dx = x^{\nu+1}J_{\nu+1}(x), \qquad \int x^{1-\nu}J_\nu(x)\,dx = -x^{1-\nu}J_{\nu-1}(x).$$

2. Show that

$$J_{\frac{3}{2}}(x) = \left(\frac{2}{\pi x}\right)^{\frac{1}{2}}\left\{\frac{\sin x}{x}-\cos x\right\},$$

$$J_{-\frac{3}{2}}(x) = \left(\frac{2}{\pi x}\right)^{\frac{1}{2}}\left\{-\sin x-\frac{\cos x}{x}\right\}.$$

3. Show that $x^{\frac{1}{2}}J_1(2x^{\frac{1}{2}})$ and $x^{\frac{1}{2}}Y_1(2x^{\frac{1}{2}})$ satisfy

$$x\frac{d^2y}{dx^2}+y = 0.$$

4. Show that $x^{\frac{1}{2}}J_{1/2\beta}(\gamma x^{\beta})$ and $x^{\frac{1}{2}}Y_{1/2\beta}(\gamma x^{\beta})$ satisfy

$$\frac{d^2y}{dx^2}+\beta^2\gamma^2x^{2\beta-2}y = 0,$$

where β and γ are positive constants.

5. Kelvin's ber and bei functions (which occur in the theory of the skin effect in alternating currents) are defined by

$$\operatorname{ber} x+i\operatorname{bei} x = I_0(x\sqrt{i}).$$

Show that

$$\operatorname{ber} x = 1-\frac{(\frac{1}{2}x)^4}{(2!)^2}+\frac{(\frac{1}{2}x)^8}{(4!)^2}-\ldots,$$

$$\operatorname{bei} x = (\tfrac{1}{2}x)^2-\frac{(\frac{1}{2}x)^6}{(3!)^2}+\frac{(\frac{1}{2}x)^{10}}{(5!)^2}-\ldots.$$

6. If $P(x)$ and $Q(x)$ are any two solutions of Bessel's equation, show that

$$P(x)Q'(x)-P'(x)Q(x) = C/x,$$

where C is a constant.

[Write down the equations satisfied by P and Q, multiply the former by Q and the latter by P, and subtract; use the form of Bessel's equation given in Ex. 10.] § 98 (34) is found in this way, the value of the constant C being found from the series expansions of $J_\nu(x)$ and $Y_\nu(x)$. Deduce from § 98 (34) that

$$J_{\nu+1}(x)Y_\nu(x)-J_\nu(x)Y_{\nu+1}(x) = 2/\pi x.$$

7. Show that

$$\exp\tfrac{1}{2}x(t-t^{-1}) = J_0(x)+\sum_{n=1}^{\infty}\{t^n+(-t)^{-n}\}J_n(x).$$

[Expand the left-hand side in the power series and collect terms in t^n.] The function is called the generating function; compare the similar result § 99 (19) for Legendre polynomials.

8. By putting $t = e^{i\phi}$ in the result of Ex. 7, show that

$$\cos(x\sin\phi) = J_0(x)+2J_2(x)\cos 2\phi+2J_4(x)\cos 4\phi+\ldots,$$

$$\sin(x\sin\phi) = 2J_1(x)\sin\phi+2J_3(x)\sin 3\phi+\ldots.$$

Replacing ϕ in these by $\frac{1}{2}\pi-\phi$, deduce two similar formulae, and also

$$\exp(ix\cos\phi) = J_0(x)+2\sum_{s=1}^{\infty}i^sJ_s(x)\cos s\phi.$$

9. By multiplying results of Ex. 8 by $\cos n\phi$ and $\sin n\phi$, and assuming that the series can be integrated term by term, show that if r is zero or

any positive integer

$$\int_0^{\pi} \cos n\phi \cos(x\sin\phi)\,d\phi = \pi J_n(x) \qquad (n = 2r),$$
$$= 0 \qquad (n = 2r+1),$$
$$\int_0^{\pi} \sin n\phi \sin(x\sin\phi)\,d\phi = 0 \qquad (n = 2r),$$
$$= \pi J_n(x) \qquad (n = 2r+1),$$
$$\int_0^{\pi} \cos(n\phi - x\sin\phi)\,d\phi = \pi J_n(x).$$

The results of this example and the last are fundamental in the theory of frequency modulation.

10. Show that $J_n(\alpha r)$ satisfies

$$\frac{1}{r}\frac{d}{dr}\left\{r\frac{dJ_n(\alpha r)}{dr}\right\}+\left(\alpha^2-\frac{n^2}{r^2}\right)J_n(\alpha r) = 0,$$

if α is any constant. Show that

$$\int_0^a rJ_n(\alpha r)J_n(\beta r)\,dr = 0,$$

where α and β are any two different roots of $J_n(ax) = 0$. [Multiply the equation for $J_n(\alpha r)$ by $J_n(\beta r)$, multiply the corresponding equation for $J_n(\beta r)$ by $J_n(\alpha r)$, subtract, and integrate by parts.] Prove that the above result holds also if α and β are roots of $J_n'(ax) = 0$ or of

$$xJ_n'(ax)+hJ_n(ax) = 0,$$

where h is a positive constant.

11. Writing $u = J_n(\alpha r)$ for shortness, show, by multiplying the first equation of Ex. 10 by $2r^2(du/dr)$, that

$$\frac{d}{dr}\left(r\frac{du}{dr}\right)^2+\alpha^2r^2\frac{du^2}{dr}-n^2\frac{du^2}{dr} = 0.$$

By integrating this equation from 0 to a show that

$$\int_0^a r\{J_n(\alpha r)\}^2\,dr = \tfrac{1}{2}a^2[J_n'(\alpha a)]^2$$

if α is a root of $J_n(ax) = 0$.

12. Assuming that $f(r)$ can be expanded in the series

$$f(r) = \sum_{n=1}^{\infty} a_n J_0(r\alpha_n),$$

where $\alpha_1, \alpha_2,\ldots$ are the positive roots of $J_0(ax) = 0$, show, by using the results of Exs. 10 and 11, that

$$a_n = \frac{2}{a^2J_1^2(a\alpha_n)}\int_0^a rJ_0(r\alpha_n)f(r)\,dr.$$

The series is called a Fourier–Bessel series; cf. Chap. XI and § 99 (28) for similar results.

13. Show that
$$1 = \frac{2}{a}\sum_{n=1}^{\infty}\frac{J_0(r\alpha_n)}{\alpha_n J_1(a\alpha_n)},$$
where $\alpha_1, \alpha_2, \ldots$ are the positive roots of $J_0(ax) = 0$.

14. Prove that if n is a positive integer
$$nP_n(x) = x\frac{dP_n(x)}{dx} - \frac{dP_{n-1}(x)}{dx}.$$
[Equate the expansions for $(1-2hx+h^2)^{-\frac{3}{2}}$ obtained by differentiating § 99 (19) with respect to x, and with respect to h.]

Using this result, and differentiating § 99 (23), show that
$$(2n+1)P_n(x) = \frac{dP_{n+1}(x)}{dx} - \frac{dP_{n-1}(x)}{dx},$$
$$\int P_n(x)\,dx = \frac{1}{2n+1}\{P_{n+1}(x) - P_{n-1}(x)\}.$$

15. Using § 99 (19), show that if n is a positive integer
$$P_n(1) = 1,$$
$$P_n(-1) = 1 \quad (\text{if } n \text{ is even}),$$
$$= -1 \quad (\text{if } n \text{ is odd}).$$
$$P_n(0) = 0 \quad (\text{if } n \text{ is odd}),$$
$$P_{2r}(0) = (-1)^r\frac{1.3\ldots(2r-1)}{2.4\ldots(2r)}.$$
Sketch the graphs of the first few $P_n(x)$ for $-1 < x < 1$.

16. Show that if $f(\theta)$ is defined for $0 \leqslant \theta \leqslant \pi$
$$f(\theta) = \tfrac{1}{2}\sum_{n=0}^{\infty}(2n+1)P_n(\cos\theta)\int_0^{\pi} f(\theta)P_n(\cos\theta)\sin\theta\,d\theta.$$
If
$$f(\theta) = 1 \quad (0 < \theta < \alpha),$$
$$= 0 \quad (\alpha < \theta < \pi),$$
show that (using the result of Ex. 14)
$$f(\theta) = \tfrac{1}{2}(1-\cos\alpha) + \tfrac{1}{2}\sum_{n=1}^{\infty}\{P_{n-1}(\cos\alpha) - P_{n+1}(\cos\alpha)\}P_n(\cos\theta).$$
And if $f(\theta) = \delta(\theta-\alpha)$, show that
$$f(\theta) = \tfrac{1}{2}\sum_{n=0}^{\infty}(2n+1)P_n(\cos\theta)P_n(\cos\alpha)\sin\alpha.$$

17. Show that $P_n(\cos\theta)$ has n zeros between $\theta = 0$ and $\theta = \pi$. Thus if it is represented on a sphere this is divided into $n+1$ zones by the zeros of $P_n(\cos\theta)$; for this reason this function is called a zonal harmonic. Show that if $P_n^m(\cos\theta)\cos m\phi$ and $P_n^m(\cos\theta)\sin m\phi$ are represented in the

same way, the sphere is divided into tesserae bounded by great circles through $\theta = 0$ and $\theta = \pi$ and small circles. Sketch the patterns for the cases $n = 1, 2, 3$. These functions are called tesseral harmonics.

18. By writing

$$(1-2h\cos\theta+h^2)^{-\frac{1}{2}} = (1-he^{i\theta})^{-\frac{1}{2}}(1-he^{-i\theta})^{-\frac{1}{2}}$$

and expanding both expressions on the right by the binomial theorem, show that

$$\tfrac{1}{2}P_n(\cos\theta) = \frac{1.3...(2n-1)}{2.4...2n}\cos n\theta+\frac{1}{2}\frac{1.3...(2n-3)}{2.4...(2n-2)}\cos(n-2)\theta+....$$

Deduce that $|P_n(\cos\theta)| \leqslant 1$ for all θ.

19. The rth Laguerre polynomial $L_r(x)$ is defined by

$$L_r(x) = e^x\frac{d^r}{dx^r}(x^re^{-x});$$

verify that it satisfies the differential equation

$$x\frac{d^2y}{dx^2}+(1-x)\frac{dy}{dx}+ry = 0.$$

Show that its sth derivative

$$L_r^{(s)}(x) = \frac{d^s}{dx^s}L_r(x)$$

satisfies

$$x\frac{d^2y}{dx^2}+(s+1-x)\frac{dy}{dx}+(r-s)y = 0.$$

Show that § 100 (3) has the polynomial solution

$$\rho^l L_{k+l}^{(2l+1)}(\rho),$$

if k is any integer greater than or equal to $l+1$.

20. If

$$H_n(x) = (-1)^ne^{x^2}\frac{d^n}{dx^n}(e^{-x^2}),$$

show that $y = H_n(x)$ satisfies

$$\frac{d^2y}{dx^2}-2x\frac{dy}{dx}+2ny = 0.$$

Show also that

$$e^{-t^2+2tx} = \sum_{n=0}^{\infty}\frac{1}{n!}H_n(x)t^n.$$

These are the Hermite polynomials which occur in Schrödinger's equation for the harmonic oscillator in the same way that the Laguerre polynomials of Ex. 19 occurred in the theory of the hydrogen atom.

21. If

$$T_0(x) = 1, \qquad T_n(x) = 2^{1-n}\cos(n\cos^{-1}x),$$

show that

$$\frac{1-t^2}{1-2tx+t^2} = \sum_{n=0}^{\infty}(2t)^nT_n(x),$$

and that $y = T_n(x)$ satisfies

$$(1-x^2)\frac{d^2y}{dx^2}-x\frac{dy}{dx}+n^2y = 0.$$

These are the Chebyshev polynomials which are of importance, e.g. in the theory of filter circuits.

22. The Laguerre, Hermite, and Chebyshev polynomials are all examples of 'orthogonal polynomials' in the sense that if two different ones are multiplied by an appropriate function and integrated over an appropriate range the integral vanishes (cf. § 88 (6), § 99 (26), and Ex. 10 above for similar results).

Show that

$$\int\limits_{-1}^{1} T_n(x)T_m(x)(1-x^2)^{-\frac{1}{2}}\,dx = 0 \quad (n \neq m),$$

$$\int\limits_{-\infty}^{\infty} e^{-x^2}H_n(x)H_m(x)\,dx = 0 \quad (n \neq m),$$

$$\int\limits_{0}^{\infty} e^{-x}L_n(x)L_m(x)\,dx = 0 \quad (n \neq m).$$

23. A particle of mass m is attached to the mid-point of a string of length l stretched to harmonically varying tension

$$T = T_0(1+k\cos\omega t).$$

Show that the system is unstable if $\omega = 2n/r$, approximately, where $r = 1, 2, 3,\ldots$, and $n^2 = 4T_0/ml$.

24. The point of support of a simple pendulum is rotated with constant angular velocity ω in a vertical circle of small radius a. Show that the motion is unstable if

$$\omega = 2n/r \quad (r = 1, 2, 3,\ldots),$$

approximately, where $n^2 = g/l$.

XIII

MATRICES

105. Introductory

In § 52 vector analysis was introduced as a tool which can be used to simplify problems, especially three-dimensional problems in statics, dynamics, and electromagnetism. Matrices play a similar role in such diverse subjects as electric network theory, stress and structural analysis, vibrations, and the study of linear control systems. Common to all of these is the study of linear systems of ordinary or differential equations. Matrix algebra provides a simple notation for the description of these systems, and offers a theoretical background for deriving general theorems about their behaviour. While linear systems involving two or three equations can be solved quite easily without matrices, and in fact have been solved without them throughout most of this book, matrix methods are indispensable for the solution of larger systems.

The use of matrices is further encouraged by the large number of matrix programs which are available for most computers. These programs perform numerically all of the more common matrix operations, and make it as easy in principle for the computer-user to solve a large system of equations as a small one.

This chapter gives a brief introduction† to matrix algebra and illustrates some of its applications, for the most part by taking problems discussed in earlier chapters and recasting them in matrix form. The numerical solution of linear equations is discussed in § 107, and the remainder of the chapter is devoted to the eigenvalue problem and to sets of linear differential equations.

106. Matrix algebra

A rectangular array of numbers arranged in m rows and n

† For a fuller discussion see Bickley and Thompson, *Matrices, Their Meaning and Manipulation* (The English Universities Press, 1964).

columns is called a matrix of order m by n. We use a single letter to denote the entire array

$$\mathbf{A} = \begin{bmatrix} a_{11} & a_{12} & . & . & . & a_{1n} \\ a_{21} & a_{22} & . & . & . & a_{2n} \\ . & & & & & \\ . & & & & & \\ . & & & & & \\ a_{m1} & a_{m2} & . & . & . & a_{mn} \end{bmatrix}^{\dagger}, \tag{1}$$

and we refer to $a_{11}, a_{12},...$ as the *elements* of the matrix. It is sometimes convenient to write $\mathbf{A} = \{a_{rs}\}$ to show that $\mathbf{A}$ is the matrix with element a_{rs} in row r and column s.

Sometimes the elements of a matrix are themselves matrices. For instance, if we introduce the order m by 1 matrices, or *column vectors*,

$$\mathbf{A}_r = \begin{bmatrix} a_{1r} \\ a_{2r} \\ . \\ . \\ . \\ a_{mr} \end{bmatrix}, \quad r = 1, 2,..., n,$$

then (1) can be written

$$\mathbf{A} = [\mathbf{A}_1 \quad \mathbf{A}_2 \quad . \quad . \quad . \quad \mathbf{A}_n]. \tag{2}$$

The *transpose* $\mathbf{A}^T$ of $\mathbf{A}$ is the matrix obtained by interchanging the rows and columns of $\mathbf{A}$.

$$\mathbf{A}^T = \begin{bmatrix} a_{11} & a_{21} & . & . & . & a_{m1} \\ a_{12} & a_{22} & . & . & . & a_{m2} \\ . & & & & & \\ . & & & & & \\ . & & & & & \\ a_{1n} & a_{2n} & . & . & . & a_{mn} \end{bmatrix} = \begin{bmatrix} \mathbf{A}_1^T \\ \mathbf{A}_2^T \\ . \\ . \\ . \\ \mathbf{A}_n^T \end{bmatrix},$$

† We will use square brackets to distinguish a matrix from a determinant. The use of Clarendon type for the single letters representing a matrix should not lead to confusion with the use of Clarendon type to denote a vector in Chapter VI, especially in view of the close relationship between vectors and column or row matrices.

or, if $\mathbf{A} = \{a_{rs}\}$, it follows that $\mathbf{A}^T = \{a_{sr}\}$.

The *sum* of two matrices is only defined if both matrices are of the same order. If $\mathbf{A} = \{a_{rs}\}$ and $\mathbf{B} = \{b_{rs}\}$ are both of order m by n, their sum $\mathbf{C}$ is also of order m by n and each element of $\mathbf{C}$ is the sum of the corresponding elements of $\mathbf{A}$ and $\mathbf{B}$. We write

$$\mathbf{C} = \mathbf{A}+\mathbf{B}$$

where $$c_{rs} = a_{rs}+b_{rs}. \tag{3}$$

If $$\mathbf{C} = \mathbf{A}+\mathbf{A}+\mathbf{A},$$

then $$c_{rs} = a_{rs}+a_{rs}+a_{rs} = 3a_{rs}$$

and we write $$\mathbf{C} = 3\mathbf{A},$$

and say that we have multiplied $\mathbf{A}$ by the scalar 3. In general, if k is a scalar,

$$k\mathbf{A} = \{ka_{rs}\}. \tag{4}$$

The product $\mathbf{AB}$ of two matrices $\mathbf{A}$ and $\mathbf{B}$ is only defined if the number of columns in $\mathbf{A}$ is equal to the number of rows in $\mathbf{B}$. If $\mathbf{A}$ is of order m by n and $\mathbf{B}$ is of order n by p, then the product $\mathbf{AB}$ exists and is defined as the matrix $\mathbf{C}$ of order m by p, where

$$\mathbf{C} = \mathbf{AB}$$

if $$c_{rs} = a_{r1}b_{1s}+a_{r2}b_{2s}+\ldots a_{rn}b_{ns} = \sum_{k=1}^{n} a_{rk}b_{ks}. \tag{5}$$

The sum $\sum_{k=1}^{n} a_{rk}b_{ks}$ is called the 'inner product' of row r of $\mathbf{A}$ with column s of $\mathbf{B}$. This definition of the product of two matrices may at first seem somewhat arbitrary, but later examples will show that it is precisely this definition that makes matrices so useful.

Notice that matrix multiplication is associative and distributive, $(\mathbf{AB})\mathbf{C} = \mathbf{A}(\mathbf{BC})$ and $(\mathbf{A}+\mathbf{B})\mathbf{C} = \mathbf{AC}+\mathbf{BC}$, but *not* commutative; in general $\mathbf{AB} \neq \mathbf{BA}$ and often one or other of $\mathbf{AB}$ or $\mathbf{BA}$ does not even exist. To keep this distinction in mind, in the product $\mathbf{AB}$ we say that $\mathbf{B}$ is pre-multiplied by $\mathbf{A}$, or $\mathbf{A}$ is post-multiplied by $\mathbf{B}$.

Ex. 1.

If $$\mathbf{A} = \begin{bmatrix} 4 & 0 & -1 \\ -4 & 0 & 1 \end{bmatrix} \quad \text{and} \quad \mathbf{B} = \begin{bmatrix} 1 & 0 \\ 2 & 3 \\ 4 & 0 \end{bmatrix},$$

then
$$\mathbf{AB} = \begin{bmatrix} 0 & 0 \\ 0 & 0 \end{bmatrix}$$

and
$$\mathbf{BA} = \begin{bmatrix} 4 & 0 & -1 \\ -4 & 0 & 1 \\ 16 & 0 & -4 \end{bmatrix}.$$

Any matrix which has all of its elements equal to zero is called a zero matrix $\mathbf{0}$. In Ex. 1, $\mathbf{AB} = \mathbf{0}$, but this does not necessarily imply that either $\mathbf{A} = \mathbf{0}$ or $\mathbf{B} = \mathbf{0}$.

Ex. 2.

The set of linear equations

$$\begin{aligned} a_{11}x_1+a_{12}x_2+\ldots+a_{1n}x_n &= y_1 \\ a_{21}x_1+a_{22}x_2+\ldots+a_{2n}x_n &= y_2 \\ &\cdot \\ &\cdot \\ &\cdot \\ a_{n1}x_1+a_{n2}x_2\ldots+a_{nn}x_n &= y_n, \end{aligned}$$

can, by the rules of matrix multiplication, be written

$$\begin{bmatrix} a_{11} & a_{12} & \cdot & \cdot & \cdot & a_{1n} \\ a_{21} & a_{22} & \cdot & \cdot & \cdot & a_{2n} \\ \cdot \\ \cdot \\ \cdot \\ a_{n1} & a_{n2} & \cdot & \cdot & \cdot & a_{nn} \end{bmatrix} \begin{bmatrix} x_1 \\ x_2 \\ \cdot \\ \cdot \\ \cdot \\ x_n \end{bmatrix} = \begin{bmatrix} y_1 \\ y_2 \\ \cdot \\ \cdot \\ \cdot \\ y_n \end{bmatrix}$$

or, simply,
$$\mathbf{AX} = \mathbf{Y} \tag{6}$$

where the coefficient matrix $\mathbf{A}$ is a square matrix of order n by n.

Ex. 3. *Matrix representation of a vector.*

The vector $OP = x_1\mathbf{i}+y_1\mathbf{j}+z_1\mathbf{k}$ can be represented by a *3* by *1* matrix $\mathbf{p} = \begin{bmatrix} x_1 \\ y_1 \\ z_1 \end{bmatrix}$ or by its transpose $\mathbf{p}^T = [x_1 \; y_1 \; z_1]$. Similarly, vector $OQ = x_2\mathbf{i}+y_2\mathbf{j}+z_2\mathbf{k}$ can be represented by $\mathbf{q} = \begin{bmatrix} x_2 \\ y_2 \\ z_2 \end{bmatrix}$ or $\mathbf{q}^T = [x_2 y_2 z_2]$.

By the rules for both matrix and vector addition, the sum $OP+OQ$ can then be represented by either $\mathbf{p}+\mathbf{q}$ or $\mathbf{p}^T+\mathbf{q}^T$.

The product $\mathbf{p}^T\mathbf{q} = x_1x_2+y_1y_2+z_1z_2$ is the scalar product of OP and OQ, while

$$\mathbf{p}^T\mathbf{p} = x_1^2+y_1^2+z_1^2 \tag{7}$$

is the square of the magnitude of OP, cf. § 52 (4) and (11).

Ex. 4. *Rotation of axes.*

The vector OP in Fig. 85 can be represented by the matrix $\mathbf{p}_1^T = [x_1\, y_1]$ with respect to the OX and OY axes, or by $\mathbf{p}_2^T = [x_2\, y_2]$ with respect to the OX' and OY' axes. It follows from the figure that

$$x_2 = x_1 \cos\theta + y_1 \sin\theta$$

and

$$y_2 = -x_1 \sin\theta + y_1 \cos\theta,$$

or

$$\begin{bmatrix} x_2 \\ y_2 \end{bmatrix} = \begin{bmatrix} \cos\theta & \sin\theta \\ -\sin\theta & \cos\theta \end{bmatrix} \begin{bmatrix} x_1 \\ y_1 \end{bmatrix}.$$

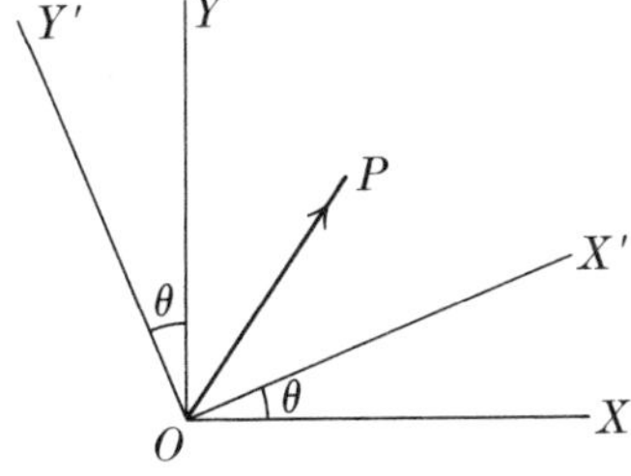

Fig. 85.

This can be written

$$\mathbf{p}_2 = \mathbf{R}^T\mathbf{p}_1, \tag{8}$$

where the matrix

$$\mathbf{R} = \begin{bmatrix} \cos\theta & -\sin\theta \\ \sin\theta & \cos\theta \end{bmatrix}$$

is known as the 'rotation matrix'. Its columns are the direction cosines of the new axes with respect to the old axes.

Successive rotations can be represented by successive matrix multiplications, using the appropriate rotation matrices. For instance, if OP has coordinates x_3 and y_3 with respect to OX'' and OY'' at an angle ϕ to the OX' and OY' axes,

$$\begin{aligned} \begin{bmatrix} x_3 \\ y_3 \end{bmatrix} &= \begin{bmatrix} \cos\phi & \sin\phi \\ -\sin\phi & \cos\phi \end{bmatrix} \begin{bmatrix} x_2 \\ y_2 \end{bmatrix} \\ &= \begin{bmatrix} \cos\phi & \sin\phi \\ -\sin\phi & \cos\phi \end{bmatrix} \begin{bmatrix} \cos\theta & \sin\theta \\ -\sin\theta & \cos\theta \end{bmatrix} \begin{bmatrix} x_1 \\ y_1 \end{bmatrix} \\ &= \begin{bmatrix} \cos(\theta+\phi) & \sin(\theta+\phi) \\ -\sin(\theta+\phi) & \cos(\theta+\phi) \end{bmatrix} \begin{bmatrix} x_1 \\ y_1 \end{bmatrix} \end{aligned}$$

on multiplying out and using the formulae for the sine and cosine of a composite angle.

If A_1, B_1, and H_1 are the moments of inertia and product of inertia (cf. § 65) of a lamina with respect to the OX and OY axes, then by arguments similar to § 65 (5) it can be shown that with respect to the

OX' and OY' axes

$$\begin{bmatrix} A_2 & -H_2 \\ -H_2 & B_2 \end{bmatrix} = \begin{bmatrix} \cos\theta & \sin\theta \\ -\sin\theta & \cos\theta \end{bmatrix}\begin{bmatrix} A_1 & -H_1 \\ -H_1 & B_1 \end{bmatrix}\begin{bmatrix} \cos\theta & -\sin\theta \\ \sin\theta & \cos\theta \end{bmatrix},$$

or

$$\mathbf{M}_2 = \mathbf{R}^T\mathbf{M}_1\mathbf{R}, \tag{9}$$

where the matrix $\mathbf{M} = \begin{bmatrix} A & -H \\ -H & B \end{bmatrix}$ is called the 'Inertia Tensor'.

Ex. 5. *Age structure in a population.*

In a more detailed study of population growth than some of the examples discussed in Chapter I, one would take into account the relative sizes of the various age groups in the population. Suppose that at zero time a population consists of x_0 pre-productive individuals, y_0 individuals in the reproducing age group, and z_0 post-reproductive individuals. If T is the time taken for an infant to reach maturity, at time T the population would consist of x_1, y_1, and z_1 individuals in the various age groups, where

$$\begin{bmatrix} x_1 \\ y_1 \\ z_1 \end{bmatrix} = \begin{bmatrix} 0 & a & 0 \\ b & c & 0 \\ 0 & d & e \end{bmatrix}\begin{bmatrix} x_0 \\ y_0 \\ z_0 \end{bmatrix}.$$

This can be written

$$\mathbf{P}_1 = \mathbf{A}\mathbf{P}_0, \tag{10}$$

where the first row of $\mathbf{A}$, for instance, shows that the number of pre-reproductive individuals at time T depends only on the number of reproducing individuals at zero time. Similarly, at time $2T$,

$$\mathbf{P}_2 = \mathbf{A}\mathbf{P}_1 = \mathbf{A}\mathbf{A}\mathbf{P}_0$$
$$= \mathbf{A}^2\mathbf{P}_0,$$

and, in general, at time nT

$$\mathbf{P}_n = \mathbf{A}^n\mathbf{P}_0.$$

Clearly the notation $\mathbf{A}^2$ or $\mathbf{A}^3$ in Ex. 5 is only meaningful if the matrix $\mathbf{A}$ is square. There are a number of other definitions and properties that only apply to square matrices, and for the remainder of this section all matrices will, unless otherwise indicated, be assumed to be square.

The diagonal that runs from the top left-hand corner to the bottom right-hand corner of a square matrix is called its 'main diagonal'. If a matrix has zero elements everywhere except on its main diagonal, then it is called a *diagonal matrix*. If, in addition, the diagonal elements are all equal to one, the matrix is called a *unit matrix*, written as $\mathbf{I}$. For example, the unit matrix of order *3* by *3* is

$$\mathbf{I} = \begin{bmatrix} 1 & 0 & 0 \\ 0 & 1 & 0 \\ 0 & 0 & 1 \end{bmatrix}.$$

The unit matrix has the property that

$$\mathbf{IA} = \mathbf{AI} = \mathbf{A}. \tag{11}$$

If the transpose of a matrix is equal to the original matrix, $\mathbf{A}^T = \mathbf{A}$, we say that $\mathbf{A}$ is *symmetric*. The inertia tensor in Ex. 4 is an example of a symmetric matrix. If $\mathbf{A}^T = -\mathbf{A}$, we say that $\mathbf{A}$ is *anti-symmetric* or *skew-symmetric*. If

$$\mathbf{A}^T\mathbf{A} = \mathbf{AA}^T = \mathbf{I}, \tag{12}$$

$\mathbf{A}$ is said to be *orthogonal*. The rotation matrix $\mathbf{R}$ of Ex. 4 is orthogonal, for

$$\mathbf{RR}^T = \begin{bmatrix} \cos\theta & -\sin\theta \\ \sin\theta & \cos\theta \end{bmatrix} \begin{bmatrix} \cos\theta & \sin\theta \\ -\sin\theta & \cos\theta \end{bmatrix} = \begin{bmatrix} 1 & 0 \\ 0 & 1 \end{bmatrix},$$

and, similarly, $\mathbf{R}^T\mathbf{R} = \mathbf{I}$.

If $\mathbf{A}$ is orthogonal then the transformation $\mathbf{P} = \mathbf{A}^T\mathbf{QA}$ is called an *orthogonal transformation*. Equation (9) is an example of an orthogonal transformation.

Associated with every square matrix $\mathbf{A}$ is a single number called its *determinant*, written as $\det \mathbf{A}$, or

$$\begin{vmatrix} a_{11} & a_{12} & \cdot & \cdot & \cdot & a_{1n} \\ a_{21} & a_{22} & & & & a_{2n} \\ \cdot & & & & & \\ \cdot & & & & & \\ \cdot & & & & & \\ a_{n1} & a_{n2} & \cdot & \cdot & \cdot & a_{nn} \end{vmatrix}.$$

Determinants have been used throughout this book and it is assumed that the reader is familiar with them and knows how to evaluate them. If $\det \mathbf{A} = 0$, $\mathbf{A}$ is said to be *singular*, while if $\det \mathbf{A} \neq 0$, $\mathbf{A}$ is *non-singular*.

Associated with every non-singular matrix $\mathbf{A}$ is a matrix $\mathbf{A}^{-1}$ which is called the *inverse* of $\mathbf{A}$ and has the property

$$\mathbf{AA}^{-1} = \mathbf{A}^{-1}\mathbf{A} = \mathbf{I}. \tag{13}$$

The solution to the set of linear equations (6)

$$\mathbf{AX} = \mathbf{Y},$$

where **X** and **Y** are column vectors, can be obtained by pre-multiplying both sides by $\mathbf{A}^{-1}$. Then

$$\mathbf{A}^{-1}\mathbf{A}\mathbf{X} = \mathbf{A}^{-1}\mathbf{Y},$$

by (13)
$$\mathbf{I}\mathbf{X} = \mathbf{A}^{-1}\mathbf{Y},$$

and by (11)
$$\mathbf{X} = \mathbf{A}^{-1}\mathbf{Y}.$$

If **A** is singular, the set of equations $\mathbf{AX} = \mathbf{Y}$ is not linearly independent and therefore does not have a unique solution.

Methods for evaluating the inverse of a matrix will be discussed in § 107. This is relatively simple in special cases. For instance, by (12) and (13) the inverse of an orthogonal matrix is its transpose. If **A** is a diagonal matrix with non-zero elements $a_{11}, a_{22},\ldots$ on its main diagonal, then $\mathbf{A}^{-1}$ is the diagonal matrix with elements $1/a_{11}$, $1/a_{22},\ldots$ on its main diagonal.

It can be shown that

$$\det(\mathbf{AB}) = (\det\mathbf{A})(\det\mathbf{B}) \tag{14}$$

and

$$(\mathbf{AB})^{-1} = \mathbf{B}^{-1}\mathbf{A}^{-1}. \tag{15}$$

107. Solution of linear equations

In this section we will discuss two of the more commonly used numerical methods for solving the set of equations

$$\mathbf{AX} = \mathbf{Y}$$

or, written out in full,

$$\begin{bmatrix} a_{11} & a_{12} & \cdot & \cdot & \cdot & a_{1n} \\ a_{21} & a_{22} & \cdot & \cdot & \cdot & a_{2n} \\ \cdot & & & & & \\ \cdot & & & & & \\ \cdot & & & & & \\ a_{n1} & a_{n2} & \cdot & \cdot & \cdot & a_{nn} \end{bmatrix} \begin{bmatrix} x_1 \\ x_2 \\ \cdot \\ \cdot \\ \cdot \\ x_n \end{bmatrix} = \begin{bmatrix} y_1 \\ y_2 \\ \cdot \\ \cdot \\ \cdot \\ y_n \end{bmatrix}. \tag{1}$$

The first method is an iteration scheme known as Gauss–Seidel iteration. It can be used whenever the matrix **A** has a 'dominant diagonal', that is, whenever the modulus of the diagonal element in every row of **A** is greater than or equal to the sum of the moduli of all the other elements in that row,

$$|a_{rr}| \geqslant |a_{r1}|+|a_{r2}|+\ldots\ |a_{rr-1}|+|a_{rr+1}|+\ldots\ |a_{rn}|$$

for $r = 1, 2,\ldots, n$.

A single iteration consists of calculating a better estimate of the solution $\mathbf{X}^{(k+1)}$ from the current estimate $\mathbf{X}^{(k)}$, using the formula

$$x_r^{(k+1)} = \Big(y_r - \sum_{s=1}^{r-1} a_{rs} x_s^{(k+1)} - \sum_{s=r+1}^{n} a_{rs} x_s^{(k)}\Big)\Big/a_{rr} \tag{2}$$

for $r = 1, 2,..., n$.

Starting with an initial guess $\mathbf{X}^{(0)}$, often taken to be the zero column vector, iterates $\mathbf{X}^{(1)}, \mathbf{X}^{(2)},...$ can be calculated by (2) until the difference between two successive iterates is less than the required accuracy of solution. The final iterate is then accepted as the solution to (1).

Ex.

$$\begin{aligned} 3x_1 - x_2 &= 1 \\ -x_1 + 3x_2 - x_3 &= 0 \\ -x_2 + 3x_3 &= 0. \end{aligned}$$

Starting with an initial guess $\mathbf{X}^{(0)} = \mathbf{0}$, the successive iterates, working to two decimal places, are, by (2),

$$\mathbf{X}^{(1)} = \begin{bmatrix} 0{\cdot}33 \\ 0{\cdot}11 \\ 0{\cdot}04 \end{bmatrix} \quad \mathbf{X}^{(2)} = \begin{bmatrix} 0{\cdot}37 \\ 0{\cdot}14 \\ 0{\cdot}05 \end{bmatrix} \quad \mathbf{X}^{(3)} = \begin{bmatrix} 0{\cdot}38 \\ 0{\cdot}14 \\ 0{\cdot}05 \end{bmatrix} = \mathbf{X}^{(4)}.$$

Convergence of the Gauss–Seidel scheme is often much slower than the above example suggests, particularly when the number of equations in (1) is large and the diagonal elements of $\mathbf{A}$ only just satisfy the conditions for dominance. The rate of convergence can often be improved quite dramatically by the use of a 'relaxation factor' ω. At each step of the calculation, $x_r^{(k+1)}$ is replaced by $\omega x_r^{(k+1)} + (1-\omega)x_r^{(k)}$ where permissible values of ω lie between 0 and 2. The case $\omega = 1$ reduces to the ordinary Gauss–Seidel method. The determination of the most suitable value for ω is often a process of trial and error (see Ex. 6 at the end of this chapter), although methods do exist for estimating the optimum value.†

If a dominant diagonal cannot be found for the set of equations (1), then they must be solved by a direct method, such as the Gaussian elimination method discussed below.

† See, for example, Varga, *Matrix Iterative Analysis* (Prentice–Hall, 1962), chapter 9.

We begin by noticing that the solution of (1) is straightforward when the matrix **A** is triangular, that is, when all the elements either above or below the main diagonal are zero. For example, if all the elements below the main diagonal are zero,

$$\begin{bmatrix} a_{11} & a_{12} & . & . & . & . & a_{1n} \\ 0 & a_{22} & . & . & . & . & a_{2n} \\ 0 & 0 & a_{33} & . & . & . & a_{3n} \\ . & . & & & & & \\ . & & & & & & \\ . & & & & & & \\ 0 & 0 & . & . & . & 0 & a_{nn} \end{bmatrix} \begin{bmatrix} x_1 \\ x_2 \\ x_3 \\ . \\ . \\ . \\ x_n \end{bmatrix} = \begin{bmatrix} y_1 \\ y_2 \\ y_3 \\ . \\ . \\ . \\ y_n \end{bmatrix}, \tag{3}$$

then from the last line of (3)

$$x_n = y_n/a_{nn},$$

from the second-to-last line

$$x_{n-1} = (y_{n-1} - a_{n-1\,n}x_n)/a_{n-1\,n-1}$$

and, in general, from line r

$$x_r = \Big(y_r - \sum_{s=r+1}^{n} a_{rs}x_s\Big)/a_{rr}.$$

Calculating $x_n, x_{n-1},\ldots x_2, x_1$ in this way is known as 'back-substitution'.

The Gaussian elimination scheme proceeds in two steps. In the first, the set of equations $\mathbf{AX} = \mathbf{Y}$ is reduced to the upper-triangular form of (3). The second step then consists of solving these equations by back-substitution. The reduction to triangular form is achieved as follows. Starting with equation (1), we subtract a_{21}/a_{11} times the first line from the second line, a_{31}/a_{11} times the first line from the third line, and so on. Writing $a'_{rs} = a_{rs} - a_{r1}a_{1s}/a_{11}$ and $y'_r = y_r - a_{r1}y_1/a_{11}$, we get

$$\begin{bmatrix} a_{11} & a_{12} & . & . & . & . & a_{nn} \\ 0 & a'_{22} & a'_{23} & . & . & . & a'_{2n} \\ 0 & a'_{32} & a'_{33} & . & . & . & a'_{3n} \\ . & & & & & & \\ . & & & & & & \\ . & & & & & & \\ 0 & a'_{n2} & a'_{n3} & . & . & . & a'_{nn} \end{bmatrix} \begin{bmatrix} x_1 \\ x_2 \\ x_3 \\ . \\ . \\ . \\ x_n \end{bmatrix} = \begin{bmatrix} y_1 \\ y'_2 \\ y'_3 \\ . \\ . \\ . \\ y'_n \end{bmatrix}, \tag{4}$$

and have introduced the requisite zeros in the first column of **A**. We then introduce zeros into the second column by a similar process of subtracting a'_{32}/a'_{22} times the second line from the third line, a'_{42}/a'_{22} times the second line from the fourth line, and so on. This process is repeated, column by column, until zeros have been introduced below the diagonal in every column of the transformed matrix **A**. The diagonal elements a_{11}, a'_{22},... of this transformed matrix are called the 'pivots' of the Gaussian elimination process, and it can be shown that the product of the pivots is the determinant of the original matrix **A**.

If $\det \mathbf{A} = 0$, there is no unique solution of (1) and no matter how the equations in (1) are arranged, the elimination process will eventually lead to a zero pivot and will break down on an attempted division by zero. If one or more of the pivots is very small compared with the magnitudes of the elements of **A**, then the subsequent elimination process can be so inaccurate that the solution obtained to the equations is meaningless. The equations are then said to be 'ill-conditioned'. (See Ex. 7 at the end of this chapter.)

There are two fairly simple ways of checking the accuracy of a solution. The first is to substitute the solution directly in (1) to calculate a 'residual vector' $\mathbf{R} = \mathbf{AX} - \mathbf{Y}$. If the elements of **R** are relatively large, then the solution is poor. On the other hand, if the elements of **R** are small, it does not necessarily follow that the solution is accurate. The second method consists of rearranging the order of the equations in (1) (see Ex. 3 at the end of the chapter) so as to get a different set of pivots. The re-ordered set of equations is then solved and the solution is compared with that of the original set.

Ex.
$$\begin{bmatrix} 3 & -1 & 0 \\ -1 & 3 & -1 \\ 0 & -1 & 3 \end{bmatrix} \begin{bmatrix} x_1 \\ x_2 \\ x_3 \end{bmatrix} \begin{bmatrix} 1 \\ 0 \\ 0 \end{bmatrix}.$$

Subtracting $-1/3$ times the first row from the second row,
$$\begin{bmatrix} 3 & -1 & 0 \\ -1 & 3 & -1 \\ 0 & -1 & 3 \end{bmatrix} \begin{bmatrix} x_1 \\ x_2 \\ x_3 \end{bmatrix} = \begin{bmatrix} 1 \\ 0 \\ 0 \end{bmatrix}.$$

and we already have a zero in the third row of the first column. We then subtract $-1/(8/3)$ times the second row from the third row

$$\begin{bmatrix} 3 & -1 & 0 \\ 0 & \frac{8}{3} & -1 \\ 0 & 0 & \frac{21}{8} \end{bmatrix} \begin{bmatrix} x_1 \\ x_2 \\ x_3 \end{bmatrix} = \begin{bmatrix} 1 \\ \frac{1}{3} \\ \frac{1}{8} \end{bmatrix}, \tag{5}$$

which is the upper-triangular form of equation (3). Back-substitution leads to the solution

$$x_3 = 0{\cdot}05, \quad x_2 = 0{\cdot}14 \quad \text{and} \quad x_1 = 0{\cdot}38.$$

The pivots are the diagonal elements of (5) and their product is $(3)(8/3)(21/8) = 21$, which is the determinant of the original coefficient matrix.

If n sets of equations of the form

$$\mathbf{AX}_1 = \begin{bmatrix} 1 \\ 0 \\ \\ 0 \\ \cdot \\ \cdot \\ 0 \end{bmatrix}, \quad \mathbf{AX}_2 = \begin{bmatrix} 0 \\ 1 \\ \\ 0 \\ \cdot \\ \cdot \\ 0 \end{bmatrix}, \quad \cdot \quad \cdot \quad \cdot \quad \mathbf{AX}_n = \begin{bmatrix} 0 \\ 0 \\ \cdot \\ \cdot \\ 0 \\ \\ 1 \end{bmatrix},$$

are solved simultaneously, they can be written in the form (cf. § 106 (2)),

$$\mathbf{A}[\mathbf{X}_1\,\mathbf{X}_2\ldots\mathbf{X}_n] = \mathbf{I}, \tag{6}$$

and the matrix of their solutions, $[\mathbf{X}_1\,\mathbf{X}_2\ldots\mathbf{X}_n]$, is the inverse of $\mathbf{A}$. Compact schemes† have been developed for solving (6), by direct methods, as economically and accurately as possible.

108. Eigenvalues and eigenvectors

These concepts are best introduced by way of an example. Consider the equation

$$\begin{bmatrix} 7 & 4 \\ 3 & 6 \end{bmatrix} \begin{bmatrix} x_1 \\ x_2 \end{bmatrix} = \lambda \begin{bmatrix} x_1 \\ x_2 \end{bmatrix}, \tag{1}$$

or $\mathbf{AX} = \lambda\mathbf{X}$, where λ is a scalar.

Equation (1) can be written

$$\begin{bmatrix} 7-\lambda & 4 \\ 3 & 6-\lambda \end{bmatrix} \begin{bmatrix} x_1 \\ x_2 \end{bmatrix} = \mathbf{0}$$

† See Bickley and Thompson, op. cit., chapter 8.

and the solution is $x_1 = x_2 = 0$, unless the determinant

$$\begin{vmatrix} 7-\lambda & 4 \\ 3 & 6-\lambda \end{vmatrix} = 0.$$

In that case, $\lambda^2-13\lambda+30 = 0$. This is known as the *characteristic equation* of $\mathbf{A}$ and its roots $\lambda_1 = 3$ and $\lambda_2 = 10$ are called the *eigenvalues* of $\mathbf{A}$. Substituting $\lambda = \lambda_1$ in (1) leads to the pair of linearly dependent equations

$$7x_1+4x_2 = 3x_1$$

and

$$3x_1+6x_2 = 3x_2.$$

The solution corresponding to $\lambda = \lambda_1$ is thus

$$\mathbf{X} = \mathbf{X}_1 = \begin{bmatrix} c \\ -c \end{bmatrix}$$

where c can take on any value. $\mathbf{X}_1$ is called the *eigenvector* of $\mathbf{A}$ corresponding to eigenvalue λ_1. It is usual to choose c so that the magnitude of $\mathbf{X}_1$ is unity. $\mathbf{X}_1 = \begin{bmatrix} 1/\sqrt{2} \\ -1/\sqrt{2} \end{bmatrix}$ is called a normalized eigenvector. Substituting $\lambda = \lambda_2$ in (1) leads in a similar way to the normalized eigenvector $\mathbf{X}_2 = \begin{bmatrix} \frac{4}{5} \\ \frac{3}{5} \end{bmatrix}$.

The general theory follows the same argument as the above example. If $\mathbf{A}$ is an n by n matrix and

$$\mathbf{AX} = \lambda\mathbf{X} \tag{2}$$

then the solution is $\mathbf{X} = \mathbf{0}$ unless

$$\det(\mathbf{A}-\lambda\mathbf{I}) = 0. \tag{3}$$

Expanding (3) leads to the characteristic equation

$$f(\lambda) = 0, \tag{4}$$

where $f(\lambda)$ is a polynomial of degree n. Its roots $\lambda_1, \lambda_2, \ldots, \lambda_n$ are the eigenvalues of $\mathbf{A}$. Each root represents a value of λ for which (2) can have a solution other than the trivial solution $\mathbf{X} = \mathbf{0}$. The solution corresponding to eigenvalue λ_r is, by (2),

$$\mathbf{AX}_r = \lambda_r \mathbf{X}_r \tag{5}$$

which is a set of linearly dependent equations. $\mathbf{X}_r$ is called the eigenvector of $\mathbf{A}$ corresponding to eigenvalue λ_r and can be specified in terms of an arbitrary constant. If this is chosen to be x_1, then (5) will give the ratios x_2/x_1, x_3/x_1,..., x_n/x_1. If x_1 is chosen such that

$$\mathbf{X}_r^T\,\mathbf{X}_r = 1, \tag{6}$$

cf. § 106 (7), the eigenvector is said to be normalized.

Let $\mathbf{Q}$ be the n by n matrix whose columns are the normalized eigenvectors of $\mathbf{A}$. Then

$$\begin{aligned} \mathbf{AQ} &= \mathbf{A}[\mathbf{X}_1 \quad \mathbf{X}_2 \quad . \quad . \quad . \quad \mathbf{X}_n] \\ &= [\mathbf{AX}_1 \quad \mathbf{AX}_2 \quad . \quad . \quad . \quad \mathbf{AX}_n] \\ \text{by (5)} \qquad &= [\lambda_1\mathbf{X}_1 \quad \lambda_2\mathbf{X}_2 \quad . \quad . \quad . \quad \lambda_n\mathbf{X}_n] \\ &= \mathbf{Q\Lambda}, \end{aligned} \tag{7}$$

where $\mathbf{\Lambda}$ is the diagonal matrix

$$\begin{bmatrix} \lambda_1 & 0 & . & . & . & . & 0 \\ 0 & \lambda_2 & 0 & . & . & . & 0 \\ . & & & & & & \\ . & & & & & & \\ . & & & & & & \\ 0 & . & . & . & . & 0 & \lambda_n \end{bmatrix}.$$

Pre-multiplying (7) by the inverse of $\mathbf{Q}$,

$$\begin{aligned} \mathbf{Q}^{-1}\mathbf{AQ} = \mathbf{Q}^{-1}\mathbf{Q\Lambda} &= \mathbf{I\Lambda} \\ &= \mathbf{\Lambda}. \end{aligned} \tag{8}$$

Equation (8) is extremely important. It shows that the eigenvectors of a matrix can be used to transform that matrix into a diagonal matrix. The non-zero elements of this diagonal matrix are just the eigenvalues of the original matrix. Since diagonal matrices are far easier to manipulate than ordinary matrices, this result leads to a great simplification in problems involving moments and products of inertia, normal modes of vibration, and the solution of sets of linear differential equations.

Two special cases are important. The first is when (4) has repeated roots. It is then not always possible to find n distinct eigenvectors, in which case $\mathbf{Q}^{-1}$ does not exist and $\mathbf{A}$ cannot be transformed into a diagonal matrix. (See Ex. 12 at the end of the chapter.) The second important case is when $\mathbf{A}$ is a symmetric matrix and all of its elements are real. It can then be shown that the eigenvalues of $\mathbf{A}$ are also real and that the matrix $\mathbf{Q}$ is orthogonal, that is

$$\mathbf{Q}^T\mathbf{Q} = \mathbf{I}$$

or

$$\left.\begin{aligned} \mathbf{X}_r^T\mathbf{X}_s &= 0 \quad \text{if } r \neq s \\ &= 1 \quad \text{if } r = s \end{aligned}\right\}. \tag{9}$$

Equation (8) then reduces to the orthogonal transformation

$$\mathbf{Q}^T\mathbf{A}\mathbf{Q} = \mathbf{\Lambda}. \tag{10}$$

Ex.

In § 106 (9) we saw that the inertia tensor $\mathbf{M}_1$ of a lamina with respect to one set of axes is related to the inertia tensor $\mathbf{M}_2$ with respect to a rotated set of axes by the orthogonal transformation

$$\mathbf{M}_2 = \mathbf{R}^T\mathbf{M}_1\mathbf{R}.$$

If the columns of $\mathbf{R}$ are the eigenvalues of $\mathbf{M}_1$, it follows from (10) that $\mathbf{M}_2$ will be a diagonal matrix, that is, a set of axes will have been found for which the product of inertia is zero. The corresponding moments of inertia, the diagonal elements of $\mathbf{M}_2$, are called the Principal Moments of Inertia of the lamina, cf. § 65.

In general, if the inertia tensor of a body with respect to a given set of axes is $\mathbf{M}$, the eigenvalues of $\mathbf{M}$ are the Principal Moments of Inertia and the eigenvectors of $\mathbf{M}$ define the direction cosines of the Principal Axis system with respect to the given set of axes. Notice that by Ex. 8, the eigenvalues of $\mathbf{M}$ are unchanged by an orthogonal transformation and hence will be independent of the orientation of the given set of axes.

Ex.

By § 106 (10), the population structure $\mathbf{P}_{n+1}$ at time $(n+1)T$ is related to the structure $\mathbf{P}_n$ at time nT by the equation $\mathbf{P}_{n+1} = \mathbf{A}\mathbf{P}_n$. We ask whether it is possible for the population to grow without changing the *relative proportions* of the various age groups in $\mathbf{P}$. In that case,

$$\mathbf{P}_{n+1} = \lambda\mathbf{P}_n$$

or

$$\mathbf{A}\mathbf{P}_n = \lambda\mathbf{P}_n.$$

It follows that the permissible values of the growth factor λ are obtained from $\det(\mathbf{A}-\lambda\mathbf{I}) = 0$, or

$$\begin{vmatrix} -\lambda & a & 0 \\ b & c-\lambda & 0 \\ 0 & d & e-\lambda \end{vmatrix} = 0.$$

The roots of this are

$$\lambda_1 = e, \quad \lambda_2 = \tfrac{1}{2}\{c+\sqrt{(c^2+4ab)}\}, \quad \text{and} \quad \lambda_3 = \tfrac{1}{2}\{c-\sqrt{(c^2+4ab)}\}.$$

Only the positive roots λ_1 and λ_2 are meaningful. The eigenvector corresponding to λ_1 or λ_2 gives the constant proportions of the various age groups in **P** for that particular growth factor.

If the eigenvalues of an n by n symmetric matrix **A** are distinct, it is possible to find scalars $c_1, c_2,..., c_n$ such that any vector **V** in an n-dimensional space can be expressed in terms of the eigenvectors of **A**. For, if

$$\mathbf{V} = \sum_{r=1}^{n} c_r \mathbf{X}_r, \tag{11}$$

then
$$\mathbf{X}_s^T \mathbf{V} = \sum_{r=1}^{n} \mathbf{X}_s^T \mathbf{X}_r c_r = c_s$$
by (9).

Pre-multiplying (11) by **A**,

$$\mathbf{AV} = \sum_{r=1}^{n} c_r \mathbf{AX}_r$$

by (5)
$$= \sum_{r=1}^{n} c_r \lambda_r \mathbf{X}_r,$$

and
$$\mathbf{A}^2\mathbf{V} = \sum_{r=1}^{n} \lambda_r \mathbf{AX}_r = \sum_{r=1}^{n} c_r \lambda_r^2 \mathbf{X}_r.$$

In general,
$$\mathbf{A}^m\mathbf{V} = \sum_{r=1}^{n} c_r \lambda_r^m \mathbf{X}_r. \tag{12}$$

Now, if the characteristic equation of **A** is $f(\lambda) = 0$, we can write down a *matrix equation* $f(\mathbf{A})$ such that, by (12),

$$f(\mathbf{A})\mathbf{V} = \sum_{r=1}^{n} c_r f(\lambda_r)\mathbf{X}_r = 0,$$

and since this holds for *any* vector **V**, it follows that

$$f(\mathbf{A}) = 0. \tag{13}$$

A matrix **A** thus satisfies its own characteristic equation. This is known as the *Cayley–Hamilton theorem*, and is true of any matrix **A** although we have only proved it for a symmetric matrix with distinct eigenvalues.

Ex.

If $\mathbf{A} = \begin{bmatrix} 7 & 4 \\ 3 & 6 \end{bmatrix}$, the characteristic equation is

$$f(\lambda) = \lambda^2 - 13\lambda + 30.$$

The corresponding matrix equation is

$$\begin{aligned} f(\mathbf{A}) &= \mathbf{A}^2 - 13\mathbf{A} + 30\mathbf{I} \\ &= \begin{bmatrix} 61 & 52 \\ 39 & 48 \end{bmatrix} - \begin{bmatrix} 91 & 52 \\ 39 & 78 \end{bmatrix} + \begin{bmatrix} 30 & 0 \\ 0 & 30 \end{bmatrix} \\ &= \begin{bmatrix} 0 & 0 \\ 0 & 0 \end{bmatrix}. \end{aligned}$$

If the inverse of **A** exists, it can be found by the Cayley–Hamilton theorem. For instance in the above example

$$\mathbf{A}^2 - 13\mathbf{A} + 30\mathbf{I} = \mathbf{0}. \tag{14}$$

Pre-multiplying by $\mathbf{A}^{-1}$,

$$\mathbf{A} - 13\mathbf{I} + 30\mathbf{A}^{-1} = \mathbf{0},$$

or

$$\mathbf{A}^{-1} = \tfrac{13}{30}\mathbf{I} - \tfrac{1}{30}\mathbf{A} = \tfrac{1}{30}\begin{bmatrix} 6 & -4 \\ -3 & 7 \end{bmatrix}.$$

The Cayley–Hamilton theorem can also be used to compute higher powers of **A**.

By (14)

$$\mathbf{A}^2 = 13\mathbf{A} - 30\mathbf{I}$$

so

$$\begin{aligned} \mathbf{A}^3 &= 13\mathbf{A}^2 - 30\mathbf{A} \\ &= 139\mathbf{A} - 390\mathbf{I}, \end{aligned}$$

and so on.

The numerical evaluation of eigenvalues and eigenvectors is an important, but except for special cases†, complex problem. A simple iterative method can be developed to find the largest (in magnitude) eigenvalue of **A** if this eigenvalue, say λ_1, is considerably larger in magnitude than the rest of the eigenvalues of **A**. Starting with an arbitrary vector **V** and pre-multiplying successively by **A**, one forms the products $\mathbf{AV}$, $\mathbf{A}^2\mathbf{V}$,... $\mathbf{A}^m\mathbf{V}$,... .

By (12)

$$\begin{aligned} \mathbf{A}^m\mathbf{V} &= \sum_{r=1}^{n} c_r \lambda_r^m \mathbf{X}_r \\ &= c_1 \lambda_1^m \mathbf{X}_1 \end{aligned} \tag{15}$$

for large m if $|\lambda_1|$ is large as compared with the other eigenvalues. For large m, then, if the vectors $\mathbf{A}^m\mathbf{V} = \mathbf{y}$ and $\mathbf{A}^{m+1}\mathbf{V} = \mathbf{z}$, we should find, by (15), that

$$\frac{z_1}{y_1} = \frac{z_2}{y_2} = \dots = \frac{z_n}{y_n} = \lambda_1.$$

† See Bickley and Thompson, op. cit., chapters 9 and 10. The authoritative work on the numerical determination of eigenvalues is Wilkinson, *The Algebraic Eigenvalue Problem* (Oxford).

The convergence of this method is therefore established by the equality of the ratios of the components of successive iterations. If $|\lambda_1|$ is only slightly larger than the next largest eigenvalue, convergence can be very slow indeed.

109. Systems of linear differential equations

If the elements b_{rs} of a matrix $\mathbf{B}$ are functions of the independent variable t, we define $\frac{d}{dt}\mathbf{B}$ as the matrix with elements $\frac{db_{rs}}{dt}$, and $\int \mathbf{B}\, dt$ as the matrix with elements $\int b_{rs}\, dt$. It follows that the set of n first-order differential equations

$$\frac{dx_1}{dt} = a_{11}\,x_1 + a_{12}\,x_2 + \ldots + a_{1n}\,x_n$$

$$\frac{dx_2}{dt} = a_{12}\,x_1 + a_{22}\,x_2 + \ldots + a_{2n}\,x_n$$

$$\vdots$$

$$\frac{dx_n}{dt} = a_{n1}\,x_1 + a_{n2}\,x_2 \ldots + a_{nn}\,x_n$$

can be written in matrix form as

$$\frac{d}{dt}\mathbf{X} = \mathbf{AX}. \tag{1}$$

This is the generalization of the pair of simultaneous first-order equations discussed in § 15.

If $\mathbf{A}$ is of order n by n, we define the matrix function $\exp(\mathbf{A})$ as the n by n matrix

$$\exp(\mathbf{A}) = \mathbf{I} + \mathbf{A} + \frac{1}{2!}\,\mathbf{A}^2 + \frac{1}{3!}\,\mathbf{A}^3 + \ldots. \tag{2}$$

It can be shown that this converges for any matrix $\mathbf{A}$. From (2), $\exp(\mathbf{A}t)$ is the n by n matrix

$$\exp(\mathbf{A}t) = \mathbf{I} + \mathbf{A}t + \frac{1}{2!}\,\mathbf{A}^2 t^2 + \frac{1}{3!}\,\mathbf{A}^3 t^3 + \ldots, \tag{3}$$

and $$\frac{d}{dt}[\exp(\mathbf{A}t)] = \mathbf{A}+\mathbf{A}^2t+\frac{1}{2!}A^3t^2+\ldots$$
$$= \mathbf{A}\exp(\mathbf{A}t)$$

if the elements of **A** are independent of t. In that case, the general solution of (1) can be written

$$\mathbf{X} = [\exp(\mathbf{A}t)]\mathbf{C} \tag{4}$$

where **C** is a column vector of n arbitrary constants $c_1, c_2,..., c_n$. If the initial condition is $\mathbf{X} = \mathbf{X}_0$ at $t = 0$, substituting in (4),

$$\mathbf{X}_0 = [\exp(\mathbf{0})]\mathbf{C} = \mathbf{IC} = \mathbf{C};$$

so, finally, $$\mathbf{X} = [\exp(\mathbf{A}t)]\mathbf{X}_0. \tag{5}$$

Equation (5) is an elegant way of writing down the solution to (1). To evaluate $\exp(\mathbf{A}t)$ in practice is, however, not an easy task. Apart from numerical methods, one approach is to use the series (3) together with the Cayley–Hamilton theorem to evaluate the higher powers of **A**.

If **A** is a diagonal matrix then $\exp(\mathbf{A}t)$ is also a diagonal matrix of the form

$$\begin{bmatrix} e^{a_{11}t} & 0 & . & . & . & . & 0 \\ 0 & e^{a_{22}t} & 0 & . & . & . & 0 \\ . & & & & & & \\ . & & & & & & \\ . & & & & & & \\ 0 & . & 0 & & & & e^{a_{nn}t} \end{bmatrix}.$$

This suggests an alternative method of solving (1).

Let **Q** be the matrix whose columns are the normalized eigenvectors of **A**, and make the substitution

$$\mathbf{X} = \mathbf{QY} \tag{6}$$

in (1). Then $\frac{d}{dt}(\mathbf{QY}) = \mathbf{AQY}$

or $$\mathbf{Q}\left(\frac{d}{dt}\mathbf{Y}\right) = \mathbf{AQY},$$

so $$\frac{d}{dt}(\mathbf{Y}) = \mathbf{Q}^{-1}\mathbf{AQY}. \tag{7}$$

But, by § 108 (8), $\mathbf{Q}^{-1}\mathbf{AQ} = \mathbf{\Lambda}$, where $\mathbf{\Lambda}$ is the diagonal matrix of the eigenvalues of **A**. Equation (7) therefore corresponds to the set of equations

$$\frac{dy_1}{dt} = \lambda_1 y_1$$

$$\frac{dy_2}{dt} = \lambda_2 y_2$$

$$\vdots$$

$$\frac{dy_n}{dt} = \lambda_n y_n,$$

and we say that we have 'decoupled' the original set (1). The solution of the decoupled equations is

$$\mathbf{Y} = \begin{bmatrix} y_1 \\ y_2 \\ \cdot \\ \cdot \\ \cdot \\ y_n \end{bmatrix} = \begin{bmatrix} c_1 e^{\lambda_1 t} \\ c_2 e^{\lambda_2 t} \\ \cdot \\ \cdot \\ \cdot \\ c_n e^{\lambda_n t} \end{bmatrix},$$

and so, by (6),

$$\mathbf{X} = \mathbf{QY} = \begin{bmatrix} q_{11} c_1 e^{\lambda_1 t} + q_{12} c_2 e^{\lambda_2 t} + \ldots + q_{1n} c_n e^{\lambda_n t} \\ q_{21} c_1 e^{\lambda_1 t} + q_{22} c_2 e^{\lambda_2 t} + \ldots + q_{2n} c_n e^{\lambda_n t} \\ \cdot \\ \cdot \\ \cdot \\ q_{n1} c_1 e^{\lambda_1 t} + q_{n2} c_2 e^{\lambda_2 t} + \ldots + q_{nn} c_n e^{\lambda_n t} \end{bmatrix}. \qquad (8)$$

Notice that for large t, the behaviour of **X** will be determined by the largest eigenvalue of **A**. If the eigenvalues of **A** are not all distinct the above argument must be modified. For instance if $\lambda_1 = \lambda_2$, the terms in $c_1 e^{\lambda_1 t}$ and $c_2 e^{\lambda_2 t}$ must be replaced by the expression

$$(c_1 + c_2 t) e^{\lambda_1 t}.$$

Ex. *The pair of simultaneous equations discussed in § 15.*

Here
$$\frac{d}{dt}\begin{bmatrix} x_1 \\ x_2 \end{bmatrix} = \begin{bmatrix} a_{11} & a_{12} \\ a_{21} & a_{22} \end{bmatrix}\begin{bmatrix} x_1 \\ x_2 \end{bmatrix}.$$

The characteristic equation of **A** is

$$\lambda^2 - (a_{11} + a_{22})\lambda + (a_{11} a_{22} - a_{12} a_{21}) = 0,$$

(cf. § 15 (6)), and if the roots of this are λ_1 and λ_2, the solution is, by (8)

$$x_1 = q_{11}c_1 e^{\lambda_1 t}+q_{12}c_2 e^{\lambda_2 t}$$
$$x_2 = q_{21}c_1 e^{\lambda_1 t}+q_{22}c_2 e^{\lambda_2 t}$$

where $\begin{bmatrix} q_{11} \\ q_{21} \end{bmatrix}$ and $\begin{bmatrix} q_{12} \\ q_{22} \end{bmatrix}$ are the normalized eigenvectors of $\mathbf{A}$.

Higher-order sets of equations can, as was seen in § 15, be reduced, by the introduction of dummy variables, to larger sets of first-order equations. The preceding analysis is therefore applicable to sets of linear equations of order greater than one. In the case of the theory of vibrations or small oscillations, however, it is simpler to solve a set of second-order differential equations directly, as in the following example.

Ex.

We consider the problem, discussed in § 33, of two equal masses connected by identical springs. Then § 33, (4) and (5) take the form

$$(D^2+2n_1^2)x_1-n_1^2 x_2 = 0$$
$$-n_1^2 x_1+(D^2+2n_1^2)x_2 = 0.$$

In matrix notation we have

$$\frac{d^2}{dt^2}\begin{bmatrix} x_1 \\ x_2 \end{bmatrix} = \begin{bmatrix} -2n_1^2 & n_1^2 \\ n_1^2 & -2n_1^2 \end{bmatrix}\begin{bmatrix} x_1 \\ x_2 \end{bmatrix}. \tag{9}$$

The eigenvalues of the coefficient matrix are the roots of

$$\lambda^2+4n_1^2\lambda+3n_1^4 = 0,$$

so $\lambda_1 = -n_1^2 \quad \text{and} \quad \lambda_2 = -3n_1^2.$

The corresponding normalized eigenvectors are $\begin{bmatrix} 1/\sqrt{2} \\ 1/\sqrt{2} \end{bmatrix}$ and $\begin{bmatrix} 1/\sqrt{2} \\ -1/\sqrt{2} \end{bmatrix}$. We therefore make the transformation

$$\begin{bmatrix} x_1 \\ x_2 \end{bmatrix} = \frac{1}{\sqrt{2}}\begin{bmatrix} 1 & 1 \\ 1 & -1 \end{bmatrix}\begin{bmatrix} y_1 \\ y_2 \end{bmatrix}, \tag{10}$$

which transforms (9) to the pair of decoupled equations

$$D^2 y_1 = -n_1^2 y_1,$$

and $$D^2 y_2 = -3n_1^2 y_2,$$

with solution $$y_1 = c_1 \sin(n_1 t+\beta_1)$$

and $$y_2 = c_2 \sin(n_1 t\sqrt{3}+\beta_2).$$

By (10) $$x_1 = \frac{c_1}{\sqrt{2}}\sin(n_1 t+\beta_1)+\frac{c_2}{\sqrt{2}}\sin(n_1 t\sqrt{3}+\beta_2)$$

and $$x_2 = \frac{c_1}{\sqrt{2}}\sin(n_1 t+\beta_1)-\frac{c_2}{\sqrt{2}}\sin(n_1 t\sqrt{3}+\beta_2),$$

which is identical to the solution § 33 (17) if we put $c_1/\sqrt{2} = A_1$ and $c_2/\sqrt{2} = A_2$.

The general many-body undamped oscillation problem leads to the matrix equation

$$\frac{d^2}{dt^2}\mathbf{X} = \mathbf{AX}.$$

From the above example it is apparent that the eigenvalues of **A** lead directly to the natural frequencies of the system, and the eigenvectors of **A** to the normal modes of oscillation.

110. Stability and control

We first consider a system of n dynamic variables $x_1, x_2,..., x_n$ which satisfy the n first-order linear differential equations

$$\frac{d}{dt}\mathbf{X} = \mathbf{AX}. \tag{1}$$

The equilibrium configuration is $\mathbf{X} = \mathbf{0}$, which is the origin in the n-dimensional phase space (cf. § 10) for the system. If the system is in equilibrium, but is subjected to a small disturbance, one of three things will occur:

(i) The values of **X** will in time return to the equilibrium configuration $\mathbf{X} = \mathbf{0}$. The phase trajectory will be a path starting at the disturbed configuration and terminating at the origin. In this case, we say that the equilibrium is *stable*.

(ii) The values of **X** will oscillate and continue to oscillate indefinitely. The phase trajectory will form a closed loop or *limit cycle*.

(iii) One or more of the variables x_r will grow with time. The phase trajectory then moves away from equilibrium at the origin, and the equilibrium is said to be *unstable*.

In § 109 (8) we saw that the general solution to (1) can be expressed completely as a sum of terms of the form $e^{\lambda_r t}$ where λ_r is an eigenvalue of **A**. It follows that the behaviour of the system depends entirely upon the eigenvalues of **A**. If the *real parts* of all the eigenvalues are negative, the solution must tend to equilibrium at $\mathbf{X} = \mathbf{0}$ no matter what the initial condition may be, and so equilibrium is stable. If the real parts of some of the eigenvalues are zero, the solution will be oscillatory and undamped, while if one or more of the eigenvalues have positive real parts, equilibrium will be unstable.

Ex. *The inverted pendulum.*

The pendulum in Fig. 86 consists of a mass m at the end of a rigid weightless rod of length l. It is free to rotate about its base point O. In an effort to stabilize the pendulum, the base point O is given a horizontal acceleration $u(t)$. This is the counterpart of balancing a pencil on the tip of one's finger, or stabilizing a rocket standing on a launching-pad.

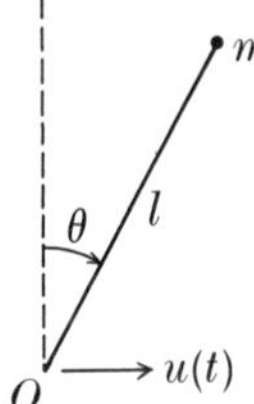

FIG. 86.

For small angles θ, the equation of motion of the pendulum is

$$lD^2\theta + u(t) = g\theta.$$

If we put $x_1 = \theta$ and $x_2 = D\theta$, the corresponding system of first-order equations is

$$\frac{d}{dt}\begin{bmatrix} x_1 \\ x_2 \end{bmatrix} = \begin{bmatrix} 0 & 1 \\ g/l & 0 \end{bmatrix}\begin{bmatrix} x_1 \\ x_2 \end{bmatrix} - \begin{bmatrix} 0 \\ 1/l \end{bmatrix} u(t). \tag{2}$$

If $u(t)$ is independent of x_1 and x_2, the homogeneous part of (2) is

$$\frac{d}{dt}\begin{bmatrix} x_1 \\ x_2 \end{bmatrix} = \begin{bmatrix} 0 & 1 \\ g/l & 0 \end{bmatrix}\begin{bmatrix} x_1 \\ x_2 \end{bmatrix},$$

and the characteristic equation is

$$\begin{vmatrix} -\lambda & 1 \\ g/l & -\lambda \end{vmatrix} = 0$$

or $\lambda^2 - g/l = 0$, so that $\lambda_1 = \sqrt{g/l}$ and $\lambda_2 = -\sqrt{g/l}$. Since one of the eigenvalues is positive the system is unstable. The only way to control the inverted pendulum must be to make the acceleration $u(t)$ depend on x_1 and/or x_2.

Suppose $u(t) = k_1 x_1 + k_2 x_2$. Then (2) becomes

$$\begin{bmatrix} \dot{x}_1 \\ \dot{x}_2 \end{bmatrix} = \begin{bmatrix} 0 & 1 \\ (g-k_1)/l & -k_2/l \end{bmatrix} \begin{bmatrix} x_1 \\ x_2 \end{bmatrix}$$

which has the characteristic equation

$$\lambda^2 + \lambda(k_2/l) - (g-k_1)/l = 0. \tag{3}$$

The roots of (3) will have negative real parts provided $k_2 > 0$ and $k_1 > g$. These are, therefore, the requirements for stability, and it is interesting to note that the controlling acceleration $u(t)$ must depend on both θ and $d\theta/dt$.

We consider next the set of non-linear equations

$$\frac{dx_1}{dt} = f_1(x_1, x_2, \ldots, x_n)$$

$$\frac{dx_2}{dt} = f_2(x_1, x_2, \ldots, x_n)$$

$$\vdots$$

$$\frac{dx_n}{dt} = f_n(x_1, x_2, \ldots, x_n), \tag{4}$$

with an equilibrium configuration $\mathbf{X} = \mathbf{X}_E$. It can be shown† that the stability of this system depends upon the eigenvalues of the matrix

$$\mathbf{M} = \begin{bmatrix} \frac{\partial f_1}{\partial x_1} & \frac{\partial f_1}{\partial x_2} & \cdots & \frac{\partial f_1}{\partial x_n} \\ \frac{\partial f_2}{\partial x_1} & \frac{\partial f_2}{\partial x_2} & \cdots & \frac{\partial f_2}{\partial x_n} \\ \vdots & & & \\ \frac{\partial f_n}{\partial x_1} & \frac{\partial f_n}{\partial x_2} & \cdots & \frac{\partial f_n}{\partial x_n} \end{bmatrix}, \tag{5}$$

† See J. L. Willems, *Stability Theory of Dynamical Systems* (Nelson, 1970), chapter 5.

where all the partial derivatives are evaluated at $\mathbf{X} = \mathbf{X}_E$. If the eigenvalues of $\mathbf{M}$ all have negative real parts, then equilibrium at $\mathbf{X}_E$ is stable, but if one or more of the eigenvalues have positive real parts, $\mathbf{X}_E$ is a position of unstable equilibrium. It is not possible to draw any definite conclusions about undamped oscillations and limit cycles from $\mathbf{M}$ alone.

Non-linear systems such as (4) are prevalent in many biological models, where the question of stability is often of the utmost importance.

EXAMPLES ON CHAPTER XIII

1. From the definition of the product, § 106 (5), prove that
$$(\mathbf{AB})^T = \mathbf{B}^T\mathbf{A}^T.$$

2. Given a right-handed system of axes OX, OY, and OZ, show that a 90° rotation about OZ, followed by a 90° rotation about OY, leads to a new system of axes OX', OY', and OZ' such that $x' = z$, $y' = -x$, and $z' = -y$. Confirm that this is not equivalent to a 90° rotation about OY followed by a 90° rotation about OZ.

3. If $\mathbf{A}$ is a *3* by *3* matrix and $\mathbf{P} = \begin{bmatrix} 1 & 0 & 0 \\ 0 & 0 & 1 \\ 0 & 1 & 0 \end{bmatrix}$, show that $\mathbf{PA}$ interchanges the second and third rows of $\mathbf{A}$, and $\mathbf{AP}$ interchanges the second and third columns of $\mathbf{A}$. Is $\mathbf{P}$ orthogonal?

4. If town r is connected to town s by a direct telephone line, element a_{rs} of a 'network matrix' $\mathbf{A}$ is set equal to one, while if there is no direct line between town r and town s, $a_{rs} = 0$. Set up the network matrix for 5 towns, given that there are direct lines between towns *1* and *3*, *1* and *4*, *1* and *5*, *2* and *3*, *2* and *4*, *2* and *5*, and *4* and *5*. Notice that $\mathbf{A}$ is symmetric. If $\mathbf{B} = \mathbf{A}^2$ confirm that b_{rr} equals the number of lines converging on town r, while b_{rs} $(r \neq s)$ equals the number of ways of routing a call from town r to town s via one other town.

5. Consider the section of a 'ladder' network in Fig. 39 (*c*). If E_0' and I_0' are the complex potential and current on the left-hand side of the section, and E_1' and I_1' are the complex potential and current on the right-hand side, show that
$$\begin{bmatrix} E_1' \\ I_1' \end{bmatrix} = \begin{bmatrix} a_{11} & a_{12} \\ a_{21} & a_{22} \end{bmatrix} \begin{bmatrix} E_0' \\ I_0' \end{bmatrix},$$
where $a_{11} = a_{22} = 1 - LC\omega^2/2$, $a_{12} = iL\omega(LC\omega^4/4 - 1)$, and $a_{21} = -iC\omega$. $\mathbf{A}$ is called the 'transmission matrix' of the network. Confirm that $\det \mathbf{A} = 1$. If a second stage is added to the network, find E_2' and I_2' in terms of E_0' and I_0'.

6. Use Gauss–Seidel iteration to solve the set of equations $\mathbf{AX} = \mathbf{Y}$, correct to four decimal places, given

$$\mathbf{A} = \begin{bmatrix} 2 & -1 & 0 & 0 \\ -1 & 2 & -1 & 0 \\ 0 & -1 & 2 & -1 \\ 0 & 0 & -1 & 2 \end{bmatrix} \quad \text{and} \quad \mathbf{Y} = \begin{bmatrix} 1 \\ 0 \\ 0 \\ 0 \end{bmatrix}.$$

Repeat the calculation using various relaxation factors ω, and by comparing the number of iterations required for convergence, show that $\omega = 1{\cdot}3$ is close to the optimum relaxation factor. Using $\omega = 1{\cdot}3$, re-solve the equations with

$$\mathbf{Y} = \begin{bmatrix} 0 \\ 1 \\ 0 \\ 0 \end{bmatrix}, \quad \begin{bmatrix} 0 \\ 0 \\ 1 \\ 0 \end{bmatrix}, \quad \text{and} \quad \begin{bmatrix} 0 \\ 0 \\ 0 \\ 1 \end{bmatrix}.$$

Hence construct $\mathbf{A}^{-1}$.

7. Write a computer program to solve the set of equations

$$\begin{aligned} 97400\,x_1+79000\,x_2+31100\,x_3 &= 207500 \\ -63100\,x_1+47000\,x_2+25100\,x_3 &= 9000 \\ 45500\,x_1+97500\,x_2+52500\,x_3 &= 195500 \end{aligned}$$

by Gaussian Elimination, using single precision. Confirm that the correct solution is $x_1 = x_2 = x_3 = 1$ and compare this with the computer results. Repeat the exercise with the following equation instead of the last equation of the set

$$45500\,x_1+97500\,x_2+42455{\cdot}01\,x_1 = 185455{\cdot}01.$$

Confirm that the correct solution is still $x_1 = x_2 = x_3 = 1$, and comment on the computed results. (Print out the pivots in both examples.)

8. If $\mathbf{R}$ is an orthogonal matrix, show that its determinant is either $+1$ or -1. If $\mathbf{B} = \mathbf{R}^T\mathbf{A}\mathbf{R}$ show that $\mathbf{A}$ and $\mathbf{B}$ have the same eigenvalues. Use § 106 (12), (14), and § 108 (3).

9. The *trace* of a square matrix is the sum of its diagonal elements, trace $\mathbf{A} = \sum_{r=1}^{n} a_{rr}$. Show that the determinant of a symmetric matrix is the product of its eigenvalues and the trace of a symmetric matrix is the sum of its eigenvalues. Use § 106 (5), (12), (14) and § 108 (10).

10. If $\mathbf{X}$ is a column vector of order n by *1*, and $\mathbf{A}$ is a symmetric matrix of order n by n, the scalar $z = \mathbf{X}^T\mathbf{A}\mathbf{X}$ is called a 'quadratic form'. If $z > 0$ for all $\mathbf{X} \neq \mathbf{0}$, $\mathbf{X}^T\mathbf{A}\mathbf{X}$ is said to be 'positive definite'.

(i) If $\mathbf{Q}$ is the matrix whose columns are the normalized eigenvectors of $\mathbf{A}$, show that the substitution $\mathbf{Y} = \mathbf{QX}$ reduces z to the form

$$z = \lambda_1 y_1^2+\lambda_2 y_2^2+...+\lambda_n y_n^2.$$

(ii) If $\mathbf{X}^T\mathbf{A}\mathbf{X}$ is positive definite, show that the eigenvalues of $\mathbf{A}$ are all positive.

11. Show that the eigenvalues of $\mathbf{A} = \begin{bmatrix} 0 & 1 & 1 \\ 1 & 0 & -1 \\ 1 & -1 & 0 \end{bmatrix}$ are $\lambda_1 = 1$, $\lambda_2 = 1$ and $\lambda_3 = -2$. Show that, corresponding to λ_2 and λ_3, $\mathbf{A}$ has eigenvectors

$$\mathbf{X}_2 = \frac{1}{\sqrt{2}}\begin{bmatrix} 1 \\ 1 \\ 0 \end{bmatrix} \quad \text{and} \quad \mathbf{X}_3 = \frac{1}{\sqrt{3}}\begin{bmatrix} -1 \\ 1 \\ 1 \end{bmatrix}.$$

Hence find an eigenvector $\mathbf{X}_1$ such that $\mathbf{Q} = [\mathbf{X}_1\,\mathbf{X}_2\,\mathbf{X}_3]$ is orthogonal, and show that

$$\mathbf{Q}^T\mathbf{A}\mathbf{Q} = \begin{bmatrix} 1 & 0 & 0 \\ 0 & 1 & 0 \\ 0 & 0 & -2 \end{bmatrix}.$$

12. If $\mathbf{A} = \begin{bmatrix} a & 1 \\ 0 & b \end{bmatrix}$, show that a transformation $\mathbf{Q}^{-1}\mathbf{A}\mathbf{Q}$ which reduces $\mathbf{A}$ to diagonal form can only be found if $a \neq b$.

13. The Maxwell stress tensor in electrostatics is

$$\mathbf{T} = \epsilon\begin{bmatrix} \tfrac{1}{2}(E_x^2 - E_y^2 - E_z^2) & E_x E_y & E_x E_z \\ E_y E_x & \tfrac{1}{2}(E_y^2 - E_x^2 - E_z^2) & E_y E_z \\ E_z E_x & E_z E_y & \tfrac{1}{2}(E_z^2 - E_x^2 - E_y^2) \end{bmatrix}$$

where ϵ is a constant and $\mathbf{E} = E_x\mathbf{i} + E_y\mathbf{j} + E_z\mathbf{k}$ is the electric field vector. Show that the eigenvalues of $\mathbf{T}$ (which are referred to as the 'principal stresses') are $\tfrac{1}{2}\epsilon E^2$, $-\tfrac{1}{2}\epsilon E^2$, and $-\tfrac{1}{2}\epsilon E^2$, where $E^2 = E_x^2 + E_y^2 + E_z^2$. Show that one of the corresponding principal axes is parallel to the electric field $\mathbf{E}$.

14. Using the matrix $\mathbf{A}$ of Ex. 11, and taking $\mathbf{v} = \begin{bmatrix} 0 \\ 0 \\ 1 \end{bmatrix}$, calculate $\mathbf{A}\mathbf{v}, \mathbf{A}^2\mathbf{v}, \ldots, \mathbf{A}^9\mathbf{v}, \mathbf{A}^{10}\mathbf{v}$ and deduce that the largest (in magnitude) eigenvalue of $\mathbf{A}$ is very nearly -2.

15. Show that the equations for the 'learning model' of § 8, Ex. 5, can be written in matrix notation as $\mathbf{P}_{n+1} = \mathbf{A}\mathbf{P}_n$, where

$$\mathbf{P}_n = \begin{bmatrix} p_n \\ q_n \end{bmatrix} \quad \text{and} \quad \mathbf{A} = \begin{bmatrix} a & b \\ 1-a & 1-b \end{bmatrix}.$$

Hence deduce that $\mathbf{P}_n = \mathbf{A}^n\mathbf{P}_0$. (This is an example of a 'Markov chain'.)

Show that the eigenvalues of $\mathbf{A}$ are 1 and $a-b$. Hence show that:

(i) if $a = b$, $\quad \mathbf{P}_n = \begin{bmatrix} a \\ 1-a \end{bmatrix}$ for $n \geqslant 1$.

(ii) if $a+b = 1$, $\quad \mathbf{P}_n \to \begin{bmatrix} 0{\cdot}5 \\ 0{\cdot}5 \end{bmatrix}$ for large n,

irrespective of the initial value $\mathbf{P}_0$. [Hint: Use the Cayley–Hamilton Theorem and § 108 (12).]

16. If
$$\mathbf{A} = \begin{bmatrix} \frac{1}{2} & \frac{1}{2}\sqrt{3} & 0 \\ -\frac{1}{2}\sqrt{3} & \frac{1}{2} & 0 \\ 0 & 0 & 1 \end{bmatrix},$$
use the Cayley–Hamilton theorem to show that $\mathbf{A}^6 = \mathbf{I}$. Interpret this result in terms of the rotation matrix **R** of § 106, Ex. 4.

17. Show that $\exp(\mathbf{A}t)\exp(\mathbf{B}t) \neq \exp\{(\mathbf{A}+\mathbf{B})t\}$ except when $\mathbf{AB} = \mathbf{BA}$.

18. Use Laplace transforms to show that the solution of the set of equations
$$\frac{d}{dt}\mathbf{X} = \mathbf{AX}+\mathbf{BU}(t)$$
is
$$\mathbf{X}(t) = [\exp(\mathbf{A}t)]\mathbf{X}(0)+\int\limits_0^t \exp\{\mathbf{A}(t-\tau)\}\mathbf{BU}(\tau)\,d\tau.$$
A and **B** are constant matrices of order n by n and n by m respectively, and $\mathbf{U}(t)$ is a known vector function of time of order m by *1*.

19. Use matrices to derive the results of Ex. 11 and Ex. 12 at the end of Chapter IV.

20. A system is described by the equation
$$\dddot{x}+a_1\ddot{x}+a_2\dot{x}+a_3 x = u(t),$$
where $u(t)$ is a control which aims to stabilize the system at $x = 0$. Use matrices to show that no control is needed if $a_1 > 0$, $a_1 a_2 - a_3 > 0$, and $a_3 > 0$. [Cf. § 47 (8).] If $a_1 > 0$, $a_2 > 0$ and $a_3 > 0$, what sort of control function will stabilize the system?

21. Find the equilibrium configuration for the competing species system § 7, Ex. 3. Write down the matrix **M** of § 110 (8) for this system and hence deduce that equilibrium is unstable if $\frac{a}{c} < \frac{e}{g}$ and $\frac{a}{b} > \frac{e}{f}$.

Compare this result with Ex. 6 at the end of Chapter I.

XIV
DIFFERENCE EQUATIONS AND THE NUMERICAL SOLUTION OF DIFFERENTIAL EQUATIONS

111. Introductory

THE foregoing chapters should have left the reader with the suspicion that explicit solutions can only be found for problems which lead to relatively simple differential equations. This suspicion is, generally speaking, well founded; applied mathematicians have, however, always been remarkably adroit at redefining a real problem and making necessary but reasonable assumptions so as to reduce it to either a linear differential equation which can be solved explicitly, or a non-linear equation that can be tackled by methods such as § 58. There is the temptation, though, either to ignore problems which cannot be solved in this way, or else to so mutilate a problem in an effort to obtain an explicit solution that the solution is, for all practical purposes, meaningless.

This temptation has been remedied by the advent of the computer and the widespread use of numerical methods for the solution of differential equations. The use of the computer leads, however, to problems and temptations of its own. The problems are related to obtaining an estimate of the accuracy of the computer solution; as we shall see, no computer solution is complete without some comment on the accuracy of the results. The temptation is to include 'everything' in the mathematical statement of the problem, thereby utterly confusing a solution that might have been revealing if the right assumptions had been made in the first place.

Modern applied mathematics relies, therefore, on a thoughtful blend of the mathematical methods developed in previous chapters with the numerical methods to be discussed in this chapter and in Chapter XVI.

The computer has also re-activated an interest in difference equations and has encouraged applied mathematicians to formulate problems in discrete rather than continuous terms. We therefore begin with a discussion of difference equations, and then proceed to the numerical solution of differential equations.

112. Linear difference equations with constant coefficients

Difference equations were introduced in Chapter I and have been solved on an *ad hoc* basis in §§ 45 and 85.

The nth-order linear difference equation with constant coefficients is

$$a_0 y_{r+n}+a_1 y_{r+n-1}+...+a_n y_r = \phi(r). \tag{1}$$

The solution of (1) is directly analogous to the solution of the nth-order differential equation with constant coefficients, as described in Chapter II.

We begin by introducing an operator E such that

$$Ey_r = y_{r+1} \tag{2}$$

$$E^2 y_r = E(Ey_r) = Ey_{r+1} = y_{r+2}$$

and in general $E^m y_r = y_{r+m}$. (3)

Equation (1) can then be written

$$(a_0 E^n+a_1 E^{n-1}+...+a_{n-1} E+a_n)y_r = \phi(r). \tag{4}$$

The solution of (4) is constructed by first solving the corresponding homogeneous equation with $\phi(r) = 0$, and then adding to it a particular solution of the inhomogeneous equation. The solution to the homogeneous equation is found in terms of the roots $\alpha_1, \alpha_2,..., \alpha_n$ of the auxiliary equation

$$a_0 \alpha^n+a_1 \alpha^{n-1}+...+a_n = 0.$$

If these roots are all distinct it can easily be verified that

$$y_r = c_1 \alpha_1^r+c_2 \alpha_2^r+...+c_n \alpha_n^r \tag{5}$$

satisfies (4) with $\phi(r) = 0$. If, for instance, $\alpha_1 = \alpha_2$ while $\alpha_3,..., \alpha_n$ are distinct, the solution with the correct number of arbitrary constants is

$$y_r = (c_1+c_2 r)\alpha_1^r+c_3 \alpha_3^r+...+c_n \alpha_n^r. \tag{6}$$

The particular solution is found by making appropriate trial solutions,† remembering that, since (4) is linear, the Principle of Superposition can be invoked.

Ex. 1. *The discrete population growth model of § 2 (2).*

The difference equation is

$$x_{r+1} = (1+k)x_r$$

or

$$\{E-(1+k)\}x_r = 0.$$

By (5), the general solution is

$$x_r = c(1+k)^r. \tag{7}$$

The arbitrary constant c is found by putting $x_r = x_1$ and $r = 1$ in (7),

$$x_1 = c(1+k) \qquad \text{or} \qquad c = x_1/(1+k),$$

so finally

$$x_r = x_1(1+k)^{r-1}.$$

Ex. 2. *The problem of § 85 (12).*

The difference equation for the bending moment at the rth support is

$$M_{r+2}+4M_{r+1}+M_r = \tfrac{1}{2}wa^2$$

or

$$(E^2+4E+1)M_r = \tfrac{1}{2}wa^2. \tag{8}$$

The roots of the auxiliary equation $\alpha^2+4\alpha+1 = 0$ are $\alpha_1 = -2+\sqrt{3}$ and $\alpha_2 = -2-\sqrt{3}$. The solution of the homogeneous equation is therefore

$$A(-2+\sqrt{3})^r+B(-2-\sqrt{3})^r.$$

To find the particular solution, try $M_r = c$ in (8), whence

$$c = wa^2/12.$$

The general solution is therefore

$$M_r = A(-2+\sqrt{3})^r+B(-2-\sqrt{3})^r+wa^2/12.$$

Ex. 3.

Let D_r be the determinant of the r by r matrix

$$\begin{bmatrix} 1 & 1 & 0 & 0 & . & . & . & . & 0 \\ 1 & 1 & 1 & 0 & . & . & . & . & 0 \\ 0 & 1 & 1 & 1 & 0 & . & . & . & 0 \\ . & & & & & & & & \\ . & & & & & & & & \\ . & & & & & & & & \\ 0 & . & . & . & . & . & 0 & 1 & 1 \end{bmatrix}.$$

Then $D_1 = 1$, $D_2 = 0$, and in general, expanding the determinant in terms of the first row

$$D_r = D_{r-1}-D_{r-2}, \tag{9}$$

or

$$(E^2-E+1)D_r = 0.$$

† For example, if $\phi(r)$ is a polynomial in r, the trial solution is also a polynomial in r. For a more formal approach see Milne-Thomson, *The Calculus of Finite Differences* (Macmillan).

The general solution is

$$D_r = A\{(1+i\surd 3)/2\}^r + B\{(1-i\surd 3)/2\}^r$$

which can be written

$$D_r = Ae^{ir\pi/3} + Be^{-ir\pi/3}.$$

As in the case of § 13 (11), this reduces to

$$D_r = P\cos(r\pi/3) + Q\sin(r\pi/3)$$

where P and Q are arbitrary. The initial conditions are

$$D_1 = 1 = P\cos(\pi/3) + Q\sin(\pi/3),$$

and

$$D_2 = 0 = P\cos(2\pi/3) + Q\sin(2\pi/3),$$

whence

$$P = 1 \quad \text{and} \quad Q = 1/(\surd 3).$$

Ex. 4. *An economic model.*

Consider an economy in which manufacturers try to control production so as to meet demand, while the retailers, through advertising and pricing, try to control demand to meet production.

If P_r and D_r are the production and demand in year r, a simplified model of the economy might be

$$P_{r+1} = P_r + k(D_r - P_r) \tag{10}$$

$$D_{r+1} = D_r + k(P_r - D_r) \tag{11}$$

where k is a positive constant.

Equations (10) and (11) could be written in terms of matrices and solved in the same way as Ex. 5 of § 106. We shall solve the equations directly by eliminating D_r in much the same way as for a pair of differential equations; cf. § 15.

From (10)

$$D_r = P_r + (P_{r+1} - P_r)/k. \tag{12}$$

Operating on (10) with E

$$P_{r+2} = P_{r+1} + k(D_{r+1} - P_{r+1})$$

by (11)

$$= P_{r+1}(1-k) + kD_r + k^2(P_r - D_r)$$

by (12)

$$= 2(1-k)P_{r+1} - (1-2k)P_r,$$

or

$$\{E^2 - 2(1-k)E + (1-2k)\}P_r = 0. \tag{13}$$

The roots of the auxiliary equation are 1 and $1-2k$, so the general solution of (13) is

$$P_r = A + B(1-2k)^r \tag{14}$$

and by (12)

$$D_r = A - B(1-2k)^r. \tag{15}$$

If $P = P_0$ and $D = D_0$ when $r = 0$, the constants are $A = (P_0 + D_0)/2$ and $B = (P_0 - D_0)/2$.

It is interesting to examine the behaviour of the solutions (14) and (15). There are three distinct cases:

(i) $0 < k \leqslant \frac{1}{2}$. Then $1-2k < 1$ and P_r and D_r approach their equilibrium values $P_r = D_r = (P_0 + D_0)/2$ monotonically.

(ii) $\frac{1}{2} < k < 1$. Then $-1 < (1-2k) < 0$ and production and demand will oscillate, with decreasing amplitude, eventually stabilizing at $P_r = D_r = (P_0+D_0)/2$, as before.

(iii) $k \geqslant 1$. Then $1-2k \leqslant -1$ and no equilibrium is possible. Production and demand will oscillate indefinitely and, if $k > 1$, with ever-increasing amplitude. The economy will therefore be unstable and, just as linear models are not representative of unstable electronic circuits (cf. § 47), the above model would not be representative of the behaviour of the economy after the first few oscillations.

An alternative mathematical model of the above economy is the pair of differential equations

$$\left.\begin{aligned} dP/dt &= c(D-P) \\ \text{and} \qquad dD/dt &= c(P-D) \end{aligned}\right\}. \tag{16}$$

If $P = P_0$ and $D = D_0$ at $t = 0$, the solution of (16) is

$$\left.\begin{aligned} P &= \tfrac{1}{2}(P_0+D_0)+\tfrac{1}{2}(P_0-D_0)e^{-2ct} \\ D &= \tfrac{1}{2}(P_0+D_0)-\tfrac{1}{2}(P_0-D_0)e^{-2ct} \end{aligned}\right\}. \tag{17}$$

No matter what the value of c in (17), P and D will always tend monotonically to equilibrium, as in (i) above. The difference equation formulation of the problem thus exhibits the possible types of behaviour (ii) and (iii) above, that cannot be derived in any way from the differential equations. This reflects the difference between an instantaneous reaction to changes in production and demand (the differential equations) and a delayed reaction from one year to the next (the difference equations). The difference equation model is clearly more realistic in this particular example.

113. Further examples of difference equations

In certain cases explicit solutions can be found to non-linear difference equations or linear difference equations with variable coefficients. For instance, one can see by writing down the first few terms of the sequence $y_{r+1} = (r+1)y_r$, that the solution of

$$\{E-(r+1)\}y_r = 0, \tag{1}$$

with $y_r = 1$ when $r = 0$,

is

$$y_r = (r!). \tag{2}$$

Ex. 1. *The 'parking lot' problem*

Suppose that a parking lot has room for N cars. Let p_r be the probability that exactly r cars are parked in the lot at a typical time of the day. It can then be argued† that

$$p_1 = \rho p_0 \tag{3}$$

and

$$(r+2)p_{r+2}-(\rho+r+1)p_{r+1}+\rho p_r = 0, \tag{4}$$

† See, for example, Haight, *Mathematical Theories of Traffic Flow* (Academic Press, 1963), chapter 6.

for $r < N-1$. The constant ρ depends on the statistical distributions assumed for the arrival process and duration of parking.

With equations (1) and (2) in mind, we make the substitution

$$p_r = u_r/r!. \tag{5}$$

Putting (5) in (4) and multiplying throughout by $(r+1)!$,

$$u_{r+2}-(\rho+r+1)u_{r+1}+\rho(r+1)u_r = 0$$

or

$$\{E^2-(\rho+r+1)E+\rho(r+1)\}u_r = 0,$$

that is

$$\{E-(r+1)\}\{E-\rho\}u_r = 0. \tag{6}$$

The general solution of (6) is, by (1) and (2),

$$u_r = Ar!+B\rho^r,$$

so that

$$p_r = u_r/r! = A+B\rho^r/r!. \tag{7}$$

When $r = 0$, by (7),

$$p_0 = A+B$$

and when $r = 1$

$$p_1 = A+B\rho = \rho p_0$$

by (3). Thus $A = 0$ and $B = p_0$ and the solution to the problem is

$$p_r = p_0\rho^r/r!.$$

This is an example of a stochastic model (cf. § 2).

Ex. 2. *The genetics problem of* § 7, *Ex. 5.*

The proportion of the rth generation born with the fatal recessive gene is p_r where

$$p_{r+1} = p_r/(1+p_r). \tag{8}$$

The substitution $p_r = 1/y_r$ reduces (8) to the linear equation

$$y_{r+1} = 1+y_r$$

or

$$(E-1)y_r = 1. \tag{9}$$

The general solution is $y_r = A+r$, so that

$$p_r = 1/(A+r). \tag{10}$$

Putting $r = 0$ in (10) gives $A = 1/p_0$, so finally,

$$p_r = p_0/(1+rp_0). \tag{11}$$

Equation (11) shows that the proportion of people born with the recessive gene will reduce only very slowly from one generation to the next if p_0 is small. For example, if $p_0 = 0{\cdot}1$, it takes 10 generations to halve the proportion of people born with the recessive gene, but if $p_0 = 0{\cdot}01$, this takes 100 generations.

Ex. 3. *A set of difference–differential equations.*

If the probability of an event happening in the time interval $(t, t+\delta t)$ is $\lambda\delta t$, independent of t, and if $P_n(t)$ is written for the probability of n events occurring in time t, it can be shown that

$$\frac{dP_0}{dt} = -\lambda P_0 \tag{12}$$

and
$$\frac{dP_r}{dt} = \lambda P_{r-1} - \lambda P_r, \tag{13}$$
for $r = 1, 2, \dots$. The initial conditions are
$$P_0(0) = 1 \quad \text{and} \quad P_r(0) = 0, \qquad r = 1, 2, \dots. \tag{14}$$
Equations (12) and (13) are a set of difference–differential equations. They can be reduced to ordinary difference equations by taking Laplace transforms with respect to time.

By (14) above and § 18 (21), we then have
$$p\bar{P}_0 - 1 = -\lambda \bar{P}_0 \tag{15}$$
and
$$p\bar{P}_r = \lambda \bar{P}_{r-1} - \lambda \bar{P}_r. \tag{16}$$
The general solution of (16) is
$$\bar{P}_r = A\left(\frac{\lambda}{p+\lambda}\right)^r,$$
where, by (15), $$A = 1/p+\lambda.$$
Thus $$\bar{P}_r = \lambda^r/(p+\lambda)^{r+1},$$
and by § 18 (12),
$$P_r(t) = (\lambda t)^r e^{-\lambda t}/r! \quad \text{for } r = 0, 1, 2 \dots.$$
The expected number of events in time t is
$$\sum_{r=0}^{\infty} rP_r = e^{-\lambda t} \sum_{r=0}^{\infty} r(\lambda t)^r/r! = \lambda t.$$
This is an important result in the theory of queues and describes what is known as a 'Poisson process'.

When suitable substitutions cannot be found for difference equations such as Ex. 1 and Ex. 2 above, the difference equation itself leads to an algorithm for finding a numerical solution on the computer. Care must, however, be taken to ensure that there is no undue accumulation of errors, as in the case of § 5 (4).

114. Numerical methods for ordinary differential equations. Equations of the first order

The general first-order equation can be written
$$dy/dx = y' = f(x, y). \tag{1}$$
Its solution contains a single arbitrary constant which can be determined from the initial condition $y = y_0$ at $x = x_0$.

Suppose we are given a function f such that (1) cannot be solved by any of the methods of Chapter III; that is, we are unable to find an explicit mathematical solution. We therefore attempt to solve the equation numerically. This means that we choose a suitable interval h and, starting with numerical values for x_0 and y_0, we attempt to construct a table of the estimated values $y_1, y_2, y_3,...$ etc., where, in general, y_r is the estimated value of y at $x = x_r = x_0+rh$.

A method of accomplishing this is suggested by Taylor's expansion

$$y(x+h) = y(x)+hy'(x)+\frac{1}{2!}h^2y''(x)+\frac{1}{3!}h^3y'''(x)+..., \tag{2}$$

or, in terms of the notation introduced above,

$$y_{r+1} = y_r+hy'_r+\frac{1}{2!}h^2y''_r+\frac{1}{3!}h^3y'''_r+.... \tag{3}$$

By the differential equation (1), we have

$$y'_r = f(x_r, y_r),$$

and so if we truncate the series (3) after the term in h, we have the approximate formula

$$y_{r+1} = y_r+hf(x_r, y_r). \tag{4}$$

This is a difference formula which can be evaluated step by step, using the given values of x_0 and y_0 to calculate y_1, then using y_1 and $x_1 = x_0+h$ to calculate y_2, and so on. This is known as *Euler's method* and is best illustrated by an example.

Ex. 1.

$$\frac{dy}{dx} = \sin(x+y), \quad \text{given } y = 0 \text{ when } x = 0.$$

Euler's formula (4) gives

$$y_{r+1} = y_r+h\sin(x_r+y_r). \tag{5}$$

Here $x_0 = y_0 = 0$. If we choose $h = 0{\cdot}5$ and successively substitute in (5) with $r = 0, 1, 2,...$ etc., we get

$$\begin{aligned} y_1 &= y_0+0{\cdot}5\sin(x_0+y_0) = 0, \\ y_2 &= y_1+0{\cdot}5\sin(x_1+y_1) = 0{\cdot}5\sin(0{\cdot}5) = 0{\cdot}240, \\ y_3 &= y_2+0{\cdot}5\sin(x_2+y_2) = 0{\cdot}24+0{\cdot}5\sin(1{\cdot}0+0{\cdot}24) \\ &= 0{\cdot}713, \end{aligned}$$

and so on. The first 10 values of y_r are shown in column 1 of Table II.

Although we have obtained a numerical 'solution' for the above example, we have as yet no way of estimating the accuracy of this solution, nor do we really know whether the chosen value of h is in fact 'suitable'. We can therefore only attach some significance to the numbers in column 1 of Table II if we can estimate the errors we might have introduced in the process of obtaining them.

Each time we apply an equation such as (4), we introduce two types of errors:

(i) A *truncation error* introduced by ignoring the terms in $h^2, h^3, \ldots$ etc., in equation (3). For Euler's method, this error is equal to $\frac{1}{2}h^2y''(\theta)$ where $x_r < \theta < x_{r+1}$, and since it is proportional to the square of the interval h we say that the truncation error is 'of order h^2', written $\mathbf{O}(h^2)$. This implies that halving the interval h will reduce the truncation error at each step by a factor of four.

(ii) A further error introduced in y_{r+1} because y_r is itself in error. The size of this error will depend on the function $f(x, y)$ and the step size h, as well as the actual error in y_r.

The above errors are introduced *at each step of the calculation*. The final entry in the table will thus contain the accumulated errors of all of the previous steps. The number of steps taken to reach a specific value of x is inversely proportional to h, and so, roughly speaking, the accumulated truncation error will be proportional to h^2/h, that is, of order h. Halving the step size thus only halves the accumulated truncation error.

The accumulation of errors of type (ii) is harder to evaluate. The errors introduced at each step could die down as the solution progresses, or, as in the calculation of § 5 (4), the errors could be magnified from one step to the next until they completely swamp the real solution. In the latter case the computation becomes meaningless and we say that the formula we are using is *unstable*. This is illustrated in the following example.

Ex. 2.

Consider the differential equations

$$dP/dt = C(D-P) \tag{6}$$

and

$$dD/dt = C(P-D) \tag{7}$$

discussed in the economic model of § 112, Ex. 4. We know that their solution is as given in § 112 (17). Here we try to reconstruct this solution using Euler's method.

Applying (4) to (6) and (7) simultaneously,

$$P_{r+1} = P_r + hC(D_r - P_r) \tag{8}$$

and

$$D_{r+1} = D_r + hC(P_r - D_r). \tag{9}$$

Given numerical values for h, C, and the initial conditions P_0 and D_0, we could use (8) and (9) to tabulate P_1 and D_1, P_2 and D_2, and so on. However, we notice that (8) and (9) are in fact the difference equations § 112 (10), (11) with $k = hC$, and we can exploit this coincidence to predict the behaviour of our numerical solution.

Referring to cases (i), (ii), and (iii) of § 112, Ex. 4, we see that if the interval h is chosen such that $0 < hC \leqslant \frac{1}{2}$, the solution obtained using Euler's method will exhibit the same behaviour as the exact solution of (6) and (7), that is, the numerical solution will also tend monotonically to the correct equilibrium values. If $\frac{1}{2} < hC < 1$, the numerical solution will contain a spurious oscillation, but this will at least be damped out as the solution progresses. If $hC > 1$, however, the spurious oscillation will grow in amplitude from one step to the next, and the numerical solution will be completely unstable.

In the above example a 'suitable' value for h is thus $hC \leqslant \frac{1}{2}$. This only ensures that the numerical solution exhibits the same sort of behaviour as the exact solution to the problem; it still does not tell us whether, in a particular calculation, we have, say, 2-, 3-, or 4-figure accuracy. The best way of determining the actual accuracy of a numerical solution is a pragmatic one; recalculate the results using a smaller value of h and compare this with the original solution. If the results coincide to n figures, we can be fairly confident that we have n-figure accuracy.

TABLE II

x	(1) *Euler's method*	(2) *'Improved' formula*	(3) *Predictor–corrector*	(4) *Taylor's method*	(5) *Runge–Kutta*
0·5	0·000	0·146	0·146	0·146	0·143
1·0	0·240	0·602	0·546	0·596	0·570
1·5	0·713	1·145	0·915	1·006	0·997
2·0	1·113	1·078	1·064	1·133	1·140
2·5	1·127	1·209	0·997	1·032	1·035
3·0	0·894	0·541	0·775	0·784	0·784
3·5	0·554	0·820	0·454	0·452	0·451
4·0	0·157	−0·384	0·075	0·070	0·069
4·5	−0·268	+0·363	−0·337	−0·343	−0·344
5·0	−0·711	−1·372	−0·771	−0·777	−0·778

115. Higher order methods for first-order equations

Euler's method is seldom used in practice because, for very little extra work, a variety of higher order formulae can be derived and applied. We shall consider four examples of the various types of higher order methods.

(i) *The 'improved' formula*

By § 114 (3),

$$y_{r+1} = y_r + hy'_r + \tfrac{1}{2}h^2y''_r + \tfrac{1}{6}h^3y'''_r + \dots \tag{1}$$

Replacing h by $-h$ in Taylor's expansion § 114 (2) leads to the equation

$$y_{r-1} = y_r - hy'_r + \tfrac{1}{2}h^2y''_r - \tfrac{1}{6}h^3y'''_r + \dots \tag{2}$$

Subtracting (2) from (1) gives the 'improved' formula

$$\begin{aligned} y_{r+1} &= y_{r-1} + 2hy'_r \\ &= y_{r-1} + 2hf(x_r, y_r), \end{aligned} \tag{3}$$

with a truncation error at each step of $\tfrac{1}{3}h^3y'''(\theta)$. The accumulated truncation error is thus $\mathbf{O}(h^2)$ as compared with $\mathbf{O}(h)$ for Euler's method.

In applying (3), the first step y_1 must be calculated by some other method, such as (iii) or (iv) below. Thereafter (3) can be applied step by step in the same way as Euler's method. The solution obtained in this way for the problem of § 114, Ex. 1, is shown in column 2 of Table II. Comparing these results with the other solutions in Table II, it is apparent that, in this case, the 'improved' method becomes unstable. In fact (3) is notoriously unstable (see Ex. 7 at the end of this chapter) and is seldom used on its own. It is, however, important as the first stage in the two-stage process discussed below.

(ii) *A predictor–corrector method*

Differentiating (1) with respect to x,

$$y'_{r+1} = y'_r + hy''_r + \tfrac{1}{2}h^2y'''_r + \dots, \tag{4}$$

and subtracting $\tfrac{1}{2}h$ times (4) from (1), leads to the formula

$$\begin{aligned} y_{r+1} &= y_r + \tfrac{1}{2}h(y'_r + y'_{r+1}) \\ &= y_r + \tfrac{1}{2}h\{f(x_r, y_r) + f(x_{r+1}, y_{r+1})\}, \end{aligned} \tag{5}$$

with a truncation error at each step of $-h^3y'''(\theta)/12$.

We cannot apply (5) directly because we require an estimate of y_{r+1} in the right-hand side before we can calculate y_{r+1} on the left-hand side. This estimate is obtained from the 'improved' formula (3), leading to a two-stage procedure in which one first *predicts* y_{r+1} using (3),

$$y^*_{r+1} = y_{r-1}+2hf(x_r, y_r), \tag{6}$$

and then *corrects* y_{r+1} using (5) with the predicted value y^*_{r+1} in the right-hand side,

$$y_{r+1} = y_r+\tfrac{1}{2}h\{f(x_r, y_r)+f(x_{r+1}, y^*_{r+1})\}. \tag{7}$$

Applying (6) and (7) at each step of the calculation leads to a method that is reasonably stable† for a wide range of problems; in addition it has the advantage that it is possible to estimate the truncation error at each step merely by calculating the difference between the corrected and predicted values. For,

$$y_{r+1}-y^*_{r+1} = -\tfrac{5}{12}h^3y'''(\theta)$$

approximately, and this is just five times the truncation error in the corrector formula.

The predictor–corrector solution of § 114, Ex. 1, is shown in column 3 of Table II. As with the 'improved' formula, some other method must be used to estimate y_1 before (6) and (7) can be applied.

(iii) *Taylor's method*

Here, the differential equation

$$y'_r = f(x_r, y_r)$$

is differentiated explicitly to obtain expressions for y''_r, y'''_r, etc., to substitute directly in (1).

Ex.

If
$$dy/dx = \sin(x+y),$$
$$y'_r = \sin(x_r+y_r),$$
$$y''_r = \cos(x_r+y_r)\{1+y'_r\},$$
and
$$y'''_r = -\sin(x_r+y_r)\{1+y'_r\}^2+y''_r\cos(x_r+y_r).$$

† For a full discussion of this as well as other predictor–corrector methods, see Hamming, *Numerical Methods for Scientists and Engineers* (McGraw-Hill), chapter 15.

Then, by (1),

$$y_{r+1} = y_r + hy'_r + \tfrac{1}{2}h^2y''_r + \tfrac{1}{6}h^3y'''_r, \tag{8}$$

with a truncation error at each step of $\mathbf{O}(h^4)$.

Results obtained by (8) for the case $x_0 = y_0 = 0$ are shown in column 4 of Table II. The above example lends itself fairly well to this approach; in more complicated examples the expressions for y''_r, y'''_r, etc., tend to be far too unwieldy to program. Taylor's method is, however, often used as a 'starting formula' to calculate y_1 as a preliminary to using a predictor–corrector method.

(iv) *Runge–Kutta formulae*

Runge–Kutta formulae are widely used in general purpose computer programs for the solution of ordinary differential equations. There are a variety of formulae, of which the two most commonly used are the 'third-order formula'

$$y_{r+1} = y_r + \tfrac{1}{6}(k_1 + 4k_2 + k_3) \tag{9}$$

where

$$\left.\begin{aligned} k_1 &= hf(x_r, y_r) \\ k_2 &= hf(x_r + \tfrac{1}{2}h, y_r + \tfrac{1}{2}k_1) \\ k_3 &= hf(x_r + h, y_r + 2k_2 - k_1) \end{aligned}\right\}, \tag{10}$$

and the 'fourth-order formula'

$$y_{r+1} = y_r + \tfrac{1}{6}(k_1 + 2k_2 + 2k_3 + k_4) \tag{11}$$

where k_1 and k_2 are as in (10) above, while

$$k_3 = hf(x_r + \tfrac{1}{2}h, y_r + \tfrac{1}{2}k_2)$$

and

$$k_4 = hf(x_r + h, y_r + k_3).$$

In both (9) and (11) the solution of $dy/dx = f(x, y)$ proceeds step by step from y_0 to y_1, y_1 to y_2, etc., as before, except that here we have to make a number of intermediate calculations at each step. The extra computational effort involved is, broadly speaking, compensated for by reasonable stability and fairly high accuracy; the derivation‡ of an nth-order Runge–Kutta formula ensures that the accumulated truncation error is of order n; the truncation error at each step is $\mathbf{O}(h^{n+1})$.

‡ See, for example, Ralston, *A First Course in Numerical Analysis* (McGraw-Hill), chapter 5.

Runge–Kutta formulae have the added advantages that they need no 'starting formula' and that it is easy to change the interval h from one step to the next. Their major disadvantage is that it is not always easy to gauge the accuracy of each step. There is, however, a simple procedure for checking the accuracy of the third-order formula. Here the truncation error in proceeding from y_r to y_{r+1} is $\mathbf{O}(h^4)$. If we then back-track and calculate y_r from y_{r+1}, that is, if we replace h by $-h$ in (9) and (10), we will again introduce a truncation error of $\mathbf{O}(h)^4$. The difference between the original y_r and the back-calculated value of y_r is therefore a measure of the truncation error. Computer routines can be written which automatically check the error in this way after every so many steps, and, depending on the error, automatically adjust the size of the interval h.

The solution of §114, Ex. 1, using the fourth-order formula (11) with an interval $h = 0{\cdot}5$ is shown in column 5 of Table II.

116. Simultaneous equations and numerical methods for higher order differential equations

The predictor–corrector and Runge–Kutta methods can both be applied to sets of simultaneous first-order differential equations.

Ex. 1. *The predator–prey model of* § 7, *Ex.* 2.

The simultaneous equations are

$$dx/dt = px - qxy \tag{1}$$

and

$$dy/dt = rxy - sy. \tag{2}$$

By analogy with § 115 (9) and (10), the third-order Runge–Kutta process for calculating x_{n+1} and y_{n+1} from x_n and y_n is as follows:

First calculate $\quad k_1 = hx_n(p - qy_n) \quad$ and $\quad l_1 = hy_n(rx_n - s)$,

then calculate $\quad k_2 = h(x_n + \frac{1}{2}k_1)\{p - q(y_n + \frac{1}{2}l_1)\}$

and $\quad l_2 = h(y_n + \frac{1}{2}l_1)\{r(x_n + \frac{1}{2}k_1) - s\}$.

Finally calculate $\quad k_3 = h(x_n + 2k_2 - k_1)\{p - q(y_n + 2l_2 - l_1)\}$

and $\quad l_3 = h(y_n + 2l_2 - l_1)\{r(x_n + 2k_2 - k_1) - s\}$,

and put $\quad x_{n+1} = x_n + \frac{1}{6}(k_1 + 4k_2 + k_3)$

and $\quad y_{n+1} = y_n + \frac{1}{6}(l_1 + 4l_2 + l_3)$.†

† Notice that in (1) and (2) the independent variable is t and so h is a time interval. It so happens that in this example t does not appear explicitly in the right-hand side of either equation. Also notice that care has to be taken not to confuse the k's and l's in the Runge–Kutta equations; k always refers to an increment in x, l to an increment in y.

The solution obtained using the values $h = 1$ year, $x_0 = 105$, $y_0 = 8$ (the units can be interpreted as, say, hundreds of animals), and $p = 0{\cdot}4$, $q = 0{\cdot}04$, $r = 0{\cdot}02$, and $s = 2{\cdot}0$, is shown in Fig. 87. This illustrates the cyclic fluctuations in population deduced in Ex. 14 at the end of Chapter III.

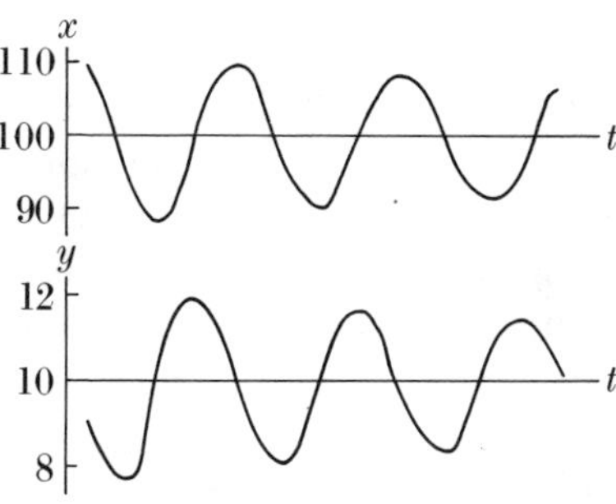

Fig. 87.

As we saw in § 15, higher order differential equations can always be reduced to a set of simultaneous equations of first order by the introduction of appropriate 'dummy' variables. The equivalent set of first-order equations can then be solved numerically by a predictor–corrector or Runge–Kutta method.

Ex. 2.
The equation for large oscillations of a simple pendulum, § 55 (13), is

$$d^2\theta/dt^2 + n^2 \sin\theta = 0, \tag{3}$$

which reduces to the pair of equations

$$d\theta/dt = v \tag{4}$$

and

$$dv/dt = -n^2 \sin\theta. \tag{5}$$

The predictor–corrector equations for (4) and (5) are, cf. § 115 (6) and (7),

$$\theta^*_{r+1} = \theta_{r-1} + 2hv_r$$

$$v^*_{r+1} = v_{r-1} - 2hn^2 \sin\theta_r,$$

and

$$\theta_{r+1} = \theta_r + \tfrac{1}{2}h(v_r + v^*_{r+1})$$

$$v_{r+1} = v_r - \tfrac{1}{2}hn^2(\sin\theta_r + \sin\theta^*_{r+1}).$$

These equations only hold for $r \geqslant 1$. In addition to the initial conditions θ_0 and v_0, we need to find θ_1 and v_1 from some 'starting formula'. If $\theta_0 = 0$, it can be verified that Taylor's method applied to (4) and (5) leads to the starting formulae

$$\theta_1 = hv_0 - \tfrac{1}{6}h^3n^2v_0$$

and

$$v_1 = v_0 - \tfrac{1}{2}h^2n^2v_0.$$

The numerical solution of a second-order equation is thus quite straightforward if we are given initial conditions, such as θ_0 and v_0 in the above example. The solution cannot be obtained quite so easily if we are given boundary conditions instead of initial conditions.

Suppose, for instance, that in Ex. 2 we are given that $\theta = 0$ at $t = 0$ and that the period of the oscillation is T, where T is specified numerically. This information can be interpreted as the pair of boundary conditions $\theta = 0$ at $t = 0$ and $\theta = 0$ at $t = \frac{1}{2}T$. We do not know $d\theta/dt = v$ at $t = 0$, and so we cannot construct the solution step by step from θ_0 and v_0 as in Ex. 2. We could, however, guess at v_0, and then proceed to solve the problem by the following iterative process:

(i) divide the interval $(0, \frac{1}{2}T)$ into m steps of size h;

(ii) use the 'guessed' value for v_0 to solve the problem as an initial value problem up to and including $t = mh$;

(iii) if θ_m is very close to zero, we have solved the boundary value problem;

(iv) if θ_m is not sufficiently close to zero, make a new guess at v_0 and repeat the procedure.

The process of 'guessing' is assisted by plotting the calculated values of θ_m against their initial velocities v_0, as in Fig. 88 (*a*).

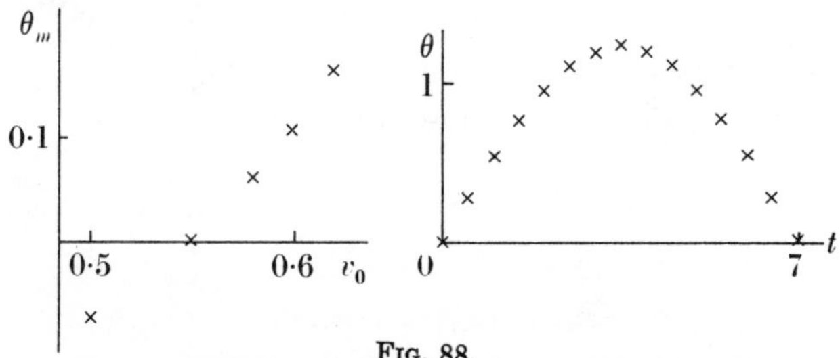

FIG. 88.

The solution obtained with the best initial guess $v_0 = 0{\cdot}55$ is shown in Fig. 88 (*b*) for the case $n = 0{\cdot}5$ and $T = 14$ seconds.

117. Interpolation, differentiation, and integration

If we have an explicit mathematical solution to a differential equation, say in the form of y as a function of x, we can evaluate y at any point that is of interest to us, and, if the integral or derivatives of y with respect to x are of practical importance, we can attempt to evaluate them explicitly too. On the other hand, if the differential equation has been solved numerically, we will only have a table of the values $y_0, y_1,..., y_n$ of y, corresponding to what we shall assume are regularly spaced values $x_0, x_1,..., x_n$ of x. To estimate y at some point between the tabulated values, or to differentiate or integrate the function, we then have to proceed numerically.

In the simplest case, we assume that successive points in the table are joined by straight lines. For instance, the line joining y_r at x_r to y_{r+1} at $x_{r+1} = x_r+h$ is

$$y(x) = y_r+(y_{r+1}-y_r)(x-x_r)/h. \tag{1}$$

We use (1) to estimate the values of y at any point x in the interval (x_r, x_{r+1}). This is known as *linear interpolation.*

The slope of the line leads to an estimate of the derivative of y with respect to x:

$$dy/dx \text{ (at } x = x_r) = (y_{r+1}-y_r)/h. \tag{2}$$

Similarly, integrating (1),

$$\int\limits_{x_r}^{x_{r+1}} y\,dx = \tfrac{1}{2}h(y_r+y_{r+1}),$$

and so we can estimate the integral of the tabulated function between x_0 and x_n from the sum

$$\int\limits_{x_0}^{x_n} y\,dx = \tfrac{1}{2}h(y_0+y_1)+\tfrac{1}{2}h(y_1+y_2)+...+\tfrac{1}{2}h(y_{n-1}+y_n)$$
$$= h(\tfrac{1}{2}y_0+y_1+y_2+...+y_{n-1}+\tfrac{1}{2}y_n). \tag{3}$$

This is the so-called 'trapezoidal rule' for numerical integration.

Linear interpolation is a special case of the more general method of *polynomial interpolation*. Here we seek a polynomial of degree k which coincides with the $k+1$ table entries $y_r, y_{r+1},..., y_{r+k}$; that is, the coefficients of the interpolating polynomial

$$P_k(x) = p_0 x^k+p_1 x^{k-1}+...+p_k\dagger \tag{4}$$

† The notation used here should not be confused with the use of $P_k(x)$ for the Legendre polynomial in Chapter XII.

are chosen so that the polynomial passes through the points

$$(x_r, y_r), \quad (x_{r+1}, y_{r+1}), \ldots, \quad (x_{r+k}, y_{r+k}).$$

There are a number of different methods for evaluating the coefficients in (4); each method leads to its own set of interpolation, differentiation, and integration formulae. However, the polynomial of degree k which passes through $k+1$ distinct points is uniquely determined by the coordinates of those points, and so the various formulae are all equivalent to each other; some formulae are well suited to computer calculation, while others are more convenient for hand-calculation. We shall derive some of the more important formulae for the case of the second-degree polynomial

$$P_2(x) = p_0 x^2 + p_1 x + p_2 \tag{5}$$

which is to be fitted to the three points (x_0, y_0), (x_1, y_1), and (x_2, y_2).

(i) *Direct determination of the coefficients*

Substituting the coordinates of the three points directly into (5),

$$p_0 x_0^2 + p_1 x_0 + p_2 = y_0$$
$$p_0 x_1^2 + p_1 x_1 + p_2 = y_1$$

and
$$p_0 x_2^2 + p_1 x_2 + p_2 = y_2.$$

These equations can be solved for p_0, p_1, and p_2. The solution is simplified by putting $x_0 = -h$, $x_1 = 0$, and $x_2 = h$, in which case

$$p_2 = y_1, \quad p_1 = (y_2 - y_0)/2h, \quad p_0 = (y_0 - 2y_1 + y_2)/2h^2,$$

and the interpolating polynomial (5) can be written

$$P_2(x) = (y_0 - 2y_1 + y_2)(x - x_1)^2/2h^2 + (y_2 - y_0)(x - x_1)/2h + y_1. \tag{6}$$

Equation (6) can be used to estimate the value of y anywhere in the interval (x_0, x_2). The first derivative of (6) is

$$dP_2/dx = (y_2 - y_0)/2h + (y_0 - 2y_1 + y_2)(x - x_1)/h^2,$$

which leads to the approximate formulae

$$dy/dx \text{ (at } x = x_0) = (4y_1 - 3y_0 - y_2)/2h, \tag{7}$$

$$dy/dx \text{ (at } x = x_1) = (y_2 - y_0)/2h, \tag{8}$$

$$dy/dx \text{ (at } x = x_2) = (y_0 - 4y_1 + 3y_2)/2h. \tag{9}$$

Equation (8) is a widely used differentiation formula, while (7) and (9) are used to estimate derivatives at the beginning and end of a table respectively.

Differentiating (6) twice with respect to x leads to the approximate formula for the second derivative

$$d^2y/dx^2 \text{ (at } x = x_1) = (y_0 - 2y_1 + y_2)/h^2, \tag{10}$$

and since

$$\int\limits_{x_0}^{x_2} P_2(x)\, dx = \tfrac{1}{3}h(y_0 + 4y_1 + y_2),$$

the integral from x_0 to x_n of the tabulated function can be estimated from

$$\int\limits_{x_0}^{x_n} y\, dx = \int\limits_{x_0}^{x_2} y\, dx + \int\limits_{x_2}^{x_4} y\, dx + \ldots + \int\limits_{x_{n-2}}^{x_n} y\, dx$$

$$= \tfrac{1}{3}h\{y_0 + 4y_1 + 2y_2 + 4y_3 + \ldots + 2y_{n-1} + y_n\} \tag{11}$$

if n is even. This is *Simpson's rule* for evaluating an integral numerically.

(ii) *Lagrangian interpolation*

The kth-degree polynomial

$$L_j(x) = \frac{(x - x_0)\ldots(x - x_{j-1})(x - x_{j+1})\ldots(x - x_k)}{(x_j - x_0)\ldots(x_j - x_{j-1})(x_j - x_{j+1})\ldots(x_j - x_k)}$$

has the property that

$$L_j(x_r) = 0 \quad \text{if } r \neq j$$
$$= 1 \quad \text{if } r = j.$$

It follows that

$$P_k(x) = \sum_{j=0}^{k} y_j\, L_j(x) \tag{12}$$

automatically satisfies the condition that the interpolating polynomial should pass through the $k+1$ points $(x_0, y_0), \ldots, (x_k, y_k)$.

Equation (12) is particularly useful when the points $x_0, \ldots, x_n$ are not spaced at regular intervals. If they *are* spaced at regular intervals h, equation (12) leads in the case $k = 2$ to differentiation and integration formulae that are identical to (7)–(11) above.

(iii) *Finite difference formulae*

If $y_0, y_1,..., y_n$ are the tabulated values, at regularly spaced intervals, of a function y, the *forward differences* of y are defined as

$$\begin{aligned}\Delta y_r &= y_{r+1} - y_r \\ \Delta^2 y_r &= \Delta y_{r+1} - \Delta y_r \\ &\;\;\vdots \\ \Delta^k y_r &= \Delta^{k-1} y_{r+1} - \Delta^{k-1} y_r,\end{aligned}$$

and a 'forward difference table' in which the first column contains the function values $y_0,..., y_n$, the second column contains the differences $\Delta y_0,..., \Delta y_{n-1}$, the third column contains the differences $\Delta^2 y_0,..., \Delta^2 y_{n-2}$, and so on, can very easily be constructed by hand.

Once a forward difference table has been constructed, a variety of formulae† can be used for interpolation between the tabulated values of y. Two important formulae are the *Gregory–Newton* formula

$$y(x_r + \theta h) = y_r + \theta \Delta y_r + \frac{\theta(\theta-1)}{2!} \Delta^2 y_r + \frac{\theta(\theta-1)(\theta-2)}{3!} \Delta^3 y_r + \dots \quad (13)$$

and the *Newton–Gauss* formula

$$y(x_r + \theta h) = y_r + \theta \Delta y_r + \frac{\theta(\theta-1)}{2!} \Delta^2 y_{r-1} + \\ + \frac{(\theta+1)\theta(\theta-1)}{3!} \Delta^3 y_{r-1} + \dots \quad (14)$$

If y is in fact a polynomial of degree k, it can be shown that

$$\Delta^m y_r = 0 \quad \text{for } m > k.$$

It follows that truncating either (13) or (14) after the term in Δ^2 is equivalent to fitting a second-degree polynomial to three

† For a full discussion of forward difference and other difference formulae, see Milne, *Numerical Calculus* (Princeton University Press); Whittaker and Robinson, *The Calculus of Observations* (Blackie).

points in the table. In the case of the Gregory–Newton formula, it can be confirmed that these are the points y_r, y_{r+1}, and y_{r+2}; while in the case of the Newton–Gauss formula, the three points are y_{r-1}, y_r, and y_{r+1}. Similarly, truncation after the term in Δ^3 is equivalent to fitting a third-degree polynomial to four points, and so on.

The interpolating variable in (13) and (14) is θ, and by a suitable choice of x_r and θ, one can use either (13) or (14) to estimate y at any point x in the interval (x_0, x_n). Differentiation and integration formulae can be derived by differentiating or integrating the interpolation formulae with respect to θ. Cf. Ex. 16 and Ex. 17 at the end of this chapter.

We have not so far discussed the accuracy of any of the formulae derived above. It can be shown, for instance by comparing formulae with the terms of a Taylor's expansion (cf. Ex. 15 at the end of this chapter), that the differentiation formula (2) is of order h, while (7)–(10) are all $\mathbf{O}(h^2)$. The trapezoidal rule for integration is $\mathbf{O}(h^2)$ and Simpson's rule is $\mathbf{O}(h^4)$.

This information encourages us, for instance, to use Simpson's rule rather than the trapezoidal rule in a particular example, but it does not tell us how accurate the solution then obtained really is. Suppose I_1 is the estimate obtained for an integral using Simpson's rule with an interval h. We could then recalculate the integral using an interval $2h$ to obtain an estimate I_2. Comparing I_1 and I_2 will give some idea of the accuracy of I_1.

We can in fact go one step further. If I^* is the true value of the integral, since Simpson's rule is $\mathbf{O}(h^4)$ we can write

$$I_1 - I^* = kh^4 \tag{15}$$

and

$$I_2 - I^* = k(2h)^4 \tag{16}$$

approximately. Eliminating kh^4 from (15) and (16) gives

$$I^* = (16I_1 - I_2)/15$$

which is a more accurate estimate of the integral than either I_2 or I_1.

EXAMPLES ON CHAPTER XIV

1. Show that the geometric series $S_r = 1+k+k^2+...+k^{r-1}$ satisfies the difference equation $S_{r+1}-S_r = k^r$. Hence show that $S_r = (k^r-1)/(k-1)$.

2. The 'inbreeding coefficient' is the probability that two genes are identical by descent. If f_r is the inbreeding coefficient after r generations of brother–sister mating, it can be shown that $f_{r+2} = \frac{1}{4}(1+2f_{r+1}+f_r)$. If $f_0 = f_1 = 0$, find f_r. Compare this with the case of self-fertilization, where $f_{r+1} = \frac{1}{2}(1+f_r)$ and $f_0 = 0$.

3. The equation $y_{r+2}-2\beta y_{r+1}+\beta y_r = c$, where c is a constant and $0 < \beta < 1$, arises in inventory analysis. Show that its general solution is

$$\beta^{r/2}(A\cos r\theta+B\sin r\theta)+c/(1-\beta),$$

which represents a damped oscillation about $c/(1-\beta)$. [$\tan\theta = (1/\beta-1)^{\frac{1}{2}}$.]

4. Suppose that it takes exactly one second to transmit a 'dot' and two seconds to transmit a 'dash' in Morse code. If N_r is the number of possible message sequences of duration r seconds, show that $N_1 = 1$, $N_2 = 2$, $N_3 = 3$, and in general $N_r = N_{r-1}+N_{r-2}$. Solve this equation and hence show that the 'capacity' of the signalling system,

$$C = \lim_{r\to\infty}\left\{\frac{\log_2 N_r}{r}\right\}, \quad \text{is } \log_2\{(1+\sqrt{5})/2\}.$$

5. In a single game a gambler wins or loses a chip with probabilities p and $q = 1-p$ respectively. The gambler has r chips to begin with and his opponent has $N-r$ chips. If P_r is the probability that the gambler will eventually be ruined, show that

$$P_r = pP_{r+1}+qP_{r-1}.$$

Hence show that $P_r = \{1-(q/p)^{N-r}\}/\{1-(q/p)^N\}$.

6. At time $t = 0$ an infected person joins a group of n susceptible and uninfected people. If $p_r(t)$ is the probability that r people will still be uninfected at time t, a stochastic epidemic model† leads to the equations

$$dp_r/dt = (r+1)(n-r)p_{r+1}-r(n-r+1)p_r \quad (r = 0, 1,..., n-1)$$

and $$dp_n/dt = -np_n.$$

The initial conditions are $p_r(0) = 0$ $(r = 0, 1, 2,..., n-1)$ and $p_n(0) = 1$. Solve these equations for the case $n = 4$ and calculate the expected number of uninfected persons $\bar{p} = \sum_{r=0}^{n} rp_r$. Compare this with the solution to the corresponding deterministic model in which the number of uninfected persons $x(t)$ is given by $dx/dt = -x(n-x+1)$ with $x(0) = n$.

7. Write down the 'improved' formula of § 115 for the equation $dy/dx = -y$ where $y = 1$ when $x = 0$. If the error in y_{r-1} is ϵ_{r-1} and the

† See Bailey, *The Mathematical Theory of Epidemics* (Griffin).

error in y_r is ϵ_r, show that these errors will produce an error

$$\epsilon_{r+1} = \epsilon_{r-1} - 2h\epsilon_r$$

in y_{r+1}. Solve this equation and hence prove that the 'improved' formula is, in this case, unstable for all values of h.

If the predictor–corrector method is applied to the same problem, show that $\epsilon_{r+1} = \epsilon_r(1-\frac{1}{2}h+h^2)-\frac{1}{2}h\epsilon_{r-1}$, and confirm that the predictor–corrector method is stable if $h \lessdot \frac{1}{2}$.

8. Show that Taylor's method (cf. § 115) applied to the equation $dy/dx = x+y$ leads to the formula

$$y_{r+1} = (1+x_r+y_r)e^h-(1+x_r+h).$$

Hence deduce that if $x_0 = y_0 = 0$, the exact solution of the differential equation is $y = e^x-x-1$.

9. Recalculate the solutions to some of the examples given in §§ 114–16. By varying the interval h and introducing checks wherever possible, comment on the accuracy of these solutions.

10. Explain how to use a predictor–corrector method to solve equation § 47 (15) for the non-linear behaviour of a tunnel–diode oscillator.

11. Solve $dy/dx = 1+y^2$ numerically, given $y = 0$ at $x = 0$. Use a third-order Runge–Kutta formula with an interval $h = 0{\cdot}1$ and check on truncation errors by 'back-tracking' at every step. Comment on your results, and explain them by solving the differential equation mathematically.

12. Write a single computer program that is capable of solving either van der Pol's equation § 58 (8) or Rayleigh's equation § 58 (9). Given $x = 0$ and $\dot{x} = 1$ at $t = 0$, use the program to solve van der Pol's equation for the cases $n = 1$, $\epsilon = 0{\cdot}1$ and $n = 1$, $\epsilon = 10{\cdot}0$. Compare your results with Figs. 49 (*b*) and (*c*). Notice that for large ϵ the solution is unstable, irrespective of the method of solution, unless very small time intervals are used. Draw phase diagrams of your solutions and notice that the van der Pol equation has an unusual limit cycle. [Hint: see Ex. 33 at the end of Chapter VII.]

13. Solve the equation $d^2y/dx^2+\omega^2y = 0$ numerically in the region $0 < x < 1$, given $y = 0$ and $dy/dx = 1$ at $x = 0$. Find the smallest value of ω for which the solution gives $y = 0$ at $x = 1$. (Notice that this is an eigenvalue problem, cf. § 87.)

14. If f is a function of x only, show that either a third- or fourth-order Runge–Kutta formula, applied to the differential equation $dy/dx = f(x)$, is equivalent to the use of Simpson's rule for evaluating $\int_0^x f(x)\,dx$.

15. Use Taylor's expansion for $y(x_1+h)$ and $y(x_1-h)$ to show that the numerical differentiation formulae § 117 (8), (10) are both of order h^2.

16. By differentiating the Gregory–Newton formula § 117 (13) with respect to θ, show that

$$hy'(x_r+\theta h) = \Delta y_r+\tfrac{1}{2}(2\theta-1)\Delta^2y_r+\dots.$$

If this series is truncated after the term in Δ^2, show that it is equivalent to the result of § 117 (7).

17. From the Newton–Gauss formula § 117 (14), show that

$$\int_{x_0}^{x_2} y(x_1+\theta h)\,dx = \tfrac{1}{3}h(y_0+4y_1+y_2)-\tfrac{1}{90}h\,\delta^4 y_1$$

where $\delta^2 y_r = \Delta y_r - \Delta y_{r-1}$. Notice that this provides a correction term for Simpson's rule.

18. Tabulate $y = \cos x$ for $0 \leqslant x \leqslant 1$ at intervals $h = 0{\cdot}1$. Hence estimate, as accurately as you can, (i) $\cos(0{\cdot}54)$; (ii) $\cos(0{\cdot}02)$; (iii) $\int_0^1 \cos x\,dx$; (iv) dy/dx at $x = 0$; (v) d^2y/dx^2 at $x = 0{\cdot}4$. Compare your results with the correct answers.

19. The error function, introduced in § 119, is defined as

$$\operatorname{erf}(x) = \frac{2}{\sqrt{\pi}}\int_0^x e^{-u^2}\,du.$$

Use Simpson's rule to tabulate $\operatorname{erf}(x)$ for $0 \leqslant x \leqslant 1$, at intervals $h = 0{\cdot}2$, working to four decimal places. Hence estimate $\operatorname{erf}(0{\cdot}7)$.
[Answer $= 0{\cdot}6778$.]

20. The four-point 'Gaussian quadrature' formula for $\int_{-1}^{1} y\,dx$ is

$$W_1\{y(x_1)+y(-x_1)\}+W_2\{y(x_2)+y(-x_2)\},$$

where $W_1 = 0{\cdot}347855$, $W_2 = 0{\cdot}652145$, $x_1 = 0{\cdot}861136$, and $x_2 = 0{\cdot}339981$. (The weights W_1, W_2 and the points x_1, x_2 are especially chosen so that this formula is exact for all polynomials up to degree seven; however, this type of formula can only be used if it is possible to evaluate y at x_1 and x_2.) Use this formula to estimate

$$\operatorname{erf}(1{\cdot}0) = \frac{1}{\sqrt{\pi}}\int_{-1}^{1} e^{-u^2}\,du,$$

and compare your answer with Ex. 19 above. [Correct answer is $0{\cdot}84270$.]

XV

PARTIAL DIFFERENTIAL EQUATIONS

118. Introductory

IN this chapter we shall give a brief account of the most important simple types of linear partial differential equations.† Naturally nothing more than a sketch can be given, but it is useful to know how such equations arise and how the methods given earlier may be used to solve them.

First we derive the most important equations in two variables, namely

$$\frac{\partial^2 v}{\partial x^2} - \frac{1}{\kappa}\frac{\partial v}{\partial t} = 0, \tag{1}$$

$$\frac{\partial^2 v}{\partial x^2} - \frac{1}{c^2}\frac{\partial^2 v}{\partial t^2} = 0, \tag{2}$$

$$\frac{\partial^2 v}{\partial x^2} + \frac{\partial^2 v}{\partial y^2} = 0, \tag{3}$$

where κ and c are constants. The first of these is *the diffusion equation*, the second *the wave equation*, and the third *Laplace's equation*. In (1) and (2), t is the time, and the equations have to be solved with *initial conditions* at $t = 0$ and *boundary conditions* at certain values of x. In (3), x and y are both space variables, and the equation has to be solved within a region in the (x, y)-plane, and with conditions at the boundary of the region.

The first point to notice is that the nature of the boundary conditions under which they can be solved, the methods of solution applicable to them, and the properties of their solutions, differ widely between the three types. The general linear second-order partial differential equation in x and y with constant coefficients is

$$a\frac{\partial^2 v}{\partial x^2} + b\frac{\partial^2 v}{\partial y^2} + 2h\frac{\partial^2 v}{\partial x \partial y} + 2g\frac{\partial v}{\partial x} + 2f\frac{\partial v}{\partial y} + cv = 0. \tag{4}$$

† The standard works are Bateman, *Partial Differential Equations* (Cambridge); Webster, *Partial Differential Equations of Mathematical Physics* (Dover); Garabedian, *Partial Differential Equations* (Wiley).

This is said to be of elliptic, parabolic, or hyperbolic type according as the conic

$$ax^2+by^2+2hxy+2gx+2fy+c = 0 \tag{5}$$

is an ellipse, parabola, or hyperbola, that is, according as $ab-h^2 \gtreqless 0$. This classification is intimately connected with the detailed theory of the nature of the solutions of the equation, and the three types have very different properties and methods of solution. It appears that (1) is of parabolic type, (2) of hyperbolic type, and (3) of elliptic type.

In §§ 119, 120, 123 the equations (1)–(3) are derived. Each of them has some simple special solutions which follow immediately from the equation and because of their simplicity and generality are of great physical importance. These are given with the equations. The equations for the uniform transmission line, which are of hyperbolic type and contain (1) and (2) as special cases, are given in § 122.

The general methods of solution are next discussed. Fourier series are applicable to all three equations if the range of the independent variable concerned is finite; if it is infinite, Fourier integrals are used in the same way. The Laplace transformation may be applied to (1) and (2) but is not very suitable for (3). But the method of conformal representation allows (3) to be solved for a wide variety of bounding surfaces.

In §§ 127 to 129 the equations corresponding to (1)–(3) but in two and three space dimensions are discussed briefly. Normally these can be solved only for regions bounded by the surfaces of some simple coordinate system, rectangular, spherical polar, cylindrical polar, elliptic, etc., in which case they can be split up into a number of equations in the separate coordinates. Fourier series and integrals and analogous expansions in Legendre and Bessel functions are used in the process of solution.

Finally the question of uniqueness must be mentioned. For each of the equations considered a *uniqueness theorem* can be proved which states that (subject, of course, to pure-mathematical restrictions on the nature of the functions involved) there is only one solution of a completely stated problem on a linear differential equation and boundary conditions. This might be regarded as obvious from the physical point of view. The importance of it is that it allows us to assert that if we can find a solution by any method, this is in fact the unique solution of the problem.

119. The equation of linear flow of heat. Simple solutions

We suppose heat to be flowing in the direction of the x-axis, the temperature being the same over any plane $x =$ constant. Let v be the temperature at the point x, and let K, ρ, and c be the thermal conductivity, density, and specific heat of the medium, which are assumed to be constant.

The fundamental assumption of the theory of conduction of heat is that the rate of flow of heat, per unit time per unit area, across the plane x is

$$-K\frac{\partial v}{\partial x}. \tag{1}$$

The differential equation is found by considering a region of unit area of the medium between the planes x and $x+\delta x$. Heat flows into this region across the plane x at the rate (1). It flows out of it across the plane $x+\delta x$ at the rate

$$-K\frac{\partial v}{\partial x}-\frac{\partial}{\partial x}\left(K\frac{\partial v}{\partial x}\right)\delta x. \tag{2}$$

Thus the region gains heat by flow across its surfaces at the rate

$$K\frac{\partial^2 v}{\partial x^2}\,\delta x \tag{3}$$

per unit time. This gain of heat causes a rise of temperature in the region, and since the thermal capacity of the region is $\rho c\,\delta x$, its rate of rise of temperature $\partial v/\partial t$ is

$$\frac{\partial v}{\partial t}=\frac{K\,\delta x}{\rho c\,\delta x}\frac{\partial^2 v}{\partial x^2}.$$

That is,

$$\frac{\partial^2 v}{\partial x^2}-\frac{1}{\kappa}\frac{\partial v}{\partial t}=0, \tag{4}$$

where

$$\kappa = K/\rho c \tag{5}$$

is called the diffusivity of the medium.

(4) is the required differential equation. It has to be solved

in some region such as $0 < x < l$, $x > 0$, or $-\infty < x < \infty$, with a given initial value of $v(x,t)$ at the instant $t = 0$, and with boundary conditions at the ends of the region. The usual boundary conditions are:

(i) Prescribed temperature v. This may be constant or a given function of the time.

(ii) Prescribed rate of flow of heat. In this case, by (1), $\partial v/\partial x$ is prescribed. If there is no flow of heat, $\partial v/\partial x = 0$.

(iii) The rate of flow of heat proportional to the temperature difference between the solid and its surroundings which are at v_0, that is

$$K\frac{\partial v}{\partial x}+H(v-v_0) = 0, \tag{6}$$

where H is a constant.

(iv) If the region extends to infinity the temperature must be finite there.

The differential equation (4) and the boundary conditions above are all linear, and it is this fact which makes it possible to go so far with the theory. In practice non-linear boundary conditions of type

$$K\frac{\partial v}{\partial x}+H(v-v_0)^n = 0 \tag{7}$$

often arise, but little can be done with these.

In this section we give some simple solutions of (4) for the infinite region $-\infty < x < \infty$, and the semi-infinite region $x > 0$, which are of great practical importance. They depend on the fact that

$$v = t^{-\frac{1}{2}}e^{-(x-x')^2/4\kappa t}, \tag{8}$$

where x' is a constant, satisfies the equation (4). This may be verified immediately by differentiation. Thus from (8)

$$\frac{\partial v}{\partial t} = \left\{-\frac{1}{2t^{\frac{3}{2}}}+\frac{(x-x')^2}{4\kappa t^{\frac{5}{2}}}\right\}e^{-(x-x')^2/4\kappa t},$$

$$\frac{\partial v}{\partial x} = -\frac{(x-x')}{2\kappa t^{\frac{3}{2}}}e^{-(x-x')^2/4\kappa t},$$

$$\frac{\partial^2 v}{\partial x^2} = \left\{-\frac{1}{2\kappa t^{\frac{3}{2}}}+\frac{(x-x')^2}{4\kappa^2 t^{\frac{5}{2}}}\right\}e^{-(x-x')^2/4\kappa t} = \frac{1}{\kappa}\frac{\partial v}{\partial t},$$

as required.

Thus (8) satisfies the equation of conduction of heat for all values of x'. We next find a physical interpretation for it. The total quantity of heat per unit area in the region $-\infty < x < \infty$ when the temperature is given by (8) is

$$Q = \rho c \int_{-\infty}^{\infty} v \, dx$$

$$= \frac{\rho c}{t^{\frac{1}{2}}} \int_{-\infty}^{\infty} e^{-(x-x')^2/4\kappa t} \, dx = 2\rho c \kappa^{\frac{1}{2}} \int_{-\infty}^{\infty} e^{-\xi^2} \, d\xi = 2\rho c(\kappa\pi)^{\frac{1}{2}}, \quad (9)$$

and thus is independent of the time.

Also, from (8), if $x \neq x'$, $v \to 0$ as $t \to 0$; but if $x = x'$, $v = t^{-\frac{1}{2}}$ and $v \to \infty$ as $t \to 0$. Therefore, multiplying (8) by a constant determined by (9), we get the result that

$$\frac{Q}{2\rho c(\pi\kappa t)^{\frac{1}{2}}} e^{-(x-x')^2/4\kappa t} \quad (10)$$

is a solution of the equation of conduction of heat which corresponds to releasing instantaneously when $t = 0$ a quantity of heat Q per unit area on the plane $x = x'$, the solid being at zero temperature when $t = 0$.

From this elementary solution many important results can be derived. For example, suppose that the solid $-\infty < x < \infty$ has temperature $f(x)$ when $t = 0$, and we wish to find its temperature subsequently. This initial temperature may be produced by liberating an amount of heat $Q = \rho c \, f(x') \, \delta x'$ per unit area in the region between each pair of planes $(x', x'+\delta x')$. Putting this value of Q in (10) and integrating with respect to x', the temperature at x in the solid at time t is found to be

$$\frac{1}{2(\pi\kappa t)^{\frac{1}{2}}} \int_{-\infty}^{\infty} f(x') e^{-(x-x')^2/4\kappa t} \, dx'. \quad (11)$$

As a second example on these elementary solutions we note that, since (8) satisfies (4), all its derivatives and integrals will do so, and in fact they all have a fundamental significance in the theory. Thus

$$\int^{x} e^{-x^2/4\kappa t}\frac{dx}{2\sqrt{(\kappa t)}} = \int^{x/2(\kappa t)^{\frac{1}{2}}} e^{-\xi^2}\,d\xi \tag{12}$$

will satisfy (4).

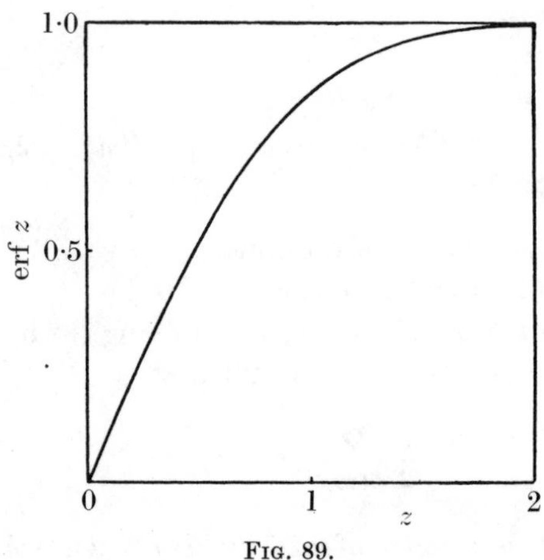

FIG. 89.

The function
$$\operatorname{erf} z = \frac{2}{\sqrt{\pi}}\int_0^z e^{-\xi^2}\,d\xi \tag{13}$$

is called the error function and is tabulated; its graph is shown in Fig. 89. Its principal properties are

$$\operatorname{erf} z \to 1 \quad \text{as } z \to \infty, \tag{14}$$

$$\operatorname{erf} 0 = 0. \tag{15}$$

It follows from (12) that

$$v = V_0 \operatorname{erf}\frac{x}{2(\kappa t)^{\frac{1}{2}}}, \tag{16}$$

where V_0 is a constant, is a solution of the equation of conduction of heat. By (14) and (15) this solution has the properties

$$v = 0, \quad \text{when } x = 0, \quad \text{for all } t > 0,$$

$$v = V_0, \quad \text{when } t = 0, \quad \text{for all } x > 0.$$

Thus (16) is the solution of the problem of *the region* $x > 0$ *with initial temperature* V_0 *and with the surface* $x = 0$ *kept at zero temperature for* $t > 0$.

In the same way, the solution of the problem of *the region* $x > 0$ *with zero initial temperature and with the surface* $x = 0$ *kept at constant temperature* V_0 *for* $t > 0$ *is*

$$v = V_0\left(1-\operatorname{erf}\frac{x}{2\sqrt{(\kappa t)}}\right). \tag{17}$$

As a final example of a different type we study the *steady periodic oscillations in the temperature in the semi-infinite solid* $x > 0$ *due to periodic surface temperature.* The results have important applications to the annual and diurnal fluctuations of soil temperature, and to temperatures in the cylinder walls of reciprocating engines.

Following the usual procedure for finding steady periodic solutions, we seek a solution of (4) of the form

$$v = V(x)e^{i\omega t}, \tag{18}$$

where $V(x)$ is a function of x only. Substituting (18) in (4) gives the differential equation for V,

$$\frac{d^2V}{dx^2}-\frac{i\omega}{\kappa}V = 0. \tag{19}$$

The general solution of (19) is

$$V = Ae^{(1+i)x(\omega/2\kappa)^{\frac{1}{2}}}+Be^{-(1+i)x(\omega/2\kappa)^{\frac{1}{2}}},$$

and, since V must be finite as $x \to \infty$, A must be zero. Therefore

$$v = Be^{-(1+i)x(\omega/2\kappa)^{\frac{1}{2}}+i\omega t}.$$

Taking the imaginary part we find that

$$v = V_0\, e^{-x(\omega/2\kappa)^{\frac{1}{2}}}\sin\left\{\omega t-x\left(\frac{\omega}{2\kappa}\right)^{\frac{1}{2}}\right\} \tag{20}$$

is the required solution for harmonic surface temperature

$$v = V_0 \sin \omega t.$$

The temperature oscillations at depth x diminish in amplitude as x increases, and they lag in phase by an increasing amount behind the surface oscillations.

Many important extensions of (4) may easily be derived in the same way. If heat is produced in the solid at a rate $A(x,t)$ per unit time per unit volume, (4) is replaced by

$$\frac{\partial^2 v}{\partial x^2}-\frac{1}{\kappa}\frac{\partial v}{\partial t}=-\frac{A(x,t)}{K}. \tag{21}$$

Clearly (4) and (21) hold for a rod with no loss of heat from its surface. If the rod is so thin that its temperature is uniform across its cross-section and it loses heat from its surface at a rate proportional to its temperature, (21) is replaced by

$$\frac{\partial^2 v}{\partial x^2}-\frac{\nu}{\kappa}v-\frac{1}{\kappa}\frac{\partial v}{\partial t}=-\frac{A(x,t)}{K}, \tag{22}$$

where ν is a constant depending on the size and material of the rod.

Differential equations of type (4) arise in many other connexions, notably in the theory of laminar motion of viscous fluid and in the theory of diffusion. In the latter case, if c is the concentration of a dissolved substance, the fundamental assumption is that the rate at which this substance crosses any plane is

$$-D\frac{\partial c}{\partial x}, \tag{23}$$

where D is the diffusion constant. Then in the same way the differential equation for c is found to be

$$\frac{\partial^2 c}{\partial x^2}-\frac{1}{D}\frac{\partial c}{\partial t}=0. \tag{24}$$

For this reason the differential equation (4) is often called the diffusion equation.

120. The wave equation in one dimension. Simple solutions

This equation appears in many connexions with different notations, for example in the longitudinal vibrations of rods, transverse vibrations of stretched strings, sound waves, water waves, etc.

We derive it first for the longitudinal vibrations of a bar of uniform cross-section. Let u be the displacement of the plane of the bar whose normal position is at x, and let $u+\delta u$ be the displacement of the plane $x+\delta x$. Let X be the stress in the bar across the plane whose normal position is at x, then, by Hooke's law,

$$X = E\,\frac{\delta u}{\delta x},$$

where E is Young's modulus. That is, in the limit as $\delta x \to 0$

$$X = E\,\frac{\partial u}{\partial x}. \tag{1}$$

We now find the equation of motion of the element of the bar whose normal position is between the planes x and $x+\delta x$. The displacement of this element is u, and its mass is $\rho\,\delta x$ per unit area of the bar, where ρ is the density of the material of the bar. The forces per unit area on the element are $-X$ on the face x, and

$$X+\frac{\partial X}{\partial x}\,\delta x$$

on the face $x+\delta x$. Thus its equation of motion is

$$\rho\,\delta x\,\frac{\partial^2 u}{\partial t^2} = \frac{\partial X}{\partial x}\,\delta x,$$

or, using (1),

$$\frac{\partial^2 u}{\partial x^2}-\frac{1}{c^2}\frac{\partial^2 u}{\partial t^2} = 0, \tag{2}$$

where

$$c^2 = E/\rho. \tag{3}$$

The common boundary conditions are:

(i) prescribed displacement u;

(ii) prescribed stress $E\,\partial u/\partial x$;

(iii) the bar attached to a mass ma, where a is the area of the bar. In this case the equation of motion of this mass gives a boundary condition

$$m\,\frac{d^2 u}{dt^2} = -X = -E\,\frac{\partial u}{\partial x}. \tag{4}$$

Before discussing the solution of (2) we derive the corresponding equation for transverse vibrations of a stretched string. Let T be the tension in the string, σ its line of density, y the displacement at the point x, and ψ the slope of the tangent at this point.

Consider an element of the string of length δs at x. The forces on this in the direction of y increasing are

$$T\sin(\psi+\delta\psi)-T\sin\psi = T\,\delta\psi, \tag{5}$$

since ψ is small. Therefore its equation of motion is

$$\sigma\,\delta s\,\frac{\partial^2 y}{\partial t^2} = T\,\delta\psi,$$

or in the limit as $\delta s \to 0$

$$\sigma\frac{\partial^2 y}{\partial t^2} = \frac{T}{\rho}, \tag{6}$$

where ρ, the radius of curvature of the string, is $\partial s/\partial\psi$. Also, since the string is nearly straight,

$$\frac{1}{\rho} = \frac{\partial^2 y}{\partial x^2},$$

very nearly, and (6) becomes

$$\frac{\partial^2 y}{\partial x^2}-\frac{1}{c^2}\frac{\partial^2 y}{\partial t^2} = 0, \tag{7}$$

where $$c^2 = T/\sigma. \tag{8}$$

We now consider some simple solutions of (2) or (7). These depend on the fact that, if $f(x)$ is any differentiable function of x, it follows immediately by differentiating that

$$f(x+ct) \tag{9}$$

satisfies (7). Physically it represents a disturbance, whose form when $t = 0$ is $f(x)$, travelling to the left with undisturbed shape and with velocity c. In the same way

$$F(x-ct) \tag{10}$$

is also a solution, and represents a wave travelling to the right with velocity c.

Thus the general solution of (7) may be regarded as a combination of motions to the left and right given by (9) and (10).

These results enable us to find the solution of (7) for an infinite string $-\infty < x < \infty$ set in motion at $t = 0$ with initial displacement $y = \phi(x)$ and initial velocity $\partial y/\partial t = \psi(x)$. Taking for the general solution

$$y = f(x+ct)+F(x-ct), \tag{11}$$

the initial conditions require

$$f(x)+F(x) = \phi(x), \tag{12}$$

$$cf'(x)-cF'(x) = \psi(x). \tag{13}$$

Integrating (13) gives

$$cf(x)-cF(x) = \int^{x} \psi(x)\, dx. \tag{14}$$

From (12) and (14),

$$f(x) = \tfrac{1}{2}\phi(x)+\frac{1}{2c}\int^{x} \psi(x)\, dx,$$

$$F(x) = \tfrac{1}{2}\phi(x)-\frac{1}{2c}\int^{x} \psi(x)\, dx.$$

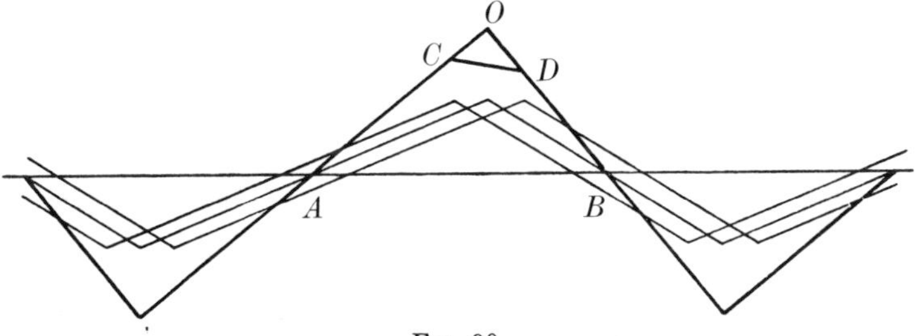

FIG. 90.

Using these values in (11), we get finally

$$y = \tfrac{1}{2}\{\phi(x+ct)+\phi(x-ct)\}+\frac{1}{2c}\int\limits_{x-ct}^{x+ct} \psi(x)\, dx. \tag{15}$$

For the case in which $\psi(x) = 0$, the solution corresponds to two waves, each of half the original wave form, propagated to the left and right respectively.

This type of solution for the infinite string may be extended to give the solution for a finite string of length l.

Suppose that the string is fixed at A and B, and is plucked at one point O as in Fig. 90. Suppose that we repeat this pattern indefinitely to make an odd periodic function of period $2l$, and regard this as an initial displacement of an infinite string. Then the general solution (15) consists of half the displacement moving to the right and half moving to the left. From the figure it appears that the sums of these displacements at A and B are both zero, so that the solution obtained in this way is a solution of (7) with $y = 0$ at A and B, and thus is the solution of our problem. The form of the string is shown in Fig. 90; it consists of three straight portions, AC, CD, DB.

Finally, a fundamental distinction between the results of this section and the last should be pointed out. Solutions of the wave equation are propagated with finite velocity c, so that if portion of the medium is initially undisturbed it remains so until the wave front reaches it. In diffusion problems, on the other hand, there is theoretically a disturbance at all points at all times, though for large distances this will be negligibly small.

121. The wave equation. Natural frequencies

It was found in Chapter IV, when studying the vibrations of a system consisting of a finite number of masses, that such a system had a finite number of natural frequencies, that with each frequency there was associated a normal mode of vibration, and that the most general motion consisted of a sum of vibrations of these types with coefficients determined by the initial conditions.

The natural frequencies and normal modes were found by seeking solutions of the equations of motion in which all quantities were proportional to $e^{i\omega t}$. The same procedure applies to the wave equation in one or more dimensions. As an example consider the vibrations of a stretched string whose ends, $x = 0$ and $x = l$, are fixed. The differential equation § 120 (7) is

$$\frac{\partial^2 y}{\partial x^2} - \frac{1}{c^2}\frac{\partial^2 y}{\partial t^2} = 0 \quad (0 < x < l), \tag{1}$$

with

$$y = 0, \text{ when } x = 0 \text{ and } x = l. \tag{2}$$

We seek a solution of this of the form

$$y = Y(x)e^{i\omega t}, \tag{3}$$

and, substituting in (1) and (2), Y must satisfy

$$\frac{d^2 Y}{dx^2} + \frac{\omega^2}{c^2} Y = 0, \tag{4}$$

with
$$Y = 0, \quad \text{when } x = 0, \tag{5}$$
$$Y = 0, \quad \text{when } x = l. \tag{6}$$

The general solution of (4) is
$$Y = A\sin\frac{\omega x}{c}+B\cos\frac{\omega x}{c}, \tag{7}$$
and (5) requires $B = 0$. Thus by (6) we must have
$$A\sin\frac{\omega l}{c} = 0, \tag{8}$$
so that either $A = 0$, which gives the trivial solution $Y = 0$, or
$$\sin\frac{\omega l}{c} = 0,$$
that is,
$$\omega = \frac{n\pi c}{l} \quad (n = 1, 2, 3,...). \tag{9}$$

The problem is an eigenvalue problem (cf. § 87). The eigenvalues (9) give the natural frequencies $(nc/2l)$, $n = 1, 2,...$, and the corresponding eigenfunctions,
$$A_n \sin\frac{n\pi x}{l} \quad (n = 1, 2,...), \tag{10}$$
are the normal modes of vibration. A combination of these, namely
$$\sum_{n=1}^{\infty}\left\{A_n\sin\frac{n\pi ct}{l}+B_n\cos\frac{n\pi ct}{l}\right\}\sin\frac{n\pi x}{l}, \tag{11}$$
gives the most general solution of (1). The A_n and B_n in (11) are determined from the initial conditions by the use of Fourier series as in § 124. But, just as in the case of the vibrations of a finite number of masses, the natural frequencies and normal modes are usually easy to find and frequently supply as much information as is needed.

The normal modes (10) are shown in Fig. 91, which may be compared with Fig. 17 for a finite number of masses.

As an example of the determination of the natural frequencies in a more complicated case we consider *transverse vibrations of a uniform beam of length l freely hinged at its ends.*

Let y be the displacement of the point x of the beam, and ρ the mass per unit length of the beam. Then the reversed effective force, § 66 (13), at x due to the motion of the beam is

$$-\frac{\rho}{g}\frac{\partial^2 y}{\partial t^2} \tag{12}$$

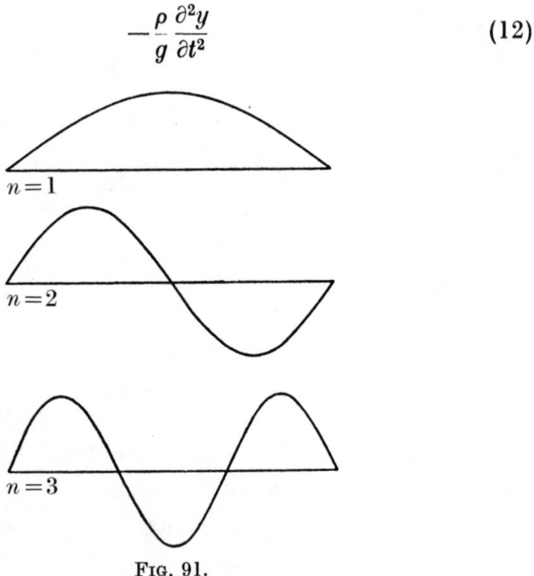

FIG. 91.

per unit length. Treating this as a load on the beam, and neglecting any static loads, § 80 (12) gives for the equation of transverse vibrations of the beam

$$EI\frac{\partial^4 y}{\partial x^4}+\frac{\rho}{g}\frac{\partial^2 y}{\partial t^2} = 0. \tag{13}$$

As before, we seek solutions of (13) of the form

$$y = Y(x)e^{i\omega t},$$

then Y has to satisfy

$$\frac{d^4 Y}{dx^4}-k^4 Y = 0, \tag{14}$$

where $k^4 = \omega^2\rho/EIg$, and

$$\dot{Y} = \frac{d^2 Y}{dx^2} = 0, \quad \text{when } x = 0 \text{ and } x = l. \tag{15}$$

This eigenvalue problem has been discussed in § 87, Ex. 3, and, using the results of that section, it appears that the eigenvalues are given by

$$kl = n\pi \quad (n = 1, 2,...). \tag{16}$$

Therefore the natural frequencies, $\omega/2\pi$, are

$$\frac{n^2\pi}{2l^2}\left(\frac{EIg}{\rho}\right)^{\frac{1}{2}} \quad (n = 1, 2,...). \tag{17}$$

Incidentally it appears that they are identical with the critical frequencies for whirling of the shaft.

122. The equations for the uniform transmission line

Suppose the line has inductance L, resistance R, capacitance C, and leakage conductance G per unit length. Let I be the current in the line and V the voltage drop across it at x, and let

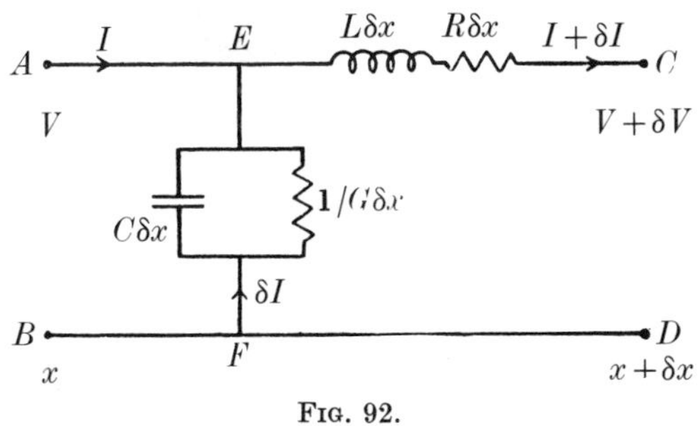

FIG. 92.

$I+\delta I$ and $V+\delta V$ be the corresponding quantities at $x+\delta x$. We may replace the portion of the line between x and $x+\delta x$ by the four-terminal network shown in Fig. 92. The circuit relations for this are

$$L\,\delta x\frac{\partial I}{\partial t}+R\,\delta x I = -\delta V, \tag{1}$$

$$C\,\delta x\frac{\partial V}{\partial t}+G\,\delta x V = -\delta I. \tag{2}$$

Dividing (1) and (2) by δx and taking the limit as $\delta x \to 0$ gives the equations

$$L\frac{\partial I}{\partial t}+RI = -\frac{\partial V}{\partial x}, \tag{3}$$

$$C\frac{\partial V}{\partial t}+GV = -\frac{\partial I}{\partial x}. \tag{4}$$

These are a pair of simultaneous linear partial differential equations of the first order for V and I. They have to be solved with given initial values of the current and voltage drop in the line as functions of x, and with boundary conditions at the ends of the line such as:

(i) prescribed voltage;

(ii) prescribed current;

(iii) a relation between I and V when the line is connected to a terminal impedance.

Either I or V may be eliminated from (3) and (4) and a second-order equation obtained. For example, eliminating I, the equation for V is

$$\frac{\partial^2 V}{\partial x^2}-LC\frac{\partial^2 V}{\partial t^2}-(LG+RC)\frac{\partial V}{\partial t}-RGV=0. \tag{5}$$

In special cases this reduces to equations which have been considered earlier:

(i) For the 'lossless' line in which $R = G = 0$, it becomes

$$\frac{\partial^2 V}{\partial x^2}-\frac{1}{c^2}\frac{\partial^2 V}{\partial t^2}=0, \tag{6}$$

with $c^2 = 1/LC$, which is the wave equation of § 120.

(ii) For the ideal submarine cable in which $L = G = 0$ it becomes

$$\frac{\partial^2 V}{\partial x^2}-\frac{1}{\kappa}\frac{\partial V}{\partial t}=0, \tag{7}$$

with $\kappa = 1/RC$, which is the diffusion equation of § 119.

(iii) For Heaviside's distortionless line in which $R/L = G/C$ it becomes

$$\frac{\partial^2 V}{\partial x^2}-LC\left(\frac{\partial}{\partial t}+\frac{R}{L}\right)^2 V=0. \tag{8}$$

In these cases solutions of transient problems may be obtained by the methods of §§ 119, 120, 124, 125. The complete solution for the general equation (5) is complicated, and here we shall consider only the steady state behaviour for this case.

Suppose that complex voltage E', in the notation of § 44, is applied to the line at $x = 0$. Let V' be the complex voltage drop at the point x of the line and let I' be the complex current there:

these are now functions of x. $V'e^{i\omega t}$ and $I'e^{i\omega t}$ will be solutions of (3), (4), and (5), and of these (5) gives

$$\frac{d^2V'}{dx^2}-\gamma^2V' = 0, \tag{9}$$

where
$$\gamma = \{(R+L\omega i)(G+C\omega i)\}^{\frac{1}{2}}, \tag{10}$$

and, for definiteness, the square root in (10) is chosen so that its real part is positive. The quantity γ is called the propagation constant of the line.

As a first example suppose that the line extends to infinity.

In the general solution

$$V' = Ae^{-\gamma x}+Be^{\gamma x} \tag{11}$$

of (9), B must be zero since V' must remain finite as $x \to \infty$. Also since $V' = E'$ when $x = 0$, $A = E'$ and we get finally

$$V' = E'e^{-\gamma x} \tag{12}$$

for the complex voltage drop at the point x of the line. The complex current I' is from (3)

$$I' = -\frac{1}{R+L\omega i}\frac{dV'}{dx} \tag{13}$$

$$= E'\left\{\frac{G+C\omega i}{R+L\omega i}\right\}^{\frac{1}{2}} e^{-\gamma x}. \tag{14}$$

The input impedance z_0 of the line, which is the ratio of E' to I' when $x = 0$, is given by

$$z_0 = \left(\frac{R+L\omega i}{G+C\omega i}\right)^{\frac{1}{2}}. \tag{15}$$

This quantity is called the characteristic impedance of the line.

Next we consider the case of the *line* $0 < x < l$, *with complex voltage* E' *applied at* $x = 0$ *as before, but terminated at* $x = l$ *by a complex impedance* z.

The boundary conditions are now

$$V' = E', \quad \text{when } x = 0, \tag{16}$$

$$zI' = V', \quad \text{when } x = l. \tag{17}$$

The solution of (9) is

$$V' = A\sinh\gamma x + B\cosh\gamma x. \tag{18}$$

The condition (16) gives $B = E'$, and (17) gives, using (13) and the notation (15),

$$-\frac{z}{z_0}\{A\cosh\gamma l + B\sinh\gamma l\} = A\sinh\gamma l + B\cosh\gamma l.$$

Solving for A and substituting in (18) gives finally

$$V' = \frac{E'\{z_0\sinh\gamma l + z\cosh\gamma l\}\cosh\gamma x - E'\{z\sinh\gamma l + z_0\cosh\gamma l\}\sinh\gamma x}{z_0\sinh\gamma l + z\cosh\gamma l}. \tag{19}$$

The input impedance is

$$\left[\frac{V'}{I'}\right]_{x=0} = z_0\frac{z_0\sinh\gamma l + z\cosh\gamma l}{z_0\cosh\gamma l + z\sinh\gamma l}. \tag{20}$$

If $z = 0$, i.e. short-circuit at $x = l$, the input impedance z_s is by (20)

$$z_s = z_0\tanh\gamma l, \tag{21}$$

while, if $z = \infty$, open circuit at $x = l$, the input impedance z_{op} is

$$z_{op} = z_0\coth\gamma l. \tag{22}$$

From (21) and (22) $\qquad z_s z_{op} = z_0^2.$

123. Laplace's equation in two dimensions. Simple solutions

Laplace's equation arises in a very large number of contexts in mathematical physics. Perhaps the most fundamental of these is the expression of continuity in steady flow. As an example of this we consider the steady flow of heat in two dimensions.

Take rectangular axes OX, OY, and consider the rectangle bounded by the planes x, $x+\delta x$ and y, $y+\delta y$. Since the flow is steady the temperature is independent of time, and therefore the net flow of heat into this rectangle must be zero.

The rates of flow into the region over the faces x, $x+\delta x$, y, and $y+\delta y$ are, respectively, by § 119 (1),

$$-K\frac{\partial v}{\partial x}\,\delta y,$$

$$-\left\{-K\frac{\partial v}{\partial x}\,\delta y-\frac{\partial}{\partial x}\left[K\frac{\partial v}{\partial x}\,\delta y\right]\delta x\right\},$$

$$-K\frac{\partial v}{\partial y}\,\delta x,$$

$$-\left\{-K\frac{\partial v}{\partial y}\,\delta x-\frac{\partial}{\partial y}\left[K\frac{\partial v}{\partial y}\,\delta x\right]\delta y\right\}.$$

Adding these we have, if K is independent of x and y,

$$\frac{\partial^2 v}{\partial x^2}+\frac{\partial^2 v}{\partial y^2}=0. \tag{1}$$

This is Laplace's equation in two dimensions in rectangular Cartesian coordinates.

Before discussing it, we derive the corresponding equation in polar coordinates. This can be done by transforming (1) into polars r, θ, or it may be obtained by applying the argument used above to the element of area bounded by the circles r and $r+\delta r$, and the rays θ and $\theta+\delta\theta$. The expression of the fact that the net rate of flow of heat into the region is zero is

$$-K\frac{\partial v}{\partial r}r\,\delta\theta-\left\{-K\frac{\partial v}{\partial r}r\,\delta\theta-\frac{\partial}{\partial r}\left[Kr\frac{\partial v}{\partial r}\,\delta\theta\right]\delta r\right\}-$$

$$-K\frac{\partial v}{r\,\partial\theta}\,\delta r-\left\{-K\frac{\partial v}{r\,\partial\theta}\,\delta r-\frac{\partial}{r\,\partial\theta}\left[-K\frac{\partial v}{r\,\partial\theta}\,\delta r\right]r\,\delta\theta\right\}=0.$$

That is,

$$r\frac{\partial}{\partial r}\left[r\frac{\partial v}{\partial r}\right]+\frac{\partial^2 v}{\partial\theta^2}=0, \tag{2}$$

or

$$\frac{\partial^2 v}{\partial r^2}+\frac{1}{r}\frac{\partial v}{\partial r}+\frac{1}{r^2}\frac{\partial^2 v}{\partial\theta^2}=0. \tag{3}$$

Laplace's equation in the form (1) has simple polynomial solutions

$$1,\ x,\ y,\ xy,\ x^2-y^2,\ x^3-3xy^2,\ldots.$$

In the form (2) it is satisfied by

$$r^n \begin{matrix}\cos\\ \sin\end{matrix} n\theta$$

for any n, and also by $\log r$.

Ex. *Radial flow of heat in the hollow cylinder $a < r < b$. $r = a$ kept at v_1, and $r = b$ at v_2.*

The differential equation (2) gives, since the solution is to be independent of θ,

$$\frac{d}{dr}\left(r\frac{dv}{dr}\right) = 0,$$

$$r\frac{dv}{dr} = A,$$

$$v = A\ln r + B,$$

where A and B are unknown constants to be found from the conditions at $r = a$ and $r = b$. These give

$$v_1 = A\ln a + B,$$

$$v_2 = A\ln b + B,$$

and therefore
$$v = \frac{v_1\ln(b/r) + v_2\ln(r/a)}{\ln(b/a)}. \tag{4}$$

The rate of flow of heat through the cylinder per unit length is

$$-2\pi rK\frac{dv}{dr} = \frac{2\pi K(v_1 - v_2)}{\ln b/a}. \tag{5}$$

Laplace's equation in two dimensions occurs also in the flow of current electricity in plane sheets, in the flow of viscous fluid between parallel planes, in the theory of torsion of shafts, in the deflexion of a soap film or a sheet of rubber, and in two-dimensional problems in hydrodynamics, electrostatics, and the flow of incompressible fluid through a porous medium.

The inhomogeneous equation

$$\frac{\partial^2 v}{\partial x^2} + \frac{\partial^2 v}{\partial y^2} = f(x, y) \tag{6}$$

is Poisson's equation in two variables. It arises, for example, in the steady flow of heat in a medium in which heat is being generated. Thus if heat is generated at a rate $f(x, y)$ per unit time per unit area in the medium, the net rate of loss of heat calculated in (1) is to be equated to this and we get

$$\frac{\partial^2 v}{\partial x^2} + \frac{\partial^2 v}{\partial y^2} = -\frac{1}{K}f(x, y). \tag{7}$$

In other fields the relation between Laplace's and Poisson's equations is the same: Poisson's equation occurs in regions where there are sources of heat or current, charges, etc., and reduces to Laplace's equation in regions free from such sources.

124. The use of Fourier series

Fourier series are applied in the same way to all the equations of §§ 119–23. A simple solution of the differential equation and some of the boundary conditions which is a product of trigonometrical or hyperbolic functions is written down; the solution of the problem is assumed to be a series of such terms, the coefficients in which are found from a Fourier series determined by the remaining boundary conditions.

To illustrate the method we first consider Laplace's equation

$$\frac{\partial^2 v}{\partial x^2}+\frac{\partial^2 v}{\partial y^2}=0, \tag{1}$$

and seek a solution of this in the rectangle $0<x<a, 0<y<b$ which satisfies the boundary conditions

$$v=0, \quad \text{when } x=0 \quad (0<y<b), \tag{2}$$

$$v=0, \quad \text{when } x=a \quad (0<y<b), \tag{3}$$

$$v=0, \quad \text{when } y=b \quad (0<x<a), \tag{4}$$

$$v=f(x), \quad \text{when } y=0 \quad (0<x<a). \tag{5}$$

We notice first that

$$\sin\frac{n\pi x}{a}\sinh\frac{n\pi(b-y)}{a} \quad (n=1,2,...) \tag{6}$$

satisfies the differential equation (1) and the boundary conditions (2), (3), and (4). This suggests that the series

$$\sum_{n=1}^{\infty} A_n\frac{\sin n\pi x}{a}\sinh\frac{n\pi(b-y)}{a}, \tag{7}$$

where the A_n are unknown, will also satisfy them. When $y=0$, (7) reduces to

$$\sum_{n=1}^{\infty} A_n\sinh\frac{n\pi b}{a}\sin\frac{n\pi x}{a}. \tag{8}$$

Now suppose that $f(x)$ is expanded in the sine series, § 89 (7),

$$f(x) = \sum_{n=1}^{\infty} b_n \sin\frac{n\pi x}{a}, \tag{9}$$

where
$$b_n = \frac{2}{a}\int_0^a f(x')\sin\frac{n\pi x'}{a}\,dx'. \tag{10}$$

Comparing coefficients between (8) and (9) gives

$$A_n \sinh\frac{n\pi b}{a} = b_n,$$

and thus, finally, the solution of the differential equation and boundary conditions is

$$v = \sum_{n=1}^{\infty} b_n \frac{\sin(n\pi x/a)\sinh n\pi(b-y)/a}{\sinh n\pi b/a}, \tag{11}$$

where b_n is given by (10).

For example, if $v = 1$ on $y = 0$, $0 < x < a$, the solution is, using § 89 (9),

$$v = \frac{4}{\pi}\sum_{r=0}^{\infty} \frac{\sin[(2r+1)\pi x/a]\sinh[(2r+1)\pi(b-y)/a]}{(2r+1)\sinh[(2r+1)\pi b/a]}. \tag{12}$$

Of course the above argument is not rigorous and needs pure-mathematical justification; this is easily supplied for all the examples of this section.

If the writing down of (6) is considered too abrupt, it may be obtained by the following method, which will also be used in § 128 for the study of equations in three variables. We seek a solution of (1) which is the product of a function of x and a function of y, say

$$v = X(x)Y(y). \tag{13}$$

Substituting (13) in (1) we get

$$\frac{1}{X}\frac{d^2X}{dx^2}+\frac{1}{Y}\frac{d^2Y}{dy^2} = 0. \tag{14}$$

This is satisfied if

$$\frac{1}{X}\frac{d^2X}{dx^2} = -k^2, \tag{15}$$

$$\frac{1}{Y}\frac{d^2Y}{dx^2} = k^2, \tag{16}$$

where k^2 is any number.

The general solution of (15) is

$$X = A\sin kx + B\cos kx. \tag{17}$$

If (17) is to satisfy the boundary conditions (2) and (3) we must have

$$B = 0, \tag{18}$$

$$A\sin ka = 0. \tag{19}$$

From (19), either $A = 0$, giving the trivial solution $v = 0$, or

$$k = \frac{n\pi}{a} \quad (n = 1, 2,...), \tag{20}$$

so that in (15) and (16), k must be $n\pi/a$ and X must have the form $A_n \sin n\pi x/a$ $(n = 1, 2,...)$. The numbers (20) are in fact the eigenvalues of the differential equation (15) with the boundary conditions (2) and (3), and the corresponding values of X are its eigenfunctions (cf. § 87).

The general solution of (16) with one of the values (20) of k may be written

$$Y = C\sinh\frac{n\pi(b-y)}{a} + D\cosh\frac{n\pi(b-y)}{a}, \tag{21}$$

and the condition (4) gives $D = 0$. Thus, finally, the solutions of (1)–(4) of type $v = X(x)Y(y)$ must be of type

$$v = A_n \sin\frac{n\pi x}{a}\sinh\frac{n\pi(b-y)}{a} \quad (n = 1, 2,...), \tag{22}$$

in agreement with (6).

The same argument may be used to derive (28) and (38) below.

Ex. 1. *Conduction of heat in the region* $0 < x < l$. *The ends* $x = 0$ *and* $x = l$ *kept at zero temperature for* $t > 0$. *The initial temperature* $f(x)$.

We have to solve § 119 (4), namely

$$\frac{\partial^2 v}{\partial x^2} - \frac{1}{\kappa}\frac{\partial v}{\partial t} = 0 \quad (0 < x < l), \tag{23}$$

with $$v = 0, \quad \text{when } x = 0 \text{ and } x = l, \quad t > 0, \tag{24}$$

and with $$v = f(x), \quad \text{when } t = 0, \quad 0 < x < l. \tag{25}$$

We suppose $f(x)$ to be expanded in the sine series § 89 (7)

$$f(x) = \sum_{n=1}^{\infty} b_n \sin\frac{n\pi x}{l}, \tag{26}$$

where $$b_n = \frac{2}{l}\int_0^l f(x')\sin\frac{n\pi x'}{l}\,dx'. \tag{27}$$

Now $$e^{-\kappa n^2\pi^2 t/l^2}\sin\frac{n\pi x}{l} \tag{28}$$

satisfies (23) and the boundary conditions (24). Thus

$$\sum_{n=1}^{\infty} A_n e^{-\kappa n^2\pi^2 t/l^2}\sin\frac{n\pi x}{l} \tag{29}$$

does so also. When $t = 0$, (29) has the value

$$\sum_{n=1}^{\infty} A_n \sin\frac{n\pi x}{l}. \tag{30}$$

Comparing coefficients between (26) and (30) we find $A_n = b_n$, and the solution is finally

$$v = \frac{2}{l}\sum_{n=1}^{\infty} e^{-\kappa n^2\pi^2 t/l^2}\sin\frac{n\pi x}{l}\int_0^l f(x')\sin\frac{n\pi x'}{l}\,dx'. \tag{31}$$

If the initial temperature is constant, V_0, this becomes

$$v = \frac{4V_0}{\pi}\sum_{r=0}^{\infty} e^{-\kappa(2r+1)^2\pi^2 t/l^2}\frac{\sin(2r+1)\pi x/l}{2r+1}. \tag{32}$$

Ex. 2. *The problem of Ex. 1 except that there is no flow of heat at* $x = 0$.

The only change is that the boundary condition at $x = 0$ is $\partial v/\partial x = 0$. To satisfy this we take as the elementary solution

$$e^{-\kappa(2n+1)^2\pi^2 t/4l^2}\cos\frac{(2n+1)\pi x}{2l}, \tag{33}$$

and expand $f(x)$ in the cosine series of Chapter XI, Ex. 17.

Ex. 3. *A string is stretched between the points* $x = 0$ *and* $x = l$ *and is set in motion at* $t = 0$ *with initial displacement* $f(x)$ *and zero initial velocity.*

The differential equation, § 120 (7), is

$$\frac{\partial^2 y}{\partial x^2}-\frac{1}{c^2}\frac{\partial^2 y}{\partial t^2}=0 \quad (0<x<l), \tag{34}$$

to be solved with

$$y=0, \quad \text{when } x=0 \text{ and } x=l, \quad t>0, \tag{35}$$

$$\frac{\partial y}{\partial t}=0, \quad 0<x<l, \quad t=0, \tag{36}$$

$$y=f(x), \quad 0<x<l, \quad t=0. \tag{37}$$

(34) and (35) are satisfied by

$$\sin\frac{n\pi x}{l}\left\{A_n\cos\frac{n\pi ct}{l}+B_n\sin\frac{n\pi ct}{l}\right\} \quad (n=1,2,...). \tag{38}$$

This has already been derived and its significance discussed in § 121. To satisfy (36) we must have $B_n = 0$ in (38). Thus (34), (35), and (36) are satisfied by

$$\sum_{n=1}^{\infty} A_n \sin\frac{n\pi x}{l}\cos\frac{n\pi ct}{l}, \tag{39}$$

and when $t = 0$ this reduces to

$$\sum_{n=1}^{\infty} A_n \sin\frac{n\pi x}{l}. \tag{40}$$

If we expand $f(x)$ in the sine series (26) and (27), comparing coefficients with (40) gives $A_n = b_n$, and the final result is

$$y=\frac{2}{l}\sum_{n=1}^{\infty}\sin\frac{n\pi x}{l}\cos\frac{n\pi ct}{l}\int_0^l f(x')\sin\frac{n\pi x'}{l}\,dx'. \tag{41}$$

Suppose, as an example, that initially the string is plucked a distance d at $x = b$, so that

$$f(x)=dx/b \quad (0<x<b), \tag{42}$$

$$f(x)=d(l-x)/(l-b) \quad (b<x<l). \tag{43}$$

Evaluating the integral in (41) we get

$$y=\frac{2dl^2}{\pi^2 b(l-b)}\sum_{n=1}^{\infty}\frac{1}{n^2}\sin\frac{n\pi b}{l}\sin\frac{n\pi x}{l}\cos\frac{n\pi ct}{l}. \tag{44}$$

125. The use of Laplace and Fourier transforms

Laplace transformations are widely used for solving equations of parabolic and hyperbolic types. We suppose the equations have to be solved for $t > 0$ with initial conditions at $t = 0$.

Suppose that $v(x,t)$ is the solution; its Laplace transform with respect to t will be

$$\bar{v} = \int_0^\infty e^{-pt} v\, dt, \tag{1}$$

which is a function of p and x. The Laplace transform of $\partial v/\partial t$ is

$$\int_0^\infty \frac{\partial v}{\partial t} e^{-pt}\, dt = [ve^{-pt}]_0^\infty + p\int_0^\infty ve^{-pt}\, dt$$

$$= -v_0(x) + p\bar{v}, \tag{2}$$

just as in § 18 (21), except that the initial value, $v_0(x)$, is now a function of x.

Similarly the Laplace transform of $\partial^2 v/\partial t^2$ is

$$\int_0^\infty \frac{\partial^2 v}{\partial t^2} e^{-pt}\, dt = -pv_0(x) - v_1(x) + p^2\bar{v}, \tag{3}$$

where $v_1(x)$ is the value of $\partial v/\partial x$ when $t = 0$.

Finally, the Laplace transform of $\partial^2 v/\partial x^2$ is

$$\int_0^\infty e^{-pt} \frac{\partial^2 v}{\partial x^2}\, dt = \frac{d^2}{dx^2}\int_0^\infty e^{-pt} v\, dt = \frac{d^2\bar{v}}{dx^2}, \tag{4}$$

assuming that the orders of differentiation and integration in (4) may be interchanged.

Now suppose we multiply our partial differential equation by e^{-pt} and integrate with respect to t from 0 to ∞. For the equation § 118 (1) we get, using (2) and (4),

$$\frac{d^2\bar{v}}{dx^2} - \frac{p}{\kappa}\bar{v} = -\frac{1}{\kappa} v_0(x). \tag{5}$$

From the equation § 118 (2) we get

$$\frac{d^2\bar{v}}{dx^2} - \frac{p^2}{c^2}\bar{v} = -\frac{p}{c^2} v_0(x) - \frac{1}{c^2} v_1(x). \tag{6}$$

And in the same way from § 122 (3) and (4) we get

$$\frac{d\bar{V}}{dx} + (Lp+R)\bar{I} = LI_0(x), \tag{7}$$

$$\frac{d\bar{I}}{dx} + (Cp+G)\bar{V} = CV_0(x), \tag{8}$$

where $I_0(x)$ and $V_0(x)$ are the values of I and V when $t = 0$.

(5), (6), and (7) and (8) are the *subsidiary equations* corresponding to the partial differential equations and their initial conditions. They have to be solved with boundary conditions which are the Laplace transforms of the given boundary conditions. When this has been done, the Laplace transform $\bar{v}$ of the solution v has been found. To find v from $\bar{v}$ two methods are available; (i) an extension of the Table of Transforms of § 18 together with the development of more theorems of the type of § 18, Theorem 1; (ii) what is in effect an extension of § 18 (28) which allows series of Fourier type for v to be written down from $\bar{v}$. To go into these points would take too long† and we merely solve one example to illustrate the way the method applies to an interesting type of problem.

Ex. 1. *A bar of length l with its end $x = l$ fixed is at rest and unstrained, when at $t = 0$ the end $x = 0$ is given a small displacement a.*

The differential equation, § 120 (2),

$$\frac{\partial^2 u}{\partial x^2} - \frac{1}{c^2}\frac{\partial^2 u}{\partial t^2} = 0 \quad (0 < x < l,\ t > 0), \tag{9}$$

has to be solved with

$$u = 0, \quad \text{when } t = 0 \quad (0 < x < l), \tag{10}$$

$$\frac{\partial u}{\partial t} = 0, \quad \text{when } t = 0 \quad (0 < x < l), \tag{11}$$

$$u = 0, \quad \text{when } x = l \quad (t > 0), \tag{12}$$

$$u = a, \quad \text{when } x = 0 \quad (t > 0). \tag{13}$$

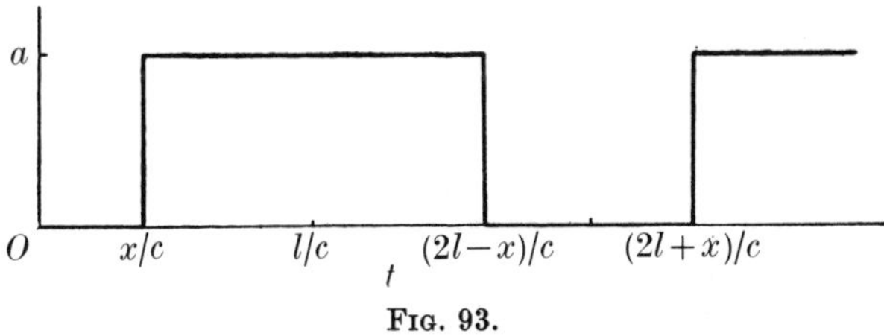

Fig. 93.

Writing $\bar{u}$ for the Laplace transform of u, the subsidiary equation for (9) with initial conditions (10) and (11) is by (6)

$$\frac{d^2\bar{u}}{dx^2} - \frac{p^2}{c^2}\bar{u} = 0. \tag{14}$$

† For further details see Carslaw and Jaeger, *Operational Methods in Applied Mathematics* (Oxford); Churchill, *Operational Mathematics* (McGraw-Hill); Gardner and Barnes, *Transients in Linear Systems* (Wiley).

Taking the Laplace transforms of (12) and (13), using § 18 (7), this has to be solved with

$$\bar{u} = 0, \quad \text{when } x = l, \tag{15}$$

$$\bar{u} = \frac{a}{p}, \quad \text{when } x = 0. \tag{16}$$

The solution of (14) satisfying (15) and (16) is

$$\bar{u} = \frac{a \sinh p(l-x)/c}{p \sinh pl/c}. \tag{17}$$

To find u from $\bar{u}$ we expand (17) in a series of negative exponentials as follows

$$\bar{u} = \frac{a\{e^{-px/c} - e^{-p(2l-x)/c}\}}{p\{1-e^{-2pl/c}\}}$$

$$= \frac{a}{p}\{e^{-px/c} - e^{-p(2l-x)/c}\}\{1+e^{-2pl/c}+e^{-4pl/c}+...\}$$

$$= \frac{a}{p}\{e^{-px/c} - e^{-p(2l-x)/c} + e^{-p(2l+x)/c} - ...\}.$$

Then, by § 18 (9),

$$u = a\left\{H\left(t-\frac{x}{c}\right) - H\left(t-\frac{2l-x}{c}\right) + H\left(t-\frac{2l+x}{c}\right) - ...\right\}. \tag{18}$$

The graph of u as a function of t is shown in Fig. 93.

Fourier transforms are used in much the same way as Laplace transforms in the solution of elliptic as well as hyperbolic and parabolic equations. Here transforms are taken with respect to one of the space variables, say x, where the region of solution is $-\infty < x < \infty$.

Ex. 2. *Conduction of heat in the infinite region* $-\infty < x < \infty$ *with an initial temperature distribution given by the delta function* $\delta(x)$.

The equation is § 119 (4)

$$\frac{\partial^2 v}{\partial x^2} - \frac{1}{\kappa}\frac{\partial v}{\partial t} = 0. \tag{19}$$

We recall from Ex. 26 at the end of Chapter XI that if $F(\omega)$ is the Fourier transform of $f(x)$, the Fourier transform of df/dx is $i\omega F(\omega)$ and that of d^2f/dx^2 is $-\omega^2 F(\omega)$. If $V(\omega)$ is the Fourier transform of $v(x,t)$ with respect to x, equation (19) therefore becomes

$$-\omega^2 V(\omega) - \frac{1}{\kappa}\frac{dV}{dt} = 0.$$

The general solution is

$$V(\omega) = Ae^{-\kappa\omega^2 t}. \tag{20}$$

The initial condition $v = \delta(x)$ when $t = 0$ transforms, by § 96 (3), into

$$V(\omega) = 1 \quad \text{when } t = 0.$$

Substituting this in (20) gives $A = 1$.

The inverse transform is obtained by Ex. 23 at the end of Chapter XI,

$$v(x,t) = \tfrac{1}{2}(\pi\kappa t)^{-\frac{1}{2}}e^{-x^2/4\kappa t}. \tag{21}$$

If the initial temperature distribution had been $\delta(x-x')$ instead of $\delta(x)$, the solution would have been

$$v(x,t) = \tfrac{1}{2}(\pi\kappa t)^{-\frac{1}{2}}e^{-(x-x')^2/4\kappa t}$$

which is essentially the result we assumed in § 119 (8). Notice that this is in fact the Green's function (cf. § 102) for this problem. We shall take this idea further in § 130.

Ex. 3. *Laplace's equation in the infinite strip* $0 < y < l$, $-\infty < x < \infty$, *with* $v = f(x)$ *on* $y = 0$ *and* $v = 0$ *on* $y = l$.

We cannot expand $f(x)$ in a Fourier series as in the problems of § 124, because the region is infinite along the x-axis. We therefore let $V(\omega)$ be the Fourier transform with respect to x of $v(x,y)$, and Laplace's equation

$$\frac{\partial^2 v}{\partial x^2}+\frac{\partial^2 v}{\partial y^2} = 0$$

transforms into

$$-\omega^2 V(\omega)+\frac{d^2V}{dy^2} = 0. \tag{22}$$

The general solution of (22) is

$$V(\omega) = Ae^{-\omega y}+Be^{\omega y}. \tag{23}$$

The boundary conditions are

$$v(x,y) = f(x) \quad \text{when } y = 0$$

or

$$V(\omega) = \int_{-\infty}^{\infty} e^{-i\omega x'}f(x')\,dx' \quad \text{when } y = 0, \tag{24}$$

and

$$v(x,y) = 0 \quad \text{when } y = l$$

or

$$V(\omega) = 0 \quad \text{when } y = l. \tag{25}$$

Substituting (24) and (25) in (23) leads to expressions for the constants A and B such that

$$V(\omega) = \frac{\sinh\omega(l-y)}{\sinh\omega l}\int_{-\infty}^{\infty} e^{-i\omega x'}f(x')\,dx',$$

and on taking inverse Fourier transforms, cf. § 95 (9), we get finally

$$v(x,t) = \frac{1}{2\pi}\int_{-\infty}^{\infty}\frac{\sinh\omega(l-y)}{\sinh\omega l}\,e^{i\omega x}\,d\omega\int_{-\infty}^{\infty} e^{-i\omega x'}f(x')\,dx'.$$

126. The use of conformal representation

While the use of Fourier series as in § 124 gave solutions of the fundamental problems on the wave equation and the diffusion equation, it only provided a solution of Laplace's equation for a region with rectangular boundaries. By the use of the theory of functions of a complex variable many two-dimensional regions can be transformed into regions with rectangular boundaries and in this way Laplace's equation solved in them.

Suppose that $\zeta = \xi + i\eta$ is a function $f(z)$ of a complex variable $z = x+iy$. We represent z by its rectangular coordinates (x, y) in the 'z-plane', and in the same way ζ by its coordinates (ξ, η) in the 'ζ-plane'. Then the relation

$$\zeta = f(z) \tag{1}$$

defines a correspondence between points in the z- and ζ-planes, and we shall suppose this to be one to one, that is, to each point of a region A in the z-plane there corresponds one point of a region B in the ζ-plane, and vice versa. The region A is then said to be mapped on the region B.

The function (1) is called an *analytic* function in a region if ζ has a definite differential coefficient with respect to z at each point of the region. That is, if

$$\lim_{\delta z \to 0} \frac{\delta \zeta}{\delta z}$$

exists and is independent of the way in which $\delta z = \delta x + i\,\delta y$ tends to zero in the (x, y)-plane. Now

$$\frac{\delta \zeta}{\delta z} = \frac{\delta(\xi+i\eta)}{\delta(x+iy)} = \frac{\left(\dfrac{\partial \xi}{\partial x}+i\,\dfrac{\partial \eta}{\partial x}\right)\delta x+\left(\dfrac{\partial \xi}{\partial y}+i\,\dfrac{\partial \eta}{\partial y}\right)\delta y}{\delta x+i\,\delta y}. \tag{2}$$

If this is to have a limit independent of $\delta y/\delta x$, the coefficient of δy in the numerator must be i times the coefficient of δx, that is

$$\frac{\partial \xi}{\partial y}+i\,\frac{\partial \eta}{\partial y} = i\left(\frac{\partial \xi}{\partial x}+i\,\frac{\partial \eta}{\partial x}\right). \tag{3}$$

Equating real and imaginary parts of (3) gives

$$\frac{\partial \xi}{\partial y} = -\frac{\partial \eta}{\partial x}, \tag{4}$$

$$\frac{\partial \eta}{\partial y} = \frac{\partial \xi}{\partial x}. \tag{5}$$

These are the Cauchy–Riemann differential equations. If they are satisfied

$$\frac{d\zeta}{dz} = \frac{\partial\xi}{\partial x} + i\frac{\partial\eta}{\partial x}. \tag{6}$$

It follows from (4) and (5) that

$$\frac{\partial^2\xi}{\partial x^2} + \frac{\partial^2\xi}{\partial y^2} = \frac{\partial^2\eta}{\partial x^2} + \frac{\partial^2\eta}{\partial y^2} = 0, \tag{7}$$

and

$$\frac{\partial\xi}{\partial x}\Big/\frac{\partial\xi}{\partial y} = -\frac{\partial\eta}{\partial y}\Big/\frac{\partial\eta}{\partial x}. \tag{8}$$

Also

$$\left|\frac{d\zeta}{dz}\right| = \left\{\left(\frac{\partial\xi}{\partial x}\right)^2 + \left(\frac{\partial\xi}{\partial y}\right)^2\right\}^{\frac{1}{2}} = \left\{\left(\frac{\partial\eta}{\partial x}\right)^2 + \left(\frac{\partial\eta}{\partial y}\right)^2\right\}^{\frac{1}{2}}. \tag{9}$$

The relation (7) states that the real and imaginary parts of any analytic function of a complex variable satisfy Laplace's equation. Also the property of satisfying Laplace's equation is preserved by the transformation (1), that is, we shall prove that if

$$\frac{\partial^2 v}{\partial\xi^2} + \frac{\partial^2 v}{\partial\eta^2} = 0, \tag{10}$$

then

$$\frac{\partial^2 v}{\partial x^2} + \frac{\partial^2 v}{\partial y^2} = 0. \tag{11}$$

To verify this we have

$$\frac{\partial v}{\partial x} = \frac{\partial v}{\partial\xi}\frac{\partial\xi}{\partial x} + \frac{\partial v}{\partial\eta}\frac{\partial\eta}{\partial x},$$

$$\frac{\partial^2 v}{\partial x^2} = \frac{\partial^2 v}{\partial\xi^2}\left(\frac{\partial\xi}{\partial x}\right)^2 + \frac{\partial^2 v}{\partial\eta^2}\left(\frac{\partial\eta}{\partial x}\right)^2 + \frac{\partial v}{\partial\xi}\frac{\partial^2\xi}{\partial x^2} + \frac{\partial v}{\partial\eta}\frac{\partial^2\eta}{\partial x^2} + 2\frac{\partial^2 v}{\partial\xi\partial\eta}\frac{\partial\xi}{\partial x}\frac{\partial\eta}{\partial x}. \tag{12}$$

Similarly

$$\frac{\partial^2 v}{\partial y^2} = \frac{\partial^2 v}{\partial\xi^2}\left(\frac{\partial\xi}{\partial y}\right)^2 + \frac{\partial^2 v}{\partial\eta^2}\left(\frac{\partial\eta}{\partial y}\right)^2 + \frac{\partial v}{\partial\xi}\frac{\partial^2\xi}{\partial y^2} + \frac{\partial v}{\partial\eta}\frac{\partial^2\eta}{\partial y^2} + 2\frac{\partial^2 v}{\partial\xi\partial\eta}\frac{\partial\xi}{\partial y}\frac{\partial\eta}{\partial y}. \tag{13}$$

Adding (12) and (13), and using (4), (5), (7), and (10), gives (11).

Since if $f(z)$ is an analytic function of z, $d\zeta/dz$ has a unique value independent of the way in which $\delta z \to 0$, an infinitesimal figure in the z-plane will be similar to the one which corresponds to it in the ζ-plane; in particular the angle at which two curves cut in the z-plane will be equal to the angle at which the corresponding curves in the ζ-plane cut. The word conformal is used to denote this property, and the transformation is said to give a conformal representation of portion of one plane on the other.

In particular, if ξ_0, ξ_1, η_0, η_1 are constants, the region bounded by the curves $\xi = \xi_0$, $\xi = \xi_1$, $\eta = \eta_0$, $\eta = \eta_1$ in the z-plane is transformed into a rectangle in the ζ-plane.

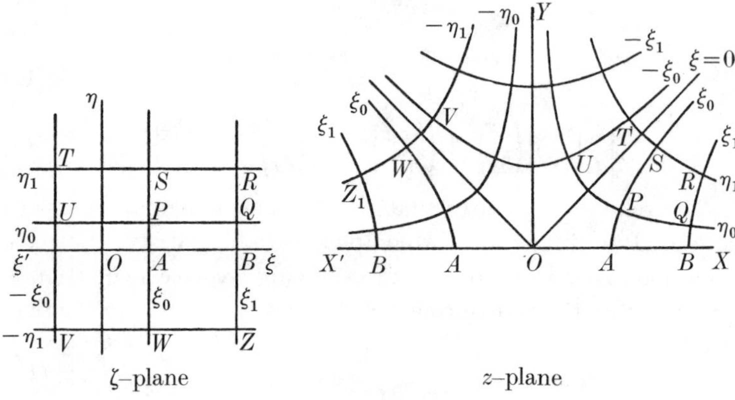

FIG. 94.

The nature of the correspondence set up by a transformation is best appreciated by the detailed study of examples. The theory of two-dimensional hydrodynamics and electrostatics is largely founded on the study of such transformations, and an account of the useful ones† is given in works on those subjects.

Ex. 1. $$\zeta = z^2. \tag{14}$$

$$\xi+i\eta = (x+iy)^2 = x^2-y^2+2ixy,$$

$$\xi = x^2-y^2, \qquad \eta = 2xy. \tag{15}$$

If $\theta = \arg z$, and $\phi = \arg \zeta$, we have $\phi = 2\theta$, and thus the upper half $0 \leqslant \theta \leqslant \pi$ of the z-plane is mapped on the whole $0 \leqslant \phi \leqslant 2\pi$ of the ζ-plane. The positive half of the x-axis, $\theta = 0$, and its negative half, $\theta = \pi$, both correspond to the positive half of the ξ-axis. The positive half of the y-axis, $\theta = \frac{1}{2}\pi$, corresponds to the negative half of the ξ-axis, $\phi = \pi$.

The line $\xi = \xi_0$ becomes the hyperbola

$$x^2-y^2 = \xi_0, \tag{16}$$

in the z-plane. $\xi = 0$ becomes the pair of lines $y = \pm x$.

† Cf. also Churchill, *Introduction to Complex Variables and Applications* (McGraw-Hill).

The line $\eta = \eta_0$ becomes the hyperbola

$$2xy = \eta_0, \tag{17}$$

and the line $\eta = 0$ becomes the pair of axes OX, OY.

The rectangle $PQRS$ in the ζ-plane corresponds to the figure $PQRS$ in the z-plane bounded by four hyperbolas, cf. Fig. 94,

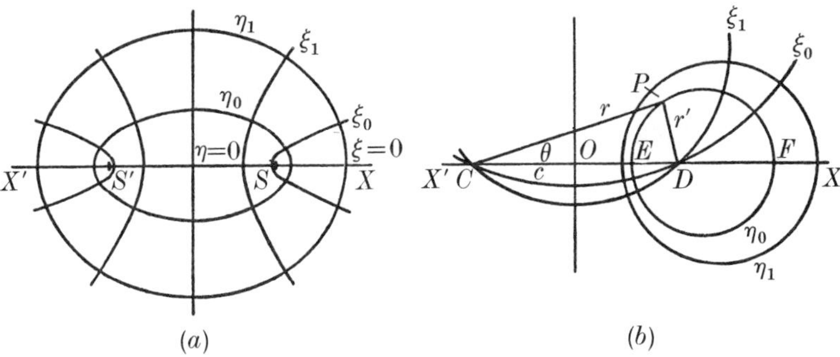

Fig. 95.

Ex. 2.
$$z = \cos\zeta. \tag{18}$$

$$x+iy = \cos(\xi+i\eta) = \cos\xi\cosh\eta+i\sin\xi\sinh\eta.$$

$$x = \cos\xi\cosh\eta, \tag{19}$$

$$y = \sin\xi\sinh\eta. \tag{20}$$

Therefore

$$\frac{x^2}{\cos^2\xi}-\frac{y^2}{\sin^2\xi} = 1, \tag{21}$$

$$\frac{x^2}{\cosh^2\eta}+\frac{y^2}{\sinh^2\eta} = 1. \tag{22}$$

The line $\xi = \xi_0$ in the ζ-plane corresponds to an hyperbola (21) in the z-plane, and the line $\eta = \eta_0$ in the ζ-plane to an ellipse (22); cf. Fig. 95 (a).

These ellipses and hyperbolas are confocal: $\eta = 0$ corresponds to the degenerate ellipse consisting of the segment SS' joining the foci, while $\xi = 0$ corresponds to the degenerate hyperbola consisting of the lines SX and $S'X'$ running outwards from the foci. Thus the transformation will be useful for studying either an isolated strip or a plane with a slit in it (for example, the hyperbolas are the stream lines for flow through a slit in a plane).

The detailed correspondence between points in the z- and ζ-planes is best followed by fixing (say) ξ in (19) and (20) and following the variations of x and y as η varies. It will be found that the whole of the z-plane is mapped on the strip $0 \leqslant \xi \leqslant \pi$, $-\infty < \eta < \infty$ of the ζ-plane.

Ex. 3.
$$\zeta = i\ln\frac{z+c}{z-c}. \tag{23}$$

Since the logarithm of a complex number z is given by

$$\ln z = \ln|z|+i\arg z,$$

it follows from (23) that

$$\xi = \theta'-\theta, \tag{24}$$

$$\eta = \ln r/r', \tag{25}$$

where r and r' are the distances of the point P, (x,y) from the points $(\mp c, 0)$, and θ and θ' are the angles PCX and PDX, Fig. 95 (*b*). The curves $\eta = \eta_0$ in the z-plane belong to a system of coaxial circles with C and D as limiting points, and the circles $\xi = \xi_0$ are a system of circles through C and D.

Now suppose that we wish to solve

$$\frac{\partial^2 v}{\partial x^2}+\frac{\partial^2 v}{\partial y^2} = 0, \tag{26}$$

in a region bounded by the curves $\xi = \xi_0$, $\xi = \xi_1$, $\eta = \eta_0$, $\eta = \eta_1$, in the (x,y)-plane which are determined by a transformation $\zeta = f(z)$, and that v is to have specified values on the boundaries. We solve

$$\frac{\partial^2 v}{\partial \xi^2}+\frac{\partial^2 v}{\partial \eta^2} = 0 \tag{27}$$

in the rectangle $\xi_0 < \xi < \xi_1$, $\eta_0 < \eta < \eta_1$, giving v at each point of the boundary the prescribed value at the corresponding point in the z-plane. This can be done as in § 124. By (10) and (11), this solution, expressed as a function of x and y, satisfies (26), and since it has the required value at each point of the boundary it is the required solution.

In most practical problems much less than this is needed. Interest is usually centred on the equipotentials on which v is constant. In such cases we take v itself to be the imaginary (or real) part of an analytic function

$$w = u+iv = f(z)$$

of z. Then, as in (7), both u and v satisfy Laplace's equation; the curves $u =$ constant, which by (8) are orthogonal to the equipotentials, also have a fundamental significance in the theory (e.g. in electrostatics they are the lines of force). As an example, suppose we consider

$$w = z^2$$

already studied in Ex. 1. The lines $v =$ constant are the hyperbolas

$$xy = \text{constant}$$

of Fig. 94: they are the equipotentials when the planes OX and OY are held at constant potential, or the lines of flow of a perfect, incompressible fluid in a right-angled corner.

127. The divergence of a vector and Laplace's equation in three dimensions

Most of the problems we have dealt with so far have involved one space variable and time, or, in the case of Laplace's equation, two space variables only, and vectors have not been used. However, in developing the general theory in three dimensions of the subjects discussed in this chapter, vectors are extremely useful; they also help to emphasize how the same mathematics underlies a number of different subjects. In all of these, the idea of the divergence of a vector is of the utmost importance.

Suppose $\mathbf{F}$ is a vector function of position in a three-dimensional space. We consider a small closed surface S in this space. Let $\mathbf{n}$ be a unit vector in the direction of the outward normal to the surface S at any point on it. Then $\mathbf{F}.\mathbf{n}$ is the normal component of $\mathbf{F}$ at that point, and if dS is an element of surface area at the point, the integral

$$\iint_S \mathbf{F}.\mathbf{n}\, dS \tag{1}$$

taken over the whole surface S, is referred to as the *flux* of the vector $\mathbf{F}$ over the surface. For example, if $\mathbf{F} = \rho\mathbf{v}$ where $\mathbf{v}$ is the velocity and ρ the density of a fluid, the integral (1) gives the net mass rate at which fluid is flowing out of the surface S.

If S encloses a point P with coordinates (x, y, z), the divergence of the vector $\mathbf{F}$, written $\operatorname{div}\mathbf{F}$, at the point P is defined as the limit of the ratio of the flux (1) to the volume δV enclosed by S, as the surface shrinks onto the point P. That is

$$\operatorname{div}\mathbf{F} = \lim_{\delta V\to 0}\frac{1}{\delta V}\iint_S \mathbf{F}.\mathbf{n}\, ds. \tag{2}$$

If this limit exists,† it will be independent of the shape of the surface S. We may therefore calculate div **F** from whatever surface is most convenient; in Cartesian coordinates this is a rectangular parallelepiped whose sides are the planes $x\pm\delta x$, $y\pm\delta y$, and $z\pm\delta z$. If F_x is the component of **F** in the x direction at (x,y,z), its value on the face $x+\delta x$ will be

$$F_x+\frac{\partial F_x}{\partial x}\delta x,$$

neglecting terms in δx^2, $\delta x\delta y$, etc., and the contribution to the integral in (2) from the face $x+\delta x$ will be

$$4F_x\,\delta y\delta z+4\frac{\partial F_x}{\partial x}\delta x\delta y\delta z.$$

The contribution from the face $x-\delta x$ will be

$$-\left\{4F_x\,\delta y\delta z-4\frac{\partial F_x}{\partial x}\delta x\delta y\delta z\right\},$$

and there will be similar results for the faces $y\pm\delta y$ and $z\pm\partial z$.

Thus the value of the integral in (2) is

$$8\partial x\partial y\partial z\left\{\frac{\partial F_x}{\partial x}+\frac{\partial F_y}{\partial y}+\frac{\partial F_z}{\partial z}\right\},$$

and dividing by the volume $8\delta x\delta y\delta z$ of the parallelepiped, we get finally

$$\operatorname{div}\mathbf{F}=\frac{\partial F_x}{\partial x}+\frac{\partial F_y}{\partial y}+\frac{\partial F_z}{\partial z}. \tag{3}$$

We now consider the important case where the vector **F** is the gradient of a scalar function of position ϕ. We write

$$\mathbf{F}=\operatorname{grad}\phi, \tag{4}$$

where, by § 72 (26),

$$F_x=\frac{\partial\phi}{\partial x},\quad F_y=\frac{\partial\phi}{\partial y},\quad F_z=\frac{\partial\phi}{\partial z}. \tag{5}$$

Then by (3)

$$\operatorname{div}\operatorname{grad}\phi=\frac{\partial^2\phi}{\partial x^2}+\frac{\partial^2\phi}{\partial y^2}+\frac{\partial^2\phi}{\partial z^2}. \tag{6}$$

† As in all such definitions, there are pure-mathematical restrictions on the nature of the function **F** in order that the limit may exist; these are simply that the components of **F** should be differentiable.

The notation
$$\nabla^2\phi = \operatorname{div}\operatorname{grad}\phi$$
$$= \frac{\partial^2\phi}{\partial x^2}+\frac{\partial^2\phi}{\partial y^2}+\frac{\partial^2\phi}{\partial z^2}, \tag{7}$$
is often used.

We can now see how the quantity $\nabla^2\phi$, which appears in so many of the problems of this chapter, usually arises in applied mathematics. Let us take conduction of heat as an example. If v is the temperature at point P in a solid, the vector
$$\mathbf{F} = -K\operatorname{grad} v$$
describes the direction and magnitude of the flow of heat at the point P. (This is the generalization of § 119 (1) and may be regarded as a fundamental assumption based on experimental evidence.) The amount of heat flowing out of a small element of volume δV surrounding P in a small time δt is
$$\delta V\,\delta t\operatorname{div}\mathbf{F},$$
and since this must be equal to the loss of heat from the volume in time δt, we get
$$-\rho c\delta V\delta v = \delta V\delta t\operatorname{div}\mathbf{F},$$
and in the limit as $\delta t \to 0$,
$$\rho c\frac{\partial v}{\partial t} = -\operatorname{div}\mathbf{F}$$
$$= \operatorname{div}(K\operatorname{grad} v). \tag{8}$$

This is the equation of conduction of heat in three dimensions. If K is constant then (8) becomes
$$\rho c\frac{\partial v}{\partial t} = K\operatorname{div}\operatorname{grad} v = K\nabla^2 v, \tag{9}$$
and if we have reached a steady state, $\partial v/\partial t = 0$ and (9) becomes
$$\nabla^2 v = 0.$$

The same type of argument applies to conservation of charge in a conducting medium, conservation of mass in a fluid, and so on. For instance, if $\mathbf{v}$ is the velocity and ρ the density at point P in a compressible fluid, the mass which flows out of a small volume δV surrounding P in time δt is
$$\delta t\delta V\operatorname{div}(\rho\mathbf{v}),$$

which, by conservation of mass, is $-\delta\rho\delta V$. Equating these we get, in the limit as $\delta t \to 0$,

$$\frac{\partial\rho}{\partial t}+\operatorname{div}(\rho\mathbf{v}) = 0, \tag{10}$$

which is called the *equation of continuity*.

For an incompressible fluid (10) becomes

$$\operatorname{div}\mathbf{v} = 0. \tag{11}$$

In certain types of motion (irrotational motion) the velocity $\mathbf{v}$ is given by minus the gradient of a scalar ϕ called the velocity potential,

$$\mathbf{v} = -\operatorname{grad}\phi,$$

and (11) becomes Laplace's equation

$$\nabla^2\phi = 0.$$

128. Solution of Laplace's equation in three dimensions

In rectangular Cartesian coordinates, Laplace's equation is

$$\nabla^2 v = \frac{\partial^2 v}{\partial x^2}+\frac{\partial^2 v}{\partial y^2}+\frac{\partial^2 v}{\partial z^2} = 0. \tag{1}$$

Assuming a solution of the form

$$v = X(x)Y(y)Z(z),$$

leads, by arguments similar to those of § 124 (13), to the elementary solution

$$\sin\frac{n\pi x}{a}\sin\frac{m\pi y}{a}\sinh\left(\frac{n^2}{a^2}+\frac{m^2}{b^2}\right)^{\frac{1}{2}}\pi z. \tag{2}$$

By superposing such solutions and using the theory of double Fourier series, (1) may be solved in the rectangular parallelepiped bounded by the planes $x = 0$, $x = a$, $y = 0$, $y = b$, $z = 0$, and $z = c$, with assigned values of v on these boundaries.

Rectangular Cartesian coordinates are suitable only when we are dealing with a problem that has to be solved in a region with rectangular boundaries. In many problems it is more convenient to use cylindrical polar coordinates or spherical polar coordinates.

In cylindrical polar coordinates, Fig. 96 (*a*), the point P is specified by (r, θ, z) and the surfaces of the coordinate system are cylinders, $r =$ constant; axial planes, $\theta =$ constant; and planes perpendicular to the axis, $z =$ constant. In these coordinates (1) becomes†

$$\frac{\partial^2 v}{\partial r^2}+\frac{1}{r}\frac{\partial v}{\partial r}+\frac{1}{r^2}\frac{\partial^2 v}{\partial \theta^2}+\frac{\partial^2 v}{\partial z^2} = 0. \tag{3}$$

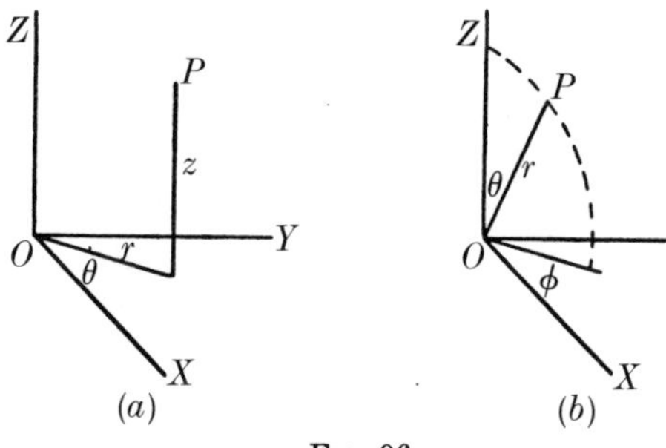

Fig. 96.

We seek an elementary solution of this of the form

$$v = R(r)\Theta(\theta)Z(z), \tag{4}$$

then, substituting (4) in (3), we get

$$\frac{1}{R}\left\{\frac{d^2R}{dr^2}+\frac{1}{r}\frac{dR}{dr}\right\}+\frac{1}{r^2\Theta}\frac{d^2\Theta}{d\theta^2}+\frac{1}{Z}\frac{d^2Z}{dz^2} = 0. \tag{5}$$

If
$$\frac{1}{Z}\frac{d^2Z}{dz^2}-m^2 = 0, \tag{6}$$

so that
$$Z = \frac{\sinh}{\cosh} mz, \tag{7}$$

and
$$\frac{1}{\Theta}\frac{d^2\Theta}{d\theta^2}+n^2 = 0, \tag{8}$$

so that
$$\Theta = \frac{\sin}{\cos} n\theta, \tag{9}$$

(5) becomes
$$\frac{d^2R}{dr^2}+\frac{1}{r}\frac{dR}{dr}+\left(m^2-\frac{n^2}{r^2}\right)R = 0. \tag{10}$$

† This result and (13) may be derived either by direct change of variables in (1) or as in Ex. 29 at the end of the chapter.

This is Bessel's equation of order n, and as in § 98 its general solution is

$$AJ_n(mr)+BY_n(mr). \tag{11}$$

Therefore

$$\begin{matrix}\cosh\\ \sinh\end{matrix} mz \begin{matrix}\cos\\ \sin\end{matrix} n\theta\{AJ_n(mr)+BY_n(mr)\} \tag{12}$$

satisfies (3) for any values of m and n. There are restrictions on the possible values of m and n caused by the boundary conditions (cf. Ex. 30).

In spherical polar coordinates, Fig. 96 (b), a point is specified by r, θ, ϕ, and the surfaces of the coordinate system are spheres, cones, and axial planes. Equation (1) becomes

$$\frac{\partial}{\partial r}\left(r^2\frac{\partial v}{\partial r}\right)+\frac{1}{\sin\theta}\frac{\partial}{\partial\theta}\left(\sin\theta\frac{\partial v}{\partial\theta}\right)+\frac{1}{\sin^2\theta}\frac{\partial^2 v}{\partial\phi^2} = 0. \tag{13}$$

As before, we seek an elementary solution of this of the form

$$v = R(r)\Theta(\theta)\Phi(\phi), \tag{14}$$

and (13) becomes

$$\frac{1}{R}\frac{d}{dr}\left(r^2\frac{dR}{dr}\right)+\frac{1}{\Theta\sin\theta}\frac{d}{d\theta}\left(\sin\theta\frac{d\Theta}{d\theta}\right)+\frac{1}{\Phi\sin^2\theta}\frac{d^2\Phi}{d\phi^2} = 0. \tag{15}$$

This is satisfied if

$$\frac{1}{R}\frac{d}{dr}\left(r^2\frac{dR}{dr}\right) = \alpha, \tag{16}$$

$$\frac{1}{\Phi}\frac{d^2\Phi}{d\phi^2}+m^2 = 0, \tag{17}$$

$$\frac{1}{\Theta\sin\theta}\frac{d}{d\theta}\left(\sin\theta\frac{d\Theta}{d\theta}\right)-\frac{m^2}{\sin^2\theta}+\alpha = 0, \tag{18}$$

where m and α are any numbers.

(16) may be written

$$r^2\frac{d^2R}{dr^2}+2r\frac{dR}{dr}-\alpha R = 0.$$

It is satisfied by $R = Ar^\nu$, where A is a constant, if

$$\nu(\nu+1)-\alpha = 0.$$

If we write

$$\alpha = n(n+1),$$

which can always be done, we have $\nu = n$ or $\nu = -(n+1)$, and

$$R = (Ar^n + Br^{-n-1}), \tag{19}$$

where A and B are arbitrary constants.

(17) is satisfied by $\quad \Phi = \begin{matrix}\cos\\ \sin\end{matrix} m\phi.$

Finally, putting $\cos\theta = \mu$, (18) becomes

$$\frac{d}{d\mu}\left\{(1-\mu^2)\frac{d\Theta}{d\mu}\right\} + \left\{n(n+1) - \frac{m^2}{1-\mu^2}\right\}\Theta = 0, \tag{20}$$

which is the associated Legendre equation § 99 (30).

In the above, m and n may have any values, but these will be restricted when the region in which the equation has to be solved is known. Thus if it is the interior of the sphere $r = a$, m must be integral, since if we add 2π to ϕ the value of the solution must not be affected. Also n must be an integer, since if n is not integral the solutions of (20) tend to infinity as $\mu \to -1$.

Thus the solution of (15) appropriate to the interior of a sphere is

$$Ar^n P_n^m(\mu) \begin{matrix}\cos\\ \sin\end{matrix} m\phi. \tag{21}$$

Ex. *To find the solution of* (1) *for the region* $0 \leqslant r < a$ *which takes the value* $f(\mu)$ *on the surface of the sphere, where* $\mu = \cos\theta$.

Since the solution is to be independent of ϕ, (21) takes the form

$$Ar^n P_n(\mu),$$

and the general solution of (1) will be

$$\sum_{n=0}^{\infty} A_n r^n P_n(\mu). \tag{22}$$

On the sphere $r = a$ this has the value

$$\sum_{n=0}^{\infty} A_n a^n P_n(\mu). \tag{23}$$

Now if we assume $f(\mu)$ to be expanded in the series § 99 (28), (29),

$$f(\mu) = \sum_{n=0}^{\infty} (n+\tfrac{1}{2}) P_n(\mu) \int_{-1}^{1} f(\mu') P_n(\mu')\, d\mu'. \tag{24}$$

Comparing coefficients between (23) and (24) gives A_n, and we get finally

$$v = \sum_{n=0}^{\infty} (n+\tfrac{1}{2}) \left(\frac{r}{a}\right)^n P_n(\mu) \int_{-1}^{1} f(\mu') P_n(\mu')\, d\mu'. \tag{25}$$

129. The wave equation in two and three dimensions

In general we can write the wave equation as

$$\nabla^2 v - \frac{1}{c^2}\frac{\partial^2 v}{\partial t^2} = 0. \tag{1}$$

This equation arises in two and three dimensions from problems such as the vibrations of a membrane, stress waves in elastic solids, water waves, and in many problems arising from Maxwell's equations.

In rectangular Cartesian coordinates (1) becomes

$$\frac{\partial^2 v}{\partial x^2}+\frac{\partial^2 v}{\partial y^2}+\frac{\partial^2 v}{\partial z^2}-\frac{1}{c^2}\frac{\partial^2 v}{\partial t^2} = 0. \tag{2}$$

A trial solution of the form

$$v = X(x)Y(y)Z(z)T(t)$$

leads to the elementary solutions

$$\begin{matrix}\cos\\ \sin\end{matrix}\left(\frac{l\pi x}{a}\right)\begin{matrix}\cos\\ \sin\end{matrix}\left(\frac{m\pi y}{b}\right)\begin{matrix}\cos\\ \sin\end{matrix}\left(\frac{n\pi z}{d}\right)\begin{matrix}\cos\\ \sin\end{matrix}(\omega t) \tag{3}$$

where

$$\left(\frac{l^2}{a^2}+\frac{m^2}{b^2}+\frac{n^2}{d^2}\right)\pi^2 = \frac{\omega^2}{c^2}. \tag{4}$$

The values of l, m, and n will be restricted by the boundary conditions; for example, if the solution has to vanish on the planes $x = 0$, $x = a$, $y = 0$, $y = b$, $z = 0$, and $z = d$, they must be integers. Each set of values of l, m, n determines a value of ω, and $\omega/2\pi$ is the corresponding natural frequency.

Ex. 1. *Maxwell's equations for a rectangular wave guide.*

The wave guide is parallel to the z-axis and consists of the rectangle bounded by the planes $x = 0$, $x = a$, $y = 0$, $y = b$. A solution of Maxwell's equations of the form

$$e^{i\omega t-\gamma z}\phi(x,y) \tag{5}$$

is sought, where $\omega/2\pi$ is the frequency of the radiation and γ is a constant. It is found that ϕ has to satisfy

$$\frac{\partial^2\phi}{\partial x^2}+\frac{\partial^2\phi}{\partial y^2}+\left(\frac{\omega^2}{c^2}+\gamma^2\right)\phi = 0, \tag{6}$$

and has to vanish on the boundaries of the rectangle. The expression

$$\sin\frac{n\pi x}{a}\sin\frac{m\pi y}{b} \tag{7}$$

vanishes on the boundaries if $n = 1, 2, 3,...$ and $m = 1, 2, 3,...,$ and satisfies (6) if

$$\left(\frac{n^2}{a^2}+\frac{m^2}{b^2}\right)\pi^2 = \frac{\omega^2}{c^2}+\gamma^2. \tag{8}$$

This determines γ in terms of m and n, and the required general solution is a sum of terms of the type

$$A_{n,m}\, e^{i\omega t-\gamma z} \sin \frac{n\pi x}{a} \sin \frac{m\pi y}{b}, \tag{9}$$

with γ given by (8). This corresponds to a disturbance propagated along the z-axis only if γ is imaginary, that is, if

$$\frac{\omega^2}{\pi^2 c^2} > \frac{n^2}{a^2}+\frac{m^2}{b^2}.$$

In practice the frequency and dimensions are usually arranged so that this is only true for $n = 1$, $m = 0$ or $n = 0$, $m = 1$, so that only this solution is propagated. For all other values of m and n, γ^2 given by (8) is positive and the corresponding solution (9) dies away exponentially.

In cylindrical polar coordinates, (1) becomes

$$\frac{\partial^2 v}{\partial r^2}+\frac{1}{r}\frac{\partial v}{\partial r}+\frac{1}{r^2}\frac{\partial^2 v}{\partial \theta^2}+\frac{\partial^2 v}{\partial z^2}-\frac{1}{c^2}\frac{\partial^2 v}{\partial t^2} = 0, \tag{10}$$

and elementary solutions of this are

$$\begin{matrix}\cos\\ \sin\end{matrix}\left(\frac{n\pi z}{l}\right)\begin{matrix}\cos\\ \sin\end{matrix}\, m\theta\{AJ_m(\alpha r)+BY_m(\alpha r)\}\begin{matrix}\cos\\ \sin\end{matrix}(\omega t), \tag{11}$$

where

$$\alpha^2+n^2\pi^2/l^2 = \omega^2/c^2, \tag{12}$$

and J_m and Y_m are Bessel functions; cf. § 98.

Here, again α and n are determined by the boundary conditions, the value ω is then determined by (12), and $\omega/2\pi$ is the corresponding natural frequency.

Ex. 2. *Symmetrical vibrations of a circular membrane of radius a.*

Since we are dealing with a thin membrane the vibrations are independent of z, and since we are told that the vibrations are symmetrical, they are also independent of θ. Equation (10) therefore reduces to

$$\frac{\partial^2 v}{\partial r^2}+\frac{1}{r}\frac{\partial v}{\partial r}-\frac{1}{c^2}\frac{\partial^2 v}{\partial t^2} = 0. \tag{13}$$

If we seek a solution of the form

$$v = R(r)\begin{matrix}\sin\\ \cos\end{matrix}(\omega t)$$

then $R(r)$ must satisfy

$$\frac{d^2R}{dr^2}+\frac{1}{r}\frac{dR}{dr}+\frac{\omega^2}{c^2}R = 0.$$

This is Bessel's equation of order zero, and the solution of it which remains finite at $r = 0$ is, as in § 98,

$$J_0\left(\frac{\omega r}{c}\right).$$

If the membrane is fixed at $r = a$, the displacement at $r = a$ must be zero for all time, and we must have

$$J_0\left(\frac{\omega a}{c}\right) = 0,$$

that is $$\omega = c\alpha_n/a, \quad n = 1, 2, \ldots, \tag{14}$$

where the α_n are the roots of $J_0(\alpha) = 0$; the first few roots are given in § 98. The values (14) of ω give the natural frequencies of the membrane, and the general solution of (13) is

$$\sum_{n=1}^{\infty} J_0\left(\frac{\alpha_n r}{a}\right)\left\{A_n \cos\frac{\alpha_n ct}{a} + B_n \sin\frac{\alpha_n ct}{a}\right\}.$$

The coefficients A_n and B_n, for given initial conditions, are found from a Fourier–Bessel series which is analogous to a Fourier series (cf. Ex. 14 on Chapter XII).

Finally, we consider the form of (1) in spherical polar coordinates. If we write μ for $\cos\theta$, this is

$$\frac{\partial^2 v}{\partial r^2} + \frac{2}{r}\frac{\partial v}{\partial r} + \frac{1}{r^2}\frac{\partial}{\partial \mu}\left\{(1-\mu^2)\frac{\partial v}{\partial \mu}\right\} + \frac{1}{r^2(1-\mu^2)}\frac{\partial^2 v}{\partial \phi^2} - \frac{1}{c^2}\frac{\partial^2 v}{\partial t^2} = 0. \tag{15}$$

We seek a solution of the form

$$v = R(r)\Theta(\theta)\Phi(\phi)\begin{matrix}\cos\\ \sin\end{matrix}\,\omega t. \tag{16}$$

Substituting (16) in (15) we find that for the region inside a sphere,

$$\Phi(\phi) = \begin{matrix}\cos\\ \sin\end{matrix}\,m\phi$$

and $$\Theta(\theta) = P_n^m(\mu),$$

where m and n are integers and P_n^m are the associated Legendre functions introduced in § 99.

The equation for R becomes

$$\frac{d^2R}{dr^2} + \frac{2}{r}\frac{dR}{dr} + R\left\{\frac{\omega^2}{c^2} - \frac{n(n+1)}{r^2}\right\} = 0. \tag{17}$$

Putting $R = (\omega r/c)^{-\frac{1}{2}}y$, (10) gives

$$\frac{d^2y}{dr^2}+\frac{1}{r}\frac{dy}{dr}+y\left\{\frac{\omega^2}{c^2}-\frac{(n+\frac{1}{2})^2}{r^2}\right\} = 0,$$

which has solutions $J_{n+\frac{1}{2}}(\omega r/c)$ and $J_{-(n+\frac{1}{2})}(\omega r/c)$. The latter of these is inadmissible in the interior of a sphere since it tends to infinity as $r \to 0$. Thus the required elementary solution of (15) is

$$\left(\frac{\omega r}{c}\right)^{-\frac{1}{2}} J_{n+\frac{1}{2}}\left(\frac{\omega r}{c}\right) P_n^m(\mu) \begin{matrix}\cos\\ \sin\end{matrix} m\phi \begin{matrix}\cos\\ \sin\end{matrix} \omega t. \tag{18}$$

As remarked in § 98 (35), the Bessel functions of half-integral order which occur in (18) can be expressed in a simpler form.

Ex. 3. *Radial vibrations of the sphere $r < a$ with $v = 0$ when $r = a$.* The solution is independent of θ and ϕ, so (18) reduces to

$$\left(\frac{\omega r}{c}\right)^{-\frac{1}{2}} J_{\frac{1}{2}}\left(\frac{\omega r}{c}\right) \begin{matrix}\cos\\ \sin\end{matrix} \omega t = \left(\frac{2}{\pi}\right)^{\frac{1}{2}} \frac{c}{\omega r} \sin\left(\frac{\omega r}{c}\right) \begin{matrix}\cos\\ \sin\end{matrix} \omega t$$

by § 98 (35).

In order to have $v = 0$ on $r = a$, we must have

$$\frac{\omega a}{c} = \pi n, \quad n = 1, 2,...,$$

and the general solution is

$$\sum_{n=1}^{\infty} \frac{1}{r} A_n \sin\left(\frac{n\pi r}{a}\right) \begin{matrix}\cos\\ \sin\end{matrix}\left(\frac{n\pi ct}{a}\right).$$

The natural frequencies are $nc/2a$, $n = 1, 2,...$.

The solution of the diffusion equation in two or three dimensions can be accomplished by following much the same procedure as in the above examples. In the next section we consider a somewhat different approach, using Green's functions.

130. Green's functions, sources, and sinks

In § 125, Ex. 2, we saw that the solution to the one-dimensional diffusion equation

$$\frac{\partial^2 v}{\partial x^2}-\frac{1}{\kappa}\frac{\partial v}{\partial t} = 0 \tag{1}$$

in the infinite region $-\infty < x < \infty$, with the initial condition $v = \delta(x-x')$ at $t = 0$, is

$$v(x,t) = \tfrac{1}{2}(\pi\kappa t)^{-\frac{1}{2}}e^{-(x-x')^2/4\kappa t}. \tag{2}$$

This is the Green's function (cf. §§ 82, 102) for equation (1) in the region $-\infty < x < \infty$. Mathematically speaking, this means that the solution to the above problem with the general initial condition $v = f(x)$ can be constructed from (2) as the integral

$$\tfrac{1}{2}(\pi\kappa t)^{-\frac{1}{2}} \int\limits_{-\infty}^{\infty} f(x')e^{-(x-x')^2/4\kappa t}\,dx'. \tag{3}$$

Physically speaking, the Green's function (2) can be interpreted, as in § 119 (10), as the temperature at x at time t due to an instantaneous heat source which liberates a quantity of heat ρc per unit area at the point x' at $t = 0$.

Both these interpretations are capable of extension, as in the following two examples.

Ex. 1. *The diffusion equation in the two-dimensional region* $x > 0$, $y > 0$ *with* $v = 0$ *on the boundaries* $x = 0$ *and* $y = 0$.

By an extension of (2), the Green's function for

$$\frac{\partial^2 v}{\partial x^2}+\frac{\partial^2 v}{\partial y^2}-\frac{1}{\kappa}\frac{\partial v}{\partial t} = 0$$

in the *infinite region* $-\infty < x < \infty$, $-\infty < y < \infty$, is

$$\frac{1}{4\pi\kappa t}\,e^{-\{(x-x')^2+(y-y')^2\}/4\kappa t}. \tag{4}$$

We can construct the Green's function for our problem in the region $x > 0$, $y > 0$ by setting up an equivalent problem in the infinite region $-\infty < x < \infty$, $-\infty < y < \infty$.

Suppose that at $t = 0$ there is an instantaneous heat source of strength ρc at the point (x', y') in the region $x > 0$, $y > 0$. Introduce a dummy heat source of strength ρc at the point $(-x', -y')$ and two dummy heat sources of strength $-\rho c$ at the points $(-x', y')$ and $(x', -y')$ in the infinite region. By symmetry, the temperature due to these four heat sources will always be zero along the lines $x = 0$ and $y = 0$. It follows that the combined effect of the four sources, or rather two sources and two sinks, is the Green's function for our problem in $x > 0$, $y > 0$. By (4), this is

$$\frac{1}{4\pi\kappa t}[e^{-\{(x-x')^2+(y-y')^2\}/4\kappa t}+e^{-\{(x+x')^2+(y+y')^2\}/4\kappa t}-$$
$$-e^{-\{(x+x')^2+(y-y')^2\}/4\kappa t}-e^{-\{(x-x')^2+(y+y')^2\}/4\kappa t}]. \tag{5}$$

If the initial temperature distribution in $x > 0$, $y > 0$ is $f(x,y)$, by analogy with (3) the solution can be obtained by first multiplying (5) by $f(x', y')$ and then integrating the resulting expression from zero to infinity with respect to both x' and y'. For the case $f(x,y) = 1$, this eventually reduces to

$$v(x,y,t) = \operatorname{erf}\frac{x}{2\sqrt{(\kappa t)}}\operatorname{erf}\frac{y}{2\sqrt{(\kappa t)}},$$

(cf. Ex. 21 at the end of this chapter).

Ex. 2.

A river flows in a straight line with velocity u. A factory at some point $x = 0$ on the river goes into production at $t = 0$ and dumps a pollutant into the river at a constant rate R (per cross-sectional area) for all $t > 0$. We want to calculate the concentration $v(y, t)$ of the pollutant at a distance y downstream from the factory.

Instead of deriving the diffusion equation in a moving medium, we solve the equivalent problem of a stationary river $-\infty < x < \infty$ with a pollutant source R moving upstream with constant velocity u. Then at time t' we have an instantaneous source of strength R at $x' = -ut'$, and its contribution to the solution of the diffusion equation (1) is, by (2)

$$\tfrac{1}{2}R\{\pi\kappa(t-t')\}^{-\frac{1}{2}}e^{-(x-x')^2/4\kappa(t-t')}. \tag{6}$$

Putting $x = y-ut$ and $x' = -ut'$ in (6), and adding the effects of all the instantaneous sources from time 0 to time t, gives us the required concentration a distance y downstream from the factory,

$$v(y,t) = \tfrac{1}{2}R\int_0^t \{\pi\kappa(t-t')\}^{-\frac{1}{2}}\, e^{-(y-ut+ut')^2/4\kappa(t-t')}dt'. \tag{7}$$

The substitution $z = u\{(t-t')/4\kappa\}^{\frac{1}{2}}$ reduces (7) to the form

$$v(y,t) = \frac{2R}{u\sqrt{\pi}}\, e^{uy/2\kappa}\int_0^{u\sqrt{(t/4\kappa)}} e^{-z^2-b^2/z^2}\, dz, \tag{8}$$

where $b^2 = u^2y^2/8\kappa^2$.

Using the result $\int_0^\infty e^{-z^2-b^2/z^2}\, dz = \frac{\sqrt{\pi}}{2}\, e^{-2b}$, for large time (8) reduces to the steady distribution

$$\begin{aligned} v(y) &= R/u \qquad \text{for } y > 0 \\ &= \frac{R}{u}\, e^{uy/\kappa} \quad \text{for } y < 0. \end{aligned}$$

Thus after a long time there will be a constant concentration of pollution downstream and the concentration will fall off exponentially upstream from the factory.

Green's functions can also be found for other types of partial differential equations. For example, the Green's function for Laplace's equation

$$\frac{\partial^2 v}{\partial x^2}+\frac{\partial^2 v}{\partial y^2} = 0 \tag{9}$$

in the region $0 < x < \infty$, $-\infty < y < \infty$, is

$$\frac{y}{\pi\{x^2+(y-y')^2\}}. \tag{10}$$

In electrostatics this gives the potential in the plate $0 < x < \infty$, $-\infty < y < \infty$, due to a point electrode of unit strength placed at $y = y'$ on the edge of the plate $x = 0$. It follows that the solution for any potential distribution $f(y)$ on the edge of the plate is

$$\frac{y}{\pi}\int_{-\infty}^{\infty} f(y')\{x^2+(y-y')^2\}^{-1}\,dy'. \tag{11}$$

For the case of a 'strip electrode', $f(y) = v_0$ for $-a < y < a$ and $f(y) = 0$ elsewhere. Then by (11)

$$\begin{aligned} v(x,y) &= \frac{v_0 y}{\pi}\int_{-a}^{a} \{x^2+(y-y')^2\}^{-1}\,dy' \\ &= \frac{v_0}{\pi}\left\{\tan^{-1}\left(\frac{y+a}{x}\right)-\tan^{-1}\left(\frac{y-a}{x}\right)\right\}. \end{aligned}$$

EXAMPLES ON CHAPTER XV

1. The region $-a < x < a$ of the infinite solid $-\infty < x < \infty$ is initially at constant temperature V, and the remainder of the solid is at zero. Show that the temperature at any point x at time t is

$$\tfrac{1}{2}V\left\{\operatorname{erf}\frac{a-x}{2\sqrt{(\kappa t)}}+\operatorname{erf}\frac{a+x}{2\sqrt{(\kappa t)}}\right\}.$$

2. Heat is supplied in the plane $x = 0$ of the infinite solid at the rate Q per unit area per unit time for $t > 0$, the solid being initially at zero temperature. Show that the temperature at the point x at time t is

$$\frac{Q}{K}\left\{\left(\frac{\kappa t}{\pi}\right)^{\frac{1}{2}} e^{-x^2/4\kappa t}-\tfrac{1}{2}x+\tfrac{1}{2}x\operatorname{erf}\frac{x}{2\sqrt{(\kappa t)}}\right\}.$$

[Combine solutions of type § 119 (10) at times from 0 to t.]

If the semi-infinite solid $x > 0$ is heated over the plane $x = 0$ at the rate Q per unit area per unit time, show that its surface temperature is

$$\frac{2Q}{K}\left(\frac{\kappa t}{\pi}\right)^{\frac{1}{2}}.$$

3. A string $0 < x < l$ of line density σ is stretched to tension $T = \sigma c^2$. If it is plucked a distance d at its middle point and then released, show by the method of Fig. 90 that for $0 < t < l/2c$ the form of the string consists of a straight portion of length $2ct$ parallel to the x-axis, which is joined to the points $x = 0$ and $x = l$ by straight portions of slope $\tan^{-1}(2d/l)$, that is, in the original direction of the string. For $t > l/2c$ this form repeats itself on the other side of the x-axis, and so on. This result may also be obtained by the Laplace transformation method of § 125. The Fourier series of § 124 (44) with $b = \frac{1}{2}l$ is the Fourier sine series for the above curve at time t.

4. A particle of mass M is attached to the end $x = l$ of a bar of length l and area a whose end $x = 0$ is fixed. If the system oscillates longitudinally, show that the natural frequencies of the system are

$$\alpha_n\left(\frac{E}{4\pi^2l^2\rho}\right)^{\frac{1}{2}},$$

where $\alpha_1, \alpha_2,...$ are the positive roots of

$$\alpha\tan\alpha = al\rho/M.$$

Show that if $m = al\rho$ is the mass of the rod, and $\lambda = aE/l$ is its stiffness regarded as a spring, the lowest natural frequency is approximately

$$\{\lambda/(M+\tfrac{1}{3}m)\}^{\frac{1}{2}}/2\pi,$$

if m/M is small.

5. Show that the natural frequencies of a uniform beam of length l and weight w per unit length, clamped at both ends, are

$$\alpha_n^2(EIg/4\pi^2\rho l^4)^{\frac{1}{2}},$$

where $\alpha_1, \alpha_2,...$ are the positive roots of $\cos\alpha\cosh\alpha = 1$.

Show that if the beam is free at both ends, the natural frequencies are the same as those given above.

Show that for the natural frequencies of a cantilever of length l, the α_n in the above expression are the roots of

$$\cos\alpha\cosh\alpha = -1.$$

Discuss graphically the nature of the roots of these equations.

6. The surface $r = a$ of a hollow cylinder $a < r < b$ is kept at temperature v_1. At $r = b$ the cylinder loses heat into a medium at temperature v_2 at a rate proportional to its temperature excess above v_2, that is

$$\frac{dv}{dr}+h(v-v_2) = 0, \quad r = b.$$

Show that under steady conditions the rate of loss of heat from the cylinder is

$$2\pi K(v_1-v_2)\frac{hb}{1+hb\ln(b/a)}.$$

Discuss the behaviour of this expression as b increases from a, and show that if $ah < 1$ it has a maximum when $b = 1/h$, that is, that in certain circumstances it is possible to increase the heat loss from a cylindrical surface by covering it with insulating material.

7. Writing $\gamma = \alpha+i\beta$, $z_s = R+iX$ in the formulae § 122 (21) for the short-circuit impedance of a uniform transmission line of length l, show that if z_0 is real

$$R = z_0\sinh\alpha l\cosh\alpha l[\cosh^2\alpha l\cos^2\beta l+\sinh^2\alpha l\sin^2\beta l]^{-1},$$

$$X = z_0\sin\beta l\cos\beta l[\cosh^2\alpha l\cos^2\beta l+\sinh^2\alpha l\sin^2\beta l]^{-1},$$

and discuss the general case in which z_0 is not real.

Find the input impedance $(R^2+X^2)^{\frac{1}{2}}$ of the line, and show that, if α

is small, it has maxima near $\beta l = (n+\frac{1}{2})\pi$ and minima near $\beta l = n\pi$.

Discuss the input impedance of the open-circuited line in the same way.

8. If the points $x = R/z_0$, $y = X/z_0$ are plotted, representing the resistance and reactance of the short-circuited line of Ex. 7, show that the curves of constant αl are circles of centre $(\coth 2\alpha l, 0)$ and radius $\operatorname{cosech} 2\alpha l$, while the curves of constant βl are circles of centre $(0, -\cot 2\beta l)$ and radius $\operatorname{cosec} 2\beta l$. Using this result, charts of coaxial circles can be drawn from which the resistance and reactance of any line can be read off.

9. A uniform string of line density σ and length l is stretched to tension $T = \sigma c^2$. It is set in motion at $t = 0$ from its equilibrium position with velocity $\phi(x)$. Show that its displacement at any time is

$$\frac{2}{\pi c}\sum_{n=1}^{\infty}\frac{1}{n}\sin\frac{n\pi x}{l}\sin\frac{n\pi ct}{l}\int_{0}^{l}\phi(x')\sin\frac{n\pi x'}{l}\,dx'.$$

If the string is set in motion by a blow of impulse P applied to a very short length of the string at $x = a$, show that the displacement is

$$\frac{2P}{\pi c\sigma}\sum_{n=1}^{\infty}\frac{1}{n}\sin\frac{n\pi x}{l}\sin\frac{n\pi a}{l}\sin\frac{n\pi ct}{l}.$$

10. Show that if $\bar{x}(p)$ is the Laplace transform of $x(t)$, then $e^{-ap}\bar{x}(p)$ is the Laplace transform of $x(t-a)H(t-a)$, that is, of the function which is zero up to time $t = a$ and subsequently has the values of $x(t)$ shifted to the right by a distance a.

11. A uniform bar of length l and unit area has its end $x = 0$ fixed. At $t = 0$, when the bar is at rest and unstrained, a constant tension T is applied at $x = l$. Show that the stress at $x = 0$ is

$$2T\left\{H\left(t-\frac{l}{c}\right)-H\left(t-\frac{3l}{c}\right)+H\left(t-\frac{5l}{c}\right)+\dots\right\},$$

and, using the result of Ex. 10, find the displacement at any point of the bar.

12. The bar of Ex. 11 is struck by a blow of impulse P at $x = l$. Show that the stress at $x = 0$ is

$$2P\left\{\delta\left(t-\frac{l}{c}\right)-\delta\left(t-\frac{3l}{c}\right)+\dots\right\}.$$

13. A bar $0 < x < l$ of density ρ is moving along the x-axis with velocity $-V$, when at $t = 0$ the point $x = 0$ is fixed. Show that the stress at $x = 0$ is

$$-\frac{EV}{c}\left\{1-2H\left(t-\frac{2l}{c}\right)+2H\left(t-\frac{4l}{c}\right)-\dots\right\}.$$

This may be regarded as the problem of the collision of two equal rods moving along the x-axis with equal speeds in opposite directions. Show that the rods separate after a time $2l/c$.

14. Voltage $V(t)$ which is any function of the time is applied at $t = 0$ to the end $x = 0$ of the semi-infinite lossless line $x > 0$, cf. § 122 (6), which has zero initial charge and current. Show that the voltage at the point x is zero up to time x/c, and is

$$V(t-x/c) \quad \text{for } t > x/c,$$

that is, it is exactly the voltage applied at $x = 0$ delayed by time x/c.

For the distortionless line § 122 (8) with the same conditions, show that the voltage at x is zero up to time x/c, and is

$$e^{-Rx/Lc}V(t-x/c) \quad \text{for } t > x/c,$$

that is, it has the same form as the applied voltage, but is attenuated by the factor $\exp(-Rx/Lc)$.

15. Given that the Laplace transform of

$$1-\operatorname{erf}[\tfrac{1}{2}x(\kappa t)^{-\frac{1}{2}}]$$

is $(1/p)\exp[-x(p/\kappa)^{\frac{1}{2}}]$, deduce the result § 119 (17).

The region $-l < x < l$ is initially at zero temperature, and for $t > 0$ its surfaces $x = -l$ and $x = l$ are kept at constant temperature V_0. Show that the temperature at the point x at time t is

$$V_0 \sum_{n=0}^{\infty} (-1)^n \left\{2-\operatorname{erf}\frac{(2n+1)l-x}{2\sqrt{(\kappa t)}}-\operatorname{erf}\frac{(2n+1)l+x}{2\sqrt{(\kappa t)}}\right\}.$$

Use the method of § 125 and the Laplace transform given above. A solution of this problem in the form of a Fourier series may be deduced from § 124 (32) or derived independently. The Fourier series converges slowly for small values of the time and the series given above is more useful.

16. Discuss the transformation of § 126, Ex. 3, in greater detail and show that the whole of the z-plane is mapped on the strip $-\pi < \xi < \pi$, $-\infty < \eta < \infty$ of the ζ-plane.

Verify, by writing down its Cartesian equation, that if η_0 is constant

$$r/r' = e^{\eta_0}$$

is a circle, and show, by writing down its values for the points E, F, Fig. 95 (b), that if this circle has radius a and its centre is at the point $(d, 0)$,

$$c^2 = d^2-a^2, \qquad \frac{d+\sqrt{(d^2-a^2)}}{a} = e^{\eta_0}.$$

17. Discuss the nature of the equipotentials $v =$ constant, where v is the real or imaginary part of the functions

(i) $\cos z$, (ii) $\ln\{(z+c)/(z-c)\}$, (iii) $1/z$.

18. Show that the imaginary part of

$$V(z+a^2/z),$$

where V and a are constants, gives a solution of Laplace's equation which vanishes on the circle $r = a$ and tends to the value Vy at large distances from the origin.

19. Show that the 'bilinear' transformation

$$\zeta = \frac{\alpha z+\beta}{\gamma z+\delta},$$

where $\alpha, \beta, \gamma, \delta$ are constants, gives a one-to-one correspondence between the z- and ζ-planes, and that there are just two points which are unchanged by the transformation.

Show that it may be built up from the three successive transformations

$$\zeta = \frac{\alpha}{\gamma}+\frac{\beta\gamma-\alpha\delta}{\gamma}z_1, \qquad z_1 = \frac{1}{z_2}, \qquad z_2 = \gamma z+\delta.$$

Show that each of these transformations possesses the property of transforming circles into circles, so that the bilinear transformation will do so also (straight lines are regarded as limiting cases of circles).

20. Show that the transformation

$$\zeta = -i\ln(z/a)$$

transforms concentric circles and radii through the origin, respectively, into the lines η = constant, and ξ = constant.

Using the result § 124 (12), show that the solution of Laplace's equation in the half-ring bounded by concentric circles of radii a and b, $b < a$, and by the portions $\theta = 0$ and $\theta = \pi$ of the x-axis, which has the value 1 on $r = a$ and vanishes on the other boundaries, is

$$\frac{4}{\pi}\sum_{s=0}^{\infty}\frac{\sin[(2s+1)\theta]\sinh[(2s+1)(\eta_0-\eta)]}{(2s+1)\sinh(2s+1)\eta_0},$$

where $\eta_0 = \ln(a/b)$, $\eta = \ln(a/r)$.

21. The rectangular corner $x > 0$, $y > 0$ is initially at unit temperature, and for $t > 0$ its surfaces $x = 0$ and $y = 0$ are kept at zero temperature. Show that

$$\operatorname{erf}\frac{x}{2\surd(\kappa t)}\operatorname{erf}\frac{y}{2\surd(\kappa t)}$$

satisfies the differential equation and boundary conditions.

22. Show that the normal modes of vibration of a rectangular membrane $0 < x < a$, $0 < y < b$, fixed at its edges, are

$$\sin\frac{n\pi x}{a}\sin\frac{m\pi y}{b}\begin{matrix}\cos\\ \sin\end{matrix}\,\omega t,$$

where

$$\pi^2\left(\frac{n^2}{a^2}+\frac{m^2}{b^2}\right) = \frac{\omega^2}{c^2}.$$

Sketch the nodal lines (lines of zero displacement) for the first few normal modes.

23. In a cylindrical wave guide of radius a, § 129 (6) in cylindrical polars has to be solved with $\phi = 0$ when $r = a$. Show that the solutions

independent of θ are of type

$$e^{i\omega t-\gamma z}J_0(r\alpha_n/a),$$

where α_n is a root of $J_0(x) = 0$, and

$$\frac{\alpha_n^2}{a^2}-\gamma^2-\frac{\omega^2}{c^2} = 0.$$

Show that this mode is propagated only if $\omega > c\alpha_n/a$.

24. The temperature of an infinite circular cylinder of radius a has the constant value V when $t = 0$, and for $t > 0$ its surface is kept at zero temperature. Show that the temperature at the radius r at time t is

$$\frac{2V}{a}\sum_{n=1}^{\infty} e^{-\kappa\alpha_n^2 t}\frac{J_0(r\alpha_n)}{\alpha_n J_1(a\alpha_n)},$$

where α_1, α_2,... are the positive roots of $J_0(ax) = 0$. [Use the result of Ex. 13 on Chap. XII.]

25. Show that

$$\frac{Q}{8\rho c(\pi\kappa t)^{\frac{3}{2}}}\exp\left\{-\frac{(x-x')^2+(y-y')^2+(z-z')^2}{4\kappa t}\right\}$$

satisfies the equation of conduction of heat and represents the temperature at (x,y,z) due to a quantity of heat Q liberated instantaneously at the point (x',y',z') at $t = 0$.

Deduce that if a quantity of heat Q per unit length is liberated at $t = 0$ along the z-axis, the temperature at (x,y) at time t will be

$$\frac{Q}{4\pi Kt}\exp\{-(x^2+y^2)/4\kappa t\}.$$

26. Show that the solution of Laplace's equation in the region $0 < x < a$, $0 < y < b$, $0 < z < c$, with $v = 1$ on the surface $z = 0$ and $v = 0$ on the other surfaces, is [using § 94 (7)]

$$\frac{16}{\pi^2}\sum_{p=0}^{\infty}\sum_{q=0}^{\infty}\frac{\sinh l(c-z)\sin[(2p+1)\pi x/a]\sin[(2q+1)\pi y/b]}{(2p+1)(2q+1)\sinh cl},$$

where
$$l^2 = \frac{(2p+1)^2\pi^2}{a^2}+\frac{(2q+1)^2\pi^2}{b^2}.$$

27. Show that, if ϵ is small, the surface

$$r = a+\epsilon P_1(\cos\theta)$$

is very nearly a sphere of radius a with its centre displaced a small distance ϵ from the origin. Show that a function v which satisfies Laplace's equation, vanishes on this sphere, and has the value unity on the sphere $r = b$, where $b < a$, is

$$\frac{b(a-r)}{r(a-b)}+\frac{\epsilon ab(r^3-b^3)}{r^2(a-b)(a^3-b^3)}P_1(\cos\theta).$$

28. Show that if F_r, F_θ, F_z are the components of a vector $\mathbf{F}$ in cylindrical coordinates (i.e. F_θ is the component in the direction in which the point specified by r, θ, z moves if θ is increased, r and z being kept constant, etc.)

$$\operatorname{div}\mathbf{F} = \frac{1}{r}\frac{\partial}{\partial r}(rF_r)+\frac{1}{r}\frac{\partial F_\theta}{\partial \theta}+\frac{\partial F_z}{\partial z}.$$

And if F_r, F_θ, F_ϕ are the components of a vector $\mathbf{F}$ in spherical polar coordinates, show that

$$\operatorname{div}\mathbf{F} = \frac{1}{r^2}\frac{\partial}{\partial r}(r^2F_r)+\frac{1}{r\sin\theta}\frac{\partial}{\partial \theta}(\sin\theta F_\theta)+\frac{1}{r\sin\theta}\frac{\partial F_\phi}{\partial \phi}.$$

29. Using the results of Ex. 28 and of Ex. 20 on Chapter IX, deduce the expressions of § 128 (3) and (13) for

$$\nabla^2 v = \operatorname{div}\operatorname{grad} v$$

in cylindrical and spherical polar coordinates.

30. Show that Laplace's equation in cylindrical coordinates, § 128 (3), has solutions of type

$$\begin{matrix}\cos\\ \sin\end{matrix} mz \begin{matrix}\cos\\ \sin\end{matrix} n\theta\{AI_n(mr)+BK_n(mr)\}$$

in addition to those specified in § 128 (12).

Discuss the choice of solutions appropriate to the finite cylinder $0 \leqslant r < a$, $0 < z < l$. Show that the solution of Laplace's equation which takes the value 1 on the curved surface of this cylinder and is zero on the plane ends is

$$\frac{4}{\pi}\sum_{s=0}^{\infty}\frac{1}{(2s+1)}\sin\left(\frac{(2s+1)\pi z}{l}\right)\frac{I_0\{(2s+1)\pi r/l\}}{I_0\{(2s+1)\pi a/l\}}.$$

31. The free vibrations of air in a sphere of radius a involve the solution of the equation of wave motion, § 129 (15), with the boundary condition

$$\frac{\partial v}{\partial r} = 0, \quad \text{when } r = a.$$

Find equations for the natural frequencies, and show that those corresponding to radial vibrations are $c\alpha_n/2\pi a$ where α_n, $n = 1, 2,\ldots$ are the positive roots of

$$\tan\alpha = \alpha.$$

32. Confirm that § 130 (10) is indeed the Green's function for Laplace's equation in the region $0 < x < \infty$, $-\infty < y < \infty$; that is, show that § 130 (10) satisfies Laplace's equation in this region and that $v = 0$ on $x = 0$ for all y except $y = y'$, where

$$\int_{-\epsilon}^{\epsilon} v(x, y-y')\,dy' = 1.$$

Hence show that for a point electrode at the origin $x = 0$, $y = 0$, the potential in the plate is constant on any circle which passes through the origin and has its centre on the x-axis.

33. Confirm that the Green's function for Laplace's equation in the three-dimensional region $0 < x < \infty$, $-\infty < y < \infty$, $-\infty < z < \infty$, is

$$(x/2\pi)\{x^2+(y-y')^2+(z-z')^2\}^{-\frac{3}{2}}.$$

Hence show that the potential distribution produced by a line electrode of strength v_0 per unit length, stretching along the entire positive y-axis, is

$$\frac{v_0 x}{2\pi(x^2+z^2)}\left\{\frac{y}{(x^2+y^2+z^2)^{\frac{1}{2}}}-1\right\}.$$

XVI

NUMERICAL METHODS FOR PARTIAL DIFFERENTIAL EQUATIONS

131. Introductory

THE partial differential equations discussed in the previous chapter were all linear equations and were solved for relatively simple boundary conditions on boundaries of regular shape. If, in a given problem, the equations are non-linear, or the boundary conditions or boundaries themselves are less simple, the chances of obtaining a mathematical solution are remote, and, if the problem is of practical importance, an approximate numerical solution must be found.

The standard methods for solving partial differential equations numerically are *finite difference methods* in which the equation is replaced by an approximate difference formula. These are derived using the differentiation formulae of § 117; the important approximations are the $\mathbf{0}(h)$ formula

$$dy/dx = (y_{n+1}-y_n)/h, \tag{1}$$

and the $\mathbf{0}(h^2)$ formulae

$$dy/dx = (y_{n+1}-y_{n-1})/2h \tag{2}$$

and

$$d^2y/dx^2 = (y_{n-1}-2y_n+y_{n+1})/h^2. \tag{3}$$

The way in which the solution is constructed depends very much on whether the partial differential equation is parabolic, hyperbolic, or elliptic; in §§ 132–4 we discuss each of these cases separately. The discussion is confined to problems in one space variable and time, or two spatial dimensions only, although the methods discussed can all be extended to two or three dimensions.

Finite difference methods are sometimes clumsy to use when boundaries and boundary conditions are particularly complex. A technique which copes with this type of problem more effectively is the finite element method, introduced briefly in § 135.

132. Finite difference methods for the diffusion equation

The diffusion equation in one spatial dimension is

$$\frac{\partial^2 v}{\partial x^2} = \frac{1}{\kappa}\frac{\partial v}{\partial t}, \tag{1}$$

which is an example of a parabolic equation.

We suppose that the boundary conditions specify v as a function of t at $x = 0$ and $x = l$, while the value of v in $0 < x < l$ at $t = 0$ is prescribed by the initial condition. We therefore have numerical values for v along the lines AB, BC, and CD in the xt-plane (see Fig. 97).

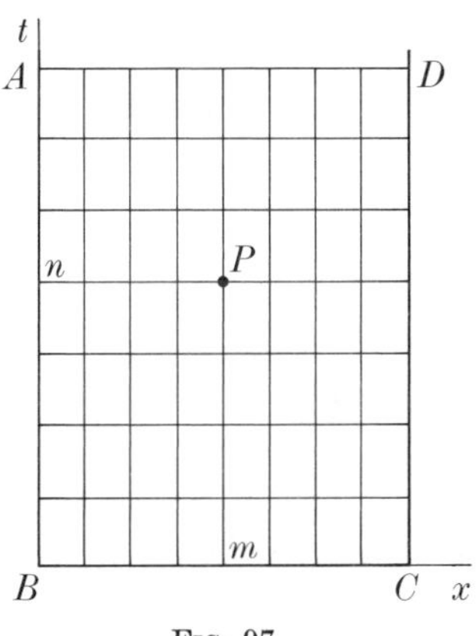

FIG. 97.

To solve (1) numerically, we first divide the t-axis into equally spaced intervals δt, and the x-axis in the region $(0, l)$ into M intervals of size h. The lines

$$x = mh \quad (m = 1, 2, \ldots, M-1)$$

and

$$y = n\delta t \quad (n = 1, 2, 3, \ldots)$$

then define a grid or mesh which spans the portion of the xt-plane contained within the lines AB, BC, and CD. A typical mesh point is P at $x = mh$ and $y = n\delta t$ in Fig. 97. We let $v_{m,n}$ denote the value of v at P. Then $v_{m+1,n}$ denotes the value of v at the mesh point immediately to the right of P, $v_{m,n+1}$ denotes the value at the point directly above P, and so on.

The next step consists of replacing the partial differential equation (1) by a difference equation at every point in the mesh. By § 131 (3), if we hold t constant, we can approximate $\partial^2 v/\partial x^2$ at P by the $\mathbf{0}(h^2)$ formula.

$$\partial^2 v/\partial x^2 = (v_{m-1,n} - 2v_{m,n} + v_{m+1,n})/h^2, \tag{2}$$

and if we hold x constant, by an equation equivalent to § 131 (1), we can approximate $\partial v/\partial t$ at P by the $\mathbf{0}(\delta t)$ formula

$$\partial v/\partial t = (v_{m,n+1} - v_{m,n})/\delta t. \tag{3}$$

Substituting (2) and (3) in (1) gives

$$(v_{m-1,n} - 2v_{m,n} + v_{m+1,n})/h^2 = (v_{m,n+1} - v_{m,n})/\kappa\delta t, \tag{4}$$

which is the difference equation we are looking for.

Rewriting (4) in the form

$$v_{m,n+1} = v_{m,n}(1-2s) + s(v_{m-1,n} + v_{m+1,n}), \tag{5}$$

where

$$s = \kappa\delta t/h^2, \tag{6}$$

suggests a straightforward algorithm for solving (1) numerically:

Putting $n = 0$ and $m = 1, 2,..., M-1$ in (5) enables us to calculate $v_{1,1}, v_{2,1},..., v_{M-1,1}$ along the line $t = \delta t$ from the known values $v_{0,0}, v_{1,0},..., v_{M,0}$ at $t = 0$. The values of $v_{0,1}$ and $v_{M,1}$ are already known by the boundary conditions. We thus have the complete solution at $t = \delta t$, which can then be used in the right-hand side of (5), with $n = 1$, to calculate the solution at $t = 2\delta t$. This, together with the boundary conditions at $t = 2\delta t$, is then used to calculate the solution at $t = 3\delta t$, and so on. The solution thus proceeds from one time step to the next, and is only terminated when we have a complete picture of how v changes with position and time.

The above method is known as the *explicit method* of solution. It is analogous to Euler's method for the solution of ordinary differential equations, and raises the same questions about accuracy and stability as were posed in connection with Euler's method in § 114.

The calculated values of v at any time step are all in error to a greater or lesser extent. Let ϵ be the absolute value of the *largest* error at time $n\delta t$. When these values of v are substituted in the right-hand side of (5), they introduce errors in the values of v at the next time step. Let ϵ' be the absolute value of the largest error introduced at time $(n+1)\delta t$ in this way. Then unless

$$\epsilon' \leqslant \epsilon \tag{7}$$

errors can grow from one time step to the next, and the numerical solution will be unstable.

By (5), we see that if $s \leqslant \frac{1}{2}$, the worst possible combination of errors in the right-hand side is

$$(1-2s)\epsilon+2s\epsilon.$$

Thus

$$\epsilon' \leqslant (1-2s)\epsilon+2s\epsilon$$

or

$$\epsilon' \leqslant \epsilon$$

and stability is guaranteed. No such guarantee exists if $s > \frac{1}{2}$, and it can in fact be shown† that for $s > \frac{1}{2}$ equation (5) is unstable. The condition for stability is therefore

$$s = \kappa\delta t/h^2 \leqslant \tfrac{1}{2}, \tag{8}$$

which restricts the choice of the intervals δt and h.

A numerical solution can be stable but inaccurate. It is much more difficult to discuss the accuracy of the solution on theoretical grounds, and the safest approach in practice is to compute a solution and then recompute it using different mesh intervals. A comparison of the two solutions should give some indication of their accuracy, and if this is unacceptable, a further solution will have to be obtained using smaller intervals. Notice, however, that if we halve the interval h, by (8) we will probably have to reduce δt by a factor of four, which would increase the computation time by a total factor of eight.

Ex. 1. *A non-linear equation.*

If a population can only migrate in the x-direction, a suggested model for the growth and migration of a population p (per unit length) is

$$\frac{\partial p}{\partial t} = \kappa\frac{\partial^2 p}{\partial x^2}+\beta p(1-p/p_0) \tag{9}$$

where κ and β are constants and p_0 is a saturation population.

An explicit formula for the solution of (9) is

$$p_{m,n+1} = p_{m,n}+s(p_{m-1,n}-2p_{m,n}+p_{m+1,n})+\beta\delta t p_{m,n}(1-p_{m,n}/p_0), \tag{10}$$

where $s = \kappa\delta t/h^2$.

† See, for example, L. Fox, *Numerical Solution of Ordinary and Partial Differential Equations* (Pergamon, 1962), chapter 19.

Equation (10) can be used to solve (9) step by step from an initial population distribution and subject to given boundary conditions on the x-axis. For stability we must at least have $s \leqslant \frac{1}{2}$, although the term in β might impose an even more stringent stability condition. This is most easily determined in practice by experimenting on the computer with the size of the time interval δt.

Ex. 2. *A non-linear boundary condition.*

We consider the solution of the diffusion equation (1) in the region $0 < x < l$ with the boundary condition of § 119 (7),

$$K\frac{\partial v}{\partial x}+H(v-v_0)^q = 0 \tag{11}$$

at $x = l$.

Mathematical solutions can sometimes be obtained for the linear case $q = 1$, but not for the non-linear case $q > 1$ which often arises in problems on heat transfer.

To solve this problem numerically, we introduce a *fictitious boundary* at $x = (M+1)h$. Using the $\mathbf{O}(h^2)$ formula § 131 (2) for $\partial v/\partial x$ at the real boundary $x = Mh$, (11) becomes

$$K(v_{M+1,n}-v_{M-1,n})/2h+H(v_{M,n}-v_0)^q = 0$$

or

$$v_{M+1,n} = v_{M-1,n}-2hH(v_{M,n}-v_0)^q/K. \tag{12}$$

Equation (12) enables one to calculate the temperature on the fictitious boundary at each time step, and the solution of the problem by the explicit method then proceeds as before. It should be noted that derivative-type boundary conditions can sometimes induce instabilities in a solution that would normally be stable. This might place a more restrictive condition on the mesh ratio $\kappa\delta t/h^2$ than equation (8).

Ex. 3.

As a final example, we discuss the solution of (1) when κ is a function of v. This often arises in metallurgical problems where v may change by a few hundred degrees during the solution of the problem, and it is unrealistic to assume that κ is constant over so wide a range of temperatures.

If we are given empirical data of the form

$$\kappa = f(v),$$

this can be incorporated in the numerical solution merely by writing

$$s = f(v_{m,n})\delta t/h^2$$

in equations (5) and (8).

The reader may have wondered why, in deriving (5), we used an $\mathbf{O}(\delta t)$ and not an $\mathbf{O}(\delta t^2)$ difference formula for $\partial v/\partial t$. The $\mathbf{O}(\delta t^2)$ formula

$$\partial v/\partial t = (v_{m,n+1}-v_{m,n-1})/2\delta t$$

leads to the equation

$$v_{m,n+1} = v_{m,n-1} + 2s(v_{m-1,n} - 2v_{m,n} + v_{m+1,n}), \tag{13}$$

in place of (5). This is analogous to the 'improved' formula for ordinary differential equations, cf. § 115, and (13) shares the unfortunate stability characteristics of the 'improved' formula. It can in fact be shown that (13) is unstable for *all* values of s.

Another $\mathbf{0}(\delta t^2)$ formula can be found for $\partial v/\partial t$ by noticing that

$$\partial v/\partial t = (v_{m,n+1} - v_{m,n})/\delta t$$

is $\mathbf{0}(\delta t)$ if we are evaluating $\partial v/\partial t$ at $t = n\delta t$, but is $\mathbf{0}(\delta t^2)$ if we are evaluating $\partial v/\partial t$ at $t = (n+\frac{1}{2})\delta t$. This is used to advantage in the *Crank–Nicholson method*, where the diffusion equation (1) is written

$$\frac{1}{\kappa}\left(\frac{\partial v}{\partial t}\right)_{(n+\frac{1}{2})\delta t} = \frac{1}{2}\left(\frac{\partial^2 v}{\partial x^2}\right)_{n\delta t} + \frac{1}{2}\left(\frac{\partial^2 v}{\partial x^2}\right)_{(n+1)\delta t} \tag{14}$$

Replacing both sides of (14) by finite difference formulae leads to the equation

$$sv_{m-1,n+1} - 2(1+s)v_{m,n+1} + sv_{m+1,n+1}$$
$$= -sv_{m-1,n} - 2(1-s)v_{m,n} - sv_{m+1,n}. \tag{15}$$

Equation (15) is an implicit formula; in order to proceed from one time step to the next, the set of simultaneous equations obtained by putting $m = 1, 2,..., M-1$ in (15) must be solved numerically. The extra computation time needed to accomplish this is usually more than compensated for by the increased accuracy and enhanced stability of (15) as compared with the explicit method of equation (5), cf. Ex. 4 at the end of this chapter.

133. A finite difference method for the wave equation

The wave equation

$$\frac{\partial^2 v}{\partial x^2} = \frac{1}{c^2}\frac{\partial^2 v}{\partial t^2} \tag{1}$$

can be solved by an explicit method in much the same way as the diffusion equation.

Setting up a grid, as in Fig. 97, we replace (1) at the point $x = mh$, $t = n\delta t$ by the difference equation

$$(v_{m-1,n} - 2v_{m,n} + v_{m+1,n})/h^2 = (v_{m,n-1} - 2v_{m,n} + v_{m,n+1})/c^2\delta t^2$$

or $$v_{m,n+1} = 2v_{m,n} - v_{m,n-1} + S(v_{m-1,n} - 2v_{m,n} + v_{m+1,n}) \tag{2}$$
where $$S = c^2\delta t^2/h^2.$$

The condition for stability is
$$S \leqslant 1,$$
and good results are often obtained with $S = 1$, in which case (2) reduces to
$$v_{m,n+1} = v_{m-1,n} + v_{m+1,n} - v_{m,n-1}. \tag{3}$$

The solution is built up from one time step to the next, using (2) or (3). The only difficulty that arises is when $n = 0$, in which case (3), for instance, becomes
$$v_{m,1} = v_{m-1,0} + v_{m+1,0} - v_{m,-1}. \tag{4}$$
$v_{m-1,0}$ and $v_{m+1,0}$ are known from an initial condition of the form $v = f(x)$ when $t = 0$. The fictitious values $v_{m,-1}$ can be found from the further initial condition
$$\partial v/\partial t = g(x) \tag{5}$$
at $t = 0$. The $\mathbf{0}(\delta t^2)$ approximation to (5) is
$$(v_{m,1} - v_{m,-1})/2\delta t = g(mh). \tag{6}$$

Substituting (6) in (4) gives the special equation for the first time step
$$v_{m,1} = \tfrac{1}{2}(v_{m-1,0} + v_{m+1,0}) + \delta t g(mh).$$

The wave equation is an example of a hyperbolic partial differential equation, and many important examples of hyperbolic equations arise in the study of waves and shock phenomena in aerodynamics and fluid mechanics. These are sometimes solved numerically using finite difference formulae such as (2) or (3) above; more often a somewhat more sophisticated technique, known as the 'method of characteristics', is used. We will not discuss this method here; the reader is referred, for example, to L. Fox, op. cit., chapter 17.

134. A finite difference method for Laplace's equation

Parabolic and hyperbolic partial differential equations are partially initial value and partially boundary value problems. This is reflected in numerical methods for their solution, which, as we saw in §§ 132, 133, incorporate the boundary conditions in a step-by-step procedure that starts with the initial conditions. Elliptic equations, on the other hand, are purely boundary value problems, and here the solution has to be constructed within a *closed* boundary subject to various conditions on that boundary. As an example, we consider the solution of Laplace's equation

$$\frac{\partial^2 v}{\partial x^2}+\frac{\partial^2 v}{\partial y^2} = 0 \tag{1}$$

in the L-shaped region of Fig. 98 (a).

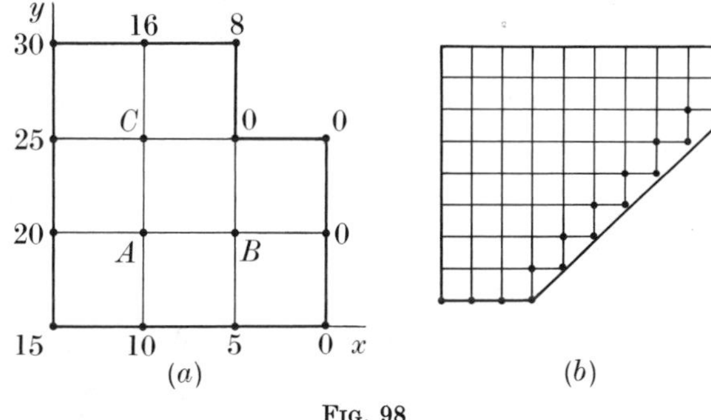

FIG. 98

As usual, we impose a grid on this region, and for simplicity we choose the same interval h along both axes. At a typical mesh point $x = mh$, $y = nh$, we replace (1) by the $\mathbf{0}(h^2)$ finite difference approximation

$$(v_{m-1,n}-2v_{m,n}+v_{m+1,n})/h^2+(v_{m,n-1}-2v_{m,n}+v_{m,n+1})/h^2 = 0,$$

or
$$-4v_{m,n}+v_{m-1,n}+v_{m+1,n}+v_{m,n-1}+v_{m,n+1} = 0. \tag{2}$$

The value of v at any interior mesh point is thus the arithmetic mean of the values of v at the four nearest adjacent mesh points.

Equation (2) is applied to all the interior mesh points. In the simplified example of Fig. 98 (a), these are the three points marked A, B, and C.

At A, $m = 1$ and $n = 1$, and (2) becomes

$$-4v_{1,1}+v_{1,2}+v_{2,1} = -(20+10). \tag{3}$$

Notice that we write the known values of v, at boundary points, on the right-hand side of the equation.

Similarly, at B, $m = 2$ and $n = 1$, and (2) becomes

$$-4v_{2,1}+v_{1,1} = -(0+5+0), \tag{4}$$

and at C,

$$-4v_{1,2}+v_{1,1} = -(25+0+16). \tag{5}$$

Equations (3), (4), and (5) can be rearranged and written in matrix form

$$\begin{bmatrix} -4 & 1 & 0 \\ 1 & -4 & 1 \\ 0 & 1 & -4 \end{bmatrix} \begin{bmatrix} v_{2,1} \\ v_{1,1} \\ v_{1,2} \end{bmatrix} = \begin{bmatrix} -5 \\ -30 \\ -41 \end{bmatrix}, \tag{6}$$

which can be solved by the Gauss–Seidel iteration scheme of § 107.

The same procedure applies to larger and more complicated problems. Equation (2) is applied at every interior mesh point and the resulting set of linear equations can be written as in (6), in the matrix form

$$\mathbf{AV} = \mathbf{C}, \tag{7}$$

where $\mathbf{V}$ is a column vector containing the unknown values of v at the interior mesh points, while $\mathbf{C}$ is a column vector which depends on the known values of v at mesh points on the boundary. A typical row of the coefficient matrix $\mathbf{A}$, row m for instance, will have non-zero elements $a_{m\,m-1} = a_{m\,m+1} = 1$, $a_{m\,m} = -4$, and at most two other non-zero elements, both of them equal to 1. The criteria for the convergence of the Gauss–Seidel process (cf. § 107) are thus satisfied and so the set of equations (7) can be solved iteratively. Computer programs can be written which take advantage of the fact that most of the elements of $\mathbf{A}$ are zero, and the actual iteration process can be accelerated by the use of appropriate relaxation factors.

When the boundary has a more complicated shape, as in Fig. 98 (*b*), the nearest grid points are assumed to lie on the boundary. It may therefore become necessary to use a very small interval h, if only to obtain a reasonably accurate representation of the boundary. There are more sophisticated ways of dealing with oddly shaped boundaries, but these are rather difficult to use, especially if the boundary conditions depend on the derivatives of v instead of v itself.†

† See L. Fox, op. cit., chapter 21.

135. The finite element method

The finite element method is widely used in the solution of the partial differential equations for stress and deformation calculations in structures and shells, and is increasingly being used for the numerical solution of more complicated versions of the problems discussed in this chapter and in Chapter XV. Here we outline the application of the method to Laplace's equation

$$\frac{\partial^2 v}{\partial x^2}+\frac{\partial^2 v}{\partial y^2}=0, \tag{1}$$

subject to prescribed values of v on the boundary of some region R.

A fundamental result of the calculus of variations states that the solution of (1) in the region R is that function v which minimizes the value of the integral

$$I=\iint_R \frac{1}{2}\left\{\left(\frac{\partial v}{\partial x}\right)^2+\left(\frac{\partial v}{\partial y}\right)^2\right\}dxdy, \tag{2}$$

subject to the same prescribed values of v on the boundary of R.

This result is analogous to the two different ways for finding the equilibrium coordinates of a system of particles; either the equations of equilibrium are written down directly, or else the potential energy V of the system is constructed and then minimized, cf. § 77 (2).

The finite element method is an approximate method for constructing and minimizing the integral (2). The region R is divided into a mesh of triangular elements,† as in Fig. 99 (a).

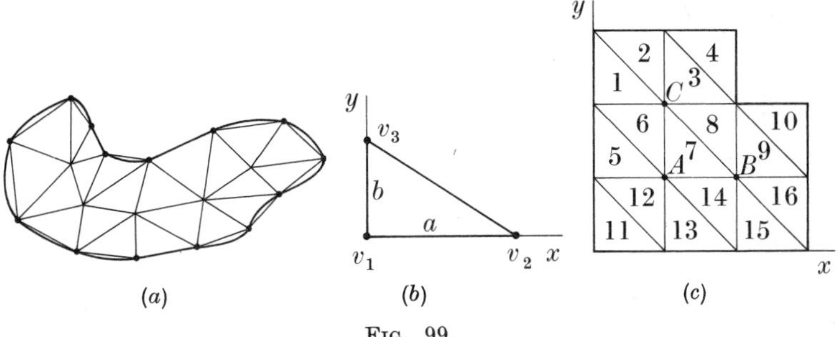

FIG. 99.

† Other types of elements can also be used; see Zienkiewicz, *The Finite Element Method in Engineering Science* (McGraw-Hill).

The mesh points are the vertices of the triangles. We number these points from 1 to N in the interior of the region and from $N+1$ to M on the boundary of R, and write v_n for the value of v at the nth mesh point.

To simplify the analysis, we consider the right-angled triangle of Fig. 99 (*b*). If we interpolate linearly between the values v_1, v_2, and v_3 at its vertices, we can assume that within this particular triangle

$$v(x,y) = v_1+(v_2-v_1)x/a+(v_3-v_1)y/b.$$

Then
$$\partial v/\partial x = (v_2-v_1)/a$$
and
$$\partial v/\partial y = (v_3-v_1)/b,$$
and the contribution of this triangle towards the integral I is

$$\tfrac{1}{2}ab\tfrac{1}{2}\left\{\left(\frac{v_2-v_1}{a}\right)^2+\left(\frac{v_3-v_1}{b}\right)^2\right\}. \tag{3}$$

Adding similar contributions towards I for all the triangles in R, and collecting like terms in v_n^2 and $v_n v_m$, eventually leads to a result of the form

$$I = \sum_{m=1}^{M} \sum_{n=1}^{M} A_{mn} v_m v_n. \tag{4}$$

The coefficients A_{mn} in (4) depend only on the topology of the mesh, and are known, as are the boundary values

$$v_n (n = N+1,..., M).$$

It remains to find the values of v_n inside the region R; these must be chosen such that I is a minimum, that is

$$\frac{\partial I}{\partial v_n} = \sum_{m=1}^{M} (A_{mn}+A_{nm})v_m = 0, \tag{5}$$

for $n = 1, 2,..., N$.

Equations (5) are a set of simultaneous linear equations for the interior mesh values $v_1, v_2,..., v_N$ in terms of the boundary values $v_{N+1},..., v_M$. These can be solved on a computer by the methods of § 107; special purpose programs can be written which take advantage of the fact that a large number of the coefficients A_{mn} are zero.

Ex. 1.

We consider the L-shaped region of Fig. 98 (a) and divide it into a number of right-angled triangles, as in Fig. 99 (c). (This is not a very fortunate example in that we need sixteen triangles to find the values v_A, v_B, and v_C at the three interior mesh points A, B, and C.)

We first calculate the contributions of each triangle towards the integral I.

By (3), if we assume $a = b = 1$, and using the boundary values specified in Fig. 98 (a), the contribution of the first triangle is

$$\tfrac{1}{4}(v_C-25)^2+\tfrac{1}{4}(30-25)^2,$$

and the contribution of triangle 6, for instance, is

$$\tfrac{1}{4}(v_A-v_C)^2+\tfrac{1}{4}(25-v_C)^2.$$

Adding similar expressions for all sixteen triangles leads to the equivalent of (4) in the form

$$I = pv_A^2+qv_B^2+rv_C^2+fv_A+gv_B+hv_C+lv_A v_B+mv_A v_C+nv_B v_C+k, \tag{6}$$

where the constants p, q,..., k are all known.

To minimize I, by (5),

$$\begin{aligned}\partial I/\partial v_A &= 2pv_A+lv_B+mv_C+f = 0\\ \partial I/\partial v_B &= lv_A+2qv_B+nv_C+g = 0\\ \partial I/\partial v_C &= mv_A+nv_B+2rv_C+h = 0,\end{aligned} \tag{7}$$

and these equations can be solved for v_A, v_B, and v_C.

The finite element method has a number of advantages, most of them arising from the flexibility with which a triangular mesh such as that of Fig. 99 (a) can be constructed. For instance, the mesh can be fitted easily and accurately to even the most oddly shaped regions. Also, the size of triangles can be varied from small triangles in regions where grad v is high, to large triangles in regions where grad v is small. In fact, the accuracy of the method is very dependent upon the size and shape of the elements chosen, which is its major disadvantage. Finally, the finite element method is particularly well suited to problems in which the properties of the medium change from point to point in the region R, as in the following example.

Ex. 2.

If the thermal conductivity K is a function of position, the equation for steady heat flow is, by § 127 (8),

$$\frac{\partial}{\partial x}\left(K\frac{\partial v}{\partial x}\right)+\frac{\partial}{\partial y}\left(K\frac{\partial v}{\partial y}\right) = 0. \tag{8}$$

The integral to be minimized is then

$$I = \iint\limits_R \frac{1}{2}\left\{K\left(\frac{\partial v}{\partial x}\right)^2 + K\left(\frac{\partial v}{\partial y}\right)^2\right\}dxdy, \tag{9}$$

and the finite element method proceeds as before, except that a different value of K is used for the contribution of each triangle towards (9); this is just the mean value of K within each triangle.

EXAMPLES ON CHAPTER XVI

1. The equation $\dfrac{\partial v}{\partial x}+k\dfrac{\partial v}{\partial t} = 0$ is often encountered in problems on flow, such as the flow of traffic along a highway. Show that the equation is parabolic and derive the explicit formula

$$v_{n,m+1} = v_{n,m}+\delta t(v_{n+1,m}-v_{n,m})/kh$$

for its solution. Show that this formula is stable if $\delta t/\kappa h \leqslant 1$.

2. Find the solution of $\dfrac{\partial^2 v}{\partial x^2}-\dfrac{1}{\kappa}\dfrac{\partial v}{\partial t} = 0$ subject to the following initial and boundary conditions: $v = 0$ for $0 < x < 1$ at $t = 0$; $v = 100$ at $x = 0$ for all $t > 0$, and $v = 0$ at $x = 1$ for all $t > 0$. Assume that $\kappa = 0{\cdot}008$. Use the explicit method with intervals $h = 0{\cdot}1$ and $\delta t = 0{\cdot}5$ and take your solution up to $t = 25$.

3. Repeat the solution of Ex. 2 with intervals $h = 0{\cdot}1$ and $\delta t = 1{\cdot}0$. Comment on your results.

4. Repeat the solution of Ex. 2 using the Crank–Nicholson method, first with $h = 0{\cdot}1$ and $\delta t = 0{\cdot}5$, and then with $h = 0{\cdot}1$ and $\delta t = 1{\cdot}0$.

5. Using the explicit method with $h = 0{\cdot}1$ and $\delta t = 0{\cdot}5$, solve the problem of Ex. 2 for the case where κ is a function of v, namely, $\kappa = 0{\cdot}006+0{\cdot}00004v$. Notice that if we ignore the variation of κ with v and replace κ by a constant 'average' value over the range $v = 0$ to $v = 100$, this problem reduces to that of Ex. 2. Compare the solution of this problem with that of Ex. 2, and comment on the importance, or otherwise, of including the variation of κ with v.

6. Solve the problem of § 132, Ex. 1, for the case $\kappa = 4$, $\beta = 1$, and $p_0 = 100$. Suppose that $p = p_0$ at $x = 0$ for all $t > 0$, and that at $x = 100$ there is a barrier to further migration, that is, $\partial p/\partial x = 0$ at $x = 100$ for $t > 0$. Assume that initially $p = p_0$ for $0 < x < 10$ and $p = 0$ for $10 < x < 100$. Study the growth and spread of the population from $t = 0$ up to $t = 40$, using intervals $h = 5$ and $\delta t = 1$.

7. Solve the problem of § 124, Ex. 3, numerically, given $c = 1$ and $d = 0{\cdot}01$, $b = 0{\cdot}5$, and $l = 1{\cdot}0$ in § 124 (42), (43). First take $h = 0{\cdot}1$ and $\delta t = 0{\cdot}1$, and then compare this solution with $h = \delta t = 0{\cdot}05$. Compare both these solutions with the Fourier series § 124 (44) for $0 < t < 2$.

8. A string of length l and mass per unit length σ is clamped at its two ends. If the tension in the string is T, it is shown in § 121 that its natural frequencies of vibration are

$$(s/2l)\sqrt{(T/\sigma)}, \quad s = 1, 2, 3 \ldots .$$

A discrete approximation to this problem is the problem of the transverse vibrations of n equally spaced particles, each of mass $m = \sigma l/n$, connected by weightless elastic strings. Using the result of Ex. 15 at the end of Chapter IV, compare the first few natural frequencies of the discrete approximation with the true solution. Consider the two cases $n = 10$ and $n = 20$.

9. Solve Laplace's equation in the L-shaped region of Fig. 98 (a) with the boundary conditions specified in the figure and using the given mesh. Then reduce the mesh interval by half and solve again. (Interpolate linearly between boundary values.) Compare the two solutions.

10. Solve Laplace's equation in the same region and subject to the same boundary conditions as in Ex. 9 above, using the finite element method with the triangular mesh of Fig. 99 (c). Compare your solution with the finite difference solutions of Ex. 9.

11. Show that an $\mathbf{O}(h^2)$ finite difference approximation to Laplace's equation in three dimensions is equivalent to the statement that the value of the function at any interior mesh point is the arithmetic mean of its values at the six nearest mesh points of a cubic grid.

12. Derive an explicit formula for the solution of

$$\frac{\partial^2 v}{\partial x^2}+\frac{\partial^2 v}{\partial y^2}-\frac{1}{\kappa}\frac{\partial v}{\partial t} = 0$$

with the same grid interval h along both the x- and y-axes. Show that this formula is stable provided $\kappa\delta t/h^2 \leqslant \frac{1}{4}$.

13. The equation for radial heat flow in a cross-section of a long pipe is

$$\frac{\partial^2 v}{\partial r^2}+\frac{1}{r}\frac{\partial v}{\partial r} = \frac{1}{\kappa}\frac{\partial v}{\partial t}.$$

A pipe has an internal radius of 4 cm and an external radius of 10 cm. It is initially at a temperature $v = 0$ throughout, and the outer boundary is kept at $v = 0$ for all $t > 0$. At $t = 0$ the temperature of the inner boundary is raised to $v = 100$ and maintained at that value for all $t > 0$.

Derive an explicit formula for the solution of this equation, using $\mathbf{O}(h^2)$ approximations on the left-hand side of the equation and an $\mathbf{O}(\delta t)$ approximation on the right-hand side. Show that this formula is stable if $\kappa\delta t/h^2 \leqslant \frac{1}{2}$. Solve the equation for the case $\kappa = 0{\cdot}02\ \mathrm{cm}^2\ \mathrm{s}^{-1}$, taking $h = 1$ cm and $\delta t = 20$ s. Confirm that the temperature tends to the steady flow distribution given by § 123 (40). Draw a graph of the temperature gradient $\partial v/\partial r$ at the outer boundary versus t. [Hint: use § 117 (9) to estimate $\partial v/\partial r$ at the outer boundary.]

14. The biharmonic equation $\nabla^2\nabla^2 v = 0$ reduces, in two dimensions to

$$\frac{\partial^4 v}{\partial x^4}+2\frac{\partial^4 v}{\partial x^2 \partial y^2}+\frac{\partial^4 v}{\partial y^4} = 0.$$

Show that this equation can be replaced by the difference formula of $\mathbf{O}(h^2)$

$$20v_{m,n}-8(v_{m-1,n}+v_{m+1,n}+v_{m,n-1}+v_{m,n+1})+$$
$$+2(v_{m-1,n-1}+v_{m+1,n-1}+v_{m-1,n+1}+v_{m+1,n+1})+$$
$$+(v_{m-2,n}+v_{m+2,n}+v_{m,n-2}+v_{m,n+2}) = 0.$$

Describe how to solve this equation in the square region $0 < x < 1$, $0 < y < 1$, given the values of v and the normal derivatives $\partial v/\partial \mathbf{n}$ on the boundaries of the region. [Hint: introduce a fictitious boundary on all sides of the region.]

15. Gourlay's 'Hopscotch' method combines the advantages of the explicit formula § 132 (5) and the Crank–Nicholson formula § 132 (15) for the solution of the diffusion equation. In a simplified form, the 'Hopscotch' algorithm for carrying the solution from $t = n\delta t$ to $t = (n+2)\delta t$ is as follows:

(i) Calculate $v_{m,n+1}$ for *even* values of m using the explicit formula.

(ii) Calculate $v_{m,n+1}$ for *odd* values of m using the Crank–Nicholson formula. Notice that this formula can now be used explicitly since two of the values on the left-hand side of § 132 (15) are known by step (i).

(iii) Calculate $v_{m,n+2}$ for *odd* values of m using the explicit formula.

(iv) Calculate $v_{m,n+2}$ for *even* values of m using the Crank–Nicholson formula.

Results are printed out at $t = 2\delta t, 4\delta t, 6\delta t,\ldots$ etc. Solve the problem of Ex. 2 using this method and compare results with Ex. 2 and Ex. 4 above.

INDEX